The Cambridge Handbook of Language and Brain

The topic of language and brain is a large and significant area of research and study, and this Handbook provides a state-of-the-art survey of the field. Bringing together contributions from an interdisciplinary team of internationally-renowned scholars, it focuses on important theoretical positions that have changed the study of language and brain in the first two decades of the 21st century. It is split into seven thematic parts, covering topics such as theoretical foundations of language and brain, neuroimaging studies of brain and language, language and cognitive development, building cognitive brain reserve and the importance of proficiency, aphasia and autism spectrum disorders, brain, language and music, and new directions and perspectives. Representing the most powerful trends in the field, it will inform new directions in the study of language and brain, cognitive neuroscience and neuroimaging, and scholars and advanced students will find this compilation an invaluable resource for years to come.

EDNA ANDREWS is Professor of Linguistics & Cultural Anthropology, Duke Institute for Brain Sciences and the Duke Center for Cognitive Neuroscience, and the Nancy & Jeffrey Marcus Distinguished Professor.

SWATHI KIRAN is a James and Cecelia Ying Professor of Neurorehabilitation in the Department of Speech and Hearing Sciences and the founding director for the Center for Brain Recovery at Boston University.

CAMBRIDGE HANDBOOKS IN LANGUAGE AND LINGUISTICS

Genuinely broad in scope, each handbook in this series provides a complete state-of-the-field overview of a major sub-discipline within language study and research. Grouped into broad thematic areas, the chapters in each volume encompass the most important issues and topics within each subject, offering a coherent picture of the latest theories and findings. Together, the volumes will build into an integrated overview of the discipline in its entirety.

Published titles

The Cambridge Handbook of Phonology, edited by Paul de Lacy
The Cambridge Handbook of Linguistic Code-switching, edited by Barbara E. Bullock and Almeida Jacqueline Toribio
The Cambridge Handbook of Child Language, Second Edition, edited by Edith L. Bavin and Letitia Naigles
The Cambridge Handbook of Endangered Languages, edited by Peter K. Austin and Julia Sallabank
The Cambridge Handbook of Sociolinguistics, edited by Rajend Mesthrie
The Cambridge Handbook of Pragmatics, edited by Keith Allan and Kasia M. Jaszczolt
The Cambridge Handbook of Language Policy, edited by Bernard Spolsky
The Cambridge Handbook of Second Language Acquisition, edited by Julia Herschensohn and Martha Young-Scholten
The Cambridge Handbook of Biolinguistics, edited by Cedric Boeckx and Kleanthes K. Grohmann
The Cambridge Handbook of Generative Syntax, edited by Marcel den Dikken
The Cambridge Handbook of Communication Disorders, edited by Louise Cummings
The Cambridge Handbook of Stylistics, edited by Peter Stockwell and Sara Whiteley
The Cambridge Handbook of Linguistic Anthropology, edited by N.J. Enfield, Paul Kockelman and Jack Sidnell
The Cambridge Handbook of English Corpus Linguistics, edited by Douglas Biber and Randi Reppen
The Cambridge Handbook of Bilingual Processing, edited by John W. Schwieter
The Cambridge Handbook of Learner Corpus Research, edited by Sylviane Granger, Gaëtanelle Gilquin and Fanny Meunier
The Cambridge Handbook of Linguistic Multicompetence, edited by Li Wei and Vivian Cook
The Cambridge Handbook of English Historical Linguistics, edited by Merja Kytö and Päivi Pahta
The Cambridge Handbook of Formal Semantics, edited by Maria Aloni and Paul Dekker
The Cambridge Handbook of Morphology, edited by Andrew Hippisley and Greg Stump
The Cambridge Handbook of Historical Syntax, edited by Adam Ledgeway and Ian Roberts
The Cambridge Handbook of Linguistic Typology, edited by Alexandra Y. Aikhenvald and R. M. W. Dixon
The Cambridge Handbook of Areal Linguistics, edited by Raymond Hickey
The Cambridge Handbook of Cognitive Linguistics, edited by Barbara Dancygier
The Cambridge Handbook of Japanese Linguistics, edited by Yoko Hasegawa
The Cambridge Handbook of Spanish Linguistics, edited by Kimberly L. Geeslin

The Cambridge Handbook of Bilingualism, edited by Annick De Houwer and Lourdes Ortega
The Cambridge Handbook of Systemic Functional Linguistics, edited by Geoff Thompson, Wendy L. Bowcher, Lise Fontaine and David Schönthal
The Cambridge Handbook of African Linguistics, edited by H. Ekkehard Wolff
The Cambridge Handbook of Language Learning, edited by John W. Schwieter and Alessandro Benati
The Cambridge Handbook of World Englishes, edited by Daniel Schreier, Marianne Hundt and Edgar W. Schneider
The Cambridge Handbook of Intercultural Communication, edited by Guido Rings and Sebastian Rasinger
The Cambridge Handbook of Germanic Linguistics, edited by Michael T. Putnam and B. Richard Page
The Cambridge Handbook of Discourse Studies, edited by Anna De Fina and Alexandra Georgakopoulou
The Cambridge Handbook of Language Standardization, edited by Wendy Ayres-Bennett and John Bellamy
The Cambridge Handbook of Korean Linguistics, edited by Sungdai Cho and John Whitman
The Cambridge Handbook of Phonetics, edited by Rachael-Anne Knight and Jane Setter
The Cambridge Handbook of Corrective Feedback in Second Language Learning and Teaching, edited by Hossein Nassaji and Eva Kartchava
The Cambridge Handbook of Experimental Syntax, edited by Grant Goodall
The Cambridge Handbook of Heritage Languages and Linguistics, edited by Silvina Montrul and Maria Polinsky
The Cambridge Handbook of Arabic Linguistics, edited by Karin Ryding and David Wilmsen
The Cambridge Handbook of the Philosophy of Language, edited by Piotr Stalmaszczyk
The Cambridge Handbook of Sociopragmatics, edited by Michael Haugh, Dániel Z. Kádár and Marina Terkourafi
The Cambridge Handbook of Task-Based Language Teaching, edited by Mohammed Ahmadian and Michael Long
The Cambridge Handbook of Language Contact: Population Movement and Language Change, Volume 1, edited by Salikoko Mufwene and Anna Maria Escobar
The Cambridge Handbook of Language Contact: Multilingualism in Population Structure, Volume 2, edited by Salikoko Mufwene and Anna Maria Escobar
The Cambridge Handbook of Romance Linguistics, edited by Adam Ledgeway and Martin Maiden
The Cambridge Handbook of Translation, edited by Kirsten Malmkjær
The Cambridge Handbook of Chinese Linguistics, edited by Chu-Ren Huang, Yen-Hwei Lin, I-Hsuan Chen and Yu-Yin Hsu
The Cambridge Handbook of Intercultural Pragmatics, edited by Istvan Kecskes
The Cambridge Handbook of Role and Reference Grammar, edited by Delia Bentley, Ricardo Mairal-Usón, Wataru Nakamura and Robert D. Van Valin Jr.
The Cambridge Handbook of Historical Orthography, edited by Marco Condorelli and Hanna Rutkowska
The Cambridge Handbook of Third Language Acquisition, edited by Jennifer Cabrelli, Adel Chaouch-Orozco, Jorge González Alonso, Eloi Puig-Mayenco, Sergio Miguel Pereira Soares and Jason Rothman
The Cambridge Handbook of Language in Context, edited by Jesús Romero-Trillo

The Cambridge Handbook of Slavic Languages, edited by Danko Šipka and Wayles Browne
The Cambridge Handbook of Gesture Studies, edited by Alan Cienki
The Cambridge Handbook of the Dictionary, edited by Edward Finegan and Michael Adams
The Cambridge Handbook of Methods in Conversation Analysis, edited by Jeffrey Robinson, Rebecca Clift, Kobin H. Kendrick and Chase Wesley Raymond
The Cambridge Handbook of Technology in Language Teaching and Learning, edited by Glenn Stockwell and Yijen Wang
The Cambridge Handbook of Bilingual Phonetics and Phonology, edited by Mark Amengual
The Cambridge Handbook of Comparative Syntax, edited by Sjef Barbiers, Norbert Corver and Maria Polinsky
The Cambridge Handbook of Construction Grammar, edited by Mirjam Fried and Kiki Nikiforidou
The Cambridge Handbook of Multilingual Education, edited by Piotr Romanowski

The Cambridge Handbook of Language and Brain

Edited by

Edna Andrews

Duke University

Swathi Kiran

Boston University

CAMBRIDGE
UNIVERSITY PRESS

CAMBRIDGE
UNIVERSITY PRESS

Shaftesbury Road, Cambridge CB2 8EA, United Kingdom

One Liberty Plaza, 20th Floor, New York, NY 10006, USA

477 Williamstown Road, Port Melbourne, VIC 3207, Australia

314–321, 3rd Floor, Plot 3, Splendor Forum, Jasola District Centre, New Delhi – 110025, India

103 Penang Road, #05-06/07, Visioncrest Commercial, Singapore 238467

Cambridge University Press is part of Cambridge University Press & Assessment, a department of the University of Cambridge.

We share the University's mission to contribute to society through the pursuit of education, learning and research at the highest international levels of excellence.

www.cambridge.org
Information on this title: www.cambridge.org/9781009202282

DOI: 10.1017/9781009202336

When citing this work, please include a reference to the DOI 10.1017/9781009202336

First published 2025

A catalogue record for this publication is available from the British Library.

Library of Congress Cataloging-in-Publication Data
Names: Andrews, Edna, 1958- editor. | Kiran, Swathi, editor.
Title: The Cambridge handbook of language and brain / edited by Edna Andrews, Duke University, Swathi Kiran, Boston University.
Description: Cambridge, United Kingdom ; New York, NY : Cambridge University Press, 2025. | Includes bibliographical references and index.
Identifiers: LCCN 2024051066 (print) | LCCN 2024051067 (ebook) | ISBN 9781009202282 (hardback) | ISBN 9781009202299 (paperback) | ISBN 9781009202336 (epub)
Subjects: LCSH: Neurolinguistics. | Psycholinguistics.
Classification: LCC QP399 .C36 2025 (print) | LCC QP399 (ebook) | DDC 612.8/233–dc23/eng/20241223
LC record available at https://lccn.loc.gov/2024051066
LC ebook record available at https://lccn.loc.gov/2024051067

ISBN 978-1-009-20228-2 Hardback

Contents

Figures

Full colour figures where available can be viewed at www.cambridge.org/andrews-kiran

Tables

Contributors

Jubin Abutalebi, Vita-Salute San Raffaele University, Milan, Italy, and The Arctic University of Norway
Edna Andrews, Duke University
Fiona Mitchel Baumer, Stanford University
Gavin M. Bidelman, Indiana University
Hayes Bierman, Duke University
David Birdsong, University of Texas at Austin
Tali Bitan, University of Haifa, Israel, and University of Toronto
Arpita Bose, University of Reading
Jane A. Brown, University of Memphis
Adam Buchwald, New York University
Lindsay Butler, University of Connecticut
Natalie V. Covington, University of Minnesota
Melissa C. Duff, Vanderbilt University Medical Center
E. Susan Duncan, City University of New York
Federico Gallo, PoLaR Lab, Center for Language, Brain and Learning, UiT – The Arctic University of Norway, Tromsø, Norway
Charlotte Gaynor, Boston University
Gerald Grant, Duke University
Teresa Gray, San Francisco State University
Harsheen Kaur, University of Utah
Juhi Kidwai, Southern Illinois University
Swathi Kiran, Boston University
Gabrielle Klassen, Utah State University
Jessica MacLean, Indiana University
Urvi Maheshwari, University of California San Diego
Erin L. Meier, Northeastern University
Meredith Pecukonis, Boston University
Christos Pliatsikas, University of Reading, and Nebrija University, Spain

Aswathy Anakkathil Pradeep, Louisiana State University
Charlotte Sophia Rammell, Minot State University
Rose Rizzi, Indiana University
Chaleece W. Sandberg, Pennsylvania State University
John W. Schwieter, Wilfrid Laurier University, Waterloo, Canada
Tamara Swaab, University of California, Davis
Alexandria Swaine, The University of North Carolina School of Medicine
Helen Tager-Flusberg, Boston University
Matthew Traxler, University of California, Davis
Sofia Vallila-Rohter, MGH Institute of Health Professions
Stephen C. Van Hedger, Huron University College at Western
Sarah Villard, Boston University
Lauryn Zipse, MGH Institute of Health Professions

Acknowledgments

We would like to express our deep gratitude to those who provided constant support during this project. There are many important people to thank. We would like to thank Rebecca Taylor for her constant and ongoing commitment to this important research volume. The tireless commitment of Cathy Lewis in working with all the contributors and with Cambridge University Press for these past two years was invaluable. Her kindness and generosity are deeply appreciated. We also want to thank Hayes Bierman for his important work on the book index and his happy demeanor during the process.

Part I

Introduction

Theoretical Foundations of Language and Brain

1

At the Crossroads of the Study of Language and Brain in Healthy Subjects and in Pathology

Deepening of the Analysis of Networks Crucial for Language Processing: Questions, Theory, Application

Edna Andrews and Swathi Kiran

It is a privilege to present the introduction to this new volume of *The Cambridge Handbook of Language and Brain.* The chapters in this volume represent important trends, methods, and central questions in research on brain and language that encompasses perspectives that include a spectrum of studies in methodology that range from healthy subjects that use one or multiple languages to neurodiversity and neurological disorders. A reader looking to come up to speed on a particular topic in language and the brain need look no further than thorough the list of contributions in this book.

The researchers who have authored chapters for this volume represent an extraordinary group of international scholars representing the multiple disciplines that are critical to the study of language and brain. The chapters represent critical findings from key international research institutes, laboratories, and universities. The clinical and research faculties which have authored contributions come from neuroscience, cognitive neuroscience, brain sciences, neuropsychology, neurolinguistics, neuroimaging, neurosurgery, bilingualism and second language acquisition, language development, specialists in autism spectrum, stroke, and dementias.

The volume is divided into five primary sections: neuroimaging studies of brain and language; language and cognitive development; brain and languages (encompassing building cognitive brain reserve, aphasia and autism spectrum disorders, memory and cognitive-communication disorders); and brain, language, music, and cross-modalities.

Section II Neuroimaging Studies of Brain and Language

In Section II, three chapters review different methodologies to map the brain. The first chapter by Gallo and Abutalebi ("Neuroimaging of Cognitive Reserve in Bilinguals") explores in detail the importance of use of multiple languages in order to change trends in cognitive aging and cognitive reserve. The authors give recent empirical research results from behavioral studies and neuroimaging studies that focus on cognitive mechanisms of the bilingual brain that can produce *cognitive reserve,* which can delay and diminish the onset of symptoms of dementias and provide special attention to the differences in brain reserve, neural reserve, and neural compensation. Conclusions include modeling of the mechanisms of bilingualism-induced reserve, the ubiquitous and equitable nature of bilingualism in the general population worldwide, the socioeconomic consequences of cognitive reserve on successful aging and quality of life.

In the second chapter, Andrews and Rammell ("Networks and Multimodalities Underlying Language Processing in the Brain: Evidence from Structural and Functional Neuroimaging") review a range of studies focusing on language and languages using invasive and noninvasive neuroimaging techniques, including CSM (cortical stimulation mapping), PET (positron emission tomography), structural (DTI) and functional MRI (including task-based and resting state), challenges in interpreting activation results across studies, and the increasing attention to including proficiency measurements to understand neural differences in processing by low vs. higher proficiency users. The role of embodied cognition and multimodalities in language processing and cognitive brain reserve is also examined.

In the third chapter of this section, Kaur, Grant, and Baumer ("TMS as a Tool for Mapping the Dynamic Properties of Language in the Brain") present the neurosurgical clinical perspective on language mapping techniques, including CSM (cortical stimulation mapping), WADA testing, MEG (magne toencephalography), fMRI and TMS (transcranial magnetic stimulation). The primary focus of the discussion is on TMS mapping procedures, especially language mapping with rTMS and error analysis. The final sections compare TMS with all the above mentioned techniques and both questions of safety and the limitations of comparative approaches.

Section III Language and Cognitive Development

Section III deals with several chapters that review the neural regions implicated in language processing in the brain. Among these chapters, several further our understanding of how recent methodological approaches have improved our understanding of language development and language disorders in adults. Also, for the first time, dedicated chapters in this book review applications of resting state fMRI analysis and resting state functional connectivity MRI (rsfc-MRI) specifically to language and brain.

In the first chapter in this section, Bitan ("Functional Connectivity and Recovery from Aphasia") examines the normalization of functional connectivity and, by inference, language processing in individuals with post-stroke aphasia. The effect of stroke and brain lesions has a devastating effect on brain processing, and contrary to old literature that belied any recovery, recent neuroimaging work has highlighted the potential for recovery and reorganization in the adult stroke brain. This chapter first explains the different methodological approaches that have addressed brain connectivity in the context of language processing in aphasia, that is, studies that have examined individuals with aphasia at single time points and those that have examined longitudinal changes in individuals. The author summarizes a convergent finding across all the studies, that after a stroke, the language network tends to normalize its functional connectivity, especially within the left hemisphere and there is concurrent evidence of compensatory connectivity within the right hemisphere and across hemispheres.

Moving on to adult language processing, the next chapter in this section, by Meier and Sandberg ("Recent Developments in Psycholinguistics and Neurolinguistics through the Lens of Aphasia"), reviews recent theories and models of language processing that have emerged from extensive research on the neurobiology of language. For instance, our understanding of semantic processing in the brain has shifted significantly from recent work that describes semantic processing as opposing views on hub-and-spoke processing versus convergent zones. Likewise, the notion of semantic control, a relatively new concept, has been substantiated by a large amount of neuroimaging research. The authors systematically cover other aspects of language processing (phonology, syntax) to bring us up to speed on the latest theoretical frameworks emergent in the research.

The next chapter in this section, addresses a slightly similar topic. Duncan, Pradeep, Kidwai, and Buchwald extend our understanding of brain connectivity beyond the language network and review the existing literature on the default mode network (DMN). Historically, the DMN has been considered a task-independent modality-free network in the brain. Duncan and colleagues review the latest evidential relationship, or the lack thereof, between the default mode network and language processing. Through studies of the healthy brain and also in aging adults as well as in individuals with stroke and neurodegenerative disease, such as dementia and primary progressive aphasia, the chapter highlights the consistent finding that connectivity in the DMN is reduced or impaired in neurological disorders and even though evidence about the link between DMN and language processing is mixed at this time, it may be a promising avenue for interventions.

Expanding beyond language processing, Villard, Zipse, and Vallila-Rohter, review recent evidence on cognitive processes such as attention, memory, working memory, and executive function. Of note, existing neuroimaging evidence has further underscored the link between cognitive and language processing and this chapter tries to link these seemingly disparate processes that also overlap with language processing through evidence from studies on aphasia. For instance, semantic memory, a form of declarative memory, relies on processing the relationship between words and their meanings. As the authors summarize cognitive functions such as memory, working memory and executive function are interconnected to language processing and impact future work on this topic.

In the final chapter of this section, Traxler and Swaab ("Predictive Processing in Language") examine the processing of "real-time" language production and comprehension through a wide range of methodological approaches. Complexities arise because of the ambiguous and multifaceted aspects of linguistic processing and the important role of different contextual outcomes. These complexities require a more robust model of ecologically valid processing and realization of speech acts. The authors challenge models of "pure bottom-up" processing and explore modeling that requires a broader range of processing, including top-down processing and hybridity.

Section IV Brain and Languages: Building Cognitive Brain Reserve and the Importance of Proficiency

Section IV deals with studies that deepen our understanding of lifelong bi/multilingualism and linguistic variation, including cross-modalities and multimodalities. In the first chapter of this section, Pliatsikas ("Bilingualism as a Dynamic Experience and Its Effects on Brain Structure, Function, and Metabolism") brings together a significant body of data that support the view that bilingualism affects domain general cognition and directly impacts both structural and functional aspects of brain function. The author explains the shift in describing bilingualism as a spectral, multivariate experience by using a *dynamic restructuring model* (DRM) and multiple stages in the bilingual experience. The result is new research that shows additional variability in empirical outcomes that were previously unknown. Topics include healthy subject research, developmental perspectives and both healthy aging and neurodegenerative diseases in aging. Structural changes are noted in gray matter and white matter regions. Functional changes are discussed in the context of resting state networks using hemodynamic and electrophysiological imaging techniques.

In the second chapter, Birdsong ("Factors in Variable Outcomes in Second Language Acquisition" (L2A)) focuses specifically on the range of multiple factors that impact second language learning outcomes, the role of working memory in L2A, learner attitudes, musical training, genetics, age of acquisition, and their interactions. Other topics include language dominance relationships and the instability of L1A and L2A learning outcomes.

In the next chapter, Andrews and Swaine ("Longitudinal fMRI Analyses of Second Language Acquisition: Expanding the Repertoire of Imaging Techniques and Proficiency Measures in Future Studies") provide a full analysis of longitudinal fMRI studies of second language acquisition conducted from 2010 to the present. All of the six existing studies use BOLD fMRI task-based protocols, and the results of these studies are compared with cross-sectional studies in order to develop methodologies and new strategies for improving research design and empirical methods used in fMRI. The authors identify important theoretically supported dynamic modeling and use variation in L1/ L2 language processing to provide the basis for the understanding of a broader array of structural and functional imaging technologies (resting state fMRI, functional connectivity, structural DTI, and fractional anisotropy measures). This study calls for more recognition of the importance of including internationally recognized proficiency measurements to enhance diversity in subject selection and improve overall research design.

Likewise, Schwieter and Klassen ("Cognitive and Neural Aspects of the Multilingual Mental Lexicon") focus on the central questions that define the research on the so-called mental lexical and neural representations in the multilingual brain. Key points include degrees and integration of lexical mental networks and how they may differ across languages. The authors consider different viewpoints of lexical-based neural processing as representations or patterns of neural activations, exploration of how this research may contribute to future knowledge about cognition, and implications for modeling the multilingual brain.

Gray and Bose contribute an important review of bilingual language processing and language control in healthy subjects. This study focuses on impairment and recovery interactions, as well as deepening understanding of how bilingual language control and cognitive control interact in general and in approaches to treatment.

In the final chapter, Pecukonis, Butler, Gaynor, and Tager-Flusberg provide a thorough review of different aspects of language processing in autism spectrum disorders (ASD). After first introducing the different methodological advances to study language processing in autism that include structural and functional imaging, the authors go on to review key findings of structural and functional differences in the brains of individuals with ASD compared to typically developing individuals. Critically, the authors summarize that ASD may be associated with reduced language connectivity and reduced volume in key frontal and posterior language regions. Finally, the authors present emerging but exciting evidence about atypical language development in infants with high-risk for ASD, when compared to their low-risk infant siblings. These early differences in reduced brain responses in key language regions and decreased functional connectivity could lead to future work on early diagnosis and early intervention.

Section V Brain, Language, and Music

Section V is devoted to the research focusing on the multi- and cross-modalities interactions between language and music and the examination of networks of regions involved in various aspects of language, music, and cognition.

The first and second chapters in this section (contributions by Bidelman and colleagues, and Van Hedger and Maheshwari) bring new perspectives to the field in a number of ways that amplify previous research on music and language, including their roles in neuroplasticity, cognitive-perceptual benefits, and specific comparative aspects of musical and linguistic structures.

First, Bidelman and colleagues focus on cross-sectional and longitudinal studies that examine the importance of music and musicianship on the *languaging* brain and potential bidirectional "transfer effects" across domains and modalities. Van Hedger and Maheshwari analyze major themes found in the research on music and language, providing compelling reasons to be cautious about overemphasis of what appear to be structural similarities between the two, and the significant differences in *meaning generation* in each. Van Hedger and Maheshwari present a general cognitive framework that can model both systems and includes both acquisition and maintenance at structural and functional neurological levels. Outcomes include elucidation of how training in one of these modalities can enhance the other at the *domain-general* system level while being distinct at the *domain-specific* level.

Section VI New Directions and Perspectives

This section focuses on new topics that intersect with the brain and language.

In the final chapter, Duff and Covington ("Hippocampal Dependent Memory Supports Communication and Language: Implications for Cognitive-Communication Disorders") review the literature that links memory abilities and language processing. Through the lens of studies performed on individuals with hippocampal lesions, the authors review the literature that dissociates between memory issues that are not linked to language impairments and those that identify common deficits in memory and language processing. Perhaps the most interesting aspect of this chapter is the implementation of novel methodological approaches that elicit behaviors of natural language processing that tap into memory processing. Ultimately, the chapter highlights the observation that hippocampal contributions to language processing are not just restricted to the formation and retrieval of new memories but occur during routine language processing tasks as well.

Summary: Questions, Theory, Application

To summarize, this book identifies key new approaches to the study of language and the brain that sets it apart from previous broad reviews on this topic. Specifically, several technological advancements and theoretical shifts in cognitive neuroscience have allowed the understanding of: (a) Expansion of neural regions implicated in research on language and brain (including cortical, subcortical, brain stem, bilaterality); (b) Examination of networks of regions involved in various aspects of language, music and cognition; (c) Examination of the implications of applying principles of embodied cognition and multimodalities; (d) Applications of resting state fMRI analysis and resting state functional connectivity MRI (rsfc-MRI) specifically to language and brain; (e) Inclusion of research focusing on cognitive reserve (GM, WM - gray matter, white matter); understanding high-end sensory-motor advantages (bilingual, music, etc.); (f) Deepening the perspectives of lifelong bi/multilingualism and linguistic variation, including cross-modalities and multi-modalities; and (g) Reimagining the role of cortical stimulation mapping for neurosurgery and language mapping in pathology and how CSM will be relevant in future research in medical and academic research. All these new advancements situate the book in contemporary human and clinical neuroscience and encourage future work on these topics through further sophisticated and impactful questioning.

Part II

Neuroimaging Studies of Brain and Language

Hemodynamic and Electrophysiological Studies

2

Neuroimaging of Cognitive Reserve in Bilinguals

Federico Gallo and Jubin Abutalebi

2.1 Introduction

Our society is witnessing a global increase in average life expectancy, and this trend is expected to continue over the coming years (Kontis et al., 2017; Raftery et al., 2013). This results not only in the obvious benefits of a protracted lifespan, but also in often overlooked drawbacks including the increased incidence of age-related cognitive impairment and dementia, currently among the major causes of decreased quality of life in the elderly (for a review, see Shearer et al., 2012). Alongside these societal repercussions come economic ones: Increasing rates of dementia will result in healthcare resource demands that are projected to progressively burden the governments of industrialized countries (Livingston et al., 2017; Prince et al., 2016; Wimo et al., 2017; Winblad et al., 2016). Concerning the purely economic consequences, recent studies show that senior citizens (aged sixty-five or above) in the United States constitute about 15 percent of the total population but they also account for 34 percent of total health spending (Mitchell, 2016). A similar proportion of senior citizens in Canada account for an even larger healthcare spending share approximating 44 percent (Canadian Institute for Health Information, 2022). Global estimates in 2016 suggested that the total costs of dementia healthcare reached nearly a trillion dollars (Xu et al., 2017). In the United States alone, they already surpass 300 billion dollars per year, and are estimated to reach one trillion dollars in the near future (Wong, 2020).

The term *cognitive aging* refers to the physiological processes of decline in cognition and brain functioning as age increases. From the behavioral viewpoint, nonpathological cognitive aging is linked with deterioration of overall processing speed and mnemonic, linguistic, visuospatial, and executive abilities. At the neural level, cognitive aging is associated with anatomical changes in gray and white matter integrity, particularly in the prefrontal cortex and hippocampus, paired with decreased functional efficiency during task performance (e.g., Dennis & Cabeza, 2008; Farokhian et al., 2017; Giorgio et al., 2010; MacPherson & Cox, 2017; Persson et al., 2006). Rates of cognitive aging are characterized by significant interindividual variability. Potentially explaining such variability, mounting evidence indicates a link between engagement with certain life experiences over the lifespan and mitigation of age-related cognitive decline (e.g., Dause & Kirby, 2019; Goh & Park, 2009; Stern, 2003, 2007; Vemuri et al., 2014). For instance, factors such as frequent physical exercise, healthy nutrition, consistent engagement with various leisure activities, and higher levels of education have all been shown to support successful cognitive aging (e.g., Hötting & Röder, 2013; Yaffe et al., 2009).

Coined to account for the significant interindividual variability observed in cognitive aging trajectories, the concept of *reserve* (see e.g., Stern, 2009) groups together all the aforementioned life experiences and activities resulting in mitigation of age-related cognitive decline. Reserve has been defined as the *discrepancy* between the severity of age-related neural deterioration and the resulting level of cognitive impairment (Stern, 2009). Hence, the protective factors mentioned above would support "successful aging" by fostering an individual's reserve. For a comprehensive overview on the concept of reserve, we refer the reader to Stern's work (e.g., Stern, 2009, 2013; Stern et al., 2020; Tucker & Stern, 2011).

Given their perspective benefits, it is crucial for research to individuate a comprehensive list of lifestyle factors with the potential to foster successful cognitive aging via reserve development. In this chapter, we examine the case of bilingualism, with a particular focus on neuroimaging evidence supporting the beneficial role of bilingualism for cognitive aging and its underlying neural mechanisms.

2.1.1 The Rationale behind the Neurocognitive Consequences of Bilingualism

Widespread experimental evidence supports the notion that bilingual experience benefits cognitive aging by promoting reserve accrual. Recent meta-analyses have reported the onset of dementia to be as much as five to seven years later for bilinguals as compared to monolinguals, despite brains of both linguistic groups showing comparable accrual of pathology-related disruption (Anderson et al., 2020; Brini et al., 2020; Paulavicius et al., 2020). Similar bilingualism-induced benefits have been observed for healthy cognitive aging, with bilingual older adults outperforming monolingual peers in an array of tasks, mainly involving executive functions (e.g., Abutalebi, Guidi, et al., 2015; Bialystok et al., 2004; Del Maschio et al., 2018; Incera & McLennan, 2018; López Zunini et al., 2019), but also related to working memory, semantic memory, and general intelligence (Arce Rentería et al., 2019; Bak et al., 2014; Bialystok, Poarch, et al., 2014; Ljungberg et al., 2013; Wodniecka et al., 2010).

Yet, which are the cognitive mechanisms making the bilingual brain more resilient against age-related decline? Bi/multilingual speakers constantly face an extra cognitive burden: unlike monolinguals they must control their languages. Provided that all languages are constantly activated in the multilingual brain, regardless of contextual need or conscious intent (e.g., Kroll et al., 2008; Spivey & Marian, 1999), potential crosslinguistic conflicts must be resolved continuously to carry out communication successfully. This is done to communicate in only one language while avoiding unwanted interferences from the language or languages not in use. To achieve this goal, bilinguals rely on *language control*, a cognitive device underpinned by a brain network overlapping with the executive functions network (Abutalebi & Green, 2007, 2016; Green & Abutalebi, 2013). Constant involvement of executive abilities and neural resources would eventually lead to the reinforcement of bi/multilinguals' executive functioning. Widespread evidence supports this argument by showing that this "training" affects domain-general executive processes and the underlying neural architecture directly and fundamentally (e.g., Baum & Titone, 2014; Bialystok et al., 2008; DeLuca, Rothman, & Pliatsikas, 2019). Given the central role played by executive functions in the aging process (for a review, see Reuter-Lorenz et al., 2021), this reinforcement is argued to lead to improved cognitive aging trajectories. Several models have schematized how the brain may structurally and functionally accommodate the dynamics of bilingual language control. Some of the most influential are reported in Figure 2.1 (adapted from Gallo, DeLuca, et al., 2022), with the right panel showing convergences and distinctions between models.

In recent years, the emergence of mixed evidence has sparked debate over the extent to which bilingualism actually results in demonstrable neurocognitive gains (e.g., Duñabeitia et al., 2014; Nichols et al., 2020; Paap et al., 2015). The lack of interstudy comparability is a significant issue in bilingualism research, which complicates the endeavor of drawing generalizable or definitive conclusions. The cause is the absence of a universal definition of bilingualism and a shared approach to qualification and quantification of bilingual experience (see Leivada et al., 2021 for discussion). Recently, an effort has been made by researchers in the field toward a more granular assessment of bilingual experience, in an attempt to increase interstudy comparability. Bilingual experience is increasingly being approached as a complex spectrum, developing along different continuous subdimensions (e.g., DeLuca, Rothman, Bialystok, et al., 2019; Gallo et al., 2021; Gullifer et al., 2018; Kuhl et al., 2016; Luo et al., 2019). This shift has further highlighted that bilingualism per se is not sufficient to result in neurocognitive benefits. Certain conditions, particularly in relation to active engagement within bilingual contexts, are required to experience crosslinguistic conflict and the resulting neurocognitive adaptations. Thus, dichotomous categorizations between bilinguals and monolinguals appear inadequate and cannot be expected to provide consistent outcomes.

Beyond behavioral evidence, results from neuroimaging studies show higher consistency. As reviewed in detail below, several investigations show structural and functional neuroplastic effects in the bilingual brain (e.g., Abutalebi et al., 2012; Dash et al., 2019; De Baene et al., 2015; DeLuca, Rothman, &

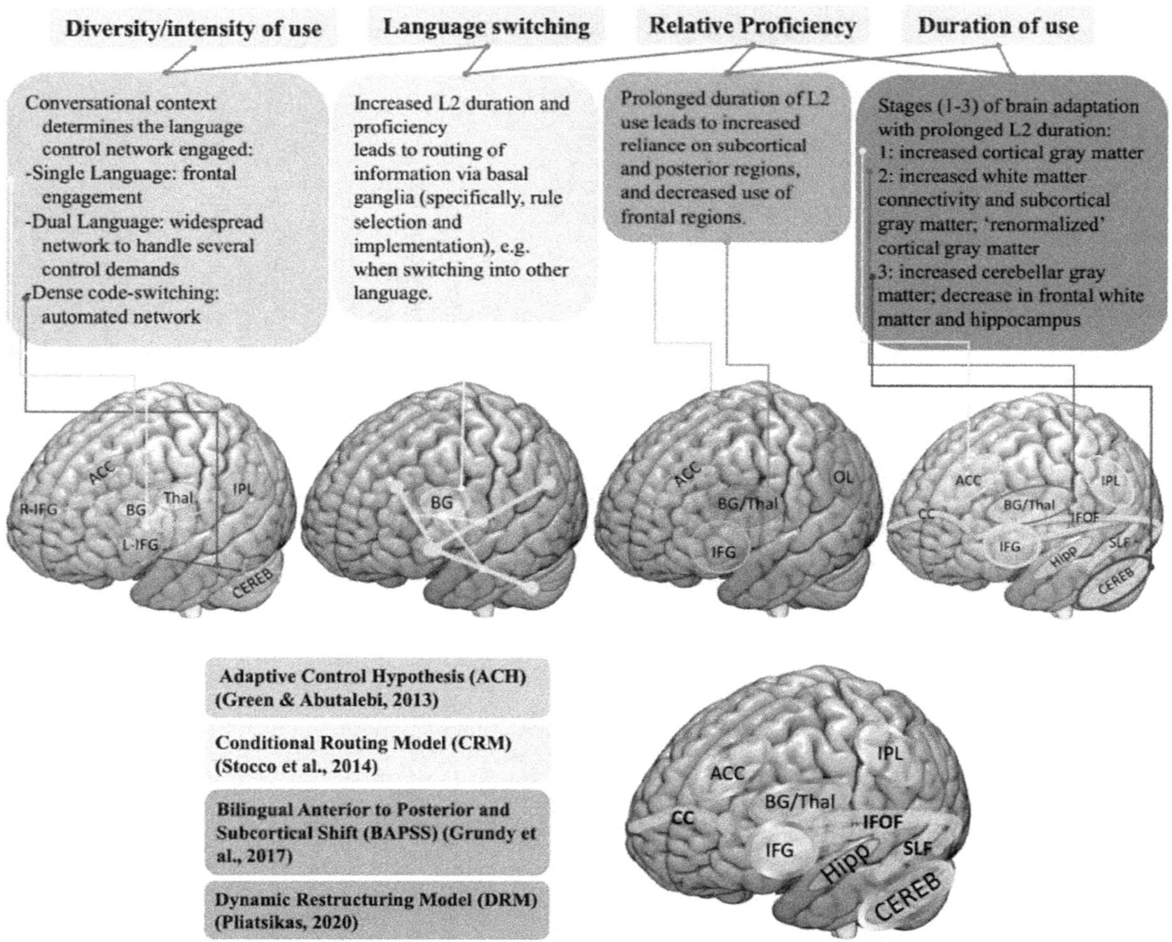

Figure 2.1 Overview of selected models of bilingualism-induced neural adaptations (Green & Abutalebi, 2013; Grundy et al., 2017; Pliatsikas, 2020; Stocco et al., 2014), and overlapping and differing brain areas in different models. Abbreviations: ACC, anterior cingulate cortex; IPL, inferior parietal lobule; BG, basal ganglia; IFG, inferior frontal gyrus; Thal, thalamus; CEREB, cerebellum; OL, occipital lobe; CC, corpus callosum; Hipp, hippocampus; IFOF, inferior fronto-occipital fasciculus; SLF, superior longitudinal fasciculus. Adapted from Gallo, F., DeLuca, V., Prystauka, Y., Voits, T., Rothman, J. & Abutalebi, J. (2022). Bilingualism and aging: Implications for (delaying) neurocognitive decline. *Frontiers of*

Pliatsikas, 2019; Gallo et al., 2021; Sulpizio et al., 2020). Neuroimaging constitutes an important asset given the issues with granularity and test–retest reliability characterizing many of the behavioral tasks commonly used in bilingualism research.

Considering the aim of this chapter, it is important to point out that the cognitive functions and brain regions argued to be enhanced by bilingualism largely overlap with those targeted by cognitive aging, mild cognitive impairment (MCI), and dementia. In other words, bilingualism-induced neural changes such as increased gray matter in those regions will render the very same regions more resistant against age-related atrophy. In what follows, we review literature on the relationship between bilingualism and cognitive aging in detail, with a particular focus on neuroimaging studies. We discuss evidence supporting the notion that bilingualism promotes different submechanisms of reserve in separate sections of this chapter. Subsequently, we discuss theoretical models proposing how the beneficial effects of bilingualism on aging might be achieved via macroscopic and microscopic neuroplastic changes, consistent with the reserve model. We conclude by highlighting the socio-economic potential of bilingualism-induced neuroprotection.

2.2 Bilingualism and Reserve: Evidence from Behavioral Studies

2.2.1 Bilingualism and Cognitive Efficiency during Healthy Aging

A first indication that bilingualism actively promotes successful aging via the development of reserve comes from studies reporting a relationship between multiple language use and behavioral enhancements in executive functioning across the lifespan (for a review, see Bialystok, 2017). Nonetheless, as mentioned above, there is debate over the consistency of these results (for a review, see Paap et al., 2015). Overall, while inconsistencies emerge regarding the benefits of bilingualism for executive functioning in samples of young adults, the picture appears much more conclusive when it comes to elderly individuals. A potential explanation for this discrepancy between age groups was offered by Valian (2015). The author highlighted how younger generations in the present digital era are exposed to a multitude of cognitively challenging activities such as musical training, intellectual exercise, videogaming, and social media, inevitably straining and thus training their cognitive resources. Hence, potential bilingual effects in young adults may be masked by their overall higher cognitive efficiency. Older individuals on the other hand, particularly those in retirement, are on average exposed to fewer cognitive challenges; therefore, the effects of dual language use on their cognitive performance may emerge more clearly. Supporting this argument, several investigations report a beneficial effect of bilingualism on the cognitive integrity of senior individuals. These benefits emerge on a number of different executive tasks that include the Simon Task, the Flanker Task, the Stroop Task, the Antisaccade Task, and various task-switching tests (e.g., Abutalebi,

Guidi, et al., 2015; Bialystok et al., 2004, 2006, 2008; Bialystok, Poarch, et al., 2014; Estanga et al., 2017; Gallo, Kubiak, et al., 2022; Gold, Kim, et al., 2013; Houtzager et al., 2017; Incera & McLennan, 2018; López Zunini et al., 2019). Bilingualism-induced effects on healthy aging, perhaps even more interestingly, seem to extend beyond executive functioning: they have also been reported for memory recall tasks (Ljungberg et al., 2013; Wodniecka et al., 2010), working memory (Bialystok, Poarch, et al., 2014; Luo et al., 2013), semantic memory (Arce Rentería et al., 2019), and even general intelligence (Bak et al., 2014). These results provide initial empirical evidence that bilingualism may promote reserve.

2.2.2 Neuroprotection during Pathological Aging

Extending Valian's (2015) argument, if the effects of bilingualism emerge for healthy aging individuals, who are not at peak cognitive efficiency, such effects should manifest even more clearly among seniors who face the ulterior burden of cognitive impairment related to pathological aging, as in the case of MCI or dementia. Indeed, several retrospective investigations have reported bilingualism-induced delays in the onset of dementia symptoms, particularly with regards to Alzheimer's Disease (AD).

A pioneering study in this regard was conducted by Bialystok and colleagues (Bialystok et al., 2007), which showed a four-year delay in AD onset in a sample of bilingual patients, as compared to monolinguals matched for gender ratio, education years, and socioeconomic status, factors that are well known to affect trajectories of dementia (Meng & D'Arcy, 2012; Sattler et al., 2012). Ever since, several studies have tried to replicate this finding. While initial support was provided by another investigation from the same group (Craik et al., 2010), two subsequent early studies raised the concern that bilingualism-induced delay of pathological aging may be restricted to specific social contexts. One such study (Gollan et al., 2011) replicated the delay effects in bilinguals with low education but not in highly educated ones. Another (Chertkow et al., 2010) reported the effects only for immigrant – but not for nonimmigrant – bilinguals. In turn, these findings were contradicted by the following studies. Regarding education, a 2017 study (Ramakrishnan et al., 2017) directly compared the independent contributions of bilingualism and education to age of MCI onset in 115 individuals (it should be noted that Gollan et al.'s study had a much smaller sample of only 44 participants). Education appeared to exert no delaying effect on MCI onset, while bilingualism delayed it by 7.4 years. A per immigrant-status study from Alladi et al. (2013) compared age of AD onset between bilinguals and monolinguals in a large sample of 648 individuals, which included both immigrants and nonimmigrants. A delay of 4.5 years emerged for nonimmigrant bilinguals as compared to monolinguals, supporting the view that bilingualism-induced neuroprotection against AD is unrelated to immigrant status. Furthermore, the study provided a more nuanced view of the concurrent effects of bilingualism and education on pathological aging. Although an independent delaying effect was observed for bilingualism, the delay increased even further (i.e., to six years) in low-educated bilinguals. These early results were then replicated by several other investigations. Bialystok and colleagues (Bialystok, Craik, et al., 2014) reported a delay of 4.7 years for AD and 7.3 years for MCI in bilingual as compared to monolingual patients. Of note, the linguistic groups were matched on a number of variables known to promote successful aging via reserve development, namely frequency of physical and leisure activities, number of social interactions, and health-related variables such as dietary, smoking, and alcohol-consumption habits. Another replication came from Woumans and colleagues (2015) who observed a 4.6-year delay in AD onset in bilinguals as compared to monolinguals matched for sociodemographic and disease-severity profiles. Finally, two more studies, by Zheng and colleagues (2018) and Mendez and colleagues (2019) reported a seven-year and a four-year AD delay for bilingual patients, respectively.

The finding of a bilingualism-induced delay in AD onset is further corroborated by meta-analyses published in recent years. Anderson and colleagues (2020) reported a moderate effect size for bilingualism-induced delays on AD (Cohen's $d = 0.4$), but only a weak effect size for lowered AD incidence among bilinguals (Cohen's $d = 0.1$; see below for further discussion on effects of bilingualism on AD incidence). Of note, Anderson and colleagues' effects survived when controlling for potential confounds from socioeconomic status, maximal educational attainment, and publication bias. Another meta-analysis (Brini et al., 2020), including data from twenty-one studies, concluded that bilinguals experience AD 4.7 years later than monolinguals, and are diagnosed 3.3 years later. Similar results emerged in another meta-analysis (Paulavicius et al., 2020) which reported a four-year delay in AD symptom onset and a two-year delay in AD diagnosis.

The literature also presents studies that aimed at delineating the differential effect of bilingualism on different variants of AD and MCI, improving the granularity of our knowledge on the relationship between bilingualism and pathological aging. For instance, Ossher and colleagues (2013) reported a bilingualism-induced 4.5-year delay in the onset of single-domain, but not multiple-domain amnestic MCI, namely the type of MCI where memory is specifically impaired. Of note, single-domain amnestic MCI – in which only one cognitive domain is impaired – is the subtype most frequently converting to AD. In a study with a similar approach, de Leon and colleagues (2020) reported a bilingualism-induced five-year delay for the logopenic variant of primary progressive aphasia (see e.g., Henry & Gorno-Tempini, 2010), but not for amnestic AD.

Besides delaying effects, some investigations also report evidence for bilingualism-induced reductions in AD and MCI incidence. For instance, Perquin and colleagues (2013) observed a negative relationship between MCI incidence and number of languages spoken in a sample of 232 individuals. Furthermore, a longitudinal study by Wilson and colleagues (Wilson et al., 2015) reported lower incidence of MCI for bilinguals as compared to monolinguals across a six-year follow up in a sample of 964 older adults. Finally, adopting a slightly different approach, Klein and colleagues (2016) reported a negative relationship between rates of multilingualism and AD incidence in a sample of ninety-three countries from across the globe.

Furthermore, some evidence suggests also a potential beneficial role of bilingualism on neurodegeneration resulting from pathologies beyond AD and MCI (Voits et al., 2020), in line with the view that bilingualism promotes the development of reserve in general. In fact, enhanced reserve has been linked to more positive outcomes in a range of neurodegenerative diseases (Hindle et al., 2017; Sumowski et al., 2009) such as: (i) diminished cognitive impairment in Huntington's disease patients as a result of better preserved brain metabolism and structural integrity in frontal brain areas (Martínez-Horta et al., 2018); (ii) delayed onset of Parkinson's disease symptoms (Ghazi-Saidi, 2019); (iii) better cognitive outcomes in patients suffering from multiple sclerosis (Aveledo et al., 2020; Soltani et al., 2018); (iv) improved post-stroke cognitive recovery (Alladi et al., 2016), and less severe symptoms of post-stroke aphasia (Paplikar et al., 2018).

However, we invite the reader to note that the studies mentioned in this section have only considered behavioral outcomes. Although this evidence suggests that bilingualism may support the development of reserve during healthy and pathological aging, it is not sufficient to gain a comprehensive view of neurocognitive mechanisms underlying these phenomena.

In the next sections, we review neuroimaging studies, that is, brain evidence, which may help us to clarify how bilingualism interacts with different sub-mechanisms of reserve to achieve neuroprotective effects against cognitive aging.

2.3 Bilingualism and Reserve: Evidence from Neuroimaging Studies

2.3.1 Sub-Mechanisms of Reserve: Brain Reserve, Neural Reserve and Neural Compensation

Reserve effects are theorized to be underpinned by two sub-mechanisms: *brain reserve* (BR) and *cognitive reserve* (CR; see Figure 2.2, adapted from Gallo et al., 2020). BR is defined as the individual's capacity for resilience to age-related neurostructural deterioration. It originates from both biological (e.g., brain size or neuronal and synaptic counts) and environmental (e.g., neuroplastic structural changes promoted by life experiences) factors. BR acts as a *passive* buffer against age-related cognitive decline, by increasing the critical threshold at which neurostructural deterioration causes cognitive impairment (Stern, 2009).

CR, instead, can be described as an *active* mechanism of compensation against age-related cognitive decline. CR manifests in two separate – although

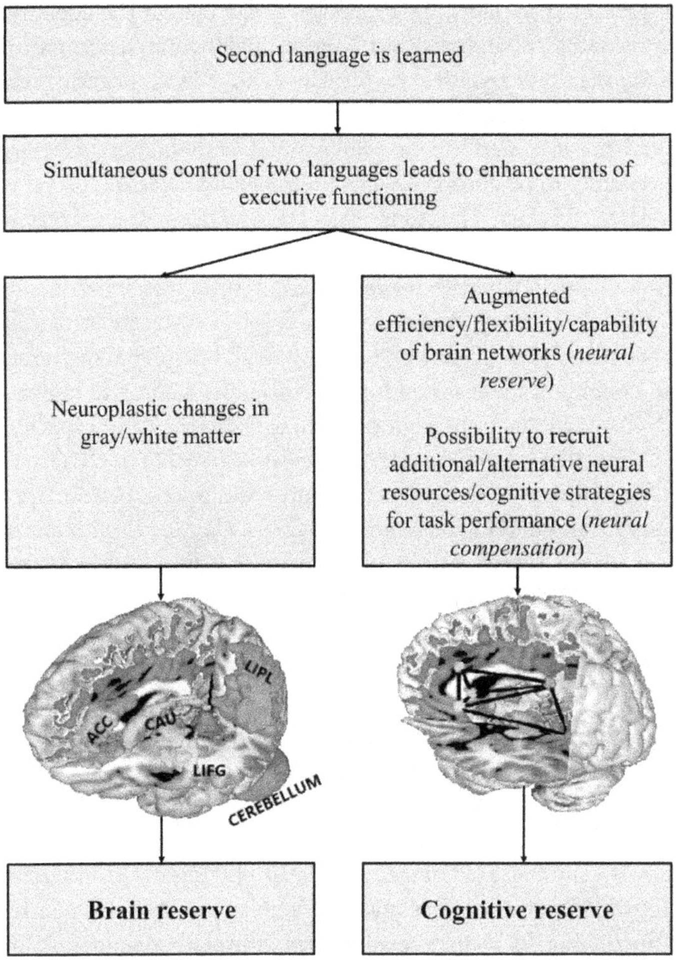

Figure 2.2 Schematic overview of the brain mechanisms underlying bilingualism-related protective effects against age-related neurocognitive decline. Abbreviations: ACC, anterior cingulate cortex; LIPL; left inferior parietal lobule; CAU, caudate nucleus; LIFG, left inferior frontal gyrus.
Adapted from Gallo, F., Myachykov, A., Shtyrov, Y., & Abutalebi, J. (2020). Cognitive and brain reserve in bilinguals: Field overview and explanatory mechanisms. *Journal of Cultural Cognitive Science*, *4*(2), 127–143. By permission of Springer Nature.

interrelated – ways (see Figure 2.2). The first sub-mechanism of CR is termed *neural reserve* (NR). It is defined as the level of efficiency, capacity, and flexibility of an individual's neural networks (Steffener et al., 2011; Stern, 2009). The term *efficiency* describes the minimum amount of neural activation required to successfully perform a cognitive task. The lower is the activation required, the higher an individual's neural efficiency. Individuals with highly efficient neural networks deal better with tasks of high difficulty, as they maintain more neural resources to be deployed when task difficulty increases. If an individual's neural efficiency is low, the maximal activation threshold is reached earlier, and further increases task difficulty leading to the deterioration of performance. How higher efficiency can benefit the cognitive aging trajectory is rather straightforward. Age-related neural deterioration poses an extra burden on cognitive performance, in a way lowering the threshold for what can be considered a "difficult" task. With a lower amount of effort sufficient for task performance, individuals with higher efficiency maintain extra resources to be deployed when their neural processing becomes taxed by age-related decline. The term *capacity* describes the maximum amount of activation a neural network can reach. Similarly to efficiency, higher network capacity allows individuals to deploy extra resources when facing increasing task difficulty. Finally, *flexibility* is the level of connectivity within a network and between different networks. Better neural pairing allows areas and networks to communicate faster and more efficiently, leading to benefits via similar mechanisms to those already described for efficiency and capacity.

The second sub-mechanism of CR is termed *neural compensation* (NC), and it describes the development of noncanonical approaches to task performance to cope with age-related neural deterioration. NC acts via the recruitment of additional and/or alternative neural resources/cognitive strategies, as compared to those typically deployed by healthy younger individuals (Steffener et al., 2011; Stern, 2009).

2.3.2 Reinforcing the Neural Substrate: Brain Reserve in Bilinguals

The accrual of BR is known to begin well before senescence, driven by lifelong experiential factors. Evidence of bilingualism-induced BR development has indeed been reported for bilinguals across the lifespan. Several neuroimaging studies have shown volumetric differences between bilinguals and monolinguals in gray and white matter regions involved in bilingual language control and executive functioning, as well as relationships between individual differences in bilingual experience and volumetric variations in such areas.

A landmark study in this sense is that by Mechelli et al. (2004), which revealed higher gray matter volumes (GMVs) in the left inferior parietal lobule (IPL) of bilingual young adults, as compared to monolingual controls. Such volumetric differences were further predicted by second language (L2) proficiency and age of acquisition. Subsequent investigations in young adult populations confirmed such findings for the IPL (Del Maschio et al., 2018; Della Rosa et al., 2013; Heim et al., 2019; Olulade et al., 2016) and provided further evidence of bilingualism-induced neuroplasticity in several other cortical and subcortical regions of the language control/executive functions network, including the anterior cingulate cortex (ACC) (Abutalebi et al., 2012; Del Maschio et al., 2018; Gallo et al., 2021), the cerebellum (Filippi et al., 2011; Pliatsikas et al., 2014), the dorsolateral prefrontal cortex (PFC) (Del Maschio et al., 2018; Olulade et al., 2016; Stein et al., 2012), Heschl's gyrus (Ressel et al., 2012), the putamen (Abutalebi et al., 2013; Pliatsikas et al., 2017), the caudate nucleus (Zou et al., 2012), the thalamus (Pliatsikas et al., 2017), and the globus pallidus (Pliatsikas et al., 2017).

Results linking bilingual experience to BR accrual in the form of neuroplastic changes in the gray matter substrate have been also reported in investigations involving elderly populations. Similar to young adults, these effects have emerged mainly in regions of the language control network, including the IPL (Abutalebi, Canini, et al., 2015; Del Maschio et al., 2018), the ACC (Abutalebi, Guidi, et al., 2015; Del Maschio et al., 2018), the PFC (Del Maschio et al., 2018), but also the temporal pole (Abutalebi et al., 2014; Olsen et al., 2015), the orbitofrontal cortex (Abutalebi et al., 2014), and the hippocampus (Voits, 2020). Importantly, some of these brain areas are targeted by age-related brain deterioration particularly early and severely, as in the case of the temporal pole (Kalpouzos et al., 2009), the IPL (Apostolova et al., 2007), and the hippocampus (Maruszak & Thuret, 2014). Thus, it is easy to individuate how bilingualism can contribute to successful aging trajectories via the development of BR in such areas. For instance, higher GMVs have been observed in the temporal pole of healthy bilingual seniors, as compared to monolingual age peers (Abutalebi et al., 2014). Interestingly, bilinguals' temporal pole volumes appeared to be associated with L2 proficiency level, suggesting a dose-response effect of bilingualism on volumetric variations. In line with this result, one subsequent investigation (Olsen et al., 2015) showed greater cortical thickness in the temporal pole of healthy senior bilinguals, as compared to monolingual age peers. Strikingly, similar results have been reported when comparing different types of bilinguals, that is, bimodal bilinguals, to monolingual controls (Li et al., 2017). Indeed, aging bimodal bilinguals (individuals who use a spoken language and a sign language) displayed increased GMVs in the left temporal pole when compared to age-matched monolingual peers. Regarding the IPL, the fact that studies report a bilingualism-induced neuroplastic trajectory not limited to senescence (Abutalebi, Canini, et al., 2015; Del Maschio et al., 2018), but already present in young adulthood (Del Maschio et al., 2018; Mechelli et al., 2004), and even childhood (Della Rosa et al., 2013), suggests that bilingualism-induced accrual of BR may begin in early life stages.

In addition to effects on gray matter, bilingualism has also been shown to be linked with BR development in the shape of white matter preservation during the aging trajectory. In a study utilizing diffusion tensor imaging (DTI), Luk and colleagues (2011) observed better preservation of white matter integrity in healthy bilingual older adults, relative to monolinguals, in a number of tracts including the corpus callosum, the superior longitudinal fasciculi, the right inferior fronto-occipital fasciculus, and the right uncinate fasciculus. Importantly, these white matter tracts are known to be severely affected by age-related deterioration (e.g., Gunning-Dixon et al., 2009). Furthermore, in an abovementioned investigation, Olsen and colleagues (2015) observed higher frontal lobe white matter volumes in bilingual as compared to monolingual seniors. Finally, Anderson and colleagues' (2018) DTI study showed a positive relationship between bilingual experience and white matter integrity in the left superior longitudinal fasciculus. Taken together, the findings reviewed above suggest that one way in which bilingualism supports successful aging is by promoting the development of BR. Yet, as mentioned above, the reserve framework posits a second mechanism through which life experiences may mitigate healthy and pathological cognitive aging, CR. In the next section, we review neuroimaging evidence supporting the notion that bilingualism triggers CR effects via the two sub-mechanisms of NR and NC.

2.3.3 Efficiency, Capacity and Flexibility: Neural Reserve in Bilinguals

Similar to BR, NR accrual is also not restricted to late life stages, but begins early in the lifespan. Indeed, evidence for bilingualism-induced NR effects comes from different age groups, spanning from youth to senescence. In an fMRI study, Luk and colleagues (2010) presented bilingual and monolingual young adults with a modified version of the Flanker task, in which a NoGo condition was inserted. Participants were required to refrain from responding to trials in the NoGo condition. Such a condition allowed the authors to distinguish between two cognitive processes usually subsumed under the umbrella term of *inhibitory control,* namely interference suppression (required for standard incongruent trials) and response inhibition (required for NoGo trials). Analyses showed that bilinguals activated a shared network for both incongruent and NoGo trials, while monolinguals showed different activation patterns for the two conditions. The authors concluded that for bilinguals, who are extensively trained in interference suppression, activation of the domain-general attentional control network sufficed to successfully process both types of trials. This result suggests that bilinguals may perform interference suppression more "automatically" and with less effort, supporting the notion that bilingual experience triggers NR effects in the form of increased neural efficiency. Providing further evidence supporting increased network efficiency in bilinguals is another fMRI-Flanker investigation in young adults by Abutalebi and colleagues (2012). The study showed that bilinguals outperformed monolinguals on the Flanker task while also showing significantly lower task-related ACC activation, in full alignment with the notion of neural efficiency and NR. The authors linked their results once again to the constant training in conflict resolution undergone by bilingual brains. The study also provided evidence for increased flexibility of bilinguals' ACC, in the form of faster adaptations to task demands. In the second run of the task, in fact, Flanker conflict effect significantly decreased for bilinguals, but not monolinguals. This pattern reflected in neuroimaging data, with bilinguals that, besides displaying lower ACC activation already in the first run, showed a further activation decrease in the second run, unlike monolinguals. Similar evidence of a more efficient use of the ACC by bilinguals, as compared to monolinguals, was reported in a subsequent investigation deploying a stop-signal task (Rodríguez-Pujadas et al., 2014).

Evidence for bilingualism-induced accrual of NR also comes from investigations conducted in senior populations. In an fMRI study, Gold and colleagues (Gold, Kim, et al., 2013) found that bilingual seniors exhibited lower switching costs – namely the difference in neural activation between trials that require switching and trials that do not – in the left dorsolateral and ventrolateral PFC and the ACC, as compared to monolinguals. Such a result indicates, once again, increased neural efficiency because of bilingual experience. Moreover, bilingual seniors outperformed monolinguals on the switching task, similarly to younger participants of the study. In similar vein, Ansaldo and colleagues (2015) reported differential task-related recruitment during a Simon Task for bilingual and monolingual healthy older adults. Indeed, despite comparable levels of task performance, during incongruent trials monolinguals activated frontal areas, while bilinguals recruited the IPL. Bilinguals appeared to avoid the typical age-related posterior-to-anterior shift in the locus of task-related activation, thought to depend on age-related deterioration of posterior-parietal neural efficiency (Davis et al., 2008; see below for further discussion). In another fMRI investigation using an Attention Network task, Dash and colleagues (2019) found a relationship between neural efficiency (as indexed by lower task-related activation) and L2 proficiency, suggesting that the effects of bilingualism are not univocal in late life stages, but rather vary as a function of L2 experiential factors, as also reported for younger age groups (e.g., DeLuca, Rothman, Bialystok, et al., 2019; Gallo et al., 2021; Li et al., 2014).

The literature also presents evidence of bilingualism-related NR development in the form of enhanced preservation of functional connections in the brain. One study providing such evidence is the abovementioned investigation by Luk and colleagues (2011), which showed better preserved anterior-to-posterior resting-state functional connectivity (rs-FC) in a sample of healthy bilingual seniors, relative to monolingual age peers. A subsequent fMRI study held similar findings, revealing increased connectivity for bilinguals in the default mode and executive control networks (Grady et al., 2015). Another fMRI study showed lower task-related recruitment needs for bilinguals during a Simon task. Healthy bilingual seniors showed task-related increases in connectivity only in visuospatial areas, while monolinguals needed to rely on increased connectivity in a much wider network including visual, motor, and executive areas. Finally, similar findings also emerged from a magnetoencephalography (MEG) investigation by de Frutos-Lucas and colleagues (2020), who reported enhanced preservation of senior bilinguals' rs-FC in several occipitoparietal clusters known to be particularly subject to AD-related disruption (Nakamura et al., 2017).

Perhaps the most interesting mechanism via which bilingualism-induced NR manifests is *modulation* of the structural brain behavior (SBB) relationship. This relationship predicts that for increasing density/integrity of structural neural resources in a given region or network, performance levels in corresponding cognitive ability will also increase. Conversely, in the face of resources deterioration, performance levels are expected to drop accordingly. Since reserve is defined as the *discrepancy between the severity of age-related neural deterioration and the resulting level of cognitive impairment,* the modulation of the SBB relationship is a clear sign of reserve. Several investigations report such modulatory effects in the relationship with bilingualism. A study by Gold and colleagues (Gold, Johnson, et al., 2013) revealed lower white matter integrity in the inferior longitudinal fasciculus, the inferior fronto-occipital fasciculus, the fornix, and parts of the corpus callosum for bilingual healthy seniors versus monolinguals with comparable levels of cognitive status and demographics. This finding supports the presence of bilingualism-induced NR effects, since the SBB relationship would have predicted greater cognitive impairment for bilinguals, who displayed lower neurostructural integrity. Thus, the relationship appeared to be mitigated by bilingualism. A subsequent study adopting the opposite approach provided consistent results: Berkes and colleagues (2021) matched their bilingual and monolingual samples for white matter integrity and demographics, and investigated cognitive performance differences. Bilingual seniors exhibited better cognitive efficiency despite matching levels of age-related white matter deterioration. Another investigation revealed that the conflict effect on the Flanker task was positively correlated with GMVs in the dorsolateral PFC healthy monolingual older adults, but not for bilingual age peers (Abutalebi, Guidi, et al., 2015). This result was later replicated and expanded in a study reporting bilingualism-induced mitigation effects on the SBB relationship in multiple regions of the language control network, including the bilateral PFC, the bilateral IPL, the left ACC, and the left caudate nucleus (Del Maschio et al., 2018). Similarly, a longitudinal investigation in bilingual and monolingual seniors revealed that bilingualism mitigated the semantic memory loss linked with enthorinal cortex thinning (Arce Rentería et al., 2019).

Taken together, these findings indicate that bilingual experience supports successful aging via NR development, by enhancing the *efficiency* and *flexibility* of the executive network. The reserve theory argues that these enhancements would result in neuroprotection from age-related cognitive decline by allowing individuals to compensate for neural deterioration during healthy and pathological aging, namely via the related mechanism of NC. In the next section, we review evidence supporting a role of bilingualism in NC development.

2.3.4 Bypassing Pathological Aging: Neural Compensation in Bilinguals

As suggested by the term, high levels of *neural compensation* allow individuals to maintain optimal or nearly optimal cognitive efficiency even in the face of neural deterioration due to healthy or pathological cognitive aging. Several neuroimaging investigations have provided evidence that bilingualism supports the development of NC, mainly by comparing neural integrity between bilingual and monolingual patients matched for cognitive status, disease severity, or disease duration. Here, the reader must keep in mind a potentially counterintuitive argument that stems from the definition of NC. For comparable levels of cognitive functioning or disease severity, the individual with higher NC is the one who displays *lower* neural integrity (i.e., reporting more extended brain damage or atrophy). In fact, notwithstanding deterioration at the neural level, this individual is able to compensate for the loss of brain structure and, hence, to preserve cognitive functioning.

The first evidence for the role of bilingualism as a promoter of NC came from an investigation by Schweizer and colleagues (2012), who used computed tomography (CT) brain scans to evaluate differences in brain atrophy between bilingual and monolingual AD patients matched for maximal educational attainment and disease severity. Bilinguals showed greater neural atrophy in the temporal horn ratio, third ventricle ratio, and the radial width of the temporal horn, well-known indicators of AD-related disruption (Frisoni, Geroldi, et al., 2002; Frisoni, Rossi, et al., 2002; Zhang et al., 2008). In another study, Gold and colleagues (Gold, Johnson, et al., 2013) compared the level of white matter integrity of bilingual and monolingual older adults matched for performance levels on an array of cognitive tests and several CR-enhancing factors including general intelligence, maximal educational attainment, and socioeconomic status. The bilingual sample displayed lower integrity in a number of white matter tracts part of the memory neural network, known to be severely targeted by AD- and MCI-related deterioration (Stebbins & Murphy, 2009). Again, this result speaks in favor of a relationship between dual language use and NC development, with bilinguals able to maintain comparable levels of cognitive functioning compared to monolingual peers, despite the lower neural resources available. It should be noted that this outcome may initially seem to contradict findings of some investigations mentioned above (i.e., Anderson et al., 2018; Luk et al., 2011; Olsen et al., 2015; see Section 2.3.2), which reported higher white matter integrity linked to bilingualism. Nonetheless, one must consider that the abovementioned studies included healthy individuals, while among Gold and colleagues' participants we can infer a higher incidence of preclinical AD, based on the specific set of white matter tracts that show deterioration, namely the corpus callosum, the inferior longitudinal fasciculus, the inferior fronto-occipital fasciculus, and the fornix (Stebbins & Murphy, 2009). It must be kept in mind that BR and CR are largely unrelated mechanisms, which may intervene at different points during the lifespan. Fluorodeoxyglucose positron emission tomography (FDG-PET) investigations also produced results consistent with a bilingualism-related contribution to NC development. Kowoll and colleagues (2016) compared glucose uptake (an AD biomarker) between bilingual and monolingual MCI patients diagnosed with probable AD and matched for age, gender ratio, and disease severity. Bilinguals displayed lower glucose uptake in frontotemporal, parietal and cerebellar brain areas. Similarly, two further FDG-PET investigations (Perani et al., 2017; Sala et al., 2022) in monolingual and bilingual AD patients matched for disease duration and cognitive status, reported more severe glucose hypometabolism for bilinguals, paired with increased functional connectivity in the executive and default mode networks. Moreover, in one of the two studies (Perani et al., 2017) the bilingual sample was found to be four and a half years older than the monolingual one, in striking resemblance with the studies mentioned in Section 2.2.2.

Further findings indicating a bilingualism-induced development of NC include mitigating effects of bilingualism on t-tau protein amounts, a cerebrospinal fluid (CSF) biomarker of AD, which were reported by Estanga and colleagues (2017) in a sample of 278 individuals matched for genetic, sociodemographic, and cognitive profiles. Finally, in a longitudinal study, Costumero and colleagues (2020) reported lower volumes of brain parenchyma (the sum of gray and white matter) for bilingual MCI patients relative to monolingual patients with comparable levels of cognitive functioning at baseline, as well as greater tissue loss and cognitive decline for monolinguals across the seven months follow-up.

Altogether, a clear pattern emerges from the evidence reviewed above, suggesting that another way in which bilingualism supports successful aging is by promoting the development of NC, which allows dual language speakers to compensate for age-related neural disruption and preserve cognitive functioning.

2.4 Modeling the Underlying Mechanisms of Bilingualism-Induced Reserve Development

As reviewed above, the literature provides widespread evidence supporting the claim that bilingual experience promotes the achievement of successful aging trajectories by aiding reserve accrual. Nonetheless, the precise neurocognitive mechanisms underlying such findings are yet to be presented in detail. Below, we discuss the main models detailing bilingualism-related neurocognitive effects leading to benefits for cognitive aging, under the light of the reserve framework.

Importantly for the argument developed herein, reserve is posited to originate from environmental influences, such as regular lifestyle activities that exert a "training" effect on the mind and brain. The evidence reviewed above presents all the typical features of observable consequences of CR and BR accrual. For example, results linking bilingualism to structural and functional neuroplastic trajectories are well in line with the notions of BR and CR respectively. Furthermore, the observed bilingualism's mitigation of the SBB relationship is perfectly overlapping with the classical definition of reserve as *the discrepancy between the severity of age-related neural deterioration and the resulting level of cognitive impairment* (Stern, 2009; Stern et al., 2020). The key question is then *how* does bilingual experience translate to structural and functional neuroplastic changes? In other words, how are BR and CR accrual triggered by bilingualism? At the macroscopic level, the rationale for these effects is the one mentioned at the beginning of this chapter, that is, the argument that the increased requirement for executive control necessary to avoid crosslinguistic interference is taxing and thus training the executive system. The logical consequence to this claim is that, similar to other lifestyle activities known to promote reserve, one should expect a dose-response effect of bilingualism on reserve accrual. This relationship is indeed emerging in the recent years, which witnessed a realization of the necessity to assess the spectrum of bilingual experience from an individual-differences standpoint rather than a monolithic dichotomous categorization between bilingualism and monolingualism. This in turn allowed researchers to show that bilingualism's neurocognitive consequences indeed surface differentially based on variations in the quantity and quality of dual-language engagement (e.g., DeLuca, Rothman, Bialystok, et al., 2019; Gallo et al., 2021; Li et al., 2014; Luk & Bialystok, 2013; Pliatsikas et al., 2020). If training of the executive system constitutes the explanation for bilingualism-induced neurocognitive effects at the *macroscopic* level, what happens in more detail in the mind and brain of bilinguals, to eventually lead to successful aging?

Several models have been proposed to map the cognitive challenges posed by bilingual experience to the outcome of developing different subtypes of reserve. For instance, Guzmán-Velez and Tranel (2015) individuate the origin of bilingualism-induced BR accrual in bilingualism-related modifications of the noradrenergic system (a neural circuitry underlying the production and circulation of the neurotransmitter norepinephrine). In turn, their account is based on the noradrenergic theory of reserve by Robertson (2013), which argues that BR and CR originate from the upregulation of noradrenergic circuits. A more efficient noradrenergic system is posited to result in BR accrual in the form of enhanced neurogenesis, angiogenesis, synaptogenesis, and increased production of the brain-derived neurotrophic factor (BDNF). Noradrenergic upregulation would also support CR development in the shape of compensatory mechanisms that mitigate the impact of age-related neuropathology. These include reduced amyloid burden, anti-inflammatory processes, and replenishment of cholinergic and dopaminergic cellular resources. The argument by Guzmán-Velez and Tranel is simple: bilingualism, not differently from other reserve-enhancing life experiences, would promote reserve accrual via upregulation of the noradrenergic system.

Models that focus on more macroscopic neural mechanisms have also been proposed. Grant and colleagues (2014) frame underlying mechanisms of bilingualism-induced successful aging in terms of the posterior-to-anterior shift in aging (PASA) (Davis et al., 2008). The PASA model stems from the observation that seniors tend to shift the locus of activation related to task performance to more anterior regions compared to younger individuals. This account claims that in the face of an age-related loss of efficiency in occipito-temporal regions, the recruitment of domain-general frontal areas would act as a compensatory mechanism to maintain optimal or nearly optimal cognitive performance. Nonetheless, since bilingualism has been shown to induce better preservation of white and gray matter resources in posterior-parietal regions (see Section 2.3.2), bilingual seniors are expected to postpone the PASA mechanism, saving extra cognitive resources to be deployed in the eventuality of an increase in task difficulty. Expanding this view, Grundy and colleagues (2017) proposed the bilingual anterior to posterior and subcortical shift (BAPSS) model, which argues that sufficiently expert bilinguals would even experience a posterior/subcortical-directed shift of the locus of activation related to executive functioning. The BAPSS builds on evidence of bilingualism-induced (i) neuroplasticity in posterior and subcortical gray and white matter (i.e., BR); (ii) lower executive-task-related frontal activation (i.e., NR in the form of increased *efficiency*); (iii) increased executive-task-related functional connectivity (i.e., NR in the form of increased *flexibility*); (iv) stronger reliance on earlier processing stages for executive task performance, as shown by EEG studies (i.e., NC in the form of alternative, more automated, strategies). This evidence led Grundy and colleagues to argue that bilingualism, via these functional and structural modifications, allows individuals to deploy more automatic, early, and economic executive processing in lieu of late, top-down mechanisms requiring a higher expenditure of neurocognitive resources. Similar to what was hypothesized for PASA, this shift would allow bilinguals to postpone the need to recruit frontal areas for compensation purposes and maintain a *fuller warehouse* of cognitive resources to be deployed in the event of more demanding task requirements.

The Dynamic Restructuring Model (DRM) (Pliatsikas, 2020) proposes arguments consistent with the abovementioned models, although describing in higher detail the correspondence between subsequent steps in the bilingual experience and progressive stages of the resulting neuroplastic trajectory. The DRM describes a stepwise mechanism of expansion and renormalization, that here we frame under the light of different reserve sub-mechanisms.

(i) At initial stages of L2 acquisition, the novel cognitive burden imposed by language control demands induces BR accrual in the form of neurostructural changes in the white and gray matter substrates, achieved via synaptogenesis.

(ii) Subsequently, progressive bilingual experience triggers NR mechanisms in the form of increased functional efficiency and flexibility of the language control network. At this stage, while part of the previously formed structural resources are consolidated, others become unnecessary for optimized language control. Thus, surplus connections are trimmed via a process called *synaptic pruning* (see e.g., Lövdén et al., 2013), leading to a partial renormalization of gray and white matter volumes toward pre-bilingualism levels. Crucially, as described by theories of synaptic reorganization (see e.g., Lövdén et al., 2013), only the *most efficient* synaptic connections survive the pruning phase (BR, NR).

(iii) As a result of this trajectory, the synaptic connectivity of the bilingual executive network undergoes a reorganization leading to higher efficiency and flexibility (BR, NR). It must be noted that cognitive aging was not the original focus of the DRM. Nonetheless, its central concept of *expansion-renormalization* seems to provide the ideal explanation for how micro- and macroscopic neuroplastic changes may lead to BR and CR accrual in bilinguals. Indeed, the DRM embodies both the notions of *quantity* and *quality* of neural resources described in the reserve theory. Senior bilinguals, as compared with monolingual peers, may not necessarily possess "more" neural resources, but potentially display similar amounts of such resources that prove nonetheless more resilient against age-related deterioration.

2.5 Conclusion: The Socioeconomic Consequences of Bilingualism

Having reviewed the theory and evidence behind the neuroprotective potential of bilingualism for aging trajectories, below we highlight the relevance of this research program on many levels.

First, we frame the body of research within the real-world context, discussing its potential benefits for multiple stakeholders for whom it may have important implications. Investigating new ways to prevent age-related cognitive decline constitutes at present a societal imperative, considering the increasing share of older individuals in societies and the ever-growing projections of life expectancy resulting from advances in healthcare (Kontis et al., 2017; Raftery et al., 2013). In this context, quality of life and economic issues are intertwined, making understanding, prediction, and treatment of cognitive aging a universal concern.

Clinical trials aimed at individuating pharmacological solutions for age-related neuropathology have, so far, produced far-from-satisfying results (Dyer et al., 2018). For instance, at present, no groundbreaking pharmacological treatment for AD is available. Until pharmacological solutions become available, individuating preventive measures to reduce the risk of age-related cognitive decline is the best option to prolong individuals' healthy cognitive life expectancy, amidst what has become a proper health crisis for many societies worldwide.

Some numbers may help us in providing an idea about the scale of the phenomenon. The WHO reported that, as of 2019, more than fifty million individuals were suffering with dementia (World Health Organization, 2019). This already significant number is likely to constitute an underestimation owing to the intrinsic complexity of the diagnosis, alongside well-known sociocultural disparities in diagnostic quality and frequency (Chin et al., 2011). Even accepting this number at face value, the resulting economic cost of dementia appears striking. As of 2016, the global estimated cost of dementia-related healthcare approximated a trillion dollars (Xu et al., 2017). This figure only considers dementia cases, and is thus only representative of a portion of the total costs resulting from cognitive aging consequences when one factors in nonpathological aging, MCI, and other neurodegenerative pathologies. Costs related to quality of life are harder to estimate, as they manifest at the individual level. Nevertheless, there is little doubt that such costs are incommensurably high for senior individuals themselves, as well as their families and communities. In this context, the investigation of lifestyle factors that may support successful aging provides hope that evidence-based recommendations, when not government-led interventions, might be deployed to mitigate aging consequences. To date, factors such as physical exercise, education years, occupational complexity, healthy diet, leisure activities, and social involvement are widely known to positively affect cognitive health in late life stages (see e.g., Cheng, 2016).

Among such factors, bilingualism as a potential contributor to successful aging stands out for several reasons (Gallo, DeLuca, et al., 2022; Gallo and Abutalebi, 2024) when compared to other successful aging factors. Below we discuss its *ubiquity*, *equity*, and *applicability*.

Bilingualism is *ubiquitous* on two different levels. (i) The majority of the global population is currently estimated to be bi-/multilingual (De Houwer, 2021). Thus, bilingualism is naturally occurring on a large scale, globally. (ii) Bilingualism is also ubiquitous in the sense of the cognitive processes it elicits in the individual. As discussed above, both languages are always active in the bilingual brain, meaning that executive and language control processes are constantly engaged to manage language processing. This distinguishes bilingualism from other reserve contributors such as, for instance, physical exercise, which are intrinsically experienced/performed intermittently.

Furthermore, it is reasonable to argue that bilingualism is more *equitable* than the vast majority of other reserve contributors. For instance, attaining higher degrees of education, maintaining a healthy diet, or performing physical exercise often require certain financial means or socioeconomic status. Bilingualism is different. Indeed, some of the most naturally diverse linguistic societies are among the most economically challenged globally. Even in monolingual dominant societies, bilingualism is often most diffused in minority migrant populations, which are often at a socioeconomic disadvantage. This latter characteristic of bilingualism acquires even greater relevance considering the well-known ethnoracial disparities in dementia diagnosis, incidence, and treatment (Chin et al., 2011).

Finally, bilingualism stands out among the majority of reserve contributors for its greater applicability in the real-world context. Learning a new language opens up novel channels for communicating, traveling and crosscultural borders. Moreover, language has an intrinsic *social* dimension. Thus, bilingualism could indirectly support successful aging by favoring new opportunities for social interactions and leisure activities, factors known to stimulate reserve accrual as a result of the cognitive demands they impose on attention, reasoning, executive functions, and processing speed (Seeman et al., 2011).

This chapter reviewed theoretical accounts, evidence, and models of bilingualism's role as a contributor to reserve. Here, we highlighted the potential of this research program from the socioeconomic standpoint. Bilingualism appears to support successful aging by affecting human neurocognition in multiple ways. This may result in ameliorations of the quality of life of senior individuals and their caregivers, as well as relief for the public health expenditure of governments all around the globe.

Notes

This project has received funding from the European Union's Horizon Europe research and innovation programme under the Marie Sklodowska-Curie grant agreement No 101106069 to Federico Gallo, and the AcqVA Aurora Center grant from UiT, the Arctic University of Norway via funding to Jubin Abutalebi.

References

Abutalebi, J., Canini, M., Della Rosa, P. A., Green, D. W., & Weekes, B. S. (2015). The neuroprotective effects of bilingualism upon the inferior parietal lobule: A structural neuroimaging study in aging Chinese bilinguals. *Journal of Neurolinguistics, 33*, 3–13. https://doi.org/10.1016/j.jneuroling.2014.09.008

Abutalebi, J., Canini, M., Della Rosa, P. A., Sheung, L. P., Green, D. W., & Weekes, B. S. (2014). Bilingualism protects anterior temporal lobe integrity in aging. *Neurobiology of Aging, 35*(9), 2126–2133. https://doi.org/10.1016/j.neurobiolaging.2014.03.010

Abutalebi, J., Della Rosa, P. A., Castro Gonzaga, A. K., Keim, R., Costa, A., & Perani, D. (2013). The role of the left putamen in multilingual language production. *Brain and Language, 125*(3), 307–315. https://doi.org/10.1016/j.bandl.2012.03.009

Abutalebi, J., Della Rosa, P. A., Green, D. W., Hernandez, M., Scifo, P., Keim, R., Cappa, S. F., & Costa, A. (2012). Bilingualism tunes the anterior cingulate cortex for conflict monitoring. *Cerebral Cortex, 22*(9), 2076–2086. https://doi.org/10.1093/cercor/bhr287

Abutalebi, J., & Green, D. (2007). Bilingual language production: The neurocognition of language representation and control. *Journal of Neurolinguistics, 20*(3), 242–275. https://doi.org/10.1016/j.jneuroling.2006.10.003

Abutalebi, J., & Green, D. W. (2016). Neuroimaging of language control in bilinguals: Neural adaptation and reserve. *Bilingualism, 19*(4), 689–698. https://doi.org/10.1017/S1366728916000225

Abutalebi, J., Guidi, L., Borsa, V., Canini, M., Della Rosa, P. A., Parris, B. A., & Weekes, B. S. (2015). Bilingualism provides a neural reserve for aging populations. *Neuropsychologia, 69*, 201–210. https://doi.org/10.1016/j.neuropsychologia.2015.01.040

Alladi, S., Bak, T. H., Duggirala, V., Surampudi, B., Shailaja, M., Shukla, A. K., Chaudhuri, J. R., & Kaul, S. (2013). Bilingualism delays age at onset of dementia, independent of education and immigration status. *Neurology, 81*(22), 1938–1944. https://doi.org/10.1212/01.wnl.0000436620.33155.a4

Alladi, S., Bak, T. H., Mekala, S., Rajan, A., Chaudhuri, J. R., Mioshi, E., Krovvidi, R., Surampudi, B., Duggirala, V., & Kaul, S. (2016). Impact of bilingualism on cognitive outcome after stroke. *Stroke, 47*(1), 258–261. https://doi.org/10.1161/STROKEAHA.115.010418

Anderson, J. A. E., Grundy, J. G., De Frutos, J., Barker, R. M., Grady, C., & Bialystok, E. (2018). Effects of bilingualism on white matter integrity in older adults. *NeuroImage, 167*, 143–150. https://doi.org/10.1016/j.neuroimage.2017.11.038

Anderson, J. A. E., Hawrylewicz, K., & Grundy, J. G. (2020). Does bilingualism protect against dementia? A meta-analysis. *Psychonomic Bulletin & Review, 27*(5), 952–965. https://doi.org/10.3758/s13423-020-01736-5

Ansaldo, A. I., Ghazi-Saidi, L., & Adrover-Roig, D. (2015). Interference control in elderly bilinguals: Appearances can be misleading. *Journal of Clinical and Experimental Neuropsychology, 37*(5), 455–470. https://doi.org/10.1080/13803395.2014.990359

Apostolova, L. G., Steiner, C. A., Akopyan, G. G., Dutton, R. A., Hayashi, K. M., Toga, A. W., Cummings, J. L., & Thompson, P. M. (2007). Three-dimensional gray matter atrophy mapping in mild cognitive impairment and mild Alzheimer disease. *Archives of Neurology, 64*(10), 1489–1495. https://doi.org/10.1001/archneur.64.10.1489

Arce Rentería, M., Casalletto, K., Tom, S., Pa, J., Harrati, A., Armstrong, N., Rajan, K. B., Manly, J., Mungas, D., & Zahodne, L. (2019). The contributions of active Spanish-English bilingualism to cognitive reserve among older Hispanic adults living in California. *Archives of Clinical Neuropsychology, 34*(7), 1235–1235. https://doi.org/10.1093/arclin/acz029.02

Aveledo, F., Higueras, Y., Marinis, T., Bose, A., Pliatsikas, C., Meldaña, A., Martínez-Guinés, M. L., García-Domínguez, J. M., Lozano-Ros, A., Cuello, J. P., & Goicochea-Briceño, H. (2020). Multiple sclerosis and bilingualism: Some initial findings. *Linguistic Approaches to Bilingualism, 10*(January), 1–30. https://doi.org/10.1075/lab.18037.ave

Bak, T. H., Nissan, J. J., Allerhand, M. M., & Deary, I. J. (2014). Does bilingualism influence cognitive aging? *Annals of Neurology, 75*(6), 959–963. https://doi.org/10.1002/ana.24158

Baum, S., & Titone, D. (2014). Moving toward a neuroplasticity view of bilingualism, executive control, and aging. *Applied Psycholinguistics, 35*(5), 857–894. https://doi.org/DOI:10.1017/S0142716414000174

Berkes, M., Calvo, N., Anderson, J. A. E., Bialystok, E., & Initiative for Alzheimer's Disease Neuroimaging. (2021). Poorer clinical outcomes for older adult monolinguals when matched to bilinguals on brain health. *Brain Structure and Function, 226*(2), 415–424. https://doi.org/10.1007/s00429-020-02185-5

Bialystok, E. (2017). The bilingual adaptation: How minds accommodate experience. *Psychological Bulletin, 143*(3), 233–262. https://doi.org/10.1037/bul0000099

Bialystok, E., Craik, F. I. M., Binns, M. A., Ossher, L., & Freedman, M. (2014). Effects of bilingualism on the age of onset and progression of MCI and AD:

Evidence from executive function tests. *Neuropsychology, 28*(2), 290–304. https://doi.org/10.1037/neu0000023

Bialystok, E., Craik, F. I. M., & Freedman, M. (2007). Bilingualism as a protection against the onset of symptoms of dementia. *Neuropsychologia, 45*(2), 459–464. https://doi.org/10.1016/j.neuropsychologia.2006.10.009

Bialystok, E., Craik, F. I. M., Klein, R., & Viswanathan, M. (2004). Bilingualism, aging, and cognitive control: Evidence from the Simon task. *Psychology and Aging, 19*(2), 290–303. https://doi.org/10.1037/0882-7974.19.2.290

Bialystok, E., Craik, F., & Luk, G. (2008). Cognitive control and lexical access in younger and older bilinguals. *Journal of Experimental Psychology: Learning, Memory, and Cognition, 34*(4), 859–873. https://doi.org/10.1037/0278-7393.34.4.859

Bialystok, E., Craik, F. I. M., & Ryan, J. (2006). Executive control in a modified antisaccade task: Effects of aging and bilingualism. *Journal of Experimental Psychology: Learning, Memory, and Cognition, 32,* 1341–1354. https://doi.org/10.1037/0278-7393.32.6.1341

Bialystok, E., Poarch, G., Luo, L., & Craik, F. I. M. (2014). Effects of bilingualism and aging on executive function and working memory. *Psychology and Aging, 29*(3), 696–705. https://doi.org/10.1037/a0037254

Brini, S., Sohrabi, H. R., Hebert, J. J., Forrest, M. R. L., Laine, M., Hämäläinen, H., Karrasch, M., Peiffer, J. J., Martins, R. N., & Fairchild, T. J. (2020). Bilingualism is associated with a delayed onset of dementia but not with a lower risk of developing it: A systematic review with meta-analyses. *Neuropsychology Review, 30*(1), 1–24. https://doi.org/10.1007/s11065-020-09426-8

Canadian Institute for Health Information. (2022). National health expenditure trends [release summary]. www.cihi.ca/en/national-health-expenditure-trends-2022-snapshot

Cheng, S. (2016). Cognitive reserve and the prevention of dementia: The role of physical and cognitive activities. *Current Psychiatry Reports, 18*(9), 85.

Chertkow, H., Whitehead, V., Phillips, N., Wolfson, C., Atherton, J., & Bergman, H. (2010). Multilingualism (but not always bilingualism) delays the onset of Alzheimer disease: Evidence from a bilingual community. *Alzheimer Disease & Associated Disorders, 24*(2). https://journals.lww.com/alzheimerjournal/Fulltext/2010/04000/Multilingualism__But_Not_Always_Bilingualism_.2.aspx

Chin, A. L., Negash, S., & Hamilton, R. (2011). Diversity and disparity in dementia: The impact of ethnoracial differences in Alzheimer disease. *Alzheimer Disease and Associated Disorders, 25*(3), 187–195. https://doi.org/10.1097/WAD.0b013e318211c6c9

Costumero, V., Marin-Marin, L., Calabria, M., Belloch, V., Escudero, J., Baquero, M., Hernandez, M., Ruiz de Miras, J., Costa, A., Parcet, M.-A., & Ávila, C. (2020). A cross-sectional and longitudinal study on the protective effect of bilingualism against dementia using brain atrophy and cognitive measures. *Alzheimer's Research & Therapy, 12*(1), 11. https://doi.org/10.1186/s13195-020-0581-1

Craik, F. I. M., Bialystok, E., & Freedman, M. (2010). Delaying the onset of Alzheimer disease: Bilingualism as a form of cognitive reserve. *Neurology, 75*(19), 1726–1729. https://doi.org/10.1212/WNL.0b013e3181fc2a1c

Dash, T., Berroir, P., Joanette, Y., & Ansaldo, A. I. (2019). Alerting, orienting, and executive control: The effect of bilingualism and age on the subcomponents of attention. *Frontiers in Neurology, 10,* 1122. www.frontiersin.org/article/10.3389/fneur.2019.01122

Dause, T. J., & Kirby, E. D. (2019). Aging gracefully: Social engagement joins exercise and enrichment as a key lifestyle factor in resistance to age-related cognitive decline. *Neural Regeneration Research, 14*(1), 39–42. https://doi.org/10.4103/1673-5374.243698

Davis, S. W., Dennis, N. A., Daselaar, S. M., Fleck, M. S., & Cabeza, R. (2008). Qué PASA? The posterior-anterior shift in aging. *Cerebral Cortex, 18*(5), 1201–1209. https://doi.org/10.1093/cercor/bhm155

De Baene, W., Duyck, W., Brass, M., & Carreiras, M. (2015). Brain circuit for cognitive control is shared by task and language switching. *Journal of Cognitive Neuroscience, 27*(9), 1752–1765. https://doi.org/10.1162/jocn_a_00817

De Frutos-Lucas, J., López-Sanz, D., Cuesta, P., Bruña, R., De La Fuente, S., Serrano, N., López, M. E., Delgado-Losada, M. L., López-Higes, R., Marcos, A., & Maestú, F. (2020). Enhancement of posterior brain functional networks in bilingual older adults. *Bilingualism, 23*(2), 387–400. https://doi.org/10.1017/S1366728919000178

De Houwer, A. (2021). *Bilingual Development in Childhood.* Cambridge University Press. https://doi.org/DOI:10.1017/9781108866002

de Leon, J., Grasso, S. M., Welch, A., Miller, Z., Shwe, W., Rabinovici, G. D., Miller, B. L., Henry, M. L., & Gorno-Tempini, M. L. (2020). Effects of bilingualism on age at onset in two clinical Alzheimer's disease variants. *Alzheimer's & Dementia, 16*(12), 1704–1713. https://doi.org/https://doi.org/10.1002/alz.12170

Del Maschio, N., Sulpizio, S., Gallo, F., Fedeli, D., Weekes, B. S., & Abutalebi, J. (2018). Neuroplasticity across the lifespan and aging effects in bilinguals and monolinguals. *Brain and Cognition, 125.* https://doi.org/10.1016/j.bandc.2018.06.007

Della Rosa, P. A., Videsott, G., Borsa, V. M., Canini, M., Weekes, B. S., Franceschini, R., & Abutalebi, J. (2013). A neural interactive location for multilingual talent. *Cortex, 49*(2), 605–608. https://doi.org/10.1016/j.cortex.2012.12.001

DeLuca, V., Rothman, J., Bialystok, E., & Pliatsikas, C. (2019). Redefining bilingualism as a spectrum of experiences that differentially affects brain structure and function. *Proceedings of the National Academy of Sciences, 116*(15), 7565–7574. https://doi.org/10.1073/pnas.1811513116

DeLuca, V., Rothman, J., & Pliatsikas, C. (2019). Linguistic immersion and structural effects on the bilingual brain: A longitudinal study. *Bilingualism: Language and Cognition, 22*(5), 1160–1175. https://doi.org/DOI:10.1017/S1366728918000883

Dennis, N. A., & Cabeza, R. (2008). Neuroimaging of healthy cognitive aging. In *The Handbook of Aging and Cognition,* 3rd ed. (pp. 1–54). Psychology Press.

Duñabeitia, J. A., Hernández, J. A., Antón, E., Macizo, P., Estévez, A., Fuentes, L. J., & Carreiras, M. (2014). The inhibitory advantage in bilingual children revisited. *Experimental Psychology, 61*(3), 234–251. https://doi.org/10.1027/1618-3169/a000243

Dyer, S. M., Harrison, S. L., Laver, K., Whitehead, C., & Crotty, M. (2018). An overview of systematic reviews of pharmacological and non-pharmacological interventions for the treatment of behavioral and psychological symptoms of dementia. *International Psychogeriatrics, 30*(3), 295–309. https://doi.org/10.1017/S1041610217002344

Estanga, A., Ecay-Torres, M., Ibañez, A., Izagirre, A., Villanua, J., Garcia-Sebastian, M., Iglesias Gaspar, M. T., Otaegui-Arrazola, A., Iriondo, A., Clerigue, M., & Martinez-Lage, P. (2017). Beneficial effect of bilingualism on Alzheimer's disease CSF biomarkers and cognition. *Neurobiology of Aging, 50*, 144–151. https://doi.org/10.1016/j.neurobiolaging.2016.10.013

Farokhian, F., Yang, C., Beheshti, I., Matsuda, H., & Wu, S. (2017). Age-related gray and white matter changes in normal adult brains. *Aging and Disease, 8*(6), 899–909. https://doi.org/10.14336/AD.2017.0502

Filippi, R., Richardson, F. M., Dick, F., Leech, R., Green, D. W., Thomas, M. S. C., & Price, C. J. (2011). The right posterior paravermis and the control of language interference. *Journal of Neuroscience, 31*(29), 10732–10740. https://doi.org/10.1523/JNEUROSCI.1783-11.2011

Frisoni, G. B., Geroldi, C., Beltramello, A., Bianchetti, A., Binetti, G., Bordiga, G., DeCarli, C., Laakso, M. P., Soininen, H., Testa, C., Zanetti, O., & Trabucchi, M. (2002). Radial width of the temporal horn: A sensitive measure in Alzheimer disease. *AJNR. American Journal of Neuroradiology, 23*(1), 35–47.

Frisoni, G., Rossi, R., & Beltramello, A. (2002). The radial width of the temporal horn in mild cognitive impairment. *Journal of Neuroimaging, 12*(4), 351–354.

Gallo, F., & Abutalebi, J. (2024). The unique role of bilingualism among cognitive reserve-enhancing factors. *Bilingualism: Language and Cognition*, 27(2), 287–294.

Gallo, F., DeLuca, V., Prystauka, Y., Voits, T., Rothman, J., & Abutalebi, J. (2022). Bilingualism and aging: Implications for (delaying) neurocognitive decline. *Frontiers in Human Neuroscience, 16*. www.frontiersin.org/article/10.3389/fnhum.2022.819105

Gallo, F., Kubiak, J., & Myachykov, A. (2022). Add bilingualism to the mix: L2 proficiency modulates the effect of cognitive reserve proxies on executive performance in healthy aging. *Frontiers in Psychology, 13*. https://www.frontiersin.org/article/10.3389/fpsyg.2022.780261

Gallo, F., Myachykov, A., Shtyrov, Y., & Abutalebi, J. (2020). Cognitive and brain reserve in bilinguals: Field overview and explanatory mechanisms. *Journal of Cultural Cognitive Science, 4*(2), 127–143. https://doi.org/10.1007/s41809-020-00058-1

Gallo, F., Novitskiy, N., Myachykov, A., & Shtyrov, Y. (2021). Individual differences in bilingual experience modulate executive control network and performance: Behavioral and structural neuroimaging evidence. *Bilingualism: Language and Cognition, 24*(2), 293–304. https://doi.org/DOI:10.1017/S1366728920000486

Ghazi-Saidi, L. (2019). Bilingual speakers postpone symptoms of cognitive deficit in Parkinson's disease. *GSA 2019 Annual Scientific Meeting, 3*, 661–662.

Giorgio, A., Santelli, L., Tomassini, V., Bosnell, R., Smith, S., De Stefano, N., & Johansen-Berg, H. (2010). Age-related changes in grey and white matter structure throughout adulthood. *NeuroImage, 51*(3), 943–951. https://doi.org/https://doi.org/10.1016/j.neuroimage.2010.03.004

Goh, J. O., & Park, D. C. (2009). Neuroplasticity and cognitive aging: The scaffolding theory of aging and cognition. *Restorative Neurology and Neuroscience, 27*, 391–403. https://doi.org/10.3233/RNN-2009-0493

Gold, B. T., Johnson, N. F., & Powell, D. K. (2013). Lifelong bilingualism contributes to cognitive reserve against white matter integrity declines in aging. *Neuropsychologia, 51*(13), 2841–2846. https://doi.org/10.1016/j.neuropsychologia.2013.09.037

Gold, B. T., Kim, C., Johnson, N. F., Kryscio, R. J., & Smith, C. D. (2013). Lifelong bilingualism maintains neural efficiency for cognitive control in aging. *Journal of Neuroscience, 33*(2), 387–396. https://doi.org/10.1523/JNEUROSCI.3837-12.2013

Gollan, T. H., Salmon, D. P., Montoya, R. I., & Galasko, D. R. (2011). Degree of bilingualism predicts age of diagnosis of Alzheimer's disease in low-education but not in highly educated Hispanics. *Neuropsychologia, 49*(14), 3826–3830. https://doi.org/10.1016/j.neuropsychologia.2011.09.041

Grady, C. L., Luk, G., Craik, F. I. M., & Bialystok, E. (2015). Brain network activity in monolingual and bilingual older adults. *Neuropsychologia, 66*, 170–181. https://doi.org/10.1016/j.neuropsychologia.2014.10.042

Grant, A., Dennis, N. A., & Li, P. (2014). Cognitive control, cognitive reserve, and memory in the aging bilingual brain. *Frontiers in Psychology, 5*(DEC), 1401. https://doi.org/10.3389/fpsyg.2014.01401

Green, D. W., & Abutalebi, J. (2013). Language control in bilinguals: The adaptive control hypothesis. *Journal of Cognitive Psychology, 25*(5), 515–530. https://doi.org/10.1080/20445911.2013.796377

Grundy, J. G., Anderson, J. A. E., & Bialystok, E. (2017). Neural correlates of cognitive processing in monolinguals and bilinguals. *Annals of the New York Academy of Sciences, 1396*(1), 183–201. https://doi.org/10.1111/nyas.13333

Gullifer, J. W., Chai, X. J., Whitford, V., Pivneva, I., Baum, S., Klein, D., & Titone, D. (2018). Bilingual experience and resting-state brain connectivity: Impacts of L2 age of acquisition and social diversity of language use on control networks. *Neuropsychologia, 117*, 123–134. https://doi.org/10.1016/j.neuropsychologia.2018.04.037

Gunning-Dixon, F. M., Brickman, A. M., Cheng, J. C., & Alexopoulos, G. S. (2009). Aging of cerebral white matter: A review of MRI findings. *International Journal of Geriatric Psychiatry, 24*(2), 109–117. https://doi.org/10.1002/gps.2087

Guzmán-Vélez, E., & Tranel, D. (2015). Does bilingualism contribute to cognitive reserve? *Cognitive and Neural Perspectives. Neuropsychology, 29*(1), 139–150. https://doi.org/10.1037/neu0000105

Heim, S., Stumme, J., Bittner, N., Jockwitz, C., Amunts, K., & Caspers, S. (2019). Bilingualism and "brain reserve": A matter of age. *Neurobiology of Aging, 81*, 157–165. https://doi.org/10.1016/j.neurobiolaging.2019.05.021

Henry, M. L., & Gorno-Tempini, M. L. (2010). The logopenic variant of primary progressive aphasia. *Current Opinion in Neurology, 23*(6), 633–637. https://doi.org/10.1097/WCO.0b013e32833fb93e

Hindle, J. V., Martin-Forbes, P. A., Martyr, A., Bastable, A. J. M., Pye, K. L., Mueller Gathercole, V. C., Thomas, E. M., & Clare, L. (2017). The effects of lifelong cognitive lifestyle on executive function in older people with Parkinson's disease. *International Journal of Geriatric Psychiatry, 32*(12), e157–e165. https://doi.org/10.1002/gps.4677

Hötting, K., & Röder, B. (2013). Beneficial effects of physical exercise on neuroplasticity and cognition. *Neuroscience & Biobehavioral Reviews, 37*(9, Part B), 2243–2257. https://doi.org/10.1016/j.neubiorev.2013.04.005

Houtzager, N., Lowie, W., Sprenger, S., & De Bot, K. (2017). A bilingual advantage in task switching? Age-related differences between German monolinguals and Dutch-Frisian bilinguals. *Bilingualism: Language and Cognition, 20*(1), 69–79. https://doi.org/10.1017/S1366728915000498

Incera, S., & McLennan, C. T. (2018). Bilingualism and age are continuous variables that influence executive function. *Aging, Neuropsychology, and Cognition, 25*(3), 443–463. https://doi.org/10.1080/13825585.2017.1319902

Jared, D., & Kroll, J. F. (2001). Do bilinguals activate phonological representations in one or both of their languages when naming words? *Journal of Memory and Language, 44*(1), 2–31. https://doi.org/10.1006/jmla.2000.2747

Kalpouzos, G., Chételat, G., Baron, J. C., Landeau, B., Mevel, K., Godeau, C., Barré, L., Constans, J. M., Viader, F., Eustache, F., & Desgranges, B. (2009). Voxel-based mapping of brain gray matter volume and glucose metabolism profiles in normal aging. *Neurobiology of Aging, 30*(1), 112–124. https://doi.org/10.1016/j.neurobiolaging.2007.05.019

Klein, R. M., Christie, J., & Parkvall, M. (2016). Does multilingualism affect the incidence of Alzheimer's disease? A worldwide analysis by country. *SSM - Population Health, 2*, 463–467. https://doi.org/10.1016/j.ssmph.2016.06.002

Kontis, V., Bennett, J. E., Mathers, C. D., Li, G., Foreman, K., & Ezzati, M. (2017). Future life expectancy in 35 industrialised countries: Projections with a Bayesian model ensemble. *The Lancet, 389*(10076), 1323–1335. https://doi.org/10.1016/S0140-6736(16)32381-9

Kowoll, M. E., Degen, C., Gorenc, L., Küntzelmann, A., Fellhauer, I., Giesel, F., Haberkorn, U., & Schröder, J. (2016). Bilingualism as a contributor to cognitive reserve? Evidence from cerebral glucose metabolism in mild cognitive impairment and Alzheimer's disease. *Frontiers in Psychiatry, 7*(APR), 15. https://doi.org/10.3389/fpsyt.2016.00062

Kroll, J. F., Bobb, S. C., Misra, M., & Guo, T. (2008). Language selection in bilingual speech: Evidence for inhibitory processes. *Acta Psychologica, 128*(3), 416–430. https://doi.org/10.1016/j.actpsy.2008.02.001

Kuhl, P. K., Stevenson, J., Corrigan, N. M., van den Bosch, J. J. F., Can, D. D., & Richards, T. (2016). Neuroimaging of the bilingual brain: Structural brain correlates of listening and speaking in a second language. *Brain and Language, 162*, 1–9. https://doi.org/10.1016/j.bandl.2016.07.004

Leivada, E., Westergaard, M., Duñabeitia, J. A., & Rothman, J. (2021). On the phantom-like appearance of bilingualism effects on neurocognition: (How) should we proceed? *Bilingualism: Language and Cognition, 24*(1), 197–210. https://doi.org/10.1017/S1366728920000358

Li, L., Abutalebi, J., Emmorey, K., Gong, G., Yan, X., Feng, X., Zou, L., & Ding, G. (2017). How bilingualism protects the brain from aging: Insights from bimodal bilinguals. *Human Brain Mapping, 38*(8), 4109–4124. https://doi.org/10.1002/hbm.23652

Li, P., Legault, J., & Litcofsky, K. A. (2014). Neuroplasticity as a function of second language learning: Anatomical changes in the human brain. *Cortex, 58*, 301–324. https://doi.org/10.1016/j.cortex.2014.05.001

Livingston, G., Sommerlad, A., Orgeta, V., Costafreda, S. G., Huntley, J., Ames, D., Ballard, C., Banerjee, S., Burns, A., Cohen-Mansfield, J., Cooper, C., Fox, N., Gitlin, L. N., Howard, R., Kales, H. C., Larson, E. B., Ritchie, K., Rockwood, K., Sampson, E. L., ... Mukadam, N. (2017). Dementia prevention, intervention, and care. *The Lancet, 390*(10113), 2673–2734. https://doi.org/10.1016/S0140-6736(17)31363-6

Ljungberg, J. K., Hansson, P., Andrés, P., Josefsson, M., & Nilsson, L.-G. (2013). A longitudinal study of memory advantages in bilinguals. *PLoS ONE*, 8(9). https://doi.org/10.1371/journal.pone.0073029

López Zunini, R. A., Morrison, C., Kousaie, S., & Taler, V. (2019). Task switching and bilingualism in young and older adults: A behavioral and electrophysiological investigation. *Neuropsychologia, 133*, 107186. https://doi.org/10.1016/j.neuropsychologia.2019.107186

Lövdén, M., Wenger, E., Mårtensson, J., Lindenberger, U., & Bäckman, L. (2013). Structural brain plasticity in adult learning and development. *Neuroscience & Biobehavioral Reviews, 37*(9, Part B), 2296–2310. https://doi.org/10.1016/j.neubiorev.2013.02.014

Luk, G., Anderson, J. A. E., Craik, F. I. M., Grady, C., & Bialystok, E. (2010). Distinct neural correlates for two types of inhibition in bilinguals: Response inhibition versus interference suppression. *Brain and Cognition, 74*(3), 347–357. https://doi.org/10.1016/j.bandc.2010.09.004

Luk, G., & Bialystok, E. (2013). Bilingualism is not a categorical variable: Interaction between language proficiency and usage. *Journal of Cognitive Psychology, 25*(5), 605–621. https://doi.org/10.1080/20445911.2013.795574

Luk, G., Bialystok, E., Craik, F. I. M., & Grady, C. L. (2011). Lifelong bilingualism maintains white matter integrity in older adults. *Journal of Neuroscience, 31*(46), 16808–16813. https://doi.org/10.1523/JNEUROSCI.4563-11.2011

Luo, D., Kwok, V. P. Y., Liu, Q., Li, W., Yang, Y., Zhou, K., Xu, M., Gao, J.-H., & Tan, L. H. (2019). Microstructural plasticity in the bilingual brain. *Brain and Language, 196*, 104654. https://doi.org/10.1016/j.bandl.2019.104654

Luo, L., Craik, F. I. M., Moreno, S., & Bialystok, E. (2013). Bilingualism interacts with domain in a working memory task: Evidence from aging. *Psychology and Aging, 28*(1), 28–34. https://doi.org/10.1037/a0030875

MacPherson, S., & Cox, S. (2017). The frontal ageing hypothesis: Evidence from normal ageing and dementia. In *Handbook of Gerontology Research Methods: Understanding Successful Ageing* (pp. 139–158). Routledge/Taylor & Francis Group.

Martínez-Horta, S., Moreu, A., Perez-Perez, J., Sampedro, F., Horta-Barba, A., Pagonabarraga, J., Gomez-Anson, B., Lozano-Martinez, G. A., Lopez-Mora, D. A., Camacho, V., Fernández-León, A., Carrió, I., & Kulisevsky, J. (2018). The impact of bilingualism on brain structure and function in Huntington's disease. *Parkinsonism & Related Disorders, 60*, 92–97. https://doi.org/10.1016/j.parkreldis.2018.09.017

Maruszak, A., & Thuret, S. (2014). Why looking at the whole hippocampus is not enough: A critical role for anteroposterior axis, subfield and activation analyses to enhance predictive value of hippocampal changes for Alzheimer's disease diagnosis. *Frontiers in Cellular Neuroscience, 8*, 95. www.frontiersin.org/article/10.3389/fncel.2014.00095

Mechelli, A., Crinion, J. T., Noppeney, U., O' Doherty, J., Ashburner, J., Frackowiak, R. S., & Price, C. J. (2004). Neurolinguistics: Structural plasticity in the bilingual brain. *Nature, 431*(7010), 757. https://doi.org/10.1038/431757a

Mendez, M. F., Chavez, D., & Akhlaghipour, G. (2019). Bilingualism delays expression of Alzheimer's clinical syndrome. *Dementia and Geriatric Cognitive Disorders, 48*(5–6), 281–289. https://doi.org/10.1159/000505872

Meng, X., & D'Arcy, C. (2012). Education and dementia in the context of the cognitive reserve hypothesis: A systematic review with meta-analyses and qualitative analyses. *PLoS ONE, 7*(6), e38268–e38268. https://doi.org/10.1371/journal.pone.0038268

Mitchell, E. M. (2016). *Concentration of Health Expenditures in the U.S. Civilian Noninstitutionalized Population, 2014*. Agency for Healthcare Research and Quality (US), Rockville (MD). http://europepmc.org/abstract/MED/28422468

Nakamura, A., Cuesta, P., Kato, T., Arahata, Y., Iwata, K., Yamagishi, M., Kuratsubo, I., Kato, K., Bundo, M., Diers, K., Fernández, A., Maestú, F., & Ito, K. (2017). Early functional network alterations in asymptomatic elders at risk for Alzheimer's disease. *Scientific Reports, 7*(1), 6517. https://doi.org/10.1038/s41598-017-06876-8

Nichols, E. S., Wild, C. J., Stojanoski, B., Battista, M. E., & Owen, A. M. (2020). Bilingualism affords no general cognitive advantages: A population study of executive function in 11,000 people. *Psychological Science, 31*(5), 548–567. https://doi.org/10.1177/0956797620903113

Olsen, R. K., Pangelinan, M. M., Bogulski, C., Chakravarty, M. M., Luk, G., Grady, C. L., & Bialystok, E. (2015). The effect of lifelong bilingualism on regional grey and white matter volume. *Brain Research, 1612*, 128–139. https://doi.org/10.1016/j.brainres.2015.02.034

Olulade, O. A., Jamal, N. I., Koo, D. S., Perfetti, C. A., LaSasso, C., & Eden, G. F. (2016). Neuroanatomical evidence in support of the bilingual advantage theory. *Cerebral Cortex, 26*(7), 3196–3204. https://doi.org/10.1093/cercor/bhv152

Ossher, L., Bialystok, E., Craik, F. I. M., Murphy, K. J., & Troyer, A. K. (2013). The effect of bilingualism on amnestic mild cognitive impairment. *The Journals of Gerontology: Series B, 68*(1), 8–12. https://doi.org/10.1093/geronb/gbs038

Paap, K. R., Johnson, H. A., & Sawi, O. (2015). Bilingual advantages in executive functioning either do not exist or are restricted to very specific and undetermined circumstances. *Cortex, 69,* 265–278. https://doi.org/10.1016/j.cortex.2015.04.014

Paplikar, A., Mekala, S., Bak, T. H., Dharamkar, S., Alladi, S., & Kaul, S. (2018). Bilingualism and the severity of poststroke aphasia. *Aphasiology, 33*(1), 58–72. https://doi.org/10.1080/02687038.2017.1423272

Paulavicius, A. M., Mizzaci, C. C., Tavares, D. R. B., Rocha, A. P., Civile, V. T., Schultz, R. R., Pinto, A. C. P. N., & Trevisani, V. F. M. (2020). Bilingualism for delaying the onset of Alzheimer's disease: A systematic review and meta-analysis. *European Geriatric Medicine, 11*(4), 651–658. https://doi.org/10.1007/s41999-020-00326-x

Perani, D., Farsad, M., Ballarini, T., Lubian, F., Malpetti, M., Fracchetti, A., Magnani, G., March, A., & Abutalebi, J. (2017). The impact of bilingualism on brain reserve and metabolic connectivity in Alzheimer's dementia. *Proceedings of the National Academy of Sciences of the United States of America, 114*(7), 1690–1695. https://doi.org/10.1073/pnas.1610909114

Perquin, M., Vaillant, M., Schuller, A.-M., Pastore, J., Dartigues, J.-F., Lair, M.-L., & Diederich, N. (2013). Lifelong exposure to multilingualism: New evidence to support cognitive reserve hypothesis. *PLoS ONE, 8*(4), e62030. https://doi.org/10.1371/journal.pone.0062030

Persson, J., Nyberg, L., Lind, J., Larsson, A., Nilsson, L.-G., Ingvar, M., & Buckner, R. L. (2006). Structure–function correlates of cognitive decline in aging. *Cerebral Cortex, 16*(7), 907–915. https://doi.org/10.1093/cercor/bhj036

Pliatsikas, C. (2020). Understanding structural plasticity in the bilingual brain: The dynamic restructuring model. *Bilingualism: Language and Cognition, 23*(2), 459–471. https://doi.org/10.1017/S1366728919000130

Pliatsikas, C., DeLuca, V., Moschopoulou, E., & Saddy, J. D. (2017). Immersive bilingualism reshapes the core of the brain. *Brain Structure and Function, 222*(4), 1785–1795. https://doi.org/10.1007/s00429-016-1307-9

Pliatsikas, C., DeLuca, V., & Voits, T. (2020). The many shades of bilingualism: Language experiences modulate adaptations in brain structure. *Language Learning, 70*(S2), 133–149. https://doi.org/10.1111/lang.12386

Pliatsikas, C., Johnstone, T., & Marinis, T. (2014). Grey matter volume in the cerebellum is related to the processing of grammatical rules in a second language: A structural voxel-based morphometry study. *Cerebellum, 13*(1), 55–63. https://doi.org/10.1007/s12311-013-0515-6

Prince, M. J., Comas-Herrera, A., Knapp, M., Guerchet, M. M., & Karagiannidou, M. (2016). World Alzheimer Report 2016 – Improving healthcare for people living with dementia: Coverage, quality and costs now and in the future. Alzheimer's Disease International. www.alz.co.uk/research/world-report-2016

Raftery, A. E., Chunn, J. L., Gerland, P., & Ševčíková, H. (2013). Bayesian probabilistic projections of life expectancy for all countries. *Demography, 50*(3), 777–801. https://doi.org/10.1007/s13524-012-0193-x

Ramakrishnan, S., Mekala, S., Mamidipudi, A., Yareeda, S., Mridula, R., Bak, T. H., Alladi, S., & Kaul, S. (2017). Comparative effects of education and bilingualism on the onset of mild cognitive impairment. *Dementia and Geriatric Cognitive Disorders, 44*(3–4), 222–231. https://doi.org/10.1159/000479791

Ressel, V., Pallier, C., Ventura-Campos, N., Díaz, B., Roessler, A., Ávila, C., & Sebastián-Gallés, N. (2012). An effect of bilingualism on the auditory cortex. *The Journal of Neuroscience, 32*(47), 16597 LP–16601. https://doi.org/10.1523/JNEUROSCI.1996-12.2012

Reuter-Lorenz, P. A., Festini, S. B., & Jantz, T. K. (2021). Chapter 5: Executive functions and neurocognitive aging. In K. W. Schaie & S. Willis (Eds.), *Handbook of the Psychology of Aging (Ninth Edition)* (pp. 67–81). Academic Press. https://doi.org/10.1016/B978-0-12-816094-7.00019-2

Robertson, I. H. (2013). A noradrenergic theory of cognitive reserve: Implications for Alzheimer's disease. *Neurobiology of Aging, 34*(1), 298–308. https://doi.org/10.1016/j.neurobiolaging.2012.05.019

Rodríguez-Pujadas, A., Sanjuán, A., Fuentes, P., Ventura-Campos, N., Barrós-Loscertales, A., & Ávila, C. (2014). Differential neural control in early bilinguals and monolinguals during response inhibition. *Brain and Language, 132*, 43–51. https://doi.org/10.1016/j.bandl.2014.03.003

Sala, A., Malpetti, M., Farsad, M., Lubian, F., Magnani, G., Frasca Polara, G., Epiney, J.-B., Abutalebi, J., Assal, F., Garibotto, V., & Perani, D. (2022). Lifelong bilingualism and mechanisms of neuroprotection in Alzheimer dementia. *Human Brain Mapping, 43*(2), 581–592. https://doi.org/10.1002/hbm.25605

Sattler, C., Toro, P., Schönknecht, P., & Schröder, J. (2012). Cognitive activity, education and socioeconomic status as preventive factors for mild cognitive impairment and Alzheimer's disease. *Psychiatry Research, 196*(1), 90–95. https://doi.org/10.1016/j.psychres.2011.11.012

Schweizer, T. A., Ware, J., Fischer, C. E., Craik, F. I. M., & Bialystok, E. (2012). Bilingualism as a contributor to cognitive reserve: Evidence from brain atrophy in Alzheimer's disease. *Cortex, 48*(8), 991–996. https://doi.org/10.1016/j.cortex.2011.04.009

Seeman, T. E., Miller-Martinez, D. M., Stein Merkin, S., Lachman, M. E., Tun, P. A., & Karlamangla, A. S. (2011). Histories of social engagement and adult cognition: Midlife in the U.S. *Study. The Journals of Gerontology: Series B, 66B* (suppl_1), i141–i152. https://doi.org/10.1093/geronb/gbq091

Shearer, J., Green, C., Ritchie, C. W., & Zajicek, J. P. (2012). Health state values for use in the economic evaluation of treatments for Alzheimer's disease. *Drugs & Aging, 29*(1), 31–43. https://doi.org/10.2165/11597380-000000000-00000

Soltani, M., Fatemeh Emami Dehcheshmeh, S., Moradi, N., Hajiyakhchali, A., Majdinasab, N., Mahmood Latifi, S., Hosseini beidokhti, M., Soltani, C. M., &

Dehcheshmeh, E. S. (2018). Comparing executive functions in bilinguals and monolinguals suffering from relapsing-remitting multiple sclerosis. *Journal of Modern Rehabilitation, 12*(2), 133–139.

Spivey, M. J., & Marian, V. (1999). Cross talk between native and second languages: Partial activation of an irrelevant lexicon. *Psychological Science, 10*(3), 281–284. https://doi.org/10.1111/1467-9280.00151

Stebbins, G. T., & Murphy, C. M. (2009). Diffusion tensor imaging in Alzheimer's disease and mild cognitive impairment. *Behavioural Neurology, 21*(1–2), 39–49.

Steffener, J., Reuben, A., Rakitin, B. C., & Stern, Y. (2011). Supporting performance in the face of age-related neural changes: Testing mechanistic roles of cognitive reserve. *Brain Imaging and Behavior, 5*(3), 212–221. https://doi.org/10.1007/s11682–011-9125-4

Stein, M., Federspiel, A., Koenig, T., Wirth, M., Strik, W., Wiest, R., Brandeis, D., & Dierks, T. (2012). Structural plasticity in the language system related to increased second language proficiency. *Cortex, 48*(4), 458–465. https://doi.org/10.1016/j.cortex.2010.10.007

Stern, Y. (2003). The concept of cognitive reserve: A catalyst for research. *Journal of Clinical and Experimental Neuropsychology, 25*(5), 589–593. https://doi.org/10.1076/jcen.25.5.589.14571

Stern, Y. (Ed.). (2007). Cognitive reserve: Theory and applications. In *Cognitive Reserve: Theory and Applications.* (pp. xxi, 344–xxi, 344). Taylor & Francis.

Stern, Y. (2009). Cognitive reserve. *Neuropsychologia, 47*(10), 2015–2028. https://doi.org/10.1016/j.neuropsychologia.2009.03.004

Stern, Y. (Ed.). (2013). *Cognitive Reserve: Theory and Applications.* Psychology Press.

Stern, Y., Arenaza-Urquijo, E. M., Bartrés-Faz, D., Belleville, S., Cantilon, M., Chetelat, G., Ewers, M., Franzmeier, N., Kempermann, G., Kremen, W. S., Okonkwo, O., Scarmeas, N., Soldan, A., Udeh-Momoh, C., Valenzuela, M., Vemuri, P., & Vuoksimaa, E. (2020). Whitepaper: Defining and investigating cognitive reserve, brain reserve, and brain maintenance. *Alzheimer's & Dementia, 16*(9), 1305–1311. https://doi.org/10.1016/j.jalz.2018.07.219

Stocco, A., Yamasaki, B., Natalenko, R., & Prat, C. S. (2014). Bilingual brain training: A neurobiological framework of how bilingual experience improves executive function. *International Journal of Bilingualism*, 18(1), 67–92.

Sulpizio, S., Del Maschio, N., Del Mauro, G., Fedeli, D., & Abutalebi, J. (2020). Bilingualism as a gradient measure modulates functional connectivity of language and control networks. *NeuroImage, 205*, 116306. https://doi.org/10.1016/j.neuroimage.2019.116306

Sumowski, J. F., Chiaravalloti, N., Wylie, G., & Deluca, J. (2009). Cognitive reserve moderates the negative effect of brain atrophy on cognitive efficiency in multiple sclerosis. *Journal of the International Neuropsychological Society, 15*(4), 606–612. https://doi.org/10.1017/S1355617709090912

Tucker, A. M., & Stern, Y. (2011). Cognitive reserve in aging. *Current Alzheimer Research, 8*(4), 354–360. https://doi.org/10.2174/156720511795745320

Valian, V. (2015). Bilingualism and cognition. *Bilingualism: Language and Cognition, 18*(1), 3–24. https://doi.org/10.1017/S1366728914000522

Vemuri, P., Lesnick, T. G., Przybelski, S. A., Machulda, M., Knopman, D. S., Mielke, M. M., Roberts, R. O., Geda, Y. E., Rocca, W. A., Petersen, R. C., & Jack Jr, C. R. (2014). Association of lifetime intellectual enrichment with cognitive decline in the older population. *JAMA Neurology, 71*(8), 1017–1024. https://doi.org/10.1001/jamaneurol.2014.963

Voits, T. (2020). *Bilingualism as a neuroprotective factor in ageing: Insights from healthy and clinical populations* (Issue September) [University of Reading]. https://doi.org/10.48683/1926.00097032

Voits, T., Pliatsikas, C., Robson, H., & Rothman, J. (2020). Beyond Alzheimer's disease: Can bilingualism be a more generalized protective factor in neurodegeneration? *Neuropsychologia, 147*(April), 107593. https://doi.org/10.1016/j.neuropsychologia.2020.107593

Voits, T., Robson, H., Rothman, J., & Pliatsikas, C. (2022). The effects of bilingualism on hippocampal volume in ageing bilinguals. *Brain Structure and Function, 227*(3), 979–994. https://doi.org/10.1007/s00429-021-02436-z

Wilson, R. S., Boyle, P. A., Yang, J., James, B. D., & Bennett, D. A. (2015). Early life instruction in foreign language and music and incidence of mild cognitive impairment. *Neuropsychology, 29*(2), 292–302. https://doi.org/10.1037/neu0000129

Wimo, A., Guerchet, M., Ali, G.-C., Wu, Y.-T., Prina, A. M., Winblad, B., Jönsson, L., Liu, Z., & Prince, M. (2017). The worldwide costs of dementia 2015 and comparisons with 2010. *Alzheimer's & Dementia: The Journal of the Alzheimer's Association, 13*(1), 1–7. https://doi.org/10.1016/j.jalz.2016.07.150

Winblad, B., Amouyel, P., Andrieu, S., Ballard, C., Brayne, C., Brodaty, H., Cedazo-Minguez, A., Dubois, B., Edvardsson, D., Feldman, H., Fratiglioni, L., Frisoni, G. B., Gauthier, S., Georges, J., Graff, C., Iqbal, K., Jessen, F., Johansson, G., Jönsson, L., ... Zetterberg, H. (2016). Defeating Alzheimer's disease and other dementias: A priority for European science and society. *The Lancet Neurology, 15*(5), 455–532. https://doi.org/10.1016/S1474-4422(16)00062-4

Wodniecka, Z., Craik, F. I. M., Luo, L., & Bialystok, E. (2010). Does bilingualism help memory? Competing effects of verbal ability and executive control. *International Journal of Bilingual Education and Bilingualism, 13*(5), 575–595. https://doi.org/10.1080/13670050.2010.488287

Wong, W. (2020). Economic burden of Alzheimer disease and managed care considerations. *The American Journal of Managed Care, 26*(Suppl 8), S177–S183. https://doi.org/10.37765/ajmc.2020.88482

World Health Organization. (2019). *Dementia*. www.who.int/news-room/fact-sheets/detail/dementia

Woumans, E., Santens, P., Sieben, A., Versijpt, J., Stevens, M., & Duyck, W. (2015). Bilingualism delays clinical manifestation of Alzheimer's disease. *Bilingualism, 18*(3), 568–574. https://doi.org/10.1017/S136672891400087X

Xu, J., Zhang, Y., Qiu, C., & Cheng, F. (2017). Global and regional economic costs of dementia: A systematic review. *The Lancet, 390*(S47). https://doi.org/10.1016/s0140-6736(17)33185-9

Yaffe, K., Fiocco, A. J., Lindquist, K., Vittinghoff, E., Simonsick, E. M., Newman, A. B., Satterfield, S., Rosano, C., Rubin, S. M., Ayonayon, H. N., & Harris, T. B. (2009). Predictors of maintaining cognitive function in older adults. *Neurology, 72*(23), 2029 LP – 2035. https://doi.org/10.1212/WNL.0b013e3181a92c36

Zhang, Y., Londos, E., Minthon, L., Wattmo, C., Liu, H., Aspelin, P., & Wahlund, L.-O. (2008). Usefulness of computed tomography linear measurements in diagnosing Alzheimer's disease. *Acta Radiologica, 49*(1), 91–97. https://doi.org/10.1080/02841850701753706

Zheng, Y., Wu, Q., Su, F., Fang, Y., Zeng, J., & Pei, Z. (2018). The protective effect of Cantonese/Mandarin bilingualism on the onset of Alzheimer disease. *Dementia and Geriatric Cognitive Disorders, 45*(3–4), 210–219. https://doi.org/10.1159/000488485

Zou, L., Ding, G., Abutalebi, J., Shu, H., & Peng, D. (2012). Structural plasticity of the left caudate in bimodal bilinguals. *Cortex, 48*(9), 1197–1206. https://doi.org/10.1016/j.cortex.2011.05.022

3

Networks and Multimodalities underlying Language Processing in the Brain

Evidence from Structural and Functional Neuroimaging

Edna Andrews, Charlotte Sophia Rammell, and Hayes Bierman

For over twenty years, neuroscientists have been actively using neuroimaging to ask questions emerging from how language is processed in the human brain. As early as 2001, Cabeza and Kingstone edited an important volume on functional neuroimaging studies of human cognition, including functional neuroimaging (specifically PET and fMRI) of language (Binder & Price, 2001; Cabeza & Kingstone, 2001). Prior to the more general availability of PET and fMRI, one of the major sources of data on language mapping (including bilingual subjects) in the brain came from Dr. Ojemann and his use of invasive direct neuronal measurements during surgeries – CSM (cortical stimulation mapping) (1978, 1979, 1991, 1993; Corina et al., 2010 and others). It is interesting that Ojemann's group included bilingualism early in the process, while the inclusion of bilingualism took longer to appear in the later neuroimaging studies of language(s).

In the period that marked the emergence of nonclinical neuroimaging studies of language (1990s), Poeppel became one of the central critical voices on analysis of neuroimaging results of PET and phonology (Poeppel, 1996), and continued to be a leader in the framing of the current field in understanding the importance of moving forward from a nineteenth-century lesion-deficit model to a more realistic view of brain function. His contributions, together with those in collaboration with Hickok, include looking for important connections between neuroscience and linguistics, deconstructing a modularity approach to language processing, and articulating how language (and languages) fit into the broader study of cognitive neuroscience (Hickok & Poeppel, 2007; Poeppel, 2008; Poeppel & Hickok, 2004).

In this chapter, we will explore the range of outcomes in functional and structural neuroimaging studies, focusing on language processing in the brain, including studies of bi- and multilingualism. We will devote special attention to recent innovative research using not only functional MRI, but Diffusion Tensor Imaging (DTI) and resting state fMRI (rfMRI). When our narrative turns to bi- and multilingualism, a variety of issues emerge from the research that challenge and redefine earlier beliefs on questions concerning age of acquisition, bilaterality, and cognitive reserve. The chapter concludes with a review of the core criticisms that have been articulated about the use of neuroimaging in cognitive studies, software and intermethod discrepancies, protocol design, ecological validity, and future directions.

3.1 Neuroimaging Studies: 1990s through 2022

One of the early and substantial publications devoted to neuroimaging of human language is found in Binder and Price (2001: 187–251). In this important overview of publications from the 1990s on language processing in the brain, the authors contrast important differences in assumptions that originally arose in connection with the lesion-deficit tradition of the late nineteenth–first half of the twentieth century. They also engage with critical theoretical questions about how to define human language across linguistic hierarchically given subsystems, speech production and perception, speech sounds and visual character perception, the interaction of linguistic meaning (including semantics) and knowledge, the ultimate challenge of understanding and interpreting *activations* found in neuroimaging studies using the hemodynamic response in both PET and fMRI studies and emergent localization patterns for language processing. An example of one of the observations made involved misleading asymmetry results due to smoothing (Binder & Price, 2001: 194). This work includes a reference section of over 200 sources focusing on the 1990s (Binder & Price, 2001: 241–251). Two important meta-analyses of neuroimaging of human language of note from these two authors include Binder and colleagues (2009) and Price (2010).

Bookheimer (2002: 151–188) reviews a corpus of PET and fMRI studies and focuses on three particular questions: IFG subdivisions and semantic processing, examination of category-specific organization of semantic information in the temporal lobe, and right hemisphere contributions to language processing. Bookheimer discusses new methods introduced by fMRI, including *event-related* design that may provide important alternatives to block design as it can provide one stimulus at a time (Bookheimer, 2002: 154–155). In her conclusions, Bookheimer returns to these three questions and explains: (1) why the classical model (based on Broca/Wernicke) is incorrect and does not correspond to actual language processing in the brain; (2) the lack of evidence in imaging data for a "single system for semantic information or an organizational structure formed along strictly categorical lines" – rather, the outcomes show associations with sensory-motor and linguistic attributes and suggest that these very different representations are linked by a frontal lobe executive system of semantic processing; and (3) the essential role of the right hemisphere in ecologically valid integration of phonological information with emergent meanings in communication acts (Bookheimer, 2002: 182–183).

Stowe and colleagues (2005: 997–1042) bring compelling evidence from studies of sentence comprehension that identify the different integral parts of the supporting network(s) and non-domain-specific components, thus requiring an alternative view of modularity in order to account for interactions across and between networks. Their conclusion calls for a rethinking of previous assumptions about the neurological basis of language processing.

Binder and colleagues' (2009: 2770) meta-analysis included 120 PET and fMRI studies from 1990 through 2007 (after an initial pool of 520) that focused on semantic processing and semantic memory. The majority of studies (81) were from the period 2002–2007 and the results suggest three major areas of the *heteromodal cortex* that are significant to the storage and retrieval of semantic knowledge (2009: 2783–2784). Of 1145 activation foci from the studies, lateralization indices show 68 percent left lateralization and 32 percent right lateralization (2009: 2771).

Binder and colleagues also critique other meta-analyses briefly to clarify how this work is different (2009: 2783) and argue for the importance of a more rigorous definition of *activations:*

> Future meta-analyses of cognitive imaging studies would benefit from the establishment of a more formal ontological system for defining the cognitive processes represented by an "activation."

Binder and colleagues (2009), as well as other meta-analyses included in this chapter (Price, 2010; Schirmer et al., 2012), used *activation likelihood estimation* (see Eickhoff et al., 2009; Turkeltaub et al., 2002, 2012), which is a multivalidated statistical technique for analyzing coordinate-based neuroimaging data. This method was updated by Turkeltaub in 2012.

Price (2010) examined 100 fMRI studies of speech comprehension and production conducted in 2009. Price concludes that these studies show significant overlap in regions supporting speech production and comprehension, and additional regions for production (2010: 82–83). Furthermore, Price mentions studies that show activations found in the cerebellum that require further elucidation, as well as bilateral activations found in superior temporal, motor, premotor, and subcortical regions. All of these trends require further study and greater "spatial resolution of anatomical conclusions" with specific and concrete suggestions for this process (2010: 83). Price clearly anticipated the neuroimaging field twenty years later in her call for more functional connectivity studies (to be discussed later in this chapter).

While there was optimism "that under ideal circumstances it is reasonable to expect very similar results from PET and fMRI" (Binder & Price, 2001: 192), the authors were clear to note the many problems associated with PET and fMRI language studies (Binder & Price, 2001: 192–198). It is here that Poeppel's (1996) critical examination of PET studies of phonological processing is mentioned (and we will return to it in a later section). In the early 2000s to the present, there has been a move away from PET and an increase in not only fMRI, but also rfMRI and structural MRI techniques.

Binder and Price conclude in noting not only the complexities of interpretation of neuroimaging data, and specifically identifying three major issues (2001: 239–241), but also a series of significant, unresolved questions (2001: 239). Issues in interpretation of neuroimaging data include: (1) the absence of a standard vocabulary to describe computational aspects of different tasks; (2) the variability across anatomical atlases and labels; (3) standard measurements of *reliability* across studies. Unresolved questions included moving beyond the so-called *classical/traditional* nineteenth-century lesion-deficit modeling of language processing in the human brain, interrelationships between different modalities of forms (visual, auditory, semantic, lexical, phonemic), and the interaction of languages in development (including monolingual and bi/multilingual users).

Other researchers who were included in the meta-analyses mentioned above, and were prominent in the 1990s and continue to be major contributors to the field of neuroimaging studies of language and languages, include Perani, Zatorre, Kotz, Friederici, and others. In a later section, we will focus on a series of contributions that began a new subfield of inquiry in neuroimaging of bi/multilingualism – cognitive brain reserve (including Abutalebi et al., 2013, 2015; Abutalebi & Green, 2016; Bialystok et al., 2007; Perani 2017; Pliatsikas 2019, 2020, 2023; Pliatsikas et al., 2017).

A series of recent studies have continued to investigate semantic networks. While semantic control was believed to occur predominantly in the left hemisphere (Jackson, 2021), its role in the language network has been investigated further by Hodgson and colleagues (2021). They used activation likelihood estimation (ALE) analyses to determine the networks responsible for both semantic control and phonological control using datasets including those from Jackson (2021) (for semantic networks) and from Vigneau and colleagues (2006, 2011), and Humphreys and Lambon (2015) (for phonological contrasts). This meta-analysis found bilateral activation for semantics, including left temporal lobe, right superior/middle temporal gyrus, left dmPFC, and right IFG/insula. Phonological activation was reported in the left precentral gyrus/IFG, posterior STG bilaterally, left inferior parietal lobule, left posterior ITG, dmPFC, and right IFG/insula. The phonological and semantic activations did overlap in some neural areas, but the authors reported that large enough datasets allow for their differentiation from each other (Hodgson et al., 2021).

3.1.1 Studies of Functional Neuroimaging (PET and fMRI) of Multilingualism

While many studies of monolingual, or nonmultilingual, language representation in the brain have focused on linguistic function, meta-analyses of multilingualism are often centered around age of acquisition and L2 proficiency as variables of interest. While L2 proficiency is certainly an important variable to consider, future works should consider the importance of factors beyond age of acquisition.

Neural representation of multiple languages in the brain is affected by the proficiency of the second language, as shown by Sebastian and colleagues (2011). In this meta-analysis, fourteen PET and fMRI studies of bilingualism addressed the activation of L1 and L2 in high- and low/mid-proficiency groups. Importantly, in high-proficiency groups, the L1 and L2 patterned similarly in activation, while in the low/mid proficiency groups, clusters were smaller and more distributed in the L2 than the L1. The authors reported that lower proficiency in a language requires wider activation to offset language demands (see also Perani et al., 1996). For both languages in highly proficient L2 individuals, activation was found in left frontal regions, including the left IFG and left superior frontal gyrus. In less proficient L2 individuals, L1 activation was still significant in left frontal regions, but L2 activation was bilateral (Sebastian et al., 2011).

Findings about activations are not in agreement about whether there are significant differences between L1 and L2 learning. For example, Birdsong (2006) shows that behaviorally and in terms of imaging, late L2 highly proficient users can be on par with L1 users across all modalities (phonology, morphology, lexicon, etc.). The meta-analysis of Cargnelutti and colleagues (2019) includes studies that claim the activations in a highly proficient L2 user may be broader than for the L1, particularly for late age of acquisition, as seen in the meta-analysis of fifty-seven fMRI and PET studies. Recruitment of more activation for the L2 than the L1 in late bilinguals has been documented also in a meta-analysis by Liu and Cao (2016). The problems with many of these studies is the lack of an empirical baseline of actual proficiency across modalities, as well as any reliable proficiency testing data to correlate with the scan data.

Advanced proficiency language users (determined by Spanish language course placement and a modified DELE proficiency test, Instituto Cervantes, n.d.) were found to use different processing strategies for L2 sentences under degradation than they did in their L1 by Rammell and colleagues (2019). While listening to L1 sentences in noise-activated executive functioning regions of the brain, listening to L2 sentences in noise required both auditory and executive processing activation, indicating a bottom-up processing strategy focused on the phonetics of the speech. These findings also have been supported by recent studies comparing high-proficiency with mid-proficiency bilinguals' activation for listening comprehension and reading comprehension across two different language pairs (Russian–English and Spanish–English); Andrews and colleagues (2019) found significant overlap in language activation across languages for high-proficiency users (B2/C1) and less overlap across languages for lower proficiency levels (B1). In Andrews and collegues (2019), proficiency was determined based on the CEFR framework. Each language had two listening and two reading comprehension tasks from the Russian version of CEFR (Andrews, 2014; North, 2000; Sobolev & Nesterova, 2014) or the Spanish DELE (Instituto Cervantes, n.d.) and for English, a modified TOEFL (ETS, 2007).

One meta-analysis that focused on PET and fMRI studies of grammatical and lexical language learning is Tagarelli and colleagues (2019). The forty-six research studies included in their analysis, were related specifically to language learning, use of written or verbal language presentation in the task, and whole brain analyses where grammar or lexical learning were subtracted from a baseline condition. ALE results revealed activation bilaterally in the IFG and in the left posterior parietal lobe in both grammatical and lexical learning, indicating that these regions may be part of a more general language learning network. Lexical language learning, according to the analysis, relies heavily on the ventral occipito-temporal cortex. In comparison, grammatical learning studies reported activation in the left putamen and left caudate nucleus (Tagarelli et al., 2019).

Another study addressing the role of lexical/semantic networks, grammar, and phonology also addressed the role of language switching. Sulpizio and colleagues (2020) reported larger activation for L1 versus L2 in cortical, subcortical, and cerebellar regions for lexical/semantic knowledge. In comparison, both grammar and phonology had fewer clusters. Activation from language switching shared activation with general, nonlinguistic control networks. Overall, the authors concluded that both L1 and L2 share a linguistic network for activation (Sulpizio et al., 2020).

Recent meta-analyses of multilingualism offer important insight into language representation and language control in the brain. Future analyses of multilingualism should carefully interpret the role of proficiency in both languages, as its role in language representation has been well documented. Language proficiency can be quantified, as in the case of Andrews and colleagues (2013) and Andrews (2014). Andrews and colleagues (2019) used Common European Framework testing models to address proficiency and furthermore demonstrated the importance of proficiency testing across all languages of the study participants based on clear, quantitative factors.

3.2 Theoretical Contributions to the Study of Brain and Language(s)

3.2.1 Discoveries, Questions, Challenges

In order to provide a broader context for understanding the latest trends in cognitive neuroimaging of language and languages, it is important to briefly review some of the key influential research that has fundamentally changed the field. While we see the roots of these changes clearly in the earlier period of fMRI and PET studies of language processing, we will consider watershed moments in the evolution of neuroimaging of language(s).

One of the most important changes to emerge is a shift in focus to more ecologically valid protocol designs and contextualization of *languaging* (use of language(s) in the sociocultural context) in general.

Poeppel and Hickok (2004: 1–12), Hickok and Poeppel (2007: 393–402), and Poeppel (2008: 1–21) articulated important points for moving forward with the neuroimaging of language(s), including: (1) rejection of the nineteenth-century classical model (including the view of modules and monolithic/homogeneous areas); (2) the need for a computational model for understanding the processing of language(s) in the brain; (3) a shared vocabulary of analysis across linguistics and the cognitive neurosciences; (4) the importance of ecological validity at all stages of experimental design and interpretation; (5) the importance of understanding that localization is not an explanation;[1] (6) the interactive nature of speech perception/comprehension and production; (7) the importance of a bilateral view of language processing in the brain; (8) system-level processing of languages; and (9) a dual-stream (dorsal/ventral) model. These principles have been generally acknowledged by the field of neuroimaging of language(s) and, in some cases, deeply incorporated into the fabric of current and ongoing research.

The meta-analysis by Schirmer and colleagues (2012) on sound processing in the temporal lobe incorporates important information from the Hickok/Poeppel (2007) model and suggests a systems-level approach to understanding how sound is processed across modalities (including vocalizations and music):

> all sounds activate an anterior stream that likely supports more general perceptual and/or discourse-level semantic processes. A posterior stream appears to contribute to sound embodiment, whereas a ventral stream appears relevant for the mapping of sounds to meaning and associated visual representations. One characteristic, common across these three streams, is that they exist in both hemispheres and are fairly bilaterally engaged especially for highly relevant sounds such as vocalizations and music. A second characteristic is that they are activated relatively independently of sound class and sound attributes such as acoustic complexity, spectral change or combinatory complexity. Instead, all streams rely on stored memory representations of perceptual, sentient or conceptual nature against which incoming sounds are matched for identification. As these representations emerge through interactions with one's auditory environment, the gross spatial organization of auditory processing within the temporal lobe is unlikely to depend on preexisting sound classes. Rather, it seems to depend on an all-purpose auditory brain system and a lifetime of sonic experiences.
>
> *(Schirmer et al., 2012: 146)*

The take-away message points to system-level networks that would include different types of ecologically valid sound processes, as well as different modalities.

3.2.2 Critiques of Neuroimaging Studies of Language(s)

Some of the most common critiques of neuroimaging studies of language(s) from the linguistics, psycholinguistics, and neurolinguistics community focus on the problematic nature of *single word studies.* Paradis (2004: 153) immediately identifies the conundrum of violating ecological language tasks in favor of a reductionist approach, the problem of interpreting activations (or the lack of activations), and challenges the validity of any results achieved. Paradis is especially concerned with the use of a single *word* approach and why it is impossible to generalize from single word to language itself (2004: 173):

> single words are the least likely candidates for investigating "language" representation, given that what makes language most specific as a cognitive function, i.e., the language system (phonology, morphology, syntax), is supported by procedural memory, whereas isolated words are supported by declarative memory and hence are less focalized in their cortical representation ...
>
> Words are not just "building blocks of sentences" that, presented in isolation, may contribute to the study of the structure of the language edifice. When they are presented out of context, they lose most of their language-specific properties.
>
> *(2004: 176)*

Paradis identifies several problematic single word tasks, including word completion tasks, the conflation of fundamental differences between grammatical and lexical categories, single-word recognition, word reading and repetition, cued word generation, word translation, synonym generation, and object/picture naming (2004: 154–174). The challenges in understanding and interpreting activations continue to be discussed, especially in attempts to understand if and when certain activations are both necessary and sufficient for the task being examined.

De Bot (2009) furthers our understanding of the notion of a system-level processing of language(s) as he moves away from essentialist categories, and rather gives a quantitatively measurable perspective about the collective and individual levels of languages used by bi/multilinguals, that is, a constant interaction and a dynamic process of change based in variation of both contact and usage (2009: 427). We see the emergence of importance research and perspectives on cognitive brain reserve during this period as perspectives on languages (bi/multilingualism, not monolingualism) become more prevalent, and active processes of acquisition, maintenance, and loss of languages is viewed across the life cycle as interactive, dynamic, and the product of multiple and changing speech communities and communities of practice interacting in what Donald (2004: 35) calls the *distributed cognitive cultural network*.[2]

In the most recent period, there have been two studies that have criticized the statistical methods used to analyze fMRI studies (Eklund et al., 2016) as well as the reproducibility problems with fMRI datasets of brain-wide association studies (Marek et al., 2022). In both instances, questions are posed for improving methods in future studies and how the results are interpreted.

In fact, there have been important changes implemented for equitable thresholding and clustering in analysis toolkits as a result of the paper by Eklund and colleagues (Cox, 2019). Suggested outcomes include better data sharing, the importance of how statistical significance is interpreted, and improved quality of both brain and behavioral data.

3.2.3 Approaches to Embodied Cognition and Irreducible Multimodalities in Neural Processing

> *Sensory modalities like vision, touch, hearing, and so on are actually integrated with each other and with motor control and planning.*
> (Gallese & Lakoff, 2005: 459)

> *... we will argue that a key aspect of human cognition is neural exploitation – the adaptation of sensory-motor brain mechanisms to serve new roles in reason and language, while retaining their original functions as well.* (Gallese & Lakoff, 2005: 456)

Gallese and Lakoff present a detailed analysis of sensory-motor embodiment of human cognition and propose an "interactionist theory of meaning" hinged on "neural exploitation – the adaptation of the sensory-motor brain mechanism to serve new roles in reason and language," building their case on the premise of multimodality (Gallese & Lakoff, 2005: 456). Demonstrating first the multimodal properties of the brain's sensorimotor system itself, they connect this to the multimodal and sensorimotor properties of language, arguing that "language exploits the pre-existing multimodal character of the sensory-motor system" (ibid.). Gallese and Lakoff support multimodality in contrast to any modular approaches, including "supramodality" in rejecting the need or existence of higher association areas in the brain for combining various sensory input modalities. They assert that sensory-motor areas themselves control and process multiple modalities including both "action and perception" (ibid.: 459). Furthermore, Gallese and Lakoff identify premotor areas as containing "neurons that integrate motor, visual, and somatosensory modalities" as well as acoustic information, "for the purpose of controlling actions in space and perceiving peri-personal space" (ibid.).

3.2.4 Defining the Importance of Nonarbitrariness in Embodied Cognition

While Gallese and Lakoff introduced a version of nonarbitrariness in their work on embodied cognition in 2005,[3] linguistic approaches to nonarbitrariness began much earlier and were part of the fundamental framework of the Prague School of Linguistics and semiotic approaches to languages and cultures. In both cases, Roman Jakobson played a central role in the early twentieth century rejection of the Saussurean binary sign defined by arbitrariness of forms and meanings. Together with Jakobson, Bolinger and C.S. Peirce were both significant voices in shifting the perspective of the linguistic sign and participating in a triadic sign complex that is always interacting with nonbinary relations defined by relative nonarbitrariness. These linguistic perspectives predicted much of the multimodal turn in twenty-first century cognitive neurolinguistics, emphasizing the bodily and cultural systems that give rise to emergent linguistic meanings. The work of Gallese and Lakoff demonstrates how embodied cognition, multimodalities, and nonarbitrariness are vitally intertwined, and according to Gallese and Lakoff (2005: 456) requires a view of the sensory-motor system underlying the neurological basis of language processing to be irreducibly multimodal:

> circuitry across brain regions links modalities, infusing each with properties of others. The sensory-motor system of the brain is thus "multimodal" rather than modular. Accordingly, language is inherently multimodal in this sense, that is, it uses many modalities linked together—sight, hearing, touch, motor actions, and so on. Language exploits the pre-existing multimodal character of the sensory-motor system.

3.2.5 Raichle's Resting State Contributions

Raichle's work (2010, 2011, 2015; Raichle & Gusnard, 2005) on intrinsic brain activity based on rfMRI has proven to be very significant in increasing the depth of understanding the baseline functional networks (FNs) of the human brain. In rfMRI, the participant does not perform a task, but is asked to lie still, stay awake, and relax. rfMRI data has been shown to be associated with a range of factors, including disease, neural response, cognitive abilities, and others. Studies have suggested that the baseline FNs of the brain are not constrained by the underlying anatomical white matter connections observed through DTI. Raichle's network discoveries provide a clear basis for conducting rfMRI analysis and understanding the set of FNs resulting from the intrinsic fluctuations in brain activity that consist of bilateral, symmetrical, and discrete cortical areas, including the default mode (DMN), the sensorimotor (SMN), auditory (AUN), visual (VSN), executive control (ECN), dorsal attention (DAN), and salience (SN) (2011).

3.2.6 Proficiency Measurements and fMRI

There have been important advancements in understanding the importance of proficiency and interactions between proficiency in age over the past two decades of neuroimaging research. As early as 2006, Birdsong showed in his important meta-analysis that it is proficiency, not age of acquisition, that is the "strongest predictor" of degrees of similarity between a user's first and second languages both in neuroimaging studies of production and in the timing components of ERP studies (even in cases where L2 acquisition began at 12 or older) (2006: 24–26). These early findings dovetail with a call for more rigorous proficiency measurements in fMRI studies (de Bot, 2006, 2008).

Abutalebi began testing for proficiency in subjects as early as 2007 using translation tasks (Abutalebi et al., 2007). In their longitudinal study of second language acquisition, Andrews and colleagues (2013) was the first fMRI study to include Common European Framework of Reference for Languages (CEFR) proficiency testing (Andrews, 2014: 162–167). All subjects were scanned, and proficiency tested multiple times across the study. Multivariate analysis of covariance (MANCOVA) was used to correlate the change in proficiency with the change in scan results across specific regions of interest (Andrews et al., 2013, Andrews, 2014: 164–191). The Andrews lab continues to use CEFR proficiency testing in all of their language studies (Andrews et al., 2019).

Abutalebi and colleagues (2013: 910) were the first to publish on the major significance of linguistic proficiency in understanding the language control network in fMRI studies:

> The importance of our data is that they show for the first time that differences in language proficiency, defined within participants, differentially modulate activity in core regions of the language control network during language switching. Switching language increases demand on a region associated with monitoring the language in use (pre-SMA/ACC) and switching to the least proficient language increases demand in the region implicated in selecting the language in use (left caudate). Such demands change with proficiency and are likely to underlie the adaptive response to gray matter in these regions.
>
> *(Abutalebi et al., 2012; Zou et al., 2012)*

These advancements set the stage for the continuation of more rigorous proficiency testing measurements in ongoing and future analyses of neuroimaging studies of language(s) processing.

3.2.7 Studies of Cognitive Brain Reserve: fMRI, DTI

Bialystok and colleagues' (2007) work on the protective effects of bilingualism in delaying the onset of symptoms of dementia was the beginning of an important shift of functional and structural imaging techniques. The early literature discusses two different types of cognitive brain reserve – neurological and behavioral (cf. Bialystok et al., 2007; Guzman-Velez & Tranel, 2015; Valenzuela & Sachdev, 2006). Some of the central contributions to the study of cognitive brain reserve in bilinguals include the work of Pliatsikas and colleagues (2015), Luk and colleagues (2011), Anderson and colleagues (2018), Coggins and collegues (2004), and Felton and colleagues (2017).

Pliatsikas and colleagues (2015) considered four different groups of bilinguals based on their age of acquisition, including (a) early bilinguals, (b) late bilinguals, (c) older bilinguals, and (d) young late bilinguals. Their findings showed that all these different groups showed increased fractional anisotropy (FA) values (compared to monolinguals) across several regions, including the genu, body, and anterior part of the splenium of the corpus callosum (CC), extending bilaterally to the inferior frontal-occipital fasciculus (IFOF), the uncinate fasciculus (UF) and the superior longitudinal fasciculus (SLF).

Coggins and colleagues (2004) and Felton and colleagues (2017) identified greater white matter (WM) volume specifically in the CC of bilinguals versus their monolingual peers. Luk and colleagues (2011) also found greater FA values in the SLF and UF of bilinguals as opposed to monolinguals. Anderson and colleagues (2018) found greater WM integrity and higher axial diffusivity (AD) values in bilinguals across the following regions: bilateral superior posterior corona radiata, the right external capsule, the midbody and splenium of the CC, the left STLF, and the anterior IFOF.

Mechanisms for the different types of brain and cognitive reserve are still not fully understood, but may include enhanced generation of neuronal, dendritic, and synaptic connections and functional reorganization of neural networks across multimodalities. For example, Grundy and colleagues (2017) argue in favor of a shift to subcortical/posterior region-based networks in bilinguals, as opposed to the more frontal regions in monolinguals, while Abutalebi and colleagues (2015) focused on increased gray matter (GM) in bilingualism as an example of cognitive reserve and potential protective effects in aging.

While multiple studies show that WM integrity decreases in normal aging (Billiet et al., 2015; Giorgio et al., 2010; Rathee et al., 2016; Tang et al., 1997; Westlye et al., 2010), recent research on cognitive reserve shows how these trends can be changed or slowed down in lifelong bilingualism and musicianship using DTI analysis of WM integrity of subcortical fiber tracts. For example, DTI analysis of white matter integrity of subcortical fiber tracts in lifelong professional musicians has shown higher fractional anisotropy (FA), an indicator of greater WM integrity, in lifelong bilingualism (Pliatsikas, 2019) and musicianship (Andrews et al., 2021, 2023). Andrews and colleagues (2023) hypothesize that the bilateral superior longitudinal fasciculus (SLF) and bilateral uncinate fasciculi (UF) could be important in building cognitive reserve through musicianship and would demonstrate greater WM integrity in aging (ibid.). Results show that FA values in bilateral SLF and bilateral UF increase in lifelong musicianship, while this increase is not found in nonmusician controls (ibid.).

3.2.8 Resting State fMRI and Functional Connectivity

There is a vibrant body of current research that shows stronger degrees of functional connectivity in *resting state networks* in bilinguals (including differences in proficiencies, use, acquisition, simultaneous/sequential, etc.) (Pliatsikas [this volume]) and in lifelong musicians (Eierud et al., 2023). Resting state fMRI analyses use ROI (region of interest) and ICA (independent component analysis) approaches and include the original networks established by Raichle (2010, 2011) and make an important contribution to understanding complex sensory-motor learning and ability as found in bi/multiligualism and musicianship. Both the bilingual and musicianship rfMRI research show increases in average functional network connectivity (FNC) in both groups in comparison with nonbilingual/nonmusician controls. For example, Eierud and colleagues (2023) examined four different resting state networks (Default Mode [DMN], Auditory [AUN], Visual [VSN] and Sensory-motor [SMN]) and found an increase in average FNC in lifelong musicians with age in the DMN-AUN, DMN-VSN, SMN-VSN and a decrease in nonmusicians with age in the DMN-AUN, DMN-SMN, AUN-SMN, and SMN-VSN.

3.3 Future Directions

Neuroimaging studies of the neural processing of language(s) have produced a rich field of inquiry that continues to expand and deepen the viability of results obtained across different types of imaging, including functional and structural MRI and more recently the inclusion of rfMRI studies of healthy subjects. It is clear that the inclusion of bi/multilingualism as an integral part of the literature and research on language processing in the brain is an important step that will deepen the applications of future studies and expand the types of questions that can be posed. One of the important outcomes of future neuroimaging of language(s) will be strengthening the connections between the fields of theoretical linguistics and the cognitive neurosciences. The paradigm shifts to full brain analysis (including GM, WM, cortical/subcortical, networks, etc.), appreciating and grappling with neurodiversity and variations across and within brains, as well as a more interactive perspective of multimodalities and collaboration of neural networks and system, will enhance the researcher's ability to achieve more profound conclusions and generate new trajectories for future research using improved technologies (hardware, software) and imaging techniques. Several papers included in this volume explore and expand on these issues in demonstrating the importance of spectral and nonlinear definitions of bilingualism as a "multivariate, multifaceted experience" (see Pliatsikas [this volume]) and interactions/correlations between subcortical WM structural changes and rfMRI functional connectivity (Pliatsikas, Gallo, and Abutalebi [this volume]).

Notes

1 Poeppel (2008: 1) states how he views the epistemological problem with localization arguments: "It appears the localization of psychological functions has become the epistemological driving force, and that brain images are assumed to constitute a privileged type of evidence that forms the basis for the explanation of mental life. In contrast, it is argued here that cartography is immensely useful but intensely limited *vis-à-vis* the explanation of function."

2 **Signification** is the human ability to collectively invent innovative and dynamic external symbols and is a key component of Donald's DCCN (distributed cognitive cultural network). Without signification as the initial and primary ability that underlies human language and all of human cognition, there can be no nonhereditary collective memory. Languages are made possible by "evolving distributed systems" (2004: 35). Donald also notes the following (ibid.): "Culture itself possesses network properties not found in individual brains. The individual mind is thus a hybrid project, partly organismic in origin, and partly ecological, shaped by a distributed network whose properties are changing." This is in keeping with Bolinger's important statement (1965: 233): "For an event to become linguistic, a great many brains must play in unison."
Signification always requires the translation from one system into another, and the process is potentially infinite and unbounded (Jakobson, 1985: 206; Peirce 4.127). Merlin Donald provides an expanded perspective on signification by identifying the key to understanding human language as a collective phenomenon:

> [t]he isolated brain does not come up with external symbols. Human brains collectively invent symbols in a creative and dynamic process.
>
> (Donald, 2004: 43)

And symbols are invented, according to Donald, by means of executive skills "that created a nervous system that invented representation out of necessity" (ibid.).

3 Mahon and Caramazza suggest a *domain-specific sensory-motor hypothesis* (Mahon & Caramazza, 2008: 69): "... the way in which concepts are organized and represented is determined by the use to which those classes of mental representations have been, and are, put." On the other hand, Mahon and Caramazza explicitly deny the possibility of any connection between linguistic forms and meanings (Mahon & Caramazza, 2008: 61): "The relation between the language effectors (e.g. mouth for spoken language) and concepts is arbitrary." This puts them irreconcilably at odds with the larger framework proposed by Gallese and Lakoff.

References

Abutalebi, J. (2008). Neural processing of second language representation and control. *Acta Psychologica*, *128*, 466–478.

Abutalebi, J., Canini, M., Della Rosa, P. A., Green, D. W., & Weekes, B. S. (2015). The neuroprotective effects of bilingualism upon the inferior parietal lobule: A structural neuroimaging study in aging Chinese bilinguals. *Journal of Neurolinguistics*, *33*, 3–13.

Abutalebi, J., Della Rosa, P. A., Ding, G., Weekes, B., Costa, A., & Green, D. W. (2013). Language proficiency modulates the engagement of cognitive control areas in multilinguals. *Cortex*, *49*, 905–911.

Abutalebi, J., Della Rosa, P. A., Green, D. W., Hernandez, M., Scifo, P., Keim, R., et al. (2012). Bilingualism tunes the anterior cingulate cortex for conflict monitoring. *Cerebral Cortex, 22*(9), 2076–2086.

Abutalebi, J., & Green, D. (2007). Bilingual language production: The neurocognition of language representation and control. *Journal of Neurolinguistics, 20,* 242–275.

Abutalebi, J., & Green, D. W. (2016). Neuroimaging of language control in bilinguals: Neural adaptation and reserve. *Bilingualism: Language and Cognition, 19*(4): 689–698.

Abutalebi, J., Guidi, L., Borsa, V., Canini, M., Della Rosa, P.A., Parris, B.A., & Weekes, B.S. (2015). Bilingualism provides a neural reserve for aging populations. *Neuropsychologia, 69,* 201–201.

Abutalebi, J., Tettamanti, M., & Perani, D. (2009). The bilingual brain: Linguistic and non-linguistic skills. *Brain and Language, 109,* 51–54.

Anderson, J. A. E., Grundy, J. G., De Frutos, J., Barker, R. M., Grady, C., & Bialystok, E. (2018). Effects of bilingualism on white matter integrity in older adults. *Neuroimage, 167,* 143–150.

Andrews, E. (2011). Language and brain: Recasting meaning in the definition of human language. *Semiotica, 184*(1/4), 11–32.

Andrews, E. (2014). *Neuroscience and Multilingualism.* Cambridge University Press.

Andrews E., Eierud C., Banks D., Harshbarger, T., Michael, A., & Rammell, C. (2021). Effects of lifelong musicianship on white matter integrity and cognitive brain reserve. *Brain Sci, 11,* 67.

Andrews, E., Eierud, C., Banks, D., Harshbarger, T., Michael, A., & Rammell, C. (2023). DTI analysis of white matter integrity and cognitive brain reserve in lifelong musicians and controls. *J Psychiatry Psychiatric Disord, 7*(2), 67–79.

Andrews, E., Frigau, L., Voyvodic-Casabo, C., Voyvodic, J., & Wright, J. (2013). Multilingualism and fMRI: Longitudinal study of second language acquisition. *Brain Sciences, 3*(2), 849–876.

Andrews, E., Harshbarger, T., & Rammell, C. S. (2019). Multilingual listening and reading: An fMRI study of Russian/English and Spanish/English bilinguals. *Glossos, 14.*

Ben Shalom, D., & D. Poeppel, D. (2008). Functional anatomic models of language: Assembling the pieces. *The Neuroscientist, 14*(1), 119–127.

Bhatia, Tej K., & Ritchie, W. C. (2006). *The Handbook of Bilingualism.* Blackwell Publishing.

Bialystok, E., Craik, F.I.M., & Freedman, M. (2007). Bilingualism as a protection against the onset of symptoms of dementia. *Neuropsychologia, 45,* 459–464.

Binder, J. R., Frost, J. A., Hammeke, T. A., Cox, R. W., Rao, S. M., & Prieto, T. (1997). Human brain language areas identified by functional magnetic resonance imaging. *Journal of Neuroscience, 17,* 353–362.

Binder, J. R., Desai, R. H., Graves, W. W., & Conant, L. L. (2009). Where is the semantic system? A critical review and meta-analysis of 120 functional neuroimaging studies. *Cerebral Cortex, 19*(12), 2767–2796.

Binder, J. R., & Price, C. J. (2001). Functional neuroimaging of language. In R. Cabeza & A. Kingstone (Eds.), *Handbook of Functional Neuroimaging of Cognition* (pp. 187–251). MIT Press.

Billiet, T., Vandenbulcke, M., Madler, B., et al. (2015). Age-related microstructural differences quantified using myelin water imaging and advanced diffusion MRI. *Neurobiology of Aging, 36,* 2107–2121.

Birdsong, D. (2006). Age and second language acquisition and processing: A selective overview. *Language Learning, 56,* 9–49.

Bolinger, D. (1949/1965). The sign is not arbitrary. Reprinted in D. Bolinger, *Forms of English: Accent, Morpheme, Order.* Edited by Isamu Abe and Tetsuya Kanekiyo. Harvard University Press.

Bolinger, D. (1986). *Intonation and Its Parts: Melody in Spoken English.* Stanford University Press.

Bolinger, D. (1989). *Intonation and Its Uses: Melody in Grammar and Discourse.* Stanford University Press.

Bookheimer, S. (2002). Function MRI of language: New approaches to understanding the cortical organization of semantic processing. *Annual Review of Neuroscience, 25,* 151–188.

Buchel, C., Price, C., & Friston, K. (1998). A multimodal language region in the ventral visual pathway. *Nature, 394,* 274–277.

Cabeza, R., & Kingstone, A. (Eds.) (2001). *Handbook of Functional Neuroimaging of Cognition* (pp. 27–48). MIT Press.

Cabeza, R., & Nyberg, L. (2000). Imaging cognition II: An empirical review of 275 PET and fMRI studies. *Journal of Cognitive Neuroscience, 12*(1), 1–47.

Calvin, W. H., & Ojemann, G. A. (1994). *Conversations with Neil's Brain: The Neural Nature of Thought and Language.* Addison-Wesley/Harper & Row.

Cargnelutti, E., Tomasino, B., & Fabbro, F. (2019). Language brain representation in bilinguals with different age of appropriation and proficiency of the second language: A meta-analysis of functional imaging studies. *Frontiers in Human Neuroscience, 13,* 154.

Coggins, P. E., Kennedy, T. J., & Armstrong, T. A. (2004). Bilingual corpus callosum variability. *Brain Lang., 89,* 69–75.

Corina, D. P., Loudermilk, B. C., Detwiler, L., Martin, R. F., Brinkley, J. F., & Ojemann, G. A. (2010). Analysis of naming erros during cortical stimulation mapping: Implications for models of language representation. *Brain and Language, 115,* 101–112.

Cox, R. (2019). Equitable thresholding and clustering: A novel method for fMRI clustering in AFNI. *Brain Connectivity, 9*(7).

Craik, F. I., Bialystok, E., & Freedman, M. (2010). Delaying the onset of Alzheimer disease: Bilingualism as a form of cognitive reserve. *Neurology, 75*(19), 1726–1729.

de Bot, K. (2006). The plastic bilingual brain: Synaptic pruning or growth? Commentary on Green et al. In M. Gullberg & P. Indefrey (Eds.), *The Cognitive Neuroscience of Second Language Acquisition* (pp. 127–132). Blackwell.

de Bot, K. (2008). The imaging of what in the bilingual mind? *Second Language Research, 24*(1), 111–133.

de Bot, K. (2009). Multilingualism and aging. In W. C. Ritchie & Tej K. Bhatia (Eds.), *The New Handbook of Second Language Acquisition* (pp. 425–442). Emerald Group Publishing Ltd.

Donald, M. (2004). The definition of human nature. In D. Rees & S. Rose (Eds.), *The New Brain Sciences: Perils and Prospects* (pp. 34–58). Cambridge University Press.

Eierud, C., Michael, A., Banks, D., & Andrews, E. (2023). Resting-state functional connectivity in lifelong musicians. *Psychoradiology, 3*, 1–8.

Eklund, A., Nichols, T. E., & Knutsson, H. (2016). Cluster failure: Why fMRI inferences for spatial extent have inflated false-positive rates. *Proceedings of the National Academy of Sciences of the United States of America, 113*(28), 7900–7905.

Eickhoff, S. B., Laird, A. R., Grefkes, C., Wang, L. E., Zilles, K., & Fox, P.T. (2009). Coordinate-based activation likelihood estimation meta-analysis of neuroimaging data: A random-effects approach based on empirical estimates of spatial uncertainty. *Human Brain Mapping, 30*(9), 2907–2926.

ETS. (2007). *The Official Guide to the New TOEFL IBT* (3rd edition). McGraw Hill.

Fabbro, F. (1999). *The Neurolinguistics of Bilingualism: An Introduction.* Psychology Press.

Fabbro, F. (2001). The bilingual brain: Cerebral representation of languages. *Brain and Language,79*(2), 211–222.

Felton, A., Vazquez, D., Ramos-Nunez, A. I., Greene, M. R., Macbeth, A., Hernandez, A. E., & Chiarello, C. (2017). Bilingualism influences structural indices of interhemispheric organization. *Journal of Neurolinguistics, 2017*(42), 1–11.

Friederici, A. D. (1998). The neurobiology of language comprehension. In A. D. Friederici (Ed.), *Language Comprehension: A Biological Perspective* (pp. 263–301). Springer.

Gallese, V., & Lakoff, G. (2005). The brain's concepts: The role of the sensory-motor system in conceptual knowledge. *Cognitive Neuropsychology, 22*(3/4), 455–479.

Giorgio, A., Santelli, L., Tomassini, V., Bosnell, R., Smith, S., De Stefano, N., & Johansen-Berg, H. (2010). Age-related changes in grey and white matter structure throughout adulthood. *Neuroimage, 51*(3), 943–951.

Grundy, J. G., Anderson, J. A. E., & Bialystok, E. (2017). Neural correlates of cognitive processing in monolinguals and bilinguals. *Annals of the New York Academy of Sciences, 1396*(1), 183–201.

Guzmán-Vélez, E., & Tranel, D. (2015). Does bilingualism contribute to cognitive reserve? Cognitive and neural perspectives. *Neuropsychology, 29,* 139–150.

Hernandez, A. (2009). Language switching in the bilingual brain: What's next? *Brain and Language, 109,* 133–140.

Hickok, G., & Poeppel, D. (2004). Dorsal and ventral streams: A framework for understanding aspects of the functional anatomy of language. *Cognition, 92,* 67–99.

Hickok, G., & Poeppel, D. (2007). The cortical organization of speech processing. *Nature Reviews, 8,* 393–402.

Hodgson, V. J., Ralph, M. A. L., & Jackson, R. L. (2021). Multiple dimensions underlying the functional organization of the language network. *NeuroImage, 241,* 118444.

Humphreys, G. F., & Lambon Ralph, M. A. (2015). Fusion and fission of cognitive functions in the human parietal cortex. *Cerebral Cortex, 25*(10), 3547–3560.

Instituto Cervantes. (n.d.). *Preparar la prueba DELE.* Retrieved from https://examenes.cervantes.es/es/dele/preparar-prueba

Jackson, R. L. (2021). The neural correlates of semantic control revisited. *NeuroImage, 224,* 117444.

Jakobson, R. (1985/1956). Metalanguage as a linguistic problem. In S. Rudy (Ed.), *Selected Writings VII* (pp. 113–121). Mouton.

Jakobson, R. (1987/1957). Linguistics and poetics. In K. Pomorska & S. Rudy (Eds.), *Language in Literature* (pp. 62–94). Belknap Press of Harvard University Press.

Jakobson, R. (1985/1967). Language and culture. In S. Rudy (Ed.), *Selected Writings VII* (pp.101–111. Mouton.

Kotz, S. A. (2009). A critical review of ERP and fMRI evidence on L2 syntactic processing. *Brain and Language, 109,* 68–74.

Kotz, S. A., & Schwartze, M. (2010). Cortical speech processing unplugged: A timely subcortico-cortical framework. *Trends in Cognitive Sciences, 14*(9), 392–399.

Lakoff, G., & Johnson, M. (1980). *Metaphors We Live By.* University of Chicago Press.

Lieberman, P. (2006). *Towards an Evolutionary Biology of Language.* Harvard University Press.

Liu H., & Cao, F. (2016). L1 and L2 processing in the bilingual brain: A meta-analysis of neuroimaging studies. *Brain and Language, 159,* 60–73.

Luk, G., Anderson, J. A., Craik, F. I., Grady, C., & Bialystok, E. (2010). Distinct neural correlates for two types of inhibition in bilinguals: Response inhibition versus interference suppression. *Brain and Cognition, 74*(3), 347–357.

Luk, G., Bialystok, E., Craik, F. I. M., & Grady, C. L. (2011). Lifelong bilingualism maintains white matter integrity in older adults. *Journal of Neuroscience, 31,* 16808–16813.

Luo, H., & Poeppel, D. (2007). Phase patterns of neuronal responses reliably discriminate speech in human auditory cortex. *Neuron, 54,* 1001–1010.

Mahon, B. Z., & Caramazza, A. (2008). A critical look at the embodied cognition hypothesis and a new proposal for grounding conceptual content. *Journal of Physiology, 102*(1–3), 59–70.

Marek, S., Tervo-Clemmens, B., Calabro, F. J., Montez, D. F., Kay, B. P., Hatoum, A. S., et al. (2022). Reproducible brain-wide association studies require thousands of individuals. *Nature, 603,* 654–660.

North, B. (2000). *The Development of a Common Framework Scale of Language Proficiency.* P. Lang.

Ojemann, G. A. (1979). Individual variability in cortical localization of language. *Brain and Language, 6*, 239–260.

Ojemann, G. A. (1983). Brain organization for language from the perspective of electrical stimulation mapping. *Behavioral and Brain Sciences, 6*(2), 189–206.

Ojemann, G. A. (1987). Surgical therapy for medically intractable epilepsy. *Journal of Neurosurgery, 66*(4), 489–499.

Ojemann, G. A. (1991). Cortical organization of language. *The Journal of Neuroscience, 11*(8), 2281–2287.

Ojemann, G. A. (1993). Functional mapping of cortical language areas in adults: Intraoperative approaches. *Advances in Neurology, 63*, 155–163.

Ojemann, G. A. (2003). The neurobiology of language and verbal memory: Observations from awake neurosurgery. *International Journal of Psychophysiology, 48*(2), 141–146.

Ojemann, G. A., Corina, D. P., Corrigan, N., Schoenfield-McNeill, J., Poliakov, A., Zamora, L., Zanos, S. (2010). Neuronal correlates of functional magnetic resonance imaging in human temporal cortex. *Brain, 33*(I), 46–59.

Ojemann, G. A., Ojemann, J., Lettich, E., & Berger, M. (1989). Cortical language localization in left, dominant hemisphere. *Journal of Neurosurgery, 71*(3), 316–326.

Ojemann, G. A., & Whitaker, H. A. (1978). The bilingual brain. *Archives of Neurology, 35*(7), 409–412.

Paradis, M. (2000). The neurolinguistics of bilingualism in the next decades. *Brain and Language, 71*, 178–180.

Paradis, M. (2004). *A Neurolinguistic Theory of Bilingualism.* John Benjamins.

Peirce, C. S. (1931–1958). *Collected Papers of Charles Sanders Peirce, 1–8.* Harvard University Press.

Peirce, C. S. (1957). *Essays in the Philosophy of Science.* Liberal Arts Press.

Perani, D., Dehaene, S., Grassi, F., Cohen, L., Cappa, S. F., Dupoux, E., et al. (1996). Brain processing of native and foreign languages. *Neuroreport, 7*(15), 2439–2444.

Perani, D., Farsad, M., Ballarini, T., Lubian, F., Malpetti, M., Fracchetti, A., Magnani, G., March, A., & Abutalebi, J. (2017) The impact of bilingualism on brain reserve and metabolic connectivity in Alzheimer's dementia. *Proceedings of the National Academy of Science of USA, 114*(7), 1690–1695.

Perani, D., Paulesu, E., Galles, N. S., Dupoux, E., Dehaene, S., Bettinardi, V., et al. (1998). The bilingual brain: Proficiency and age of acquisition of the second language. *Brain, 121*, 1841–1852.

Perani, D., Saccuman, M. C., Scifo, P., Anwander, A., Spada, D., Baldoli, C., et al. (2011). Neural language networks at birth. *Proceedings of the National Academy of Sciences of the United States of America, 108*(38), 16056–16061.

Pliatsikas, C. (2019). Multilingualism and brain plasticity. In J. W. Schwieter & M. Paradis (Eds.), *The Handbook of the Neuroscience of Multilingualism* (pp. 230–251). Wiley Publishers.

Pliatsikas, C. (2020). Understanding structural plasticity in the bilingual brain: The Dynamic Restructuring Model. *Bilingualism: Language and Cognition, 23*(2), 459–471.

Pliatsikas, C. (2023). Bilingualism and brain structure: Insights from healthy ageing and progressive neurodenegerative disease. In G. Luk et al. (Eds.), *Understanding Language and Cognition through Bilingualism: In Honor of Ellen Bialystok* (pp. 301–317). John Benjamins.

Pliatsikas, C., DeLuca, V., Moschopoulou, E. et al. (2017). Immersive bilingualism reshapes the core of the brain. *Brain Structure and Function, 222,* 1785–1795.

Pliatsikas, C., Moschopoulou, E., & Saddy, J. D. (2015). The effects of bilingualism on the white matter structure of the brain. *Proceedings of the National Academy of Sciences of the United States of America, 112,* 1334–1337.

Poeppel, D. (1996). A critical review of PET studies of phonological processing. *Brain and Language, 55*(3), 317–351.

Poeppel, D. (2008). The cartographic imperative: Confusing localization and explanation in human brain mapping. In H. Bredekamp, M. Bruhn, & G. Werner (Eds.), *Bildwelten des Wissens (Ikonographie des Gehirns)* (pp. 1–21) Akademie Verlag.

Poeppel, D., & Embick, D. (2005). Defining the relation between linguistics and neuroscience. In A. Cutler (Ed.), *Twenty-First Century Psycholinguistics* (pp. 103–111). Four Cornerstones.

Poeppel, D., & Hickok, G. (2004). Towards a new functional anatomy of language. *Cognition, 92,* 1–12.

Price, C. J. (2000). The anatomy of language: Contributions from functional neuroimaging. *Journal of Anatomy, 197,* 335–359.

Price, C. (2010). The anatomy of language: A review of 100 fMRI studies published in 2009. *Annals of the New York Academy of Sciences, 1191,* 62–88.

Raichle, M. (2001). Functional neuroimaging: A historical and physiological perspective. In R. Cabeza & A. Kingstone (Eds.), *Handbook of Functional Neuroimaging of Cognition* (pp. 3–26). MIT Press.

Raichle, M. (2006). The brain's dark energy. *Science, 314,* 1249–1250.

Raichle, M. (2010). Two views of brain function. *Trends in Cognitive Sciences, 14* (4), 180–190.

Raichle, M. (2011). The restless brain. *Brain Connectivity, 1,* 3–12.

Raichle, M. (2015). The brain's default mode network. *Annual Review of Neuroscience, 38,* 433–447.

Raichle, M. E., & Gusnard, D. A. (2005). Intrinsic brain activity sets the stage for expression of motivated behavior. *Journal of Comparative Neurology, 493,* 167–176.

Rammell, C. S., Cheng, H., Pisoni, D. B., & Newman, S. D. (2019). L2 speech perception in noise: An fMRI study of advanced Spanish learners. *Brain Research, 1720,* 146316.

Rathee, R., Rallabandi, V. P., & Roy, P. K. (2016). Age-related differences in white matter integrity in healthy human brain: Evidence from structural MRI and diffusion tensor imaging. *Magnetic Resonance Insights, 9,* 9–20.

Savan, D. (1980). Abduction and semiotics. In I. Rauch & G. Carr (Eds.), *The Signifying Animal* (pp. 252–262). Indiana University Press.

Schirmer, A., Fox, P. M., & Grandjean, D. (2012). On the spatial organization of sound processing in the human temporal lobe: A meta-analysis. *Neuroimage, 63* (1), 137–147.

Sebastian, R., Laird, A., & Kiran, S. (2011). Meta-analysis of the neural representation of first language and second language. *Applied Psycholinguistics, 32*(4), 799–819.

Sobolev, O., & Nesterova, T. (2014). Oral communication in the framework of cognitive fluency: Developing and testing spoken Russian within the TORFL system. *Language Learning in Higher Education: Journal of the European Confederation of Language Centres in Higher Education (CercleS),3*(2), 271–282.

Stowe, L.A., Haverkort, M., & Zwarts, F. (2005). Rethinking the neurological basis of language. *Lingua, 115,* 997–1042.

Sulpizio, S., Del Maschio, N., Fedeli, D., & Abutalebi, J. (2020). Bilingual language processing: A meta-analysis of functional neuroimaging studies. *Neuroscience & Biobehavioral Reviews, 108,* 834–853.

Tagarelli, K. M., Shattuck, K. F., Turkeltaub, P. E., & Ullman, M. T. (2019). Language learning in the adult brain: A neuroanatomical meta-analysis of lexical and grammatical learning. *Neuroimage, 193,* 178–200.

Tang, Y., Nyengaard, J. R., Pakkenberg, B., & Gundersen, H. J. G. (1997). Age-induced white matter changes in the human brain: A stereological investigation. *Neurobiology of Aging, 18*(6), 609–615.

Turkeltaub, P. E., Eden, G. F., Jones, K. M., & Zeffiro, T. A. (2002). Meta-analysis of the functional neuroanatomy of single-word reading: Method and validation. *Neuroimage, 16,* 765–780.

Turkeltaub, P. E., Eickhoff, S. B., Laird, A. R., Fox, M., Wiener, M., & Fox, P. (2012). Minimizing within-experiment and within-group effects in activation likelihood estimation meta-analyses. *Human Brain Mapping, 33*(1), 1–13.

Valenzuela, M. J., & Sachdev, P. (2006). Brain reserve and dementia: A systematic review. *Psychological Medicine, 36,* 441–454.

Vigneau, M., Beaucousin, V., Hervé, P. Y., Duffau, H., Crivello, F., Houde, O., et al. (2006). Meta-analyzing left hemisphere language areas: Phonology, semantics, and sentence processing. *Neuroimage, 30*(4), 1414–1432.

Vigneau, M., Beaucousin, V., Hervé, P. Y., Jobard, G., Petit, L., Crivello, F., et al. (2011). What is right-hemisphere contribution to phonological, lexico-semantic, and sentence processing? Insights from a meta-analysis. *Neuroimage, 54*(1), 577–593.

Westlye, L.T., Walhovd, K. B., Dale, A. M., Bjørnerud, A., Due-Tønnessen, A., Engvig, A., et al. (2010). Life-span changes of the human brain white matter: Diffusion tensor imaging (DTI) and volumetry. *Cerebral Cortex, 20*(9), 2055–2068.

Zatorre, R. J. (1989). On the representation of multiple languages in the brain: Old problems and new directions. *Brain and Language, 36*(1), 127–147.

Zatorre, R. J., Meyer, E., Gjedde, A., & Evans, A. C. (1996). PET studies of phonetic processing of speech: Review, replication, and reanalysis. *Cerebral Cortex, 6,* 21–30.

Zou, L., Guosheng, D., Abutalebi, J., Shu, H., & Peng, D. (2012). Structural plasticity of left caudate in bimodal bilinguals. *Cortex, 48*(9), 1197–1206.

4

TMS as a Tool for Mapping the Dynamic Properties of Language in the Brain

Harsheen Kaur, Gerald Grant, and Fiona Baumer

4.1 Introduction: Mapping the Language Regions

It has been long appreciated that specific areas in the brain are critical for language. In the second half of the nineteenth century, studies of patients with brain lesions helped identify important, predominantly left-hemispheric language regions, including Broca's area in the left inferior frontal gyrus; Wernicke's area in the posterior superior temporal gyrus; the left angular gyrus as well as the insula (Ardila et al., 2016). Neurolinguistic research has further advanced the in-depth understanding of functional neuroanatomy of language (Poeppel & Hickok, 2004). Over the last century, noninvasive neuroimaging and neurophysiology techniques capable of measuring whole-brain activity have allowed us to study language in much greater depth, not only in neurologic patients but also healthy individuals. Through this foundational work, we have come to understand that in addition to these canonical regions, language function depends on distributed and bilateral networks throughout the brain (Ferstl et al., 2008; Hickok, 2009; Hickok & Poeppel, 2007), including white matter tracts like the superior longitudinal fasciculus and arcuate fasciculus (Madhavan et al., 2014). In contrast to the nuanced information from these studies, the typical focus of *clinical* language mapping is to identify "eloquent cortex," (Chang et al., 2011) – brain regions to which damage from neurosurgery would lead to devastating losses of language.

In clinical language mapping, we often discuss both the *lateralization* and *localization* of language function. Lateralization refers to the hemisphere (left vs. right) most important for language function. Localization refers to identifying more specific regions within a hemisphere critical for specific language functions (i.e., expressive vs. receptive language). For approximately 90 percent of the healthy population, language lateralizes to the left hemisphere. Left lateralization is more consistent in right-handed individuals, 90–95 percent of whom are left lateralized. In contrast, only 70–85 percent of left-handed individuals have left-hemispheric dominance (Packheiser et al., 2020), whereas 14 percent have shown to have symmetric language distribution and 8 percent have right hemispheric dominance (Szaflarski et al., 2002) In most healthy individuals, *receptive* language is localized in perisylvian regions, including the supramarginal, angular gyri, superior and middle temporal gyri while *expressive* language localizes to the middle and inferior frontal gyri, including the pars opercularis and pars triangularis (Ardila et al., 2016; Binder et al., 1997).

In contrast to healthy individuals, patients with neurological disorders can lateralize and localize language in atypical and sometimes surprising ways (Amirian et al., 2010). Children's receptive and expressive language skills develop during early years, which in turn affect their cognitive development. Childhood neurologic disorders affect the brain during a critical period of language development and thus can lead to an atypical language network (Jacola et al., 2006). Certain patterns are often seen with particular pathologies. For example, children who experience a destructive left-hemispheric insult, such as a stroke, in the first few years of life, tend to lateralize language to the right hemisphere (Guzzetta et al., 2008). Children with left-sided developmental lesions, such as cortical dysplasias, tend to retain left-hemispheric language, even if they have significant seizures arising from that hemisphere (Duchowny et al., 1996). Patients who acquire cortical pathology later in life, including tumors or strokes, may reorganize their language due to neuroplasticity (Pasquini et al., 2022), especially if the pathology is close to eloquent language regions (Gibb et al., 2020).

Lateralizing and localizing language function becomes especially clinically relevant for patients with neurologic disorders such as epilepsy and brain tumors who require neurosurgical intervention. The goal of neurosurgical procedures is to optimize the surgical resection in order to remove abnormal tissue while preserving eloquent cortex, language function, and quality of life. In this setting, understanding the spatial boundaries of a patient's critical language areas is crucial for optimal surgical planning (Hamberger & Cole, 2011; Ille et al., 2019). There are several technologies available to lateralize and localize eloquent language cortex in such patients. In this chapter, we will discuss a noninvasive method for mapping language cortex called transcranial magnetic stimulation (TMS), which is a particularly useful tool for evaluating pediatric and adult neurosurgical candidates who cannot cooperate with other tests. We will first compare TMS with other modalities available for language mapping. We will then review the technique for performing TMS and summarize the results as well as discuss the benefits and drawbacks for TMS.

4.2 Common Language Mapping Techniques

Several options exist for identifying brain eloquent language cortex, all of which have unique advantages and drawbacks (see Table 4.1 for a summary).

Table 4.1 *Available modalities for language mapping*

Modalities	Technique	Language Task	Advantages	Disadvantages
INVASIVE:				
Wada test or intracarotid amobarbital procedure (IAP)	One hemisphere is anesthetized temporarily while language and memory function are tested to determine the ability of each hemisphere to independently support language and memory.	Overt tasks (i.e., naming and remembering objects)	Most sensitive and specific test	1 percent risk of stroke. Invasive procedure. Requires significant cooperation.
Direct cortical stimulation	Specific cortical electrodes or areas of cortex are stimulated while the patient participates in a language task to determine regions that produce speech arrest, anomia, or other aspects of aphasia.	Overt tasks (i.e., naming, counting)	Sensitive and specific if electrode is in the right location and if the stimulation is in the right location.	Mapping is limited to location where electrodes have been implanted or where surgeon stimulates during awake surgery. Invasive procedure. Seizure induction. Interpretation limited in patients with after-discharges.
NONINVASIVE:				
Magnetoencephalography (MEG)	The magnetic field generated by the electrical activity of neurons is measured while the patient performs a language task. Task-induced changes in magnetic field are measured.	Covert tasks (i.e., object naming; verb generation).	High spatial and temporal resolution.	Movement creates artifact. Technically difficult, only available in certain centers.
Functional MRI (fMRI)	Changes in oxygenation concentration (blood oxygen level dependent) are measured. Task activation induces different neural states in the brain as the visual, auditory, or other stimulus is manipulated during the scan, and activation maps are obtained by comparing the signals recorded during the different states (Glover, 2011).	Covert tasks to limit movement artifact.	High spatial resolution.	Low temporal resolution. Brain lesions may alter vascular dynamics and tissue oxygenation, thus skewing interpretation in nearby regions. Requires significant patient cooperation.
TMS	A strong magnetic field applied to the scalp, induces an intracranial electric current in the underlying cortex. When applied in brief repetitive pulses, TMS transiently disrupts cortical function creating a "virtual lesion."	Overt (i.e., picture naming).	Cost-effective. Well tolerated in children.	Cannot map deep frontal, temporomesial or basal brain structures. Pain can limit the spatial extent of rTMS language mapping, especially in frontal regions.

We begin with a brief overview of the most frequently used mapping modalities and will then discuss TMS more fully in the following sections. The two invasive techniques reviewed – the intracarotid amobarbital procedure (IAP or Wada test, explained below) and cortical stimulation mapping (CSM) – are typically considered the gold standards (Jeltema et al., 2021), but carry higher risks of adverse events and provide spatially limited information. The non-invasive mapping options, functional magnetic resolution imaging (fMRI) and magnetoencephalography (MEG), are safe and can map the entire brain at once, thus helping in localization in addition to lateralization but can be intolerable or uninterpretable for certain patients.

4.2.1 Wada Test

During the Wada test, a single hemisphere is temporarily anesthetized via intracarotid injection of a sedative, sodium amobarbital. Appropriate sedation is confirmed by EEG recording indicating unilateral brain slowing. During this period, the patient is asked a battery of questions to determine if that hemisphere supports language or memory function. Though it is the most sensitive test, Wada carries a risk of stroke approaching 1 percent (Rossi et al., 2009). Furthermore, the patient must be able to cooperate with instructions and perform language and memory tasks in a stressful environment – while laying on an operating room table – making it unfeasible for most young children or patients with intellectual disability (Papanicolaou et al., 2014).

4.2.2 Cortical Stimulation Mapping

Cortical stimulation mapping (CSM) is a second invasive language mapping method that can be performed when intracranial electrodes have been implanted, typically in the context of epilepsy surgery, or open during awake tumor resection. During CSM, patients engage in a language task (i.e., reading, counting) while brief trains of stimulation are applied to specific intracranial electrodes to disrupt cortical function. The provider performing the mapping monitors the patient for expressive language errors, including aphasia/speech arrest and paraphasic errors (e.g., semantic or production errors). If stimulation consistently provokes an error on multiple trials, the electrode is in a critical language region. A primary drawback of CSM is that mapping is limited to regions with implanted electrodes, but electrodes may not cover canonical language regions. A second issue with CSM is safety. Seizures are reported in up to 58 percent of patients undergoing CSM (Cuello Oderiz et al., 2019). Seizures carry inherent safety risks and often disrupt the rest of the mapping session (Chou et al., 2018). Moreover, CSM can also provoke something called "after-discharges," electrical brain activity that outlasts the stimulation and can occur both in the stimulated electrode and in remote brain regions. When after-discharges occur, they compromise interpretation of the stimulation mapping, because it is difficult to determine whether the language error is related to disruption of the stimulated cortex vs. wider brain regions. Furthermore, CSM can be particularly challenging in children, because children typically require higher intensity stimulation to induce errors than adults (Aungaroon et al., 2017) and thus are more prone to after-discharges and seizures.

4.2.3 Magnetoencephalography

M agnetoencephalography (MEG) measures changes in the magnetic field associated with postsynaptic electrical activity of the dendritic pyramidal cells in response to external stimul (Kim & Davis, 2021). This magnetic field change is measured by special sensors placed on the scalp. Mathematical analyses are then used to identify the "source" of these signals within the brain. To identify language regions, patients perform specific language tasks during the MEG. Typically, MEG mapping is done using *covert language tasks*, in which the patient is asked to complete an activity in his or her head. *Overt language tasks*, in which the patient responds out loud, are avoided to avoid motion artifacts and limit motor system activation. Verb generation and object naming tasks are commonly used with MEG mapping. In the verb generation task, the patient sees a picture of an object or hears a noun played out loud and must think of a verb associated with the stimulus. In the object naming task, the patient is shown a picture of an object and is asked to think of the name (Bowyer et al., 2020). Changes in the magnetic fields induced by these tasks are measured and used to lateralize and localize language (Papanicolaou et al., 2004). Moreover, MEG has very high spatial and temporal resolution, and provides information about *when* specific brain regions become involved in the language task (Dikker et al., 2020). An important challenge of MEG is that the patient must remain very still to achieve an adequate signal to noise ratio (Papadelis & Chen, 2020). Furthermore, MEG is expensive and requires highly specialized equipment, typically including a magnetically shielded room (Singh, 2014).

4.2.4 Functional Magnetic Resonance Imaging

Functional magnetic resonance imaging (fMRI) measures hemodynamic changes in the brain induced by performance of a language task (Cascino, 2002), specifically, changes in the blood oxygen level dependent (BOLD) activity. Results are presented as images overlayed on structural brain scans, thus illustrating lateralization and localization of eloquent language cortex (Szaflarski et al., 2017). Typically, covert language tasks are also used during fMRI mapping, though passive listening (Starbuck et al., 2000) has also been tried. Moreover, fMRI has excellent spatial resolution but poor temporal resolution. Similar to MEG, the patient must remain very still. Furthermore, the patient must be able to tolerate being inside an MRI scanner for the duration of test, which can be challenging in children and those with neurologic or psychiatric disorders.

Typically, MEG and fMRI require active participation by patients. This can be challenging in young children, especially those below the age of three years or with patients who have significant developmental delays or anxiety (Narayana et al., 2021). To overcome these limitations, sedated fMRI and MEG studies have been developed. In these studies, auditory and visual cortical activation is measured during "passive" tasks (i.e., listening to mother's voice). Some promising results have been reported from sedated studies in younger patients as well as patients who are unable to cooperate during the test (Bernal et al., 2012; Birg et al., 2013), though this is still not standard of care.

4.2.5 Transcranial Magnetic Stimulation

One newer language mapping modality is transcranial magnetic stimulation (TMS). Most analogous to intracranial stimulation, TMS works by causing a "virtual lesion" in the brain but does it via an electromagnetic coil placed superficial to the scalp. It is considered a noninvasive functional language mapping tool which is safe and well tolerated and can be utilized even in younger children (Narayana et al., 2021). The remainder of this chapter will discuss TMS language mapping more fully.

4.3 Mechanisms of TMS Language Mapping

4.3.1 Overview

TMS operates based on Faraday's law of electromagnetic induction (Wagner et al., 2009). An electric current is passed through a magnetic coil placed superficially on the skull. This stimulation creates a fluctuating magnetic field which can pass through the skull and create an electric field (E-field) in the brain (Groppa et al., 2012). These intracranial electric fields create brief depolarizations of the cortical cells.

Two types of TMS stimulation are used for functional mapping. *Single pulses* of TMS are used to identify the motor cortex and establish a resting motor threshold, which are important initial steps for both motor and language mapping. When single pulses are applied to the motor cortex, they depolarize pyramidal neurons and induce muscle twitches on the contralateral side of the body (Sondergaard et al., 2021). In contrast, *repetitive TMS* (rTMS) (trains of pulses applied in quick succession at a specific frequency, which is typically between 5 and 10 times per second) transiently inhibits the function of the stimulated cortex and thus creates a "virtual lesion." During language mapping, rTMS is applied as the patient completes a language task. If the rTMS induces a speech error, one can infer that the stimulated cortex is important for expressive language (Pascual-Leone et al., 2000).

4.3.2 Mapping Procedure

In the United States, currently only the Nexstim Neuronavigation System has FDA clearance for functional brain mapping, and hence the procedures discussed below, and accompanying figures, reflect this system's workflow.

4.3.2.1 Neuronavigation

The use of TMS for language mapping was first tried nearly three decades ago (Cohen & Hallett, 1988; Pascual-Leone et al., 1991), but has been clinically developed since the invention of the neuro-navigation system (Krings et al., 2001) in the year 2000. The currently available optically tracked, frameless stereotaxic neuronavigational system combines anatomical MRI data with TMS, which allows the user to focus the stimulation on relevant language regions (Sparing et al., 2008). First, a three-dimensional model is created from a patient's brain MRI. The patient wears an optical tracking unit which allows the MRI-based model to be registered to the patient's head using scalp landmarks (see Figure 4.1).

In addition to tracking the person in space, the neuronavigation system tracks the TMS coil. Coils vary in size and shape. In general, coils with larger dimensions produce broader electric fields and smaller coils produce more localized and superficial fields (Deng et al., 2013). The most commonly utilized coil for language mapping is a figure-of-eight coil, in which the wires overlap in the center, allowing for a fairly focal E-field that has been modified

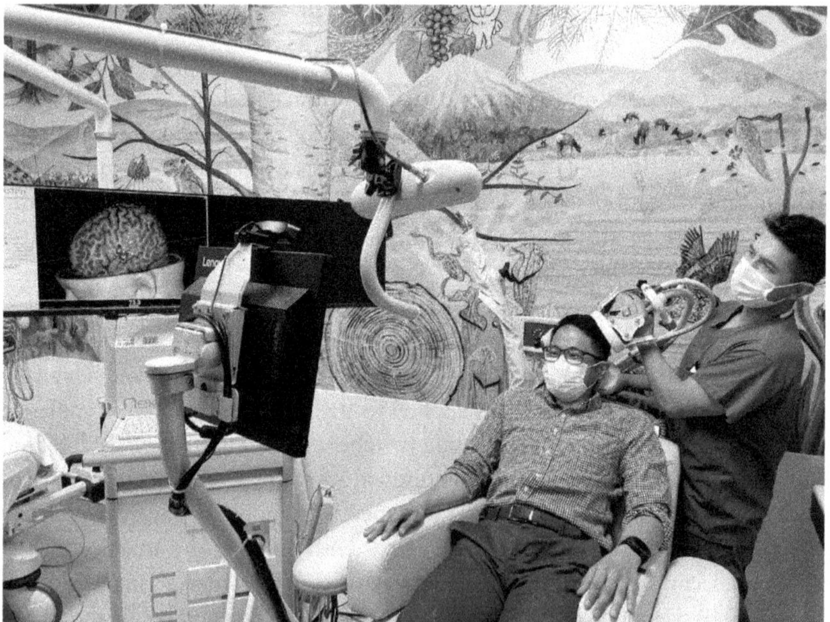

Figure 4.1 The patient sits in a chair, watching the screen where images are played for 1000 ms. The operator moves the TMS coil along the left side of the head, targeting repetitive trains of stimulation to perisylvian and frontal regions as guided by the neuronavigation system.

to optimize depth and focality of the stimulation (Ueno & Sekino, 2021). The depth of stimulation of the figure-of-eight coil varies from 0.9 to 3.4 cm and the surface area stimulated is up to 1–2 cm^2 (Narayana et al., 2021).

The neuronavigation system paired with the figure-of-eight coil provides precise and systematic placement of stimuli during the study, with millimeter accuracy (Sparing et al., 2008). Moreover, TMS coil positioning can be achieved by this system using functional and anatomical imaging (Lancaster et al., 2004). Furthermore, the neuronavigational system allows the TMS provider to calculate the expected E-field generated at the cortical surface, thus estimating the amount of stimulation the person is receiving with each pulse (Ruohonen & Karhu, 2010). This E-field calculation can guide the required stimulation intensity needed for mapping, as discussed below.

4.3.2.2 Stimulation Intensity

The next step is to define the intensity of stimulation required to perform the language mapping. First, the resting motor threshold (rMT) (Sondergaard et al., 2021), which is the minimum amount of stimulation needed to induce a muscle twitch of at least 50 mV in 5 out of 10 trials (Rothwell et al., 1999), is defined. Muscle twitches induced by TMS of the motor cortex are measured via electromyogram electrodes (Thickbroom et al., 1998) placed over intrinsic hand muscles such as the abductor pollicis brevis (APB) or first dorsal interosseous muscle (FDI). The rMT is identified for each hemisphere. Language mapping is typically performed using stimulation intensities ranging from 80 to 120 percent of the rMT (Sollmann et al., 2018). Alternatively, for patients with high rMT – such as children and patients on antiseizure medications – the stimulation intensity can be decided based on the E-field. Studies in both pediatric and adult patients (Picht et al., 2013; Rezaie et al., 2020; Schiller et al., 2020) suggest that an E-field of 80–100 V/m at the cortical surface is adequate for mapping.

4.3.3.3 Establishing Baseline Language Performance

Patients begin the language task by establishing a reliable baseline without stimulation. Patients are sequentially shown a series of pictures depicting discrete common objects which they must quickly name while video is recorded. To avoid erroneous interpretations, objects without synonyms are presented. Pictures that are not named accurately are removed from the deck as the goal is to identify prompts that elicit a reliable and rapid response (Krieg et al., 2017). For patients who might have difficulty with object naming, including developmentally delayed individuals and younger children, simpler tasks such as color naming have been used as a successful alternative (Schiller et al., 2020). In addition, there has been research into other language tasks such as pseudoword reading, verb generation, and action naming, though these are not part of the standard TMS language mapping protocols (Hauck et al., 2019).

4.3.3.4 Language Mapping with rTMS

After baseline performance without TMS is established, patients name the same set of pictures while the brain is stimulated to identify brain regions where TMS induces naming errors. Stimulation is focused on perisylvian (including the supramarginal and angular gyri and superior and middle temporal gyri) and frontal (middle and inferior frontal gyri, including the pars opercularis and pars triangularis) as well as sites that may be surgically resected (Cattaneo, 2013). The cortex is stimulated approximately every 10 mm, though denser 3–5 mm coverage is performed in areas of surgical interest (Krieg et al., 2017). Picture presentation is timed with the stimulation. Typically, pictures are shown for 1000 ms (Narayana et al., 2021) during which the patient receives five pulses of rTMS at a frequency of 5Hz (Krieg et al., 2014; Narayana et al., 2021) (equivalent to 1000 ms of stimulation). These stimulation parameters can be modified on a case-by-case basis to meet individual patient needs. For example, patients who demonstrate delayed responses during the baseline period may require longer trains of rTMS to ensure the task is disrupted. If a language error is noted during mapping, then that cortical region is stimulated several times to ensure consistency of the response. In general, a two thirds error rule is applied: if two out of three stimulations to a region induce a naming error, then that area is considered "positive" for language (Sanai et al., 2008; Sollmann et al., 2016). Typically, both the left and right hemisphere are mapped during a session.

4.3.3.5 Analysis of Errors

Though the TMS operator monitors for language errors during the mapping procedure, the whole session is recorded on video for formal scoring. The provider interpreting the study compares videos taken during the baseline to those recorded during the TMS mapping session to determine whether an error occurred, and if so, what type of error was induced.

The four basic types of errors include (Hernandez-Pavon et al., 2014; Mueller-Sarnowski et al., 2022; Pascual-Leone et al., 1991):

1. Speech arrest/no response: Speech arrest is described as complete absence of naming. This is differentiated from anomia where a patient can usually start the phrase but is unable to name the object (Tarapore et al., 2013).
2. Performance errors: Slurring, stuttering or improper articulation of the word. Dysarthria and oral motor speech disorder such as apraxia of speech are included in this category. Performance errors can also include (Corina et al., 2010):
   a. Hesitation errors: A delay is noted, and the response is slower than baseline.
   b. Phonological paraphasia: Unintentional replacement of the word with a phonetically different, but somewhat similar sounding word, for example, WAGON → DRAGON.

3. Semantic paraphasia: Replacement with word that has some association and/or belongs to the same category of the word of interest, for example, Strawberry → "fruit." Specific categories of semantic paraphasia include:
   a. Neologisms: A new word that does not exist but sounds like the original word, for example, SPOON → LOON.
   b. Circumlocution errors: A descriptor indicating the function of the object is given in place of the name, for example, LADDER → CLIMB.
4. Muscle activation: The TMS can activate the facial nerve because it passes through the scalp regions stimulated. This is particularly a problem when frontal brain regions or anterior temporal regions are stimulated. This stimulation can induce jaw twitching that can be painful and interrupt speech, and thus can confound the results (Holmes & Meteyard, 2018). Muscle twitching must be accounted for in the interpretation.

4.3.3.6 Creating the Language Map and Transferring to Other Imaging Modalities

After responses recorded during each stimulation have been reviewed and scored, the software plots each scored response on the brain MRI (Figure 4.2) to create a map of "language positive" and "negative" regions. This is typically done for both hemispheres. In addition, the language mapping data can be exported using DICOM (Digital Imaging and Communications in Medicine) representations and overlaid on other imaging for surgical planning. One such overlay is performed with diffusion tensor imaging fiber tracking (DTI-FT), which traces white matter tracts and provides further information about pertinent cortical and subcortical language processing areas, critical for surgical planning (Krieg et al., 2017). These maps can be used to help delineate "safe" surgical margins that avoid eloquent language cortex.

4.3.3.7 Determination of Laterality Index

The *laterality index* measures whether the patient is left or right hemisphere dominant for language. It is calculated by comparing errors induced by right vs. left hemisphere stimulation (Narayana et al., 2021). The most basic laterality index is derived by counting all errors in the left hemisphere and right hemispheres separately, and calculating the following:

$$\frac{\left[\text{Errors}_{\text{Left}} - \text{Errors}_{\text{Right}}\right] \times 100}{\left[\text{Errors}_{\text{Left}} + \text{Errors}_{\text{Right}}\right]}$$

Laterality indices of 100 to ≤ 10 are typically interpreted as indicating left hemispheric dominance, -100 to ≤ -10 as indicating right hemispheric dominance, and -10 to $+10$ as indicating bilateral representation of language (Rezaie et al., 2020). However, there is ambiguity regarding this method. Should absolute number of errors be considered or the percentage of stimuli that caused errors? Should all types of errors "count" the same amount? Several studies have utilized variable weightage for different errors. For

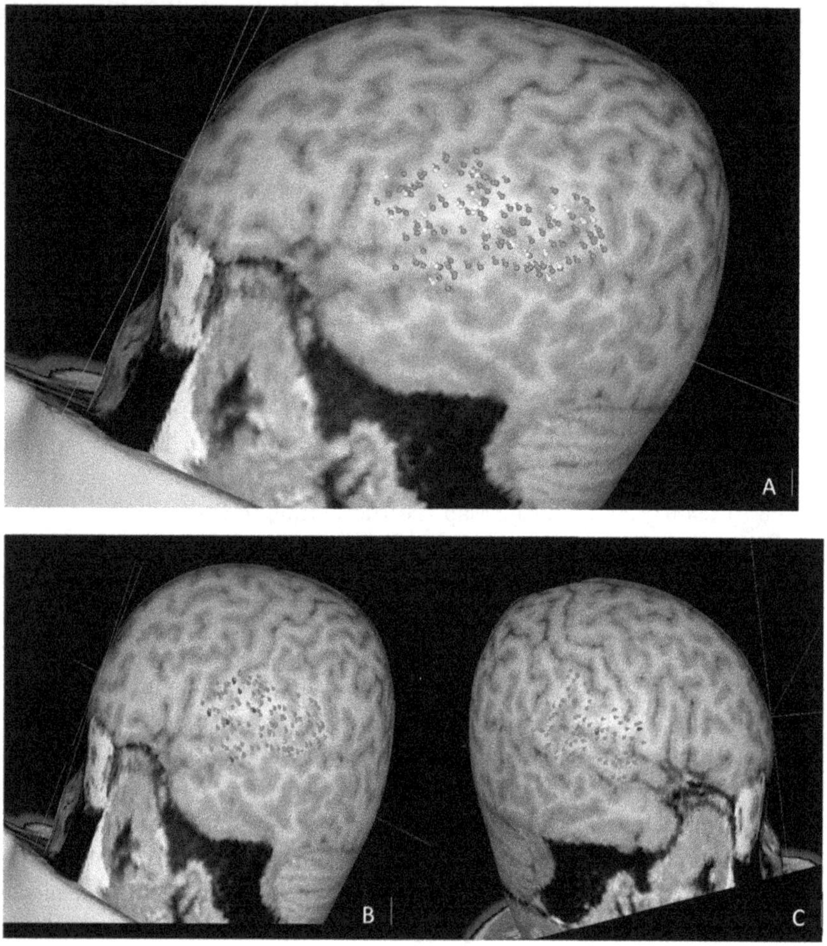

Figure 4.2 (A) Language map showing language positive areas represented by gray pegs and language negative areas represented by white pegs. Left (B) and Right (C) hemisphere mapping of specific error types (Gray = No Error; Blue = Semantic Error; Green = Performance Error; White = Speech Arrest; Red = Unclear Response). More errors are seen in the left vs. right hemisphere, though this highlights that speech errors are induced bilaterally. Performance errors (green) are more widely distributed in the superior temporal, supramarginal, and frontal regions. Semantic errors (blue) are concentrated in the frontal regions. A black and white version of this figure will appear in some formats. For the color version, refer to the online version.

example, weighing the speech arrests more than semantic errors, and semantic errors more than performance errors (Narayana et al., 2021). The vagaries of this method are still an area of research, as is establishing good interrater reliability in error analysis across centers (Sollmann et al., 2013).

4.4 Advantages of TMS Language Mapping

Functional mapping with TMS offers several practical advantages compared to other invasive and noninvasive mapping modalities. One major advantage of TMS is its utility in mapping motor and language functions in the pediatric population. In contrast to fMRI and MEG, TMS can be performed in an unrestricted environment and does not require the patient to lie still. Instead, patients sit in a comfortable semi-reclined chair during the procedure and can easily take breaks or move around if needed. Young children can be positioned on their parent's lap for the duration of the mapping and support staff, such as child life specialists or patient advocates, can safely be in the room to assist. With these environmental modifications, TMS can map expressive language in children as young as five to six years of age (Braden et al., 2022; Narayana et al., 2015, 2021) who otherwise would require sedation for imaging (Tarapore et al., 2012). For example, a high volume pediatric TMS center reported that TMS successfully mapped nearly 90 percent of children for whom other mapping modalities failed (Narayana et al., 2021). For similar reasons, TMS can also be an excellent option for adult patients with intellectual disability or anxiety.

A second feature of TMS is that expressive language tasks are *overtly* performed, meaning that patients respond, "out loud." In contrast, language tasks during MEG and fMRI tend to be *covertly* performed, with patients thinking of but not verbalizing responses. Covert performance in such studies is used to avoid motion artifacts (Narayana et al., 2021) and activation of the motor system. The overt responses during TMS allow the investigator to confirm that the patient is actively engaged throughout the language task. This can be particularly useful for patients who may not always be able to follow directions for sustained periods. Furthermore, as overt and covert tasks (termed "executed" vs. "imagined" in the clinical neuroscience literature) elicit different patterns of brain activity during motor tasks (Kilteni et al., 2018; Kline et al., 2020; Lotze et al., 1999; Neuper et al., 2005) and language (Proix et al., 2022), TMS mapping may uncover different activity than is seen by the covert language mapping methods.

TMS in children can provide an index of maturation of the cortex and corticospinal tracts and can help in differentiating static and dynamic nature of a disease process (Frye et al., 2008). Involvement of white matter tracts in certain pathologies can help differentiate them from others and provide a better idea about prognosis and treatment options. Following brain injury, reorganization of eloquent cortex can also be assessed by TMS (Najib et al., 2011). Over the course of a disease process, such as a brain tumor, plastic changes can be identified not only for motor tracts but also for language which can be of prognostic value to the neurosurgeons and neurologists (Raffa et al., 2018). Although imaging techniques such as diffusion tensor imaging (DTI) and fMRI provide similar data, the dynamic neurophysiological aspects of the maturing and healing brain are better delineated by TMS.

TMS has advantages mapping individuals with highly vascularized lesions that interfere with fMRI interpretation. More specifically, fMRI depends on the blood oxygen level-dependent (BOLD) signal and intact neurovascular coupling. Neurovascular coupling is the relationship between the neuronal discharges and associated blood flow that occurs with changing demands during any performed task. In patients with vascular abnormalities, strokes or tumors, neurovascular uncoupling can confound the results for fMRI (Pak et al., 2017; Tieleman et al., 2009). TMS is less affected by brain lesions when compared to fMRI because it is not affected by blood oxygen level and is based on virtual lesioning of the area being tested (Ille et al.; 2015; Jeltema et al., 2021).

A final advantage of TMS is that it is highly accessible and portable. Compared with the expensive equipment and dedicated, magnetically shielded hospital space required for fMRI or MEG, the TMS system is relatively inexpensive and can be performed in a normal procedure room or even at the patient's bedside (Singh, 2014). This is especially beneficial when a patient is clinically unstable and cannot be moved away from an ICU setting or if a patient is being considered for an urgent surgical procedure.

4.5 Limitations of TMS Language Mapping

4.5.1 Procedural Limitations

From a practical perspective, the repetitive TMS pulses administered during language mapping can be uncomfortable. Stimulation aimed at the frontal language regions and more anterior temporal targets often stimulates facial and trigeminal nerves and can induce muscle twitching of the eye and jaw (Maizey et al., 2013). When this occurs, it can be difficult to distinguish a true language error from dysarthria caused by the jaw jerking. Furthermore, response/reaction time during tasks performance can be affected due to this discomfort leading to false errors (Meteyard & Holmes, 2018).

Lower stimulation intensities can be tried at sensitive areas to reduce discomfort, but this must be balanced against using an intensity that could reasonably be expected to disrupt cortical function. As discussed in Section 4.3.2.2, the resting motor threshold is used to guide the stimulation intensity, but this is not feasible for patients with high motor thresholds, such as children or patients on anti-seizure medications (Narayana et al., 2021). In these cases, using the calculated E-field of 80–100 V/m can be an appropriate alternative (Picht et al., 2013; Rezaie et al., 2020; Schiller et al., 2020).

4.5.2 Limited Specificity

TMS often identifies more bilateral language representation than other mapping modalities (Rezaie et al., 2020; Schiller et al., 2020). Generally, TMS language maps are thought to have lower specificity and identify more "false positives" than other modalities. The specificity of TMS vs. other common modalities is generally estimated between 50 and 60 percent (Krieg et al., 2014; Sollmann et al., 2017; Tarapore et al., 2013). Several factors may contribute to this reduced specificity. As previously described in Section

4.3.3.6, TMS maps are drawn by considering the effect of focal cortical stimulation on language output; it is assumed that when a language disturbance occurs, it is because that specific stimulated cortical region is critical for language. This "virtual lesion" model however is likely too simplistic. First, the magnetic field created by TMS coil can stimulate a cortical region of up to 1–2 cm^2 and thus has somewhat limited spatial specificity (Narayana et al., 2021). Second, as with invasive direct cortical stimulation, TMS can cause transsynaptic excitation of downstream (and, possibly, upstream) neuronal units in functionally connected but geographically distant brain regions. TMS may thus identify false positive language regions (Tarapore et al., 2013). Another interpretation of the bilateral language representation is that TMS detects regions important for language that may be missed by other modalities, but has been described in the cognitive neuroscience literature (Hickok & Poeppel, 2007; Luo & Poeppel, 2007; Poeppel & Hickok, 2004). Alternatively, it is possible that the overt test used in TMS recruits more bilateral regions, as an fMRI study comparing overt and covert versions of the same language task showed bilateral activation with the overt task (Berro et al., 2021; Partovi et al., 2012). Regardless of the physiologic region, "false positive" language areas can have important clinical consequences during surgical planning as it can inappropriately limit needed resection of pathological areas. In contrast, given its very high sensitivity, TMS is considered an excellent modality for determining "true negatives," areas that would be safe to resect without risking language deficits (Tarapore et al., 2013). Since a human is operating the TMS and evaluating the errors and giving a score, there is a sense of subjectivity which also depends on the operator's experience with the test and subjects of varying ages. It is important to speak to caregivers to get a sense of what the patient is comfortable with during the procedure to set expectations (Lin et al., 2022).

4.6 Comparison between Available Techniques for Language Mapping

Several authors have compared TMS language mapping to other language modalities. It is important to note that these studies use the other language modality as the "gold standard," and provide test characteristics (sensitivity, specificity, negative and positive predictive value) for TMS relative to the other modality. While direct cortical stimulation mapping (CSM) and the Wada test are considered the "gold standard" for language mapping, it is less clear that fMRI or MEG are superior to TMS. Before reviewing the comparative literature, we will briefly review the terms used to describe and compare test accuracy (Monaghan et al., 2021; Schiller et al., 2020).

Imagine a TMS session involving stimulation of 200 regions with 100 regions being biologically important for language (true positives) and 100 being irrelevant (true negatives). *Sensitivity* refers to the ability of TMS to identify all true positive regions. If TMS only classified 98 of 100 true positive regions as "positive," then the sensitivity of TMS would be 98 percent. *Specificity* refers to the ability of TMS to classify true negative regions. If TMS classified 80 true negative regions as "negative" and the other 20 as "positive," the specificity of TMS would be 80 percent. In contrast, *positive and negative predictive value* refer to the probability that areas classified as "language positive" or "language negative" by TMS are truly positive/negative for language respectively. Based on the performance described above, the positive predictive value would be 83 percent ([98 True Positives Identified]/[98 True Positives + 20 False Positives]) and the negative predictive value would be 97.5 percent ([80 True Negatives]/[80 True Negatives + 2 False Negatives]).

Accuracy of a testing modality is the ability to differentiate true positives and true negatives out of all the results (Baratloo et al., 2015). In other words, it is a measurement of how close a result is to it's true value, be it negative or positive. It is calculated as [True positive + True negative/True positive + False positive + True negative + False negative]. Using the above mentioned numbers, accuracy for TMS would be 89 percent [98 True Positives Identified + 80 True Negative Identified/98 True Positives + 20 False Positives + 80 True Negatives + 2 False Negatives]. In an ideal situation, it is desirable to utilize a modality that is 100 percent accurate (Baratloo et al., 2015).

Concordance between two testing modalities describes the degree of agreement between the two modalities (Kwiecien et al., 2011). It is determined by comparing the regions that TMS determines to be "language positive" to those marked "positive" by the other modality. For example, when comparing TMS with fMRI, the concordance of presence of language in specific regions is measured by comparing whether the error-sites in TMS were similar to the activated clusters found in fMRI results in each region (Schiller et al., 2020).

4.6.1 TMS vs. Cortical Stimulation Mapping

Studies comparing TMS to cortical stimulation mapping (CSM) have been conducted both in patients with tumors and epilepsy. The earliest study by Tarapore and colleagues (Tarapore et al., 2013) focused on 12 adult patients with brain tumors around the language area and found a good correlation between the two testing modalities. The sensitivity of TMS was 90%, specificity was 98%, the positive predictive value was 69%, and the negative predictive value was 99% when compared with intraoperative direct cortical stimulation. A second study that included both adults and children (14 pediatric and 6 adult epilepsy patients, ranging from 9 to 32 years) (Lehtinen et al., 2018) noted a statistically significant association ($p < 0.001$) between TMS and CSM. No statistically significant differences were detected in sensitivity, specificity, or predictive values between adult and pediatric patients. This study concluded that TMS language mapping was similarly effective in epilepsy as in tumor patients and that performance was comparable in the pediatric and adult populations.

A more recent systematic review of ten studies comparing TMS and CSM language mapping in pediatric and adult patients (Jeltema et al., 2021) found large variability in test performance of TMS. Sensitivity of TMS ranged from 10 to 100 percent and specificity ranged from 13.3 to 98 percent. The positive and negative predictive values ranged from 17 to 75 percent and 57–100 percent, respectively. The authors suggest several potential sources of variability. One important factor is variability between different centers in criteria used to determine if a cortical area is considered positive or negative for language function. Some institutes employ the two thirds error rule as previously mentioned (Section 4.3.3.4) and others use varying error rate thresholds as a cut off. This suggests a need for consensus building across centers in order to optimize TMS language mapping protocols for all age groups.

A second source of variability in how TMS and CSM compare may be due to the cortical area being mapped. In the previously mentioned study of predominantly pediatric epilepsy patients (Lehtinen et al., 2018), TMS was more sensitive in frontal (86%) than temporoparietal (53%) regions, and it was more specific in temporoparietal (80%) than frontal (68%) sites. The positive predictive value of TMS was low at 27% (36% for frontal, 20% for temporoparietal sites) but negative predictive value was high at 95% (96% for frontal, 94% for temporoparietal sites) across regions. A study of adult tumor patients (Picht et al., 2013) suggested that TMS performed better for mapping frontal than posterior language regions. In this study of 20 patients with tumors in or close to left-sided language eloquent regions, there was excellent sensitivity and negative predictive value around Broca's area but moderate values (sensitivity, negative predictive value) around posterior regions of language mapping (Picht et al., 2013). Another study of 32 adult patients with tumors (Krieg et al., 2014), noted that timing of the TMS administration relative to picture presentation could be region-specific; posterior areas were better mapped if stimulation coincided precisely with picture presentation whereas frontal mapping was improved when there was a 300ms delay between picture presentation and stimulation onset. These findings may reflect the processing time it takes to transmit information from "receptive" posterior regions to "expressive" frontal regions.

In summary, the literature suggests that compared to CSM, TMS shows similar sensitivity in identifying areas important to speech but worse specificity. TMS induces language errors in regions where CSM does not. In contrast, TMS has excellent negative predictive value, so one can be confident that resection of regions that do not elicit language errors can be safely performed.

4.6.2 TMS vs. Wada

In contrast to CSM or TMS, which aim to localize specific language regions within a hemisphere, the goal of the Wada test is to identify more generally whether the left or right hemisphere is language dominant. An early study of seventeen epilepsy surgery patients found that TMS and Wada were concordant in 12 (70 percent) patients (Epstein et al., 2000). However, this study was conducted before neuronavigated TMS was readily available and thus language regions may not have been specifically stimulated. In contrast, a more recent study using navigated TMS (Han et al., 2021) with nine right-handed epilepsy patients found perfect concordance (100 percent) between TMS and Wada lateralization, even though language lateralization was atypical in 57 percent of patients. Although, in this study, only inferior frontal pars triangularis and superior temporal regions with their corresponding sites on the contralateral hemisphere were stimulated, whereas other areas associated with language (maybe more specific), including supra-marginal or angular gyri, were not. In summary, small studies have shown the concordance between TMS and Wada varies between 70 and 100 percent, with neuronavigated TMS improving consistency of findings between the two modalities.

4.6.3 TMS vs. fMRI

Several studies have compared fMRI to TMS and overall have identified modest concordance between the two techniques when lateralizing hemispheric dominance for language. The choice of language task used during the mapping procedure affects the concordance between fMRI and TMS results. One study compared standard TMS object-naming mapping to fMRI maps obtained during two different tasks (object naming and verb generation) in a mixed group of pediatric (n=26) and adult (n=14) patients with brain tumors and/or epilepsy. Out of this cohort, 12 patients completed the object naming task, and 34 completed the verb generation task, and 6 patients completing both tasks. This study found that TMS language maps were more concordant with fMRI maps generated during the object naming paradigm (concordant in 8/12, 67%) than the verb generation (concordant in 16/34, 47%) protocols. TMS performed better in children than adults. In particular, in children, TMS maps agreed more with fMRI object naming (sensitivity 86%, specificity 71%, accuracy 79%) than fMRI verb generation (sensitivity 74%, specificity 43%, accuracy 59%) maps. Additionally, in all patients, TMS showed higher concordance and accuracy with functional MRI results for left-handed (60% concordance; 80% accuracy) than right-handed (51% concordance; 66% accuracy) patients (Schiller et al., 2020).

A second quite novel study tested if concordance between TMS and fMRI language maps could be improved by utilizing different language tasks (Hauck et al., 2019). Typically, TMS language mapping is performed only with the object naming task previously described (Section 4.3.3.4), but other language tasks do exist, and it is not clear that object naming is the optimal task for mapping. One involved 19 healthy, young, right-handed adults who performed four different language tasks while mapping the left hemisphere with both TMS and with fMRI: object naming, pseudoword reading, verb generation, and action naming. The study found that TMS maps and fMRI maps derived during the pseudoword reading task had good concordance while there was moderate to good concordance in verb generation and action naming. There was moderate concordance noted with object naming. While the results of this study were preliminary, they highlight the fact that TMS language mapping might be improved by optimizing task selection.

In addition to direct comparisons between fMRI and TMS, several studies have obtained fMRI, TMS, and CSM in the same patients and have used CSM as the "gold standard" against which they compared the performance of both fMRI and TMS. One study of 28 adult patients (age range 25–71 years) with brain tumors (Muir et al., 2022) found that TMS was more sensitive (83%) and specific (67%), and had a higher negative predictive value (99%) than fMRI (sensitivity 71%, specificity 40% negative predictive value 80%), but had a worse positive predictive value (TMS 16% vs. fMRI 29%). Poor cognitive baseline was associated with decreased TMS accuracy results. A second study of 35 adult patients (ages 24–74 years) with left perisylvian brain tumors (Ille et al., 2015) showed that rTMS was a more sensitive but less specific tool for preoperative language mapping than fMRI. Combining fMRI and TMS led to improved prediction of the language positive areas identified by CSM.

In summary, overall results of fMRI and rTMS are concordant, although the degree of concordance varies based on the language task performed. Additionally, TMS may be more concordant with fMRI mapping in children than adults and in left-handed vs. right-handed individuals. Task optimization may be an important future direction for improving the utility of TMS language mapping.

4.6.4 TMS vs. MEG

Only one large-scale study compared TMS to MEG, and it specifically examined how well each technique determined the dominant hemisphere for language (Rezaie et al., 2020). Of all the studies comparing TMS to other mapping modalities, this study was unique in that most patients were pediatric (54/67 were less than age 19 years, range 5–18 years). There was a 63% concordance between techniques. Compared to MEG, TMS had a sensitivity of 74% and specificity of 72%, and moderate positive (78%) and negative (68%) predictive values. Another group of researchers determined in their proof of nTMS technique and validation study (Tarapore et al., 2013) that with MEG, reliable language lateralization is obtained and with TMS, precise language areas can be mapped and, although the results for both these techniques are concordant, MEG and TMS can be used as complementary studies.

4.6.5 Limitations of Comparative Studies

Most of the patients in these cohorts had brain tumors. Although, tumors are an important indication for presurgical mapping, the results from these studies might not be generalizable to patients with other forms of neuropathology, for example, epilepsy. For example, the language literature describes that destructive lesions to the left hemisphere, such as in pediatric stroke (Lidzba et al., 2017) as well as infiltrative lesions such as tumors (Ille et al. 2019; Krieg et al., 2013), lead to remapping of language to the other hemisphere. In contrast, developmental lesions in the left hemisphere present since birth, such as cortical dysplasia, can induce reorganization of language function within the same hemisphere but do not necessarily shift language contralaterally (Gaillard et al., 2007). Furthermore, the comparative studies largely focused on adults or on mixed-age populations. Pediatric neurologic disorders can interfere with critical windows of language network development whereas pathology that begins in adulthood disturbs stably organized and lateralized language networks. Given this, results of studies performed on mixed age groups may miss important, age-specific phenomena. Another important and clinically relevant point is that typically the more data we have to help us with surgical decision making, the better our outcomes will be. In the real world, we use multiple modalities to help us map out the eloquent cortex and create surgical margins for the neurosurgeons before they enter the operating room. If the surgery will be proximate to a "high stakes" region, we often attend the surgery in order to refine the mapping with intraoperative CSM.

4.7 Safety and Tolerability of TMS Language Mapping

TMS was first introduced in 1985 (Barker et al., 1985) and has been used for noninvasive functional mapping of neurosurgical patients since 1990. The International Federation of Clinical Neurophysiology updated their Expert Guidelines regarding safe practices for TMS in 2021 (Rossi et al., 2021). Overall, TMS has an excellent safety profile with mild acute side effects and no expected long-term side effects (Hadley et al., 2011; Tarapore et al., 2016). The most common adverse event induced by TMS functional mapping is pain and discomfort at the stimulation site (Narayana et al., 2021). To improve tolerability of mapping, patients are actively monitored for discomfort throughout the study and offered breaks if needed. Many patients benefit from having a family member or friend present for support; young children can sometimes be mapped while sitting on a parent's lap. We have found that in pediatric patients, having a Child Life Specialist present to provide age- and developmentally appropriate support and incentives can improve tolerability and facilitate collection of clinically useful data. Notably, it is important that distractions be minimized during the mapping itself. Other potential adverse events include short term headache, lightheadedness, abnormal scalp sensation, or transient visual defect (Bae et al., 2007; Hett et al., 2021).

One of the primary concerns with brain stimulation is seizure induction. Seizures are a major concern for both safety and feasibility reasons (as seizures will prematurely end a mapping session). The risk of seizures during a TMS session is quite low (Bae et al., 2007; Schrader et al., 2004), even in patients with epilepsy. A study of a large group of 733 brain tumor patients, ages 8–86 years, found that none had seizures during language or motor mapping, even though 50 percent had preexisting seizures secondary to the lesion (Tarapore et al., 2016). A second, smaller study of 15 healthy controls and 50 brain tumor patients showed no seizures during language mapping with TMS (Rösler et al., 2014). In adult epilepsy patients, a crude seizure risk from 2.9 to 3.6 percent (Pereira et al., 2016; Schrader et al., 2004), has been estimated for single-, paired-pulse, and rTMS protocols. Seizures induced during TMS of epilepsy patients usually are similar to patient's "typical" seizure semiology. In a mixed pediatric cohort of patients with tumors or refractory epilepsy, 20 percent of patients were noted to have seizures during or immediately after motor mapping but none had seizures during or after language mapping (Narayana et al., 2021). Of note, the seizures during motor mapping were all typical semiology; brief (<1 minute), self-resolving and did not prevent continuation of mapping or affect quality of results. The authors concluded that these seizures occurred during the mapping coincidentally rather than being provoked by the stimulation; they drew this conclusion because the patients had very frequent seizures at baseline due to their refractory epilepsy. Though no serious consequences of seizures during TMS have been reported, it is generally recommended that epilepsy patients bring their typical rescue medication to mapping sessions in case of prolonged or repeated seizures (Narayana et al., 2021).

There are certain absolute contraindications for TMS. Ferromagnetic objects implanted in the head or neck areas in close proximity to the TMS coil magnetic fields (i.e., cochlear implants, implanted brain stimulators or electrodes, aneurysm clips) are not considered safe because the Eddy currents induced by the TMS magnetic field in the metal objects can cause the objects to heat and pose a risk of thermal injury (McClintock et al., 2018). Patients with tattoos with ferromagnetic-containing ink are also contraindicated for this procedure (Laux et al., 2016). It is important to note that many of the recent implanted neurosurgical plates and clips are *not* ferromagnetic and thus are TMS compatible. Therefore, the TMS provider needs to check specific devices. Nonferromagnetic hardware (e.g., braces, implants, fillings, cranial titanium plates, certain programmable ventriculo-peritoneal shunts) are also typically considered safe. Implanted metal below the head and neck are generally considered safe (e.g., hip implants) (Deng et al., 2013). There is paucity of data regarding VNS and cardiac pacemakers (McClintock et al., 2018).

TMS safety guidelines as suggested by the International Federation of Clinical Neurophysiology (Rossi et al., 2021) provide guidance on safe/recommended: stimulation frequencies, number of stimuli, and magnitude of induced E-field. The guidelines also discuss risk of seizures and use of TMS in patients with implanted devices.

4.8 Summary

Language mapping is an essential part of presurgical work up in neurosurgical patients. TMS is a quick, safe, well-tolerated language mapping modality that can be used in adult and pediatric patients. It is especially useful for lateralizing language or identifying eloquent cortex around proposed resections in patients who cannot tolerate other mapping modalities. Compared with other techniques, TMS has a high sensitivity and an excellent negative predictive value. Resection of regions that are "language negative" on TMS mapping typically will not lead to language deficits postoperatively. In contrast, TMS has relatively poor specificity and hence there is room to optimize the procedure, including potentially by altering the language paradigms used during mapping. Ongoing research is underway to improve TMS language mapping techniques and to include TMS as a routine part of the preneurosurgical work up for patients with epilepsy or tumors in eloquent cortex.

References

Amirian, E., Liu, Y., Scheurer, M. E., El-Zein, R., M. R. Gilbert M. R., & Bondy, M. L. (2010). Genetic variants in inflammation pathway genes and asthma in glioma susceptibility. *Neuro Oncol, 12*(5), 444–452.

Ardila, A., Bernal, B., & Rosselli, M. (2016). How localized are language brain areas? A review of Brodmann areas involvement in oral language. *Arch Clin Neuropsychol, 31*(1), 112–122.

Aungaroon, G., Zea Vera, A., Horn, P. S., Byars, A. W., Greiner, H. M., Tenney, J. R., Arthur, T. M., Crone, N. E., Holland, K. D., Mangano, F. T., & Arya, R. (2017). After-discharges and seizures during pediatric extra-operative electrical cortical stimulation functional brain mapping: Incidence, thresholds, and determinants. *Clin Neurophysiol, 128*(10), 2078–2086.

Bae, E. H., Schrader, L. M., Machii, K., Alonso-Alonso, M., Riviello, J. J., Pascual-Leone A., & Rotenberg, A. (2007). Safety and tolerability of repetitive transcranial magnetic stimulation in patients with epilepsy: A review of the literature. *Epilepsy Behav, 10*(4), 521–528.

Baratloo, A., Hosseini, M., Negida, A., & El Ashal, G. (2015). Part 1: Simple definition and calculation of accuracy, sensitivity and specificity. *Emerg (Tehran), 3*(2), 48–49.

Barker, A. T., Jalinous, R., & Freeston, I. L. (1985). Non-invasive magnetic stimulation of human motor cortex. *Lancet, 1*(8437), 1106–1107.

Bernal, B., Grossman, S., Gonzalez, R., & Altman, N. (2012). FMRI under sedation: What is the best choice in children? *J Clin Med Res, 4*(6), 363–370.

Berro, D. H., Lemée, J. M., Leiber, L. M., Emery, E., Menei, P., & Ter Minassian, A. (2021). Overt speech critically changes lateralization index and did not allow determination of hemispheric dominance for language: An fMRI study. *BMC Neurosci, 22*(1), 74.

Binder, J. R., Frost, J. A., Hammeke, T. A., Cox, R. W., Rao, S. M., & Prieto, T. (1997). Human brain language areas identified by functional magnetic resonance imaging. *J Neurosci, 17*(1), 353–362.

Birg, L., Narayana, S., Rezaie, R., & Papanicolaou, A. (2013). Technical tips: MEG and EEG with sedation. *Neurodiagn J, 53*(3), 229–240.

Bowyer, S. M., Zillgitt, A., Greenwald, M., & Lajiness-O'Neill, R. (2020). Language mapping with magnetoencephalography: An update on the current state of clinical research and practice with considerations for clinical practice guidelines. *J Clin Neurophysiol, 37*(6), 554–563.

Braden, A. A., Weatherspoon, S. E., Boardman, T., Williard, T., Adkins, A., Gibbs, S. K., Wheless, J. W., & Narayana, S. (2022). Image-guided TMS is safe in a predominately pediatric clinical population. *Clin Neurophysiol, 137*, 193–206.

Cascino, G. (2002). Functional MRI for language localization. *Epilepsy Curr, 2*(6), 178–179.

Cattaneo, L. (2013). Language. *Handb Clin Neurol, 116*, 681–691.

Chang, E. F., Clark, A., Smith, J. S., Polley, M. Y., Chang, S. M., Barbaro, N. M., Parsa, A. T., McDermott, M. W., & Berger, M. S. (2011). Functional mapping-guided resection of low-grade gliomas in eloquent areas of the brain: Improvement of long-term survival. Clinical article. *J Neurosurg, 114*(3), 566–573.

Chou, N., Serafini, S., & Muh, C. R. (2018). Cortical language areas and plasticity in pediatric patients with epilepsy: A review. *Pediatr Neurol, 78*, 3–12.

Cohen, L. G., & Hallett, M. (1988). Noninvasive mapping of human motor cortex. *Neurology, 38*(6), 904–909.

Corina, D. P., Loudermilk, B. C., Detwiler, L. R., Martin, F., Brinkley, J. F., & Ojemann, G. (2010). Analysis of naming errors during cortical stimulation mapping: Implications for models of language representation. *Brain Lang, 115*(2), 101–112.

Cuello Oderiz, C., von Ellenrieder, N., Dubeau, F., Eisenberg, A., Gotman, J., Hall, J., Hincapié, A. S., Hoffmann, D., Job, A. S., Khoo, H. M., Minotti, L., Olivier, A., Kahane, P., & Frauscher, B. (2019). Association of cortical stimulation-induced seizure with surgical outcome in patients with focal drug-resistant epilepsy. *JAMA Neurol, 76*(9), 1070–1078.

Deng, Z. D., Lisanby, S. H., & Peterchev, A. V. (2013). Electric field depth-focality tradeoff in transcranial magnetic stimulation: Simulation comparison of 50 coil designs. *Brain Stimul, 6*(1), 1–13.

Dikker, S., Assaneo, M. F., Gwilliams, L., Wang, L., & Kösem, A. (2020). Magnetoencephalography and language. *Neuroimaging Clin N Am, 30*(2), 229–238.

Duchowny, M., Jayakar, P., Harvey, A. S., Resnick, T., Alvarez, L., Dean, P., & Levin, B. (1996). Language cortex representation: Effects of developmental versus acquired pathology. *Ann Neurol, 40*(1), 31–38.

Epstein, C. M., Woodard, J. L., Stringer, A. Y., Bakay, R. A., Henry, T. R., Pennell, P. B., & Litt, B. (2000). Repetitive transcranial magnetic stimulation does not replicate the Wada test. *Neurology, 55*(7), 1025–1027.

Ferstl, E. C., Neumann, J., Bogler, C., & von Cramon, D. Y. (2008). The extended language network: A meta-analysis of neuroimaging studies on text comprehension. *Hum Brain Mapp, 29*(5), 581–593.

Frye, R. E., Rotenberg, A., Ousley, M., & Pascual-Leone, A. (2008). Transcranial magnetic stimulation in child neurology: Current and future directions. *J Child Neurol, 23*(1), 79–96.

Gaillard, W. D., Berl, M. M., Moore, E. N., Ritzl, E. K., Rosenberger, L. R., Weinstein, S. L., Conry, J. A., Pearl, P. L., Ritter, F. F., Sato, S., Vezina, L. G., Vaidya, C. J., Wiggs, E., Fratalli, C., Risse, G., Ratner, N. B., Gioia, G., & Theodore, W. H. (2007). Atypical language in lesional and nonlesional complex partial epilepsy. *Neurology, 69*(18), 1761–1771.

Gibb, W. R., Kong, N. W., & Tate, M. C. (2020). Direct evidence of plasticity within human primary motor and somatosensory cortices of patients with glioblastoma. *Neural Plast, 2020*, 8893708.

Glover, G. H. (2011). Overview of functional magnetic resonance imaging. *Neurosurg Clin N Am, 22*(2), 133–139, vii.

Groppa, S., Oliviero, A., Eisen, A., Quartarone, A., Cohen, L. G., Mall, V., Kaelin-Lang, A., Mima, T., Rossi, S., Thickbroom, G. W., Rossini, P. M., Ziemann, U., Valls-Solé, J., & Siebner, H. R. (2012). A practical guide to diagnostic transcranial magnetic stimulation: Report of an IFCN committee. *Clin Neurophysiol, 123*(5), 858–882.

Guzzetta, A., Pecini, C., Biagi, L., Tosetti, M., Brizzolara, D., Chilosi, A., Cipriani, P., Petacchi E., & Cioni, G. (2008). Language organisation in left perinatal stroke. *Neuropediatrics, 39*(3): 157–163.

Hadley, D., Anderson, B. S., Borckardt, J. J., Arana, A., Li, X., Nahas, Z., & George, M. S. (2011). Safety, tolerability, and effectiveness of high doses of adjunctive daily left prefrontal repetitive transcranial magnetic stimulation for treatment-resistant depression in a clinical setting. *J ECT, 27*(1), 18–25.

Hamberger, M. J., & Cole, J. (2011). Language organization and reorganization in epilepsy. *Neuropsychol Rev, 21*(3), 240–251.

Han, Y., Tong, X., Wang, X., Teng, F., Deng, Q., Zhou, J., Guan, Y., Yan, Z., Chen, L., Luan, G., & Wang, M. (2021). A concordance study determining language dominance between navigated transcranial magnetic stimulation and the Wada test in patients with drug-resistant epilepsy. *Epilepsy Behav, 117*, 107711.

Hauck, T., Probst, M., Zimmer, C., Ringel, F., Meyer, B., Wohlschlaeger A., & Krieg, S. M. (2019). Language function shows comparable cortical patterns by functional MRI and repetitive nTMS in healthy volunteers. *Brain Imaging Behav, 13*(4), 1071–1092.

Hernandez-Pavon, J. C., Mäkelä, N., Lehtinen, H., Lioumis, P., & Mäkelä, J. P. (2014). Effects of navigated TMS on object and action naming. *Front Hum Neurosci, 8*, 660.

Hett, D., Rogers, J., Humpston, C., & Marwaha, S. (2021). Repetitive transcranial magnetic stimulation (rTMS) for the treatment of depression in adolescence: A systematic review. *J Affect Disord, 278*, 460–469.

Hickok, G. (2009). The functional neuroanatomy of language. *Phys Life Rev, 6*(3), 121–143.

Hickok, G., & Poeppel, D. (2007). The cortical organization of speech processing. *Nat Rev Neurosci, 8*(5), 393–402.

Holmes, N. P., & Meteyard, L. (2018). Subjective discomfort of TMS predicts reaction times differences in published studies. *Front Psychol, 9*, 1989.

Ille, S., Engel, L., Albers, L., Schroeder, A., Kelm, A., Meyer, B., & Krieg, S. M. (2019). Functional reorganization of cortical language function in glioma patients: A preliminary study. *Front Oncol, 9*, 446.

Ille, S., Sollmann, N., Hauck, T., Maurer, S., Tanigawa, N., Obermueller, T., Negwer, C., Droese, D., Boeckh-Behrens, T., Meyer, B., Ringel, F., & Krieg, S. M. (2015). Impairment of preoperative language mapping by lesion location: A functional magnetic resonance imaging, navigated transcranial magnetic stimulation, and direct cortical stimulation study. *J Neurosurg, 123*(2), 314–324.

Ille, S., Sollmann, N., Hauck, T., Maurer, S., Tanigawa, N., Obermueller, T., Negwer, C., Droese, D., Zimmer, C., Meyer, B., Ringel, F., & Krieg, S. M. (2015). Combined noninvasive language mapping by navigated transcranial magnetic stimulation and functional MRI and its comparison with direct cortical stimulation. *J Neurosurg, 123*(1), 212–225.

Jacola, L. M., Schapiro, M. B., Schmithorst, V. J., Byars, A. W., Strawsburg, R. H., Szaflarski, J. P., Plante, E., & Holland, S. K. (2006). Functional magnetic resonance imaging reveals atypical language organization in children following perinatal left middle cerebral artery stroke. *Neuropediatrics, 37*(1), 46–52.

Jeltema, H. R., Ohlerth, A. K., de Wit, A., Wagemakers, M., Rofes, A., Bastiaanse, R., & Drost, G. (2021). Comparing navigated transcranial magnetic stimulation mapping and "gold standard" direct cortical stimulation mapping in neurosurgery: A systematic review. *Neurosurg Rev, 44*(4), 1903–1920.

Kilteni, K., Andersson, B. J., Houborg, C., & Ehrsson, H. H. (2018). Motor imagery involves predicting the sensory consequences of the imagined movement. *Nat Commun, 9*(1), 1617.

Kim, J. A., & Davis, K. D. (2021). Magnetoencephalography: Physics, techniques, and applications in the basic and clinical neurosciences. *J Neurophysiol, 125*(3), 938–956.

Kline, A., Pittman, D., Ronsky, J., & Goodyear, B. (2020). Differentiating the brain's involvement in executed and imagined stepping using fMRI. *Behav Brain Res, 394*, 112829.

Krieg, S. M., Lioumis, P., Mäkelä, J. P., Wilenius, J., Karhu, J., Hannula, H., Savolainen, P., Lucas, C. W., Seidel, K., Laakso, A., Islam, M., Vaalto, S., Lehtinen, H., Vitikainen, A. M., Tarapore, P. E., & Picht, T. (2017). Protocol for motor and language mapping by navigated TMS in patients and healthy volunteers: Workshop report. *Acta Neurochir (Wien), 159*(7), 1187–1195.

Krieg, S. M., Sollmann, N., Hauck, T., Ille, S., Foerschler, A., Meyer, B., & F. Ringel, F. (2013). Functional language shift to the right hemisphere in patients with language-eloquent brain tumors. *PLoS ONE, 8*(9), e75403.

Krieg, S. M., Tarapore, P. E., Picht, T., Tanigawa, N., Houde, J., Sollmann, N., Meyer, B., Vajkoczy, P., Berger, M. S., Ringel, F., & Nagarajan, S. (2014). Optimal timing of pulse onset for language mapping with navigated repetitive transcranial magnetic stimulation. *Neuroimage, 100,* 219–236.

Krings, T., Chiappa, K. H., Foltys, H., Reinges, M. H., Cosgrove, G. R., & Thron, A. (2001). Introducing navigated transcranial magnetic stimulation as a refined brain mapping methodology. *Neurosurg Rev, 24*(4), 171–179.

Kwiecien, R., Kopp-Schneider, A., & Blettner, M. (2011). Concordance analysis: Part 16 of a series on evaluation of scientific publications. *Dtsch Arztebl Int, 108*(30), 515–521.

Lancaster, J. L., Narayana, S., Wenzel, D., Luckemeyer, J., Roby, J., & P. Fox, P. (2004). Evaluation of an image-guided, robotically positioned transcranial magnetic stimulation system. *Hum Brain Mapp, 22*(4), 329–340.

Laux, P., Tralau, T., Tentschert, J., Blume, A., Dahouk, S. A., Bäumler, W., Bernstein, E., Bocca, B., Alimonti, A., Colebrook, H., de Cuyper, C., Dähne, L., Hauri, U., Howard, P. C., Janssen, P., Katz, L., Klitzman, B., Kluger, N., Krutak, L., Platzek, T., Scott-Lang, V., Serup, J., Teubner, W., Schreiver, I., Wilkni?, E., & Luch, A. (2016). A medical-toxicological view of tattooing. *Lancet, 387*(10016), 395–402.

Lehtinen, H., Mäkelä, J. P., Mäkelä, T., Lioumis, P., Metsähonkala, L., Hokkanen, L., Wilenius, J., & Gaily, E. (2018). Language mapping with navigated transcranial magnetic stimulation in pediatric and adult patients undergoing epilepsy surgery: Comparison with extraoperative direct cortical stimulation. *Epilepsia Open, 3*(2), 224–235.

Lidzba, K., Küpper, H., Kluger, G., & Staudt, M. (2017). The time window for successful right-hemispheric language reorganization in children. *Eur J Paediatr Neurol, 21*(5), 715–721.

Lin, Y. Y., Chen, R. S., & Huang, Y. Z. (2022). Impact of operator experience on transcranial magnetic stimulation. *Clin Neurophysiol Pract, 7,* 42–48.

Lotze, M., Montoya, P., Erb, M., Hülsmann, E., Flor, H., Klose, U., Birbaumer, N., & Grodd, W. (1999). Activation of cortical and cerebellar motor areas during executed and imagined hand movements: An fMRI study. *J Cogn Neurosci, 11*(5), 491–501.

Luo, H., & Poeppel, D. (2007). Phase patterns of neuronal responses reliably discriminate speech in human auditory cortex. *Neuron, 54*(6), 1001–1010.

Madhavan, K. M., McQueeny, T., Howe, S. R., Shear, P., & Szaflarski, J. (2014). Superior longitudinal fasciculus and language functioning in healthy aging. *Brain Res, 1562,* 11–22.

Maizey, L., Allen, C. P., Dervinis, M., Verbruggen, F., Varnava, A., Kozlov, M., Adams, R. C., Stokes, M., Klemen, J., Bungert, A., Hounsell, C. A., & Chambers, C. D. (2013). Comparative incidence rates of mild adverse effects to transcranial magnetic stimulation. *Clin Neurophysiol, 124*(3), 536–544.

McClintock, S. M., Reti, I. M., Carpenter, L. L., McDonald, W. M., Dubin, M., Taylor, S. F., Cook, I. A., O'Reardon, J., Husain, M. M., Wall, C., Krystal, A. D., Sampson, S. M., Morales, O., Nelson, B. G., Latoussakis, V., George, M. S., Lisanby, S. H., N. N. o. D. C. r. T. Group and A. P. A. C. o. R. T. F. o. N. B.

a. Treatments (2018). Consensus recommendations for the clinical application of repetitive transcranial magnetic stimulation (rTMS) in the treatment of depression. *J Clin Psychiatry, 79*(1).

Meteyard, L., & Holmes, N. P. (2018). TMS SMART: Scalp mapping of annoyance ratings and twitches caused by transcranial magnetic stimulation. *J Neurosci Methods, 299,* 34–44.

Monaghan, T. F., Rahman, S. N., Agudelo, C. W., Wein, A. J., Lazar, J. M., K. Everaert, K., & Dmochowski, R. R. (2021). Foundational statistical principles in medical research: Sensitivity, specificity, positive predictive value, and negative predictive value. *Medicina (Kaunas), 57*(5).

Mueller-Sarnowski, F., Sollmann, N., Schröder, A., Houri, L., Ille, S., Grimmer, T., Krieg, S. M., & Diehl-Schmid, J. (2022). Neuronavigated repetitive transcranial magnetic stimulation as novel mapping technique provides insights into language function in primary progressive aphasia. *Brain Imaging Behav, 16*(3), 1208–1216.

Muir, M., Patel, R., Traylor, J., de Almeida Bastos, D. C., Prinsloo, S., Liu, H. L., Noll, K., Wefel, J., Tummala, S., Kumar, V., & Prabhu, S. (2022). Validation of non-invasive language mapping modalities for eloquent tumor resection: A pilot study. *Front Neurosci, 16,* 833073.

Najib, U., Bashir, S., Edwards, D., Rotenberg, A., & Pascual-Leone, A. (2011). Transcranial brain stimulation: Clinical applications and future directions. *Neurosurg Clin N Am, 22*(2), 233–251, ix.

Narayana, S., Gibbs, S. K., Fulton, S. P., McGregor, A. L., Mudigoudar, B., Weatherspoon, S. E., Boop, F. A., & Wheless, J. W. (2021). Clinical utility of transcranial magnetic stimulation (TMS) in the presurgical evaluation of motor, speech, and language functions in young children with refractory epilepsy or brain tumor: Preliminary evidence. *Front Neurol, 12,* 650830.

Narayana, S., Papanicolaou, A. C., McGregor, A., Boop, F. A., & Wheless, J. W. (2015). Clinical applications of transcranial magnetic stimulation in pediatric neurology. *J Child Neurol, 30*(9), 1111–1124.

Neuper, C., Scherer, R., Reiner, M., & Pfurtscheller, G. (2005). Imagery of motor actions: Differential effects of kinesthetic and visual-motor mode of imagery in single-trial EEG. *Brain Res Cogn Brain Res, 25*(3), 668–677.

Packheiser, J., Schmitz, J., Arning, L., Beste, C., Güntürkün, O., & S. Ocklenburg, S. (2020). A large-scale estimate on the relationship between language and motor lateralization. *Sci Rep, 10*(1), 13027.

Pak, R. W., Hadjiabadi, D. H., Senarathna, J., Agarwal, S., Thakor, N. V., Pillai, J. J., & Pathak, A. P. (2017). Implications of neurovascular uncoupling in functional magnetic resonance imaging (fMRI) of brain tumors. *J Cereb Blood Flow Metab, 37*(11), 3475–3487.

Papadelis, C., & Chen, Y. H. (2020). Pediatric magnetoencephalography in clinical practice and research. *Neuroimaging Clin N Am, 30*(2), 239–248.

Papanicolaou, A. C., Rezaie, R., Narayana, S., Choudhri, A. F., Wheless, J. W., Castillo, E. M., Baumgartner, J. E., & Boop, F. A. (2014). Is it time to replace the Wada test and put awake craniotomy to sleep? *Epilepsia, 55*(5), 629–632.

Papanicolaou, A. C., Simos, P. G., Castillo, E. M., Breier, J. I., Sarkari, S., Pataraia, E., Billingsley, R. L., Buchanan, S., Wheless, J., Maggio, V., & Maggio, W. W. (2004). Magnetocephalography: A noninvasive alternative to the Wada procedure. *J Neurosurg, 100*(5), 867–876.

Partovi, S., Konrad, F., Karimi, S., Rengier, F., Lyo, J. K., Zipp, L., Nennig, E., & Stippich, C. (2012). Effects of covert and overt paradigms in clinical language fMRI. *Acad Radiol, 19*(5), 518–525.

Pascual-Leone, A., Gates, J. R., & Dhuna, A. (1991). Induction of speech arrest and counting errors with rapid-rate transcranial magnetic stimulation. *Neurology, 41*(5), 697–702.

Pascual-Leone, A., Walsh, V., & Rothwell, J. (2000). Transcranial magnetic stimulation in cognitive neuroscience: Virtual lesion, chronometry, and functional connectivity. *Curr Opin Neurobiol, 10*(2), 232–237.

Pasquini, L., Di Napoli, A., Rossi-Espagnet, M. C., Visconti, E., Napolitano, A., A. Romano, A., Bozzao, A., Peck, K. K., & Holodny, A. I. (2022). Understanding Language reorganization with neuroimaging: How language adapts to different focal lesions and insights into clinical applications. *Front Hum Neurosci, 16*, 747215.

Pereira, L. S., Müller, V. T., da Mota Gomes, M., A. Rotenberg, A., & Fregni, F. (2016). Safety of repetitive transcranial magnetic stimulation in patients with epilepsy: A systematic review. *Epilepsy Behav, 57*(Pt A), 167–176.

Picht, T., Krieg, S. M., Sollmann, N., Rösler, J., Niraula, B., Neuvonen, T., Savolainen, P., Lioumis, P., Mäkelä, J. P., Deletis, V., Meyer, B., Vajkoczy, P., & Ringel, F. (2013). A comparison of language mapping by preoperative navigated transcranial magnetic stimulation and direct cortical stimulation during awake surgery. *Neurosurgery, 72*(5), 808–819.

Poeppel, D., & Hickok, G. (2004). Towards a new functional anatomy of language. *Cognition, 92*(1–2), 1–12.

Proix, T., Delgado Saa, J., Christen, A., Martin, S., Pasley, B. N., Knight, R. T., Tian, X., Poeppel, D., Doyle, W. K., Devinsky, O., Arnal, L. H., Mégevand, P., & Giraud, A. L. (2022). Imagined speech can be decoded from low- and cross-frequency intracranial EEG features. *Nat Commun, 13*(1), 48.

Raffa, G., Quattropani, M. C., Scibilia, A., Conti, A., Angileri, F. F., Esposito, F., Sindorio, C., Cardali, S. M., Germanò, A., & Tomasello, F. (2018). Surgery of language-eloquent tumors in patients not eligible for awake surgery: The impact of a protocol based on navigated transcranial magnetic stimulation on presurgical planning and language outcome, with evidence of tumor-induced intra-hemispheric plasticity. *Clin Neurol Neurosurg, 168*, 127–139.

Rezaie, R., Schiller, K. K., Embury, L., Boop, F. A., Wheless, J. W., & Narayana, S. (2020). The clinical utility of transcranial magnetic stimulation in determining hemispheric dominance for language: A magnetoencephalography comparison Study. *J Clin Neurophysiol, 37*(2), 90–103.

Rösler, J., Niraula, B., Strack, V., Zdunczyk, A., Schilt, S., Savolainen, P., Lioumis, P., Mäkelä, J., Vajkoczy, P., Frey, D., & Picht, T. (2014). Language mapping in healthy volunteers and brain tumor patients with a novel navigated TMS system: Evidence of tumor-induced plasticity. *Clin Neurophysiol, 125*(3), 526–536.

Rossi, S., Antal, A., Bestmann, S., Bikson, M., Brewer, C., Brockmöller, J., Carpenter, L. L., Cincotta, M., Chen, R., Daskalakis, J. D., Di Lazzaro, V., Fox, M. D., George, M. S., Gilbert, D., Kimiskidis, V. K., Koch, G., Ilmoniemi, R. J., Lefaucheur, J. P., Leocani, L., Lisanby, S. H., Miniussi, C., Padberg, F., Pascual-Leone, A., Paulus, W., Peterchev, A. V., Quartarone, A., Rotenberg, A., Rothwell, J., Rossini, P. M., Santarnecchi, E., Shafi, M. M., Siebner, H. R., Ugawa, Y., Wassermann, E. M., Zangen, A., Ziemann, U. Hallett, M., & F. o. T. S. The basis of this article began with a Consensus Statement from the IFCN Workshop on "Present, Ethical Guidelines", Siena, October 17–20, 2018, updating through April 2020 (2021). Safety and recommendations for TMS use in healthy subjects and patient populations, with updates on training, ethical and regulatory issues: Expert Guidelines. *Clin Neurophysiol, 132*(1), 269–306.

Rossi, S., Hallett, M., Rossini, P. M., Pascual-Leone, A., & S. o. T. C. Group (2009). Safety, ethical considerations, and application guidelines for the use of transcranial magnetic stimulation in clinical practice and research. *Clin Neurophysiol, 120*(12), 2008–2039.

Rothwell, J. C., Hallett, M., Berardelli, A., Eisen, A., Rossini, P., & Paulus, W. (1999). Magnetic stimulation: Motor evoked potentials. The International Federation of Clinical Neurophysiology. *Electroencephalogr Clin Neurophysiol Suppl, 52*, 97–103.

Ruohonen, J., & Karhu, J. (2010). Navigated transcranial magnetic stimulation. *Neurophysiol Clin, 40*(1), 7–17.

Sanai, N., Mirzadeh, Z., & Berger, M. S. (2008). Functional outcome after language mapping for glioma resection. *N Engl J Med, 358*(1), 18–27.

Schiller, K., Choudhri, A. F., Jones, T., Holder, C., Wheless, J. W., & Narayana, S. (2020). Concordance between transcranial magnetic stimulation and functional magnetic resonance imaging (MRI) derived localization of language in a clinical cohort. *J Child Neurol, 35*(6), 363–379.

Schrader, L. M., Stern, J. M., Koski, L., Nuwer, M. R., & Engel, J. (2004). Seizure incidence during single- and paired-pulse transcranial magnetic stimulation (TMS) in individuals with epilepsy. *Clin Neurophysiol, 115*(12), 2728–2737.

Singh, S. P. (2014). Magnetoencephalography: Basic principles. *Ann Indian Acad Neurol, 17*(Suppl 1), S107–112.

Sollmann, N., Fuss-Ruppenthal, S., Zimmer, C., Meyer, B., & Krieg, S. M. (2018). Investigating stimulation protocols for language mapping by repetitive navigated transcranial magnetic stimulation. *Front Behav Neurosci, 12*, 197.

Sollmann, N., Hauck, T., Hapfelmeier, A., Meyer, B., Ringel, F., & Krieg, S. M. (2013). Intra- and interobserver variability of language mapping by navigated transcranial magnetic brain stimulation. *BMC Neurosci, 14*, 150.

Sollmann, N., Ille, S., Boeckh-Behrens, T., Ringel, F., Meyer, B., & Krieg, S. M. (2016). Mapping of cortical language function by functional magnetic resonance imaging and repetitive navigated transcranial magnetic stimulation in 40 healthy subjects. *Acta Neurochir (Wien), 158*(7), 1303–1316.

Sollmann, N., Ille, S., Negwer, C., Boeckh-Behrens, T., Ringel, F., Meyer, B., & Krieg, S. M. (2017). Cortical time course of object naming investigated by repetitive navigated transcranial magnetic stimulation. *Brain Imaging Behav, 11*(4), 1192–1206.

Sondergaard, R. E., Martino, D., Kiss, Z. H. T., & Condliffe, E. G. (2021). TMS motor mapping methodology and reliability: A structured review. *Front Neurosci, 15*, 709368.

Sparing, R., Buelte, D., Meister, I. G., Paus, T., & Fink, G. R. (2008). Transcranial magnetic stimulation and the challenge of coil placement: A comparison of conventional and stereotaxic neuronavigational strategies. *Hum Brain Mapp, 29*(1), 82–96.

Starbuck, V. N., Kay, G. G., Platenberg, R. C., Lin, C. S., & Zielinski, B. A. (2000). Functional magnetic resonance imaging reflects changes in brain functioning with sedation. *Hum Psychopharmacol, 15*(8), 613–618.

Szaflarski, J. P., Binder, J. R., Possing, E. T., McKiernan, K. A., Ward, B. D., & Hammeke, T. A. (2002). Language lateralization in left-handed and ambidextrous people: fMRI data. *Neurology, 59*(2), 238–244.

Szaflarski, J. P., Gloss, D., Binder, J. R., Gaillard, W. D., Golby, A. J., Holland, S. K., Ojemann, J., Spencer, D. C., Swanson, S. J., French, J. A., & Theodore, W. H. (2017). Practice guideline summary: Use of fMRI in the presurgical evaluation of patients with epilepsy: Report of the guideline development, dissemination, and implementation subcommittee of the American Academy of Neurology. *Neurology, 88*(4), 395–402.

Tarapore, P. E., Findlay, A. M., Honma, S. M., Mizuiri, D., Houde, J. F., Berger, M. S., & Nagarajan, S. S. (2013). Language mapping with navigated repetitive TMS: Proof of technique and validation. *Neuroimage, 82*, 260–272.

Tarapore, P. E., Picht, T., Bulubas, L., Shin, Y., Kulchytska, N., Meyer, B., Berger, M. S., Nagarajan, S. S., & Krieg, S. M. (2016). Safety and tolerability of navigated TMS for preoperative mapping in neurosurgical patients. *Clin Neurophysiol, 127*(3), 1895–1900.

Tarapore, P. E., Tate, M. C., Findlay, A. M., Honma, S. M., Mizuiri, D., Berger, M. S., & Nagarajan, S. S. (2012). Preoperative multimodal motor mapping: A comparison of magnetoencephalography imaging, navigated transcranial magnetic stimulation, and direct cortical stimulation. *J Neurosurg, 117*(2), 354–362.

Thickbroom, G. W., Sammut, R., & Mastaglia, F. L. (1998). Magnetic stimulation mapping of motor cortex: Factors contributing to map area. *Electroencephalogr Clin Neurophysiol, 109*(2), 79–84.

Tieleman, A., Deblaere, K., Van Roost, D., Van Damme, O., & Achten, E. (2009). Preoperative fMRI in tumour surgery. *Eur Radiol, 19*(10), 2523–2534.

Ueno, S., & Sekino, M. (2021). Figure-eight coils for magnetic stimulation: From focal stimulation to deep stimulation. *Front Hum Neurosci, 15*, 805971.

Wagner, T., Rushmore, J., Eden, U., & Valero-Cabre, A. (2009). Biophysical foundations underlying TMS: Setting the stage for an effective use of neurostimulation in the cognitive neurosciences. *Cortex, 45*(9), 1025–1034.

Part III

Language and Cognitive Plasticity and Processing

5

Functional Connectivity and Recovery from Aphasia

Tali Bitan

5.1 Functional Connectivity in the Study of Plasticity Following Brain Damage

One of the most intriguing questions related to brain plasticity concerns the neural mechanisms underlying the recovery and rehabilitation following brain lesions. Throughout the last decades, as the cognitive neuroscience pendulum swings closer toward distributional and connectionist approaches, and away from the very localistic explanations, there is a growing appreciation of the importance of functional connectivity (FC) measures for understanding recovery processes. The focus of this chapter is on the role of FC among brain regions, as measured in fMRI, in the neural processes involved in the recovery from aphasia.

Aphasia typically occurs following left hemisphere (LH) stroke in the territory of the middle cerebral artery. A large study that measured *structural* connectivity in 818 stroke patients, showed that the integrity of specific white matter pathways did *not* predict patients' language skills beyond lesion location (Hope et al., 2018). In contrast, measures of *functional* connectivity reflect a wide range of neural mechanisms, including polysynaptic connections, and thus go beyond structural measures of white matter pathways (Deco et al., 2011). A recent study (Keator et al., 2021), with 97 post-stroke patients, tested the role of resting-state functional connectivity (rs-FC), when controlling for local measures of lesion properties and structural connectivity measured with diffusion tensor imaging. It showed that FC among left frontotemporal and temporoparietal regions was correlated with standardized measures of auditory comprehension, even after controlling for lesion location and structural connectivity. These findings suggest that the neural coherence in the network, measured in FC, provides a useful lens for looking at brain plasticity underlying language recovery processes.

5.2 Normalization and Compensation Processes

Questions about the neural recovery from brain lesions have particularly interesting implications in the language domain. Language is a high-level cognitive skill that relies on a bilaterally distributed brain network (Fedorenko & Thompson-Schill, 2014; Siegel et al., 2016), while it is also asymmetrically organized. An important question about rehabilitation is to what extent behavioral recovery involves the "normalization" of neural processing, that is, the restoration of dysfunctional perilesional regions. The alternative is "compensation," which is functional reorganization of the network, associated with recruiting new brain areas that were not previously involved in the task, or enhancing of existing connections. Due to the asymmetric distribution of typical language processes, this question is often investigated through the role of the right hemisphere (RH) in recovery. While many neuroimaging studies, looking at local activation measures, showed enhanced recruitment of RH regions following LH lesion, supporting the notion of compensation and reorganization (Abel et al., 2015; Abo et al., 2004; Meltzer et al., 2013), others point to the importance of normalization processes in LH perilesional areas (Fridriksson, 2010; Heiss et al., 1997; Thiel et al., 2001), especially in chronic stages (Saur et al., 2006). A related hypothesis, that received a lot of attention, is that of a maladaptive role of the RH in recovery (Fridriksson et al., 2009; Postman-Caucheteux et al., 2010). This idea is based on the notion of "trans-callosal suppression" between homotopic language areas, suggesting that intact RH regions inhibit LH lesioned areas, thus hindering recovery (Heiss et al., 2003; Naeser et al., 2004; Price & Crinion, 2005). These hypotheses underlie the focus of many FC studies on interhemispheric connectivity, and on comparing intra-hemispheric connectivity between hemispheres.

The distinction between normalization and compensation is also reflected in questions about the specificity of the neural changes underlying recovery. While normalization processes necessarily involve changes which are specific to the impaired language functions, compensation may involve the recruitment of domain general processes, such as cognitive control and attention, which could be shared across different types of therapeutic approaches. Functional connectivity measures are useful for studying these cognitive questions. In addition to looking at the connectivity between pairs of regions, studies can also examine the integrity of the language network and its connections with domain general networks, using both local and global network efficiency measures.

5.3 Methodological Approaches to the Research of FC and Language Recovery

This chapter summarizes results of fMRI studies that used FC measures to examine the neural mechanisms underlying recovery from aphasia. Generally speaking, two types of study designs have been used to examine the role of FC in language recovery, differentiated by the number of times FC is measured, and the control over the treatment patients receive. Studies that test patients before and after undergoing a specific language therapy (pre-/post-designs) can directly measure changes in connectivity between the two timepoints and control the length and type of treatment patients received. In other studies, patients are tested at a single timepoint, and the correlation between their neural FC measures and the severity of their language symptoms is computed. Studies with the later design can recruit larger sample sizes, because they do not provide therapy, and can inform us about the connections that contribute to adequate language performance. Nevertheless, this design cannot distinguish between connections that are spared by the lesion, and those undergoing plastic changes during recovery. This design also does not control for the type and number of interventions patients received since the stroke, and thus do not provide information about the specificity of neural changes underlying different treatment approaches. In the current chapter we will discuss these two types of study designs separately.

Another important methodological difference is between studies using task-based FC, and those using resting-state functional connectivity (rs-FC). Measuring FC while participants perform a language task inside the scanner enables connectivity changes, which are *specific* to the treated stimuli or task, to be distinguished from those supporting generalization of treatment effects to new or untreated stimuli or tasks. In contrast, rs-FC reflects intrinsic network properties, which are relatively stable as they are evident even in the absence of any task (Zhu et al., 2014). Moreover, rs-FC also enables studying the role of more distributed connections, including domain general networks and global measures of connectivity. Finally, rs-FC avoids the confounding effects of different levels of task performance between timepoints and between different participants, and enables testing even severely affected patients who cannot perform language tasks (Klingbeil et al., 2019). Therefore, the current chapter will separately address task-based FC and rs-FC.

Many other methodological factors are important in studies of aphasia recovery and can affect their conclusions. The heterogeneity of patients with aphasia in terms of aphasia type, lesion size and location, patients' age, and the time since onset implies that there is often a tradeoff between sample size and sample homogeneity. Specifically with regards to the time post-lesion onset, studies testing patients in acute and early subacute stages can typically identify larger and more robust recovery effects, which rely on restoration of temporary dysfunctional perilesional regions, while longer-term recovery is thought to reflect neuroplasticity through functional reorganization (Wilson et al., 2019). Moreover, testing of chronic patients pre- and post-treatment can better dissociate spontaneous recovery from the effect of specific interventions. Importantly, studies measuring FC also differ in terms of the type of analysis methods they use, which include: correlations between predefined regions of interest (ROI-to-ROI) (e.g., Bitan et al., 2018), and between ROIs and the whole brain (i.e., seed to voxel, e.g., van Hees et al., 2014); effective

Table 5.1 *The methodological dimensions on which the chapter is organized, and the advantages for each category*

Study design:	Single timepoint	Pre-/post-treatment
	Larger samples	Control over treatment type and parameters
	Can identify the critical network	Can identify plastic network changes
Type of scan:	**Task-based FC**	**Resting-state FC**
	Differentiates treated stimuli and tasks from generalization	A stable intrinsic measure
		Independence of differences in task performance

connectivity using dynamic causal modeling (DCM, e.g., Chu et al., 2018; Meier et al., 2019; Truzman et al., 2021); voxel mirrored homotopic connectivity (VMHC, e.g., Yang et al., 2017); amplitudes of low frequency fluctuations (ALFF, e.g., van Hees et al., 2014); independent component analysis (ICA, e.g., Duncan & Small, 2018); eigenvector centrality mapping (ECM, e.g., Gili et al., 2017); graph theoretical measures (Baliki et al., 2018; Johnson et al., 2020); and more, each with its own strengths and weaknesses.

The following sections will give examples from studies in the four categories described above, differentiated based on the study design (pre-/post-treatment or a single timepoint) and based on the use of task-based or rs-FC, in light of the questions regarding normalization vs. compensation (see Table 5.1). It should be noted that this chapter is not intended as a comprehensive review, and that studies using other neuroimaging modalities (not fMRI), children, or studies testing the effect of neurostimulation techniques are not included.

5.4 Single Timepoint Studies

Studies that measure FC in patients with aphasia at a single timepoint test the association between FC measures and language performance at that point in time, to reveal the patterns of connections and networks associated with adequate language function. Comparing the connectivity in patients to that of healthy controls (HC) can indicate whether high-performing patients show similar connectivity to HC, which could be described as "normalization," or whether they are characterized by unique patterns of connectivity, interpreted as "compensatory."

5.4.1 Studies Using Resting-State Functional Connectivity

Many studies measured rs-FC at a single timepoint and examined its association with aphasia severity. Klingbeil and colleagues (2019) reviewed 10 such studies (published between 2013 and 2017), including, altogether, 176 stroke patients with aphasia. The authors point out that despite the mostly small sample sizes, heterogeneous groups of patients, and diverse analytic strategies, they could still converge on several interesting conclusions. Most studies showed reduced connectivity in the language network in patients with aphasia, compared to HC, and a positive correlation between language function and connectivity, both between hemispheres (New et al., 2015; Sandberg, 2017; Siegel et al., 2016; Yang et al., 2017; Zhu et al., 2014) and within the lesioned hemisphere (Harvey et al., 2013; Sandberg, 2017; Siegel et al., 2016). Altogether, these findings point to a process of "normalization," whereby better language performance among patients is associated with the typical connectivity pattern also evident in HC, rather than with new compensatory connectivity. The results of a more recent study with ninety-seven post-stroke patients (Keator et al., 2021), showing that left intrahemispheric connectivity in the language network was associated with speech comprehension, are also consistent with this conclusion of normalization. While in many of these studies reduced connectivity in patients compared to HC is associated with poor language performance, some of the results show that *enhanced* connectivity in patients compared to HC is correlated with poor performance. Such results were found in a study by Yang and colleagues (Yang et al., 2017) who examined interhemispheric connectivity using voxel-mirrored homotopic connectivity (VMHC) in thirty-seven patients with aphasia and twenty-eight HC. They showed increased interhemispheric connectivity in patients compared to HC in the superior temporal gyri (STG), which was associated with poor language performance. These results still indicate that the typical connectivity is associated with adequate performance and are thus considered as "normalization."

In contrast, another recent study (Xu et al., 2020), examined twelve patients two weeks after suffering stroke to the basal-ganglia that caused aphasia. All patients regained fluent speech by the time of scanning and were compared to seventeen HC. The results showed increased intrahemispheric FC in the language network in patients compared to HC. The testing of language function at two timepoints (during admission to the hospital and immediately before scanning), and the early phase of recovery, when all patients showed significant improvement, may suggest that this increased intrahemispheric connectivity reflects a recovery process. However, because this study did not measure correlation with language performance, we do not know whether these differences in the language network indeed supported language function. This study also found reduced *interhemispheric* FC in patients compared to HC, a finding replicated by another study that tested (twenty) patients in early stages of recovery (one month post-onset) (Zhang et al., 2021). In Zhang and colleagues' study, this interhemispheric connectivity was correlated with auditory verbal comprehension, making it consistent with normalization processes.

An important feature of rs-FC over task-based measures is the possibility to efficiently examine the effects of other, domain general networks. It can test the role of domain general cognitive processes in the compensation for the damaged function. Some support for the relevance of domain general networks comes from the above study (Zhang et al., 2021) that showed reduced coupling between the left language network and posterior default model network (DMN) in patients compared to HC. However, because this coupling was not associated with language performance it is hard to conclude to what extent it was relevant to recovery of function.

5.4.2 Studies Measuring Task-Based Functional Connectivity

Measuring task-based FC enables one to distinguish between connections relevant to different aspects of language processing. Several studies used dynamic causal modeling (DCM) to assess *effective* connectivity, which is the directional influence that one brain region exerts on another (Friston et al., 2003). In one of these studies (Chu et al., 2018), we examined inter and intrahemispheric connectivity between bilateral inferior frontal gyri (IFG) and primary auditory cortices (A1) in fifteen patients with chronic post-stroke aphasia while they performed a sentence comprehension task, and compared them to HC. The results showed that the same inter and intrahemispheric connections which were significant in HC, were associated with better performance on standard language tests in patients, consistent with the notion of "normalization" as a mechanism for improved language performance in patients. There was no evidence for the emergence of a new compensatory connection emanating from the RH. Nevertheless, there was also no evidence for maladaptive homotopic transcallosal suppression from right to left hemisphere regions. Altogether, the results point to the critical role of the balanced coordination between regions in both hemispheres for language function.

Two more studies that also examined effective connectivity using DCM, have shown mixed results with regards to normalization vs. compensation processes. They included large sample sizes and examined FC during a semantic feature decision task. In one of these studies (Meier et al., 2019), thirty-five patients with aphasia were compared to twenty-one HC who showed stronger left intrahemispheric connectivity compared to the patients. Within the patient group, connectivity depended on lesion location, where damage to left superior frontal structures resulted in greater right intrahemispheric connectivity (compensation), whereas damage to left ventral temporal structures resulted in stronger modulation of left frontal regions (normalization). Left intrahemispheric connectivity in patients was also correlated with processing speed in the scanner (also normalization), but overall aphasia severity was only predicted by the amount of tissue spared from the lesion. Another study from the same group (Meier et al., 2018) compared the left intrahemispheric connectivity in twenty-five patients to eighteen HC. They showed that connectivity within the left hemisphere (left posterior middle temporal gyrus to LIFG pars triangularis) differentiated between patients and HC, and its strength was positively associated with lexical-semantic abilities measured inside and outside the scanner, also supporting a process of normalization.

A different approach was taken by a study that examined FC in fifty-three patients with aphasia while they performed a picture description task and compared them to twenty-four HC (Geranmayeh et al., 2016). Although this was a task-based scan, they examined the connectivity between the language and domain general networks such as the DMN, showing that language performance was associated with the *difference* in activation between the language network and the DMN, even though activity within each of these networks did not predict language performance by itself. Specifically, good speech performance in the scanner was associated with the combination of higher activity in the left language networks and low activity in the DMN, while the difference between right language network and DMN was negatively associated with performance. These findings are consistent with studies in other domains, showing that decreased segregation between networks is associated with degenerative states (Chan et al., 2014; Tinaz et al., 2016), and suggest that normalization of segregation and specialization contributes to language recovery.

5.5 Pre-/Post-treatment Studies

The benefit of pre-/post-treatment designs is that they can identify specific neural *changes* underlying the recovery process, and dissociate them from useful connections that were spared by the lesion and do not reflect plasticity. When such pre-/post-treatment studies include a control group of healthy participants, who serve as a benchmark for the healthy language network, the results can indicate whether treatment-related changes reflect normalizing or compensatory processes. Moreover, when these studies include a control group of patients or HC who are scanned twice without receiving treatment, they can indicate which plastic changes can be attributed to the treatment itself. Finally, correlating the behavioral treatment outcomes with the changes in connectivity provides the strongest evidence for the relevance of changes in connectivity to the effect of treatment. It should be noted that although all studies in this section included at least two testing timepoints, some did not identify neural changes in FC between the two assessments or only reported the connectivity patterns pre-treatment that predicted good treatment outcomes.

5.5.1 Studies Measuring Resting-State Functional Connectivity

Many of the rs-FC studies show treatment-related changes in the language network. In some of these, the changes are specifically related to the treatment strategy, or the language aspects emphasized during treatment.

In one case study (Bitan et al., 2018), we tested the effect of forty-eight sessions of melody-based treatment, in which the patient repeats phrases with exaggerated prosody. The basis for this therapy is the observation that production of melodic speech is often intact, even when standard speech is impaired, in patients with nonfluent aphasia following LH lesion (Sparks et al., 1974). Based on the association of prosodic processing and the RH (Callan et al., 2006), the underlying assumption of this therapy is that recruitment of RH regions can compensate for LH lesions (Albert et al., 1973). In this case study, a chronic patient with aphasia was scanned at three timepoints: baseline, pre-therapy and post-therapy, and was compared to a control patient who underwent similar testing without therapy. The treated patient improved in several language production tests and maintained this improvement during the follow-up testing. Interestingly, only the treated patient showed increase in rs-FC between RH frontal language areas and motor speech control areas during the treatment interval. This change, which was specific for the RH, was greater than any changes that occurred during the baseline interval before treatment. These results are compatible with a compensatory role played by RH language areas when the intervention was specifically designed to recruit these regions. They demonstrate that rs-FC measures can show changes in specific connections, relevant to the treated task.

The review by Klingbeil and colleagues (2019) found four rs-FC studies using the pre-/post-design, which included altogether thirty-nine patients with aphasia. They found mixed results regarding evidence for normalization vs. compensation processes in the language network. For example in one study (van Hees et al., 2014), eight chronic patients with aphasia received twelve treatment sessions aimed at improving object naming, with half of the sessions focused on phonological and half on semantic strategies with different items. Only the improvement in the phonological treatment was correlated in with neural changes, and specifically with a shift from reliance on right middle temporal gyrus (MTG) pre-treatment to left MTG post-treatment. Furthermore, left intrahemispheric connectivity between left MTG and STG, was reduced compared to HC pre-treatment, but not post-treatment, consistent with a process of normalization. On the other hand, left intrahemispheric connectivity between left MTG and IFG was stronger in patients compared to HC post-treatment, consistent with the compensatory mechanism.

Other rs-FC studies using pre-/post-treatment designs show connectivity changes outside the language network, tying these changes to specific treatment approaches. A study that tested a phonological picture-naming treatment in ten patients (Masson-Trottier et al., 2021) showed changes in connectivity between left temporal language areas and bilateral occipital visual areas following treatment, suggesting that this was due to the visual requirements of the picture-naming task. However, no correlation was found with treatment outcome, and no control group was used. Another study examined four patients who learned sensorimotor strategies to trigger action verb naming (Durand et al., 2021). They showed increase in connections within and between visuomotor and language areas, and between areas involved in action processing. These studies point to connectivity changes among regions presumed to be relevant to the specific treatment strategy.

Other studies demonstrate the role of domain general networks to language recovery. They examine how global measures of network integration (i.e., coactivation of large sets of regions and networks), or segregation (i.e., separation between networks; Cohen & D'Esposito, 2016), contribute to recovery. In one study (Baliki et al., 2018), eight patients, four to thirty-two months post stroke, were scanned before and after an intensive four-week comprehensive language and communication therapy. Rs-FC across the whole brain was assessed using graph theoretical measures. Although this study does not report changes in connectivity between the two timepoints, they found that measures of network *integration*, such as global efficiency, as well as interhemispheric connectivity before treatment predicted treatment outcome, while the modularity of the network did not. When examining connectivity in separate networks, the study found that connectivity between the DMN and auditory regions, and between the SN and visual areas, positively predicted improved language recovery.

In contrast, several other studies point to the positive role of *segregation* between networks, and measures of network modularity to recovery during treatment. A seminal study (Siegel et al., 2018) used rs-FC to study recovery from stroke in 132 post-stroke patients, 33 of them with aphasia. Although they did not control the type of treatment participants received, they tested patients at three timepoints during the recovery process (two weeks, three months, and one-year post stroke) measuring changes in brain network modularity, which is the degree of integration within a network, and segregation between different networks in the brain. The results showed reduced network modularity in patients at the subacute phase compared with HC, with partial recovery of modularity over time, consistent with a normalization mechanism. Interestingly, the improvement in network modularity was associated with recovery in language, attention, and spatial memory, but not with recovery of visual or motor skills. The authors suggest that the recovery process involves simultaneous local reorganization processes (evident in other studies), and more global normalization processes. A similar conclusion on the positive effect of network modularity on language recovery is drawn from a study in which twelve patients with aphasia received imitation-based aphasia therapy (Duncan & Small, 2018), and rs-FC was measured at several timepoints before and after treatment. The results showed an association between the increase in patients' narrative abilities from baseline to a follow-up assessment and increase in the time of minimal connectivity between separate networks. These findings suggest that segregation between networks may facilitate language function following stroke.

5.5.2 Task-Based Functional Connectivity

Fewer studies examined changes in task-based FC before and after language treatment, and those mainly utilized effective connectivity analysis with DCM. In one recent study (Truzman et al., 2021), we examined connectivity changes within the RH in four patients with chronic aphasia before and after undergoing a phonological naming treatment. The patients and eight HC were scanned at two timepoints, while performing phonological and semantic judgments with object pictures. Moreover, DCM enables to dissociate baseline connectivity, across all perceptual and linguistic conditions in which objects were presented, from connectivity specific to the phonological or semantic tasks. The treatment-related changes in the baseline connectivity, which reflect silent object name retrieval, enhanced in its similarity to HC after treatment, reflecting a normalization process. In contrast, treatment-related changes in connectivity specific to the phonological task decreased in the resemblance to HC, presumably reflecting emerging of compensatory reorganization in RH connectivity to support phonological processing. No specific changes were associated with the semantic task. These findings suggest that both normalization and compensatory processes can simultaneously play a role in the recovery of the language network.

Another study that used DCM to measure treatment-related changes in task-based connectivity in the language network included twenty treated and ten untreated patients with chronic aphasia, further divided into treatment responders and nonresponders, and compared to a group of HC (Johnson et al., 2021). Patients received twenty-one sessions of semantic naming treatment and were scanned during a naming task. The results in a bilateral network of naming-related regions showed that a set of interhemispheric and left intrahemispheric connections was abnormally weak in patients compared to HC before treatment. These connections were strengthened following treatment only in treatment responders, providing strong evidence for normalization as a recovery process evident in the network of regions targeted by the treatment. Another study, conducted by the same group with mostly the same patients, performing a semantic judgment task (Johnson et al., 2020), provided more evidence for the relevance of global graph theoretical measures to the rehabilitation process. The results from a large network of regions associated with semantic processing showed that global integration of the network before treatment (as measured by the average network strength and global efficiency), was higher for treatment responders compared to nonresponders, and was correlated with treatment outcomes. No such effects were found for measures of network segregation (such as the clustering coefficient). These results support the contribution of global integration of the network for the language network recovery.

Finally, one study in this category also showed the relevance of domain general networks. This study, with nine patients with chronic aphasia and ten HC showed increase in the strength of the DMN, measured during naming, following semantic treatment (Marcotte et al., 2013).

5.6 Summary

This chapter reviewed studies that examined the patterns of FC involved in language recovery in aphasia, many of which were published in very recent years. These studies use diverse methodological approaches in terms of the activity during scanning (task-based vs. rs-FC), the study design (single time-point vs. pre-/post-treatment), the stage of recovery (acute, subacute, or chronic), the type of aphasia, the specific regions, and the analysis method for computing FC. Moreover, many of the studies have small sample sizes. Although these differences clearly affect the results, there are also some shared findings across the different categories. For example, as concluded by the review of Klingbeil and colleagues (2019), there is strong evidence that one of the mechanisms contributing to language recovery is the normalization of reduced left intrahemispheric connectivity within language network (Harvey et al., 2013; Johnson et al., 2021; Meier et al., 2018; Meier et al., 2019; Sandberg, 2017; Siegel et al., 2016; van Hees et al., 2014). Nevertheless, there is also evidence for the emergence of compensatory connectivity in the RH or in interhemispheric connections (Bitan et al., 2018; Meier et al., 2019; Truzman et al., 2021), or even within the LH (van Hees et al., 2014). Beyond the language network, studies from different categories point to the role of domain general networks, such as the DMN in the process of recovery from aphasia (Geranmayeh et al., 2016; Marcotte et al., 2013; Zhang et al., 2021). When looking at a larger scale, with network measures encompassing all (or most of) the brain, there is mixed evidence regarding the contribution of integration across networks (Baliki et al., 2018; Johnson et al., 2020) or the segregation and specialization of networks (Duncan & Small, 2018; Geranmayeh et al., 2016; Siegel et al., 2018) to language recovery. Altogether, these findings suggest that there is still much need for research into the FC mechanisms underlying brain plasticity in the context of language recovery.

References

Abel, S., Weiller, C., Huber, W., Willmes, K., & Specht, K. (2015). Therapy-induced brain reorganization patterns in aphasia. *Brain, 138*, 1097–1112.

Abo, M., Senoo, A., Watanabe, S., Miyano, S., Doseki, K., Sasaki, N., Kobayashi, K., Kikuchi, Y., & Yonemoto, K. (2004). Language-related brain function during word repetition in post-stroke aphasics. *Neuroreport, 15*(12), 1891–1894.

Albert, M. L., Sparks, R. W., & Helm, N. A. (1973). Melodic intonation therapy for aphasia. *Archives of Neurology, 29*(2), 130–131.

Baliki, M. N., Babbitt, E. M., & Cherney, L. R. (2018). Brain network topology influences response to intensive comprehensive aphasia treatment. *Neurorehabilitation, 43*(1), 63–76.

Bitan, T., Simic, T., Saverino, C., Jones, C., Glazer, J., Collela, B., Wiseman-Hakes, C., Green, R., & Rochon, E. (2018). Changes in resting-state connectivity following melody-based therapy in a patient with aphasia. *Neural Plasticity, 2018*, 13.

Callan, D. E., Tsytsarev, V., Hanakawa, T., Callan, A. M., Katsuhara, M., Fukuyama, H., & Turner, R. (2006). Song and speech: Brain regions involved with perception and covert production. *Neuroimage, 31*(3), 1327–1342.

Chan, M. Y., Park, D. C., Savalia, N. K., Petersen, S. E., & Wig, G. S. (2014). Decreased segregation of brain systems across the healthy adult lifespan. *Proceedings of the National Academy of Sciences, 111*(46), E4997–E5006.

Chu, R., Meltzer, J. A., & Bitan, T. (2018). Interhemispheric interactions during sentence comprehension in patients with aphasia. *Cortex, 109*, 74–91.

Cohen, J. R., & D'Esposito, M. (2016). The segregation and integration of distinct brain networks and their relationship to cognition. *Journal of Neuroscience, 36* (48), 12083–12094.

Deco, G., Jirsa, V. K., & McIntosh, A. R. (2011). Emerging concepts for the dynamical organization of resting-state activity in the brain. *Nature Reviews Neuroscience, 12*(1), 43–56.

Duncan, E. S., & Small, S. L. (2018). Changes in dynamic resting state network connectivity following aphasia therapy. *Brain Imaging and Behavior, 12*(4), 1141–1149.

Durand, E., Masson-Trottier, M., Sontheimer, A., & Ansaldo, A. I. (2021). Increased links between language and motor areas: A proof-of-concept study on resting-state functional connectivity following personalized observation, execution and mental imagery therapy in chronic aphasia. *Brain and Cognition, 148*.

Fedorenko, E., & Thompson-Schill, S. L. (2014). Reworking the language network. *Trends in Cognitive Sciences, 18*(3), 120–126.

Fridriksson, J. (2010). Preservation and modulation of specific left hemisphere regions is vital for treated recovery from anomia in stroke. *Journal of Neuroscience, 30*(35), 11558–11564.

Fridriksson, J., Baker, J. M., & Moser, D. (2009). Cortical mapping of naming errors in aphasia. *Human Brain Mapping, 30*(8), 2487–2498.

Friston, K. J., Harrison, L., & Penny, W. (2003). Dynamic causal modelling. *Neuroimage, 19*(4), 1273–1302.

Geranmayeh, F., Leech, R., & Wise, R. J. S. (2016). Network dysfunction predicts speech production after left hemisphere stroke. *Neurology, 86*(14), 1296.

Gili, T., Fiori, V., De Pasquale, G., Sabatini, U., Caltagirone, C., & P. Marangolo, P. (2017). Right sensory-motor functional networks subserve action observation therapy in aphasia. *Brain Imaging Behav, 11*(5), 1397–1411.

Harvey, D. Y., Wei, T., Ellmore, T. M., Hamilton, A. C., & Schnur, T. T. (2013). Neuropsychological evidence for the functional role of the uncinate fasciculus in semantic control. *Neuropsychologia, 51*(5), 789–801.

Heiss, W. D., Karbe, H., WeberLuxenburger, G., Herholz, K., Kessler, J., Pietrzyk, U., & Pawlik, G. (1997). Speech-induced cerebral metabolic activation reflects recovery from aphasia. *Journal of Neuroscience, 145*(2), 213–217.

Heiss, W. D., Thiel, A., Kessler, J., & Herholz, K. (2003). Disturbance and recovery of language function: Correlates in PET activation studies. *Neuroimage, 20*, Supplement 1, S42–S49.

Hope, T. M. H., Leff, A. P., & Price, C. J. (2018). Predicting language outcomes after stroke: Is structural disconnection a useful predictor? *Neuroimage-Clinical, 19*, 22–29.

Johnson, J. P., Meier, E. L., Pan, Y., & Kiran, S. (2020). Pre-treatment graph measures of a functional semantic network are associated with naming therapy outcomes in chronic aphasia. *Brain and Language, 207.*

Johnson, J. P., Meier, E. L., Pan, Y., & Kiran, S. (2021). Abnormally weak functional connections get stronger in chronic stroke patients who benefit from naming therapy. *Brain and Language, 223.*

Keator, L. M., Yourganov, G., Basilakos, A., Hillis, A. E., Hickok, G., Bonilha, L., Rorden, C., & Fridriksson, J. (2021). Independent contributions of structural and functional connectivity: Evidence from a stroke model. *Network Neuroscience, 5*(4), 911–928.

Klingbeil, J., Wawrzyniak, M., Stockert, A., & Saur, D. (2019). Resting-state functional connectivity: An emerging method for the study of language networks in post-stroke aphasia. *Brain and Cognition, 131*, 22–33.

Marcotte, K., Perlbarg, V., Marrelec, G., Benali, H., & Ansaldo, A. I. (2013). Default-mode network functional connectivity in aphasia: Therapy-induced neuroplasticity. *Brain and Language, 124*(1), 45–55.

Masson-Trottier, M., Sontheimer, A., Durand, E., & Ansaldo, A. I. (2021). Resting-state functional connectivity following phonological component analysis: The combined action of phonology and visual orthographic cues." *Brain sciences, 11*(11), 1458.

Meier, E. L., Johnson, J. P., & Kiran, S. (2018). Left frontotemporal effective connectivity during semantic feature judgments in patients with chronic aphasia and age-matched healthy controls. *Cortex, 108*, 173–192.

Meier, E. L., J. P. Johnson, Y. Pan & S. Kiran (2019). "A lesion and connectivity-based hierarchical model of chronic aphasia recovery dissociates patients and healthy controls." *Neuroimage-Clinical*, **23**.

Meltzer, J. A., Wagage, S., Ryder, J., Solomon, B., & Braun, A. R. (2013). Adaptive significance of right hemisphere activation in aphasic language comprehension. *Neuropsychologia, 51*(7), 1248–1259.

Naeser, M. A., Martin, P. I., Baker, E. H., Hodge, S. M., Sczerzenie, S. E., Nicholas, M., Palumbo, C. L., Goodglass, H., Wingfield, A., Samaraweera, R., Harris, G., Baird, A., Renshaw, P., & Yurgelun-Todd, D. (2004). Overt propositional speech in chronic nonfluent aphasia studied with the dynamic susceptibility contrast fMRI method. *Neuroimage, 22*(1), 29–41.

New, A. B., Robin, D. A., Parkinson, A. L., Duffy, J. R., McNeil, M. R., O. Piguet, O., Hornberger, M., Price, C. J., Eickhoff, S. B., & Ballard, K. J. (2015). Altered resting-state network connectivity in stroke patients with and without apraxia of speech. *NeuroImage: Clinical, 8*, 429–439.

Postman-Caucheteux, W. A., Birn, R. M., Pursley, R. H., Butman, J. A., Solomon, J. M., Picchioni, D., McArdle, J., & Braun, A. R. (2010). Single-trial fMRI shows contralesional activity linked to overt naming errors in chronic aphasic patients. *Journal of Cognitive Neuroscience, 22*(6), 1299–1318.

Price, C. J., & Crinion, J. (2005). The latest on functional imaging studies of aphasic stroke. *Current Opinion in Neurology, 18*(4), 429–434.

Sandberg, C. W. (2017). Hypoconnectivity of resting-state networks in persons with aphasia compared with healthy age-matched adults. *Frontiers in Human Neuroscience, 11.*

Saur, D., Lange, R., Baumgaertner, A., Schraknepper, V., Willmes, K., Rijntjes, M., & Weiller, C. (2006). Dynamics of language reorganization after stroke. *Brain, 129*(6), 1371–1384.

Siegel, J. S., Ramsey, L. E., Snyder, A. Z., Metcalf, N. V., Chacko, R. V., Weinberger, K., Baldassarre, A., Hacker, C. D., Shulman, G. L., & Corbetta, M. (2016). Disruptions of network connectivity predict impairment in multiple behavioral domains after stroke. *Proceedings of the National Academy of Sciences, 113*(30), E4367–E4376.

Siegel, J. S., Seitzman, B. A., Ramsey, L. E., Ortega, M., Gordon, E. M., Dosenbach, N. U. F., Petersen, S. E., Shulman, G. L., & Corbetta, M. (2018). Re-emergence of modular brain networks in stroke recovery. *Cortex, 101,* 44–59.

Sparks, R., Helm, N., & Albert, M. (1974). Aphasia rehabilitation resulting from melodic intonation therapy. *Cortex, 10*(4), 303–316.

Thiel, A., Herholz, K., Koyuncu, A., Ghaemi, M., Kracht, L. W., Habedank, B., & Heiss, W. D. (2001). Plasticity of language networks in patients with brain tumors: A positron emission tomography activation study. *Annals of Neurology, 50*(5), 620–629.

Tinaz, S., Lauro, P., Hallett, M., & Horovitz, S. G. (2016). Deficits in task-set maintenance and execution networks in Parkinson's disease. *Brain Structure and Function, 221*(3), 1413–1425.

Truzman, T., Rochon, E., Meltzer, J., Leonard, C., & Bitan, T. (2021). Simultaneous normalization and compensatory changes in right hemisphere connectivity during aphasia therapy. *Brain Sciences, 11*(10), 1330.

van Hees, S., McMahon, K., Angwin, A., de Zubicaray, G., Read, S., & Copland, D. A. (2014). A functional MRI study of the relationship between naming treatment outcomes and resting state functional connectivity in post-stroke aphasia. *Human Brain Mapping, 35*(8), 3919–3931.

Wilson, S. M., Eriksson, D. K., Yen, M., Demarco, A. T., Schneck, S. M., & Lucanie, J. M. (2019). Language mapping in aphasia. *Journal of Speech, Language, and Hearing Research, 62*(11), 3937–3946.

Xu, L., Huang, L., Cui, W., & Yu, Q. (2020). Reorganized functional connectivity of language centers as a possible compensatory mechanism for basal ganglia aphasia. *Brain Injury, 34*(3), 430–437.

Yang, H. Q., Bai, L., Zhou, Y., Kang, S., Liang, P. P., Wang, L. H., & Zhu, Y. F. (2017). Increased inter-hemispheric resting-state functional connectivity in acute lacunar stroke patients with aphasia. *Experimental Brain Research, 235* (3), 941–948.

Zhang, C., Xia, Y. Y., Feng, T., Yu, K., Zhang, H. Y., Sami, M. U., Xiang, J., & Xu, K. (2021). Disrupted functional connectivity within and between resting-state networks in the subacute stage of post-stroke aphasia. *Frontiers in Neuroscience, 15.*

Zhu, D., Chang, J., Freeman, S., Tan, Z., Xiao, J., Gao, Y., & Kong, J. (2014). Changes of functional connectivity in the left frontoparietal network following aphasic stroke. *Frontiers in Behavioral Neuroscience, 8*.

Zhu, L., Fan, Y., Zou, Q., Wang, J., Gao, J.-H., & Niu, Z. (2014). Temporal reliability and lateralization of the resting-state language network. *PLoS ONE*, *9*(1).

6

Recent Developments in Psycholinguistics and Neurolinguistics through the Lens of Aphasia

Erin L. Meier and Chaleece W. Sandberg

Language is considered a uniquely human trait (Bolhuis et al., 2014). While the power of language for communication is obvious, Friederici and colleagues (2017) argue that the purpose of language is to generate complex thought. Language can be expressed in spoken, written, and signed form and comprises the linguistic subsystems of phonology (i.e., sound), semantics (i.e., meaning), and morphosyntax (i.e., structure of words and sentences). This chapter focuses on recent advances in the fields of psycholinguistics and neurolinguistics, with a heavier emphasis on the latter, due to the recent tendency to combine psycholinguistic and neurolinguistic aspects into a single model. The field of psycholinguistics is concerned with the psychological processes that mediate language behaviors – including during the development of language and when a fully intact language system is disrupted with an acquired brain injury – while neurolinguistics focuses on where these processes occur in the brain. While many psycholinguists cite Chomsky's work in the 1950s and 60s as the origin of the field, both psycholinguistics and neurolinguistics have roots in work started by aphasiologists in the mid nineteenth century (Levelt, 2012). Therefore, this chapter begins with a historical overview of the neurobiology of language and aphasia before turning to developments in these fields within the last twenty years.

6.1 An Historical Overview of the Neurobiology of Language

French physician Paul Broca and German neuroanatomist and neurologist Carl Wernicke are widely recognized as the first to link brain damage to certain parts of the left hemisphere – namely the inferior frontal gyrus and superior temporal gyrus – with respective deficits in speech production and speech comprehension (Broca, 1861; Wernicke, 1874). Fierce debates about the nature of language representation in the brain arose following the presentation of Broca and Wernicke's seminal aphasia cases. Several neurobiological models of language were proposed in the mid to late nineteenth century, most notably the first network models by Wernicke, later specified and illustrated by Ludwig Lichtheim (Lichtheim, 1885; Wernicke, 1874). Wernicke proposed that primary sensorimotor cortical areas mediate foundational sensory and motor functions but that higher order thinking is the consequence of connections between these regions. Wernicke is often considered the godfather of disconnection theory (Catani & ffytche, 2005) and was the first to propose another aphasia syndrome, conduction aphasia, characterized primarily by repetition deficits due to a disconnection of regions responsible for speech comprehension (Wernicke's area) and speech production (Broca's area). Lichtheim extended Wernicke's disconnection proposal into describing other aphasia syndromes (Lichtheim, 1885). He proposed the existence of an anatomically nonspecific "concept center" that when disconnected from either Broca's or Wernicke's areas, resulted in, respectively, transcortical motor and transcortical sensory aphasia.

Criticisms regarding the clinical–anatomical association approach followed the Wernicke–Lichtheim model, partially because Lichtheim's aphasia syndrome framework was theoretical and based on clinical observations of patients with aphasia without evidence of anatomical correlates of these syndromes (Catani & ffytche, 2005). In addition to a lack of evidence, some took issue with the fact that Wernicke's school was purely disconnectionist, meaning that proponents of disconnection theory believed disorders arose purely from damage to the connections between regions (white matter tracts) without affording a particular role in language processing to the cortical regions themselves. At the time, the school of associationism also became mired with a plethora of cortical specialization maps that harkened back to phrenology, resulting in anti-associationism and anti-localizationist views.

By the late nineteenth century, interest in the neural correlates of behavior had dwindled. Associationism also fell out of favor within the field of neurology until it was rejuvenated by Norman Geschwind in the mid 1960s (Devinsky, 2009). Geschwind proposed an updated language model, often coined the Wernicke–Lichtheim–Geschwind model, wherein information flows from cortical conceptual centers to the posterior auditory language region (i.e., Wernicke's area in the posterior superior temporal lobe) and to the anterior motor language area (i.e., Broca's area in the inferior frontal gyrus) via the arcuate fasciculus (Geschwind, 1965a, 1965b, 1970). Geschwind's major contributions to the neurobiology of language included an emphasis on association cortex as necessary relays of language information, theories regarding the role of the inferior parietal lobe in language processing, and the proposal that in addition to white matter damage, disconnection syndromes (e.g., conduction aphasia) can arise from damage to critical cortical hubs (Catani & ffytche, 2005). However, Geschwind did not distinguish the effects of cortical lesions to association cortex from lesions to white matter tracts nor did he assign specialized functions to specific association regions. Despite these drawbacks, the Wernicke–Lichtheim–Geschwind model (and its various iterations, collectively often referred to as the "Classic Model") have remained in widespread clinical and scientific use through the turn of the twenty-first century.

In recent years, however, additional criticisms and evidence against the utility of both the Classic Model aphasia syndrome classifications have been presented. Tremblay and Dick (2016) emphasized that the focus of the Classic Model on Broca's and Wernicke's areas excludes several other cortical and subcortical structures that modern neuroimaging has indicated are critical for language processing. They specifically highlight evidence that several white matter tracts have been implicated in different language and linguistic processes (see Section 6.2) and state that the focus on the arcuate fasciculus as the sole language "pathway" in the Geschwind-weighted Classic Model even overlooks Wernicke's proposal from over a 100 years ago that a second, ventral language pathway exists. Tremblay and Dick (2016) also stress the lack of consensus in the literature and between language neuroscientists regarding the precise location of Broca's and Wernicke's areas and argue for the use of more precise neuroanatomical terms.

Relatedly, accumulating evidence from before (e.g., Mohr et al., 1978; Willmes & Poeck, 1993) and after (e.g., Fridriksson et al., 2007; Kasselimis et al., 2017; Landrigan et al., 2021; Yourganov et al., 2015) the turn of the twenty-first century suggests that there is often low correspondence between aphasia syndromes and classic lesion sites per the Lichtheim framework. This is even true when the brain of Broca's classic patient, Monsieur Tan, was reexamined with modern neuroimaging tools, revealing a much larger area of damage than merely a circumscribed Broca's area lesion (Dronkers et al., 2007). Forty years ago, Schwartz (1984) argued against aphasia classification and discussed how aphasia syndromes follow a polytypic classification structure, meaning that people with the same aphasia syndrome can exhibit deficits unshared by other people with the same aphasia type and that people with different aphasia syndromes can confusingly share the same attributes. For example, Blumstein (2016) highlighted that patients diagnosed with Broca's, Wernicke's, and conduction aphasia all often produce phonological errors so the presence or absence of these types of errors is not a useful clinical feature for determining an aphasia classification. Furthermore, in speech-language pathology clinical practice, aphasia syndromes are often determined by deficits in fluency, auditory comprehension, and repetition, yet Landrigan and colleagues (2021) found that impairments in semantics and phonology – rather than comprehension and production – distinguished different subgroups of patients with aphasia. As such, many aphasiologists (e.g., Blumstein, 2016; Caplan, 1993; Charidimou et al., 2014; Kasselimis et al., 2017; McNeil & Pratt, 2001; Schwartz, 1984) have proposed focusing on deficits in specific linguistic domains and understanding which brain regions are responsible for those processes is a more useful endeavor than aphasia classification per the Classic Model.

The pioneering work of the late nineteenth and twentieth centuries shaped modern views of the neurobiology of language. However, taking a forward-thinking perspective, the remainder of this chapter presents evidence for modern language neurobiology models based on neuroimaging evidence (rather than the Classic Model), includes precise neuroanatomical terms (when possible), and reviews evidence from aphasia (without an overreliance on aphasia syndromes).

6.2 Contemporary Neurolinguistic and Psycholinguistic Models and Current Evidence

The focus of current neuroimaging research of language continues to be the extent to which linguistic processes can be localized within the brain. As shown in Figure 6.1, brain regions responsible for semantics, phonology,

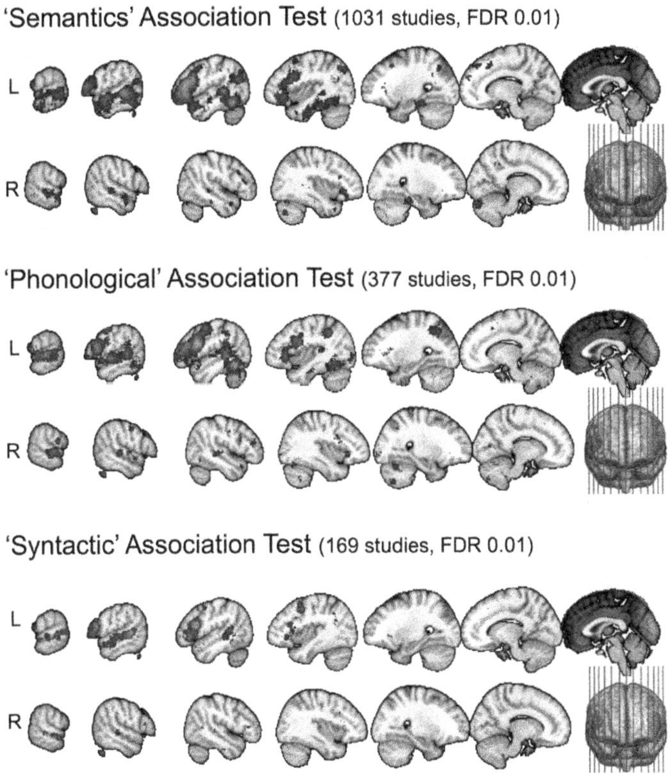

Figure 6.1 Meta-analytic fMRI results to date in Neurosynth (Yarkoni et al., 2011) for the linguistic domains of semantics, phonology, and syntax. Association test activation maps (corrected for multiple comparisons at an FDR of 0.01) for the search terms "semantic" (1031 studies), "phonological" (377 studies), and "syntactic" (169 studies) are overlaid in stereotactic space. L = left, R = right.

and syntax lie predominantly within the left cerebral hemisphere. While some spatial separation of these processes is apparent, there is also a high degree of overlap in certain parts of the cortex.

Many contemporary models of the neurobiology of language propose a dual-route architecture for the organization of semantics, phonology, and syntax in the brain. In one such model, Hickok and Poeppel (2004, 2007) framed the neural correlates of component processes within the context of word-level repetition and comprehension. They proposed spectrotemporal analysis of the incoming acoustic signal is processed bilaterally in the dorsal superior temporal gyrus (STG) followed by processing of phonological word features within bilateral (although perhaps weakly left-dominant) representation regions in the superior temporal sulcus (STS). For word repetition, processing diverges to a left-lateralized dorsal route wherein the Sylvian parietal temporal area (area Spt) serves as a critical sensorimotor hub and interface between regions involved in spectrotemporal analysis and regions within the articulatory network that retrieve syllable structure and word forms (left posterior inferior frontal gyrus [IFG]) and activate motor programs (primary motor cortex) for production. For word comprehension, processing diverges to the ventral route, wherein phonological representations are mapped onto lexical conceptual representations within the bilateral posterior middle temporal gyrus (MTG) and inferior temporal gyrus (ITG) and sulcus. Left-lateralized cortex anterior to these regions within the temporal lobe gives rise to a widely distributed semantic combinatorial network through which conceptual representations are used to impart linguistic meaning.

Other dual-stream models include a similar dorsal/ventral division of processes with some important distinctions from the Hickok and Poeppel (2004, 2007) model. For example, Friederici and Gierhan (2013) framed their model within more general language processing contexts (i.e., not specific to single word repetition or comprehension) and specified the role of white matter structures in linguistic functions. They posited that two subroutes comprise the dorsal pathway, with one subroute spanning the posterior temporal to premotor cortices critical for sensorimotor mapping for speech repetition and a second subroute from posterior IFG to posterior STG (directly connected via the arcuate fasciculus), important for processing syntactically complex sentences. Friederici and Gierhan (2013) proposed crucial white matter structures – including the uncinate and inferior frontooccipital fasciculi – connect inferior frontal and temporal regions of the ventral pathway. They also suggested a distinction between regions important for lexical-semantic processing (left MTG) versus sentence-level semantic processes (left angular gyrus [AG]) and proposed that the ventral stream is responsible for processing syntactically simple sentences in addition to semantic processes.

Not all modern neurobiological language models adhere to a dual-route composition or focus solely on linguistic processes. For instance, Hagoort (2013) described a model in which language processing occurs through retrieval of linguistic knowledge stored within memory areas of the temporal and inferior parietal cortices, unification (assembly) of retrieval linguistic structures within the inferior frontal gyrus, and control operations mediated by Brodmann area 46 (middle frontal gyrus [MFG]) and the anterior cingulate cortex (ACC). Other model work has aimed to disentangle brain networks responsible for language and nonlinguistic processes (see e.g., Fedorenko & Thompson-Schill, 2014 for review). The following sections focus on advances pertaining to the building blocks of language: semantics, phonology, and syntax.

6.2.1 Semantics

Semantics is the branch of psycholinguistics that is concerned with how humans transform their sensory, motor, and emotional experiences into meaning, generalize those experiences to new concepts and events, and access context-dependent semantic information as needed. Semantic knowledge accumulates across the lifespan and is critical for verbal and nonverbal behaviors alike. Major advances in the last twenty years in the study of semantics have centered on how components of the semantic system are organized and represented in the brain.

6.2.1.1 Semantic Representation and Localization of Semantic Processing

Semantic knowledge is thought to be stored as a network of conceptual nodes that are connected via semantic similarity and that activation can spread from an activated concept to related concepts (Collins & Loftus, 1975). Traditionally, there has been a debate as to whether semantic knowledge is stored as abstracted, amodal information (e.g., Fodor, 1975) or as modality-specific information based on sensory, motor, and emotional systems per embodied cognition theories (e.g., Barsalou, 2008). Related to these debates is the fact that individuals with acquired brain damage – most notably individuals with post-stroke aphasia – can exhibit category-specific difficulties. The most well-studied double dissociation is between animals and tools. Individuals with a specific difficulty with animal concepts tend to have damage to the left ventral and medial temporal lobes (e.g., Gainotti, 2010). Those with a specific difficulty with manipulable (tool) concepts tend to have damage to left posterior temporoparietal regions (e.g., Riccardi et al., 2020). Historically, there were two general opposing theories regarding category specific deficits – one that focused on the relative importance of different types of features (e.g., visual/perceptual vs. functional/associative, Warrington & Mccarthy, 1987), and one that focused on the evolutionary importance of these domains (domain-specific hypothesis, Caramazza & Shelton, 1998). More recently, these two ideas have been brought together in the distributed domain-specific hypothesis (see Mahon & Caramazza, 2009 for a review). The distributed, domain-specific hypothesis (Mahon & Caramazza, 2009, 2011) proposes that the functional connectivity of different brain regions within the semantic network drives how concepts and categories are spatially represented, leading to a distributed network without a single amodal storage hub or zone.

However, critics of this hypothesis (e.g., Lambon Ralph et al., 2017) have emphasized that a transmodal hub is necessary for forming coherent concepts that can be generalized to new ideas and situations. As such, other models (e.g., Binder & Desai, 2011) propose that regions located between modal systems – that is, mid middle and inferior temporal and inferior parietal cortex – serve as convergence zones that store abstract representations of event and object knowledge received from modality-specific representation areas within primary sensory, motor, and affective cortices. A related model that has gained popularity in recent years is the hub-and-spoke controlled semantic cognition (CSC) model (Lambon Ralph et al., 2017). In this model, there are modality-specific "spokes," located in modality-specific cortices, where the modality-specific information that has been encoded for each concept is stored/processed. The "hub," located in the anterior temporal lobes (ATLs), with a leftward bias for verbal information and a rightward bias for visual information, combines the modality-specific information from the spokes into a coherent concept. For example, the concept "dog" is comprised of visual information (fur, tail), tactile information (texture of fur), auditory information (bark, growl), olfactory information (breath, wet fur), and even emotional information (joy, fear), and/or experiential/idiosyncratic information (e.g., *my dog likes carrots*) that all combine into a conceptual representation that supports our ability to think and talk about dogs.

There are many similar features of the hub-and-spoke and convergence zone models, but one key difference is often the area(s) highlighted as the central hub(s)/zone(s) (i.e., posterior temporal and inferior parietal areas versus the anterior temporal lobe). In recent years, lesion symptom mapping (LSM) has been widely used to demonstrate the roles of these regions for processes that engage the semantic system, such as single word comprehension and production and judgments regarding semantic relationships between words and pictures. For example, in a series of papers, Hillis and colleagues (Cloutman et al., 2009; DeLeon et al., 2007; Hillis & Heidler, 2002; Hillis et al., 2006; Hillis, Kane, et al., 2001; Hillis, Wityk, et al., 2001) linked hypoperfusion within regions of interest to deficits and recovery of word comprehension and/or production in stroke survivors within the first few days of left hemisphere stroke. Across several studies, they reported that hypoperfusion within the left superior temporal gyrus (STG), corresponding to Brodmann area (BA) 22 was associated with impairments of spoken word comprehension and production whereas hypoperfusion within the left posterior inferior temporal (pITG)/fusiform gyri – corresponding to BA 37 – was implicated in impaired oral naming with spared word comprehension. Along the same vein, improvement of word-level lexical semantics during comprehension and production tasks in the two to three days after stroke was linked to reperfusion of tissue within left STG/BA 22, while improved naming was associated with reperfusion of BA 37 (Hillis et al., 2006; Hillis, Kane, et al.,

2001). In a later study by the same group employing different perfusion methods, Meier and colleagues (2021) found that semantic production errors – classified as semantic paraphasias (e.g., "apple" for *grapes*, "milk" for *cow*), circumlocutions, and mixed errors (e.g., "cat" for *rat*) – during object naming was associated with left parietal hypoperfusion. In sum, these findings from acute stroke primarily implicate posterior brain regions in semantic processes but link damage to pSTG/Wernicke's area in semantic comprehension and pITG/fusiform gyrus in semantic production.

In chronic aphasia, anterior brain regions – particularly the left ATL – have been implicated in semantic processes. In the first of a series of papers, Schwartz and colleagues (2009) found that in 64 people with aphasia with mixed (but predominantly chronic) stroke chronicity, production of semantic errors during object naming was linked to damage to the left temporal pole, anterior half of the middle temporal gyrus (MTG), the posterior MTG, and portions of the left inferior frontal gyrus, pars triangularis (IFGtri) and middle frontal gyrus (MFG). When the authors controlled for nonverbal semantic abilities, only voxels within the ATL remained significant. In subsequent studies by this group, the link between lexical semantic production errors and left ATL damage remained, particularly in the anterior half of MTG and ITG. These findings persisted even when the definition of semantic errors was expanded (Walker et al., 2011), a larger sample and more extensive testing battery was employed (Mirman, Chen, et al., 2015), and multivariate support vector regression over univariate voxel-based LSM methods were utilized (Mirman, Zhang, et al., 2015). In contrast to semantic errors made during naming, deficits in semantic recognition – which reflected impairments in nonverbal and lexical semantics – were associated with damage to the frontal white matter "bottleneck" where the inferior fronto-occipital fasciculus (IFOF), uncinate fasciculus (UF), and anterior thalamic radiations converge (Mirman, Chen, et al., 2015; Mirman, Zhang, et al., 2015). These findings led Mirman, Zhang, and colleagues (2015) to conclude that the role of white matter connections had not been incorporated adequately into models of semantic processing that emphasized the importance of gray matter hubs (either the ATL or inferior parietal/posterior superior temporal regions e.g., Binder & Desai, 2011; Lambon Ralph et al., 2017).

Indeed, LSM study findings from other groups demonstrate the importance not only of the left ATL and temporoparietal regions in semantic processing but also white matter tracts that connect these regions. For example, in another series of studies by a different group with overlapping subsets of individuals with chronic aphasia, Lambon Ralph and colleagues interrogated the neural correlates of a semantic factor obtained from omnibus principal component analysis (PCA) across a battery of nonverbal and lexical-semantic tasks (Alyahya et al., 2020; Butler et al., 2014; Halai et al., 2017, 2018; Schumacher et al., 2019; Woollams et al., 2018; Zhao et al., 2020). They consistently found that lower semantic factor scores (reflecting more impaired semantic abilities) were most strongly associated with damage to left ATL but also damage to posterior cortical regions such as pITG, pMTG, and the superior lateral occipital cortex and underlying white matter tracts (e.g., IFOF, ILF, and UF) that connect anterior and posterior cortices. These findings accord with other studies that incorporated connectome LSM (Fridriksson et al., 2018) or diffusion tensor imaging (DTI) (Han et al., 2013; Ivanova et al., 2016; Rolheiser et al., 2011) to interrogate cortico-cortical connections or white matter tract damage implicated in semantic deficits, respectively.

These collective findings contributed to the creation of a recent dual-hub model for combining semantic information (Mirman et al., 2017). In this model, rather than one hub responsible for integrating all types of semantic information, there are two hubs: one that combines information regarding taxonomic relationships (based on semantic similarity, e.g., both dogs and cats are pets), located in the ATLs, and one that combines information regarding thematic relationships (based on co-occurrence in a context, e.g., dog and leash are related in the context of walking the dog), located in the angular gyrus (AG).

6.2.1.2 Semantic Control

In recent years, increased attention has also been given to how semantic information is accessed and manipulated in a controlled fashion, often termed semantic control (Lambon Ralph et al., 2017). Specific to storage, the distributed domain-specific hypothesis (Mahon & Caramazza, 2011) does not explain how semantic information is accessed or manipulated. In their model, Binder and Desai (2011) highlighted the role of dorsolateral and prefrontal regions in goal-directed access of information stored within temporoparietal convergence zones. Unlike many other frameworks of semantic cognition, the CSC model (Lambon Ralph et al., 2017) specifies a set of brain regions implicated in semantic control – including the left inferior frontal gyrus (IFG), posterior middle temporal gyrus (pMTG), and inferior parietal sulcus (IPS) – that exerts top-down situation/context-dependent access and retrieval of semantic information stored within the ATLs. One limitation of the LSM literature described above is that the studies rarely separated semantic representation from control. However, functional neuroimaging meta-analyses in neurologically healthy adults (Diveica et al., 2021; Hodgson et al., 2021; Noonan et al., 2013) support the role of IFG and pMTG in particular in semantic control.

Perhaps the strongest evidence for a distinction between semantic representation and semantic control comes from comparison of clinical disorders. In stroke-based aphasia, it is generally accepted that semantic knowledge is not impaired, though there can be difficulty accessing semantic information. Semantic aphasia (SA) has been presented as a subtype of aphasia in which individuals exhibit particular difficulty with semantic control and often have damage to left inferior frontal and/or posterior temporal regions. Moreover, SA has been contrasted with the semantic variant of primary progressive aphasia (PPA-S) – sometimes referred to as semantic dementia (SD) – in which there is a gradual loss of conceptual representations (Jefferies & Lambon Ralph, 2006). In SD, the primary locus of atrophy is within the ATL (e.g., Gorno-Tempini et al., 2011) whereas inferior frontal and posterior temporal regions often remain spared, at least until later stages of the disease. Notably, Chapman and colleagues (2020) argue that the mapping of semantic control to SA and semantic storage to SD is not absolute and suggest that the use of syndromic categories may mask alternative explanations that lesion-based studies could help uncover. Regardless, as semantic knowledge forms the basis for language and thought, the effect of difficulty with semantic control versus deficits in semantic knowledge on language production and comprehension may be very different. For example, an issue with control may result in semantic errors that show a preservation of semantic knowledge (e.g., saying *lion* instead of *tiger*, the feeling of knowing what one wants to say), whereas an issue with semantic knowledge may result in semantic errors that show a degradation of semantic knowledge (e.g., saying *animal* instead of *tiger*, neglecting characteristic semantic features – like the hump of a camel, empty speech).

Distinctions in Word Class Representations

In addition to the debate regarding the organization of semantic knowledge in general, there are debates regarding the specific psycholinguistic and neurolinguistic organization of concepts for which behavior tends to dissociate. For example, individuals with aphasia typically exhibit more difficulty with verbs than nouns, but there are cases in which nouns are more difficult than verbs (e.g., Hillis & Caramazza, 1995). The increased difficulty with verbs may result from the semantic distinction that nouns represent objects while verbs represent actions or events (see Kemmerer, 2014 for a review), that nouns have higher imageability than verbs (e.g., Bird et al., 2003), that nouns and verbs have different morphological processing demands (Tsapkini et al., 2002), and/or that verbs with more complex argument structures are harder to retrieve (argument structure complexity hypothesis; Thompson, 2003). However, Kemmerer (2014) argues that the morphological/syntactic distinction between nouns and verbs is not clear-cut in all languages. Further, Alyahya and colleagues (2018) found that forty-eight chronic stroke survivors with aphasia exhibited no behavioral differences in noun versus verb single-word production and comprehension when stimulus lists were carefully balanced on several psycholinguistic properties (i.e., age-of-acquisition, imeagibility, familiarity, frequency, word length, visual complexity).

Two opposing theories exist regarding the neural correlates of noun versus verb processing, with one view proposing overlapping cortical regions supporting both noun and verb processing, and the other view suggesting noun and verb networks are dissociable. Current neuroimaging evidence suggests both views may be partially correct. For example, using voxel-based correlation methods, Alyahya and colleagues (2018) reported nearly 80 percent overlap in regions implicated in noun and verb naming deficits in participants with chronic aphasia, including the left AG, pSMG, superior lateral occipital cortex extending to pMTG and pITG, temporal fusiform, planum polare to aMTG and temporal pole, and underlying white matter corresponding to the AF, ILF, and internal capsule. Verb (but not noun) naming was also associated with voxels in IFG, medial frontal cortex, frontal pole, and PreCG, but a direct comparison of nouns and verbs revealed no significant clusters. They found less overlap for noun versus verb single-word comprehension, but still common regions included left AG, inferior lateral occipital cortex, temporal occipital fusiform cortex extending along MTG and ITG, and underlying white matter, including ILF, posterior segment of the AF, fornix, and CST. Controlling for noun comprehension, parietal regions (e.g., PostCG, aSMG, and anterior segment of AF) were implicated solely in verb comprehension. Consistent with these findings, a review of functional neuroimaging work has shown preferential activation in ventral temporal areas for nouns and a fronto-temporoparietal network for verbs but still some overlap in activations between word classes (Kemmerer, 2014).

Similarly, neurologically intact adults and those with aphasia exhibit more difficulty with abstract words than concrete words (referred to as the concreteness effect), with aphasia exaggerating this effect (e.g., Sandberg & Kiran, 2014). However, some individuals with primary progressive aphasia of the semantic type (PPA-S) – also referred to as semantic dementia (SD) – have shown a reversal of the concreteness effect, in which abstract words appear to be easier to use (e.g., Warrington, 1975; however, see Hoffman & Lambon Ralph, 2011 who suggest that this is the exception rather than the rule and that damage to the ATLs should affect both abstract and concrete words). Many theories have been proposed to account for this double dissociation, including those that propose two separate systems for the storage and processing of abstract and concrete concepts (e.g., the dual coding hypothesis, Paivio, 1971) and those that propose the same system, but different types and/or amounts of information (e.g., the context availability hypothesis, Schwanenflugel & Shoben, 1983). Neuroimaging work has shown preferential activation for abstract words in left anterior areas, such as IFG and anterior superior temporal cortex, and preferential activation for concrete words in posterior areas, such as fusiform and posterior cingulate cortex (e.g., Binder et al., 2009; Bucur & Papagno, 2021; Wang et al., 2010); however, note that a graded representation for abstract and concrete words has been found in the anterior temporal lobes (e.g., Hoffman et al., 2015).

While there is not yet one agreed-upon theory, one recent theory suggests that while concrete words are grounded in sensory experience, abstract words are grounded in internal perception and metacognition and may be considered to be social tools (the words as social tools account by Borghi et al., 2019). Similarly, Desai and colleagues (2018) suggest that abstract and concrete concepts are not categorically distinct, but rely differentially on semantic dimensions (e.g., event-related, interoception, introspection, and sensory-motor). Thus, there appears to be a general move toward framing both abstract and concrete concept organization within the semantic system using a grounded cognition or embodied approach (e.g., Barsalou, 2008; Conca et al., 2021). This move is supported by neuroimaging work showing activation in areas related to numbers, theory of mind, morality, and emotion for abstract words (see Desai et al., 2018 for a meta-analysis). Less attention has been paid to how abstract and concrete words may be connected within the semantic network, though Troche and colleagues (2014) have examined the topography of abstract and concrete words within a multidimensional semantic space, finding that while their topographies are unique, abstract and concrete words can cluster together (e.g., father and love cluster together, partly because they both have high loadings for the emotion dimension).

6.2.2 Phonology

Phonology is the system of speech sounds or signs that are the basic building blocks of a spoken or signed language, respectively. Individuals with aphasia can show deficits in the production of phonemes in the form of phonemic paraphasias (e.g., "ocopus" for *octopus*) and/or perception of phonemes causing comprehension errors (e.g., interpreting "my chair is missing" as a claim of baldness).

6.2.2.1 Production of Phonology

Phonemic paraphasias may consist of sound substitutions (e.g., cake vs. take), additions (e.g., bread vs. bed), and/or deletions (e.g., bed vs. bread). As with speech errors in nonimpaired individuals, phonemic paraphasias can be perseverative (e.g., brass bread vs. brass bed), anticipatory (e.g., bass bed vs. brass bed), swaps (e.g., cakes the take vs. takes the cake), or blends (e.g., incesticide = insecticide + pesticide). In the Lemma Model of speech production, Levelt (1989) proposes that sound speech errors occur during the stage of retrieving morphemic phonological codes, which occurs after lemma[1] selection and prior to creating the prosodic and syllabic structure. Indefrey and Levelt, (2004) propose that this process occurs in the left posterior STG/STS, the left posterior MTG, the left anterior insula, and the right SMA. In the interactive activation model of speech production (Dell & O'Seaghdha, 1992), phonology is represented in both the lexical (word form) network level and the phonological network level, and speech errors are discussed in terms of the competition that exists within levels, influenced by activation that can feed forward or backward through the levels. As described by Martin (2022), phonemic paraphasias can occur either at the phonological network level or the lexical network level. Errors that occur at the lexical level can be influenced by feedforward connections from the semantic network and feedback connections from the phonological network, which produces lexical competition from items that are both semantically and phonetically similar (i.e., mixed errors; e.g., dice vs. slice) or just feedback connections from the phonological network, which produces lexical competition from items that are just phonetically similar (e.g., slide vs. slice). Phonemic paraphasias that result in nonwords (e.g., slike vs. slice) are thought to result from competition only within the phonological network as phonological encoding for speech is taking place – though issues with phonological encoding can also produce phonologically similar real words. Many models incorporate some sort of working memory component to the phonological encoding process, as the word form must be kept active while the phonemes are sequenced (e.g., phonological output buffer in Ellis & Young, 1988). This may be an especially salient aspect of phonemic encoding in aphasia, as connectionist models of word retrieval have shown that decay of information leads to phonemic paraphasias (e.g., Dell et al., 1997).

Many LSM studies of phonological production have focused on the neural correlates of phonemic paraphasias. In 106 people with aphasia, Schwartz and colleagues (2012) reported that phonemic paraphasias were associated with damage primarily to the postcentral gyrus (PostCG) (with extension into STG), the supramarginal gyrus (SMG), and the insula when controlling for total lesion volume. Contrary to their predictions, the authors found that only a few voxels implicated in phonological production errors were located in pSTG and the Sylvian-parietal-temporal (Spt) region. In a different sample of 102 participants with aphasia, Henseler and colleagues (2014) similarly found that phonemic paraphasias produced in connected speech were localized to SMG, whereas Halai and colleagues (2018) found that phonological errors during naming were associated with more anterior damage to the precentral gyrus (PreCG). Despite some differences in the specific implicated regions, these collective findings localize phonemic paraphasias to the dorsal stream.

6.2.2.2 Comprehension of Phonology

Individuals with aphasia can also exhibit deficits in comprehension that are influenced by phonology. For example, it may be difficult to choose the correct target (e.g., bear) from an array that includes a phonological distractor (e.g., pear), or to reject a nonword differing by one phoneme from a real word (e.g., gat). Notably, it is rare to have a pure deficit in phoneme perception, called pure word deafness or word sound deafness, which appears to occur with bilateral damage to the superior temporal gyrus (STG) (Binder, 2016). Rogalsky and colleagues (2008) argue that the auditory comprehension deficits observed in aphasia are most likely post-phonemic – that is, lexical-semantic – as left hemisphere damage, even in the acute phase, results in consistently better phonemic than semantic discrimination. One example of a purely psycholinguistic model of speech recognition that accounts for lexical involvement in phonemic comprehension is Shortlist B (Norris & McQueen, 2008), which proposes that listeners use Bayesian decision-making during speech comprehension. In other words, listeners select the most probable word (based on their experience) for a given phonemic input. Within the Dual Stream Model, proposed by Hickok and Poeppel (2007), phonological processing is proposed to take place in the posterior superior temporal sulcus bilaterally, while the process of mapping sound to meaning occurs in posterior middle and inferior temporal lobe, with a slight leftward bias. Thus, the left posterior temporal damage so often associated with auditory comprehension deficits in aphasia is thought to be related to this process of mapping sound to meaning.

In contrast to their findings for phonemic paraphasias, Schwartz and colleagues (2012) found that phonological recognition errors (as measured by an auditory discrimination task) were linked to damage to the entire STG, the posterior planum temporale and area Spt. Later studies by this group that either mapped parameters from the two-step model of lexical access (Dell et al., 2013) or included a more robust set of neuropsychological tests (Mirman, Chen, et al., 2015; Mirman, Zhang, et al., 2015) corroborated their prior findings, reporting an inferior–superior split between phonological production deficits (localized to SMG, PostCG, PreCG, insula, and premotor areas) and phonological recognition impairments (localized to anterior, medial, posterior STG, the posterior portion of the planum temporale and Spt). Several groups (Butler et al., 2014; Halai et al., 2017; Lacey et al., 2017; Schumacher et al., 2019; Zhao et al., 2020) have employed PCA data reduction on assessments requiring phonological recognition and production and have found that the resulting phonological factor is associated with a mix of the aforementioned regions. Schwartz and colleagues (2012) additionally determined that the conjunction of phonological production and comprehension deficits was localized to PreCG, PostCG, and SMG, suggesting that these regions in particular are important for both phonological input and output processes.

6.2.2.3 Repetition

Another task in which phonological processing plays an important role is repetition, especially repetition of nonwords, as the semantic system cannot support nonword repetition. However, repetition tasks should be paired with both auditory and verbal tasks to better understand the root of the repetition difficulty. In some individuals with aphasia, repetition is much more difficult than would be expected based on their comprehension and production abilities, though production may contain phonological errors. These individuals often have circumscribed damage to the left temporoparietal junction and underlying white matter. Within the Dual Stream Model (Hickok & Poeppel, 2007), area Spt (Sylvian-parietal-temporal region of the left hemisphere) is responsible for the sensorimotor integration of phonological information and has a vital role in mediating the storage and maintenance components of verbal working memory. Thus, the authors suggest that verbal working memory may be impacted in anyone with damage to area Spt. Historically, within Wernicke's original work (Wernicke, 1874) and Geschwind's classic model (Geschwind, 1970), repetition deficits have been linked to damage to the arcuate fasciculus (AF). Indeed, several studies employing DTI methods have found that repetition impairments in patients with post-stroke are related to poorer integrity of the AF (Breier et al., 2008, 2011; Geva et al., 2015; Kummerer et al., 2013; Yourganov et al., 2016). Repetition impairments have also been linked to damage to primary sensorimotor cortices, inferior parietal, and posterior temporal regions also implicated in phonological production deficits (Fridriksson et al., 2010; Kummerer et al., 2013).

6.2.2.4 Phonological Processing in Reading and Writing

Speech sounds are expressed in written form via orthography, though whether the written characters represent phonetic features, single phonemes, syllables, or whole words varies by language (Kemmerer, 2022). Thus, difficulty with phonological processing can also affect reading and writing. In the Dual Route Cascaded (DRC) model (Coltheart et al., 2001), there are two routes for reading: the lexical route, which is subdivided into the semantic route (most important for irregularly spelled words like yacht) and the nonsemantic route, and the grapheme-to-phoneme conversion (GPC) route (most important for nonwords). Of specific interest to phonological processing is the GPC route, as difficulty with phonological processing will inevitably affect grapheme-to-phoneme conversion (i.e., translating letters into sounds). Individuals with aphasia who have difficulty with the GPC route exhibit phonological alexia (Madden et al., 2022). Unsurprisingly, there is evidence that the neural correlates of phonological segmentation and oral reading abilities overlap. For example, in forty participants with chronic aphasia, Pillay and colleagues (2014) found that deficits in written rhyme judgment correlated with damage to a region localized to the junction of SMG and AG with extension into pSTG and the planum temporale. These findings did not change when the authors controlled for potential input phonological deficits (captured by an auditory word-to-picture matching task), but controlling for single-word oral reading deficits removed all significant voxels.

6.2.3 Syntax

Syntax is the arrangement of linguistic units (e.g., words) into meaningful structures (e.g., sentences) and is an essential element of sentence processing. Individuals with aphasia can have a variety of difficulties during sentence production and comprehension, and while some difficulties may stem from semantic or phonological difficulties, others appear to be rooted in syntax or other cognitive functions required for sentence processing (e.g., memory). Along this vein, many brain regions implicated in a variety of linguistic and cognitive processes have been linked to sentence processing in general and syntactic operations specifically. For example, in a recent meta-analysis of selected PET and fMRI studies, Walenski and colleagues (2019) found that general sentence processing elicited 784 foci from 996 participants in 60 studies within the left IFG (triangularis and opercularis), MFG, SFG, precentral gyrus, SMA, insula, temporal pole, MTG, posterior STG, AG, posterior fusiform and parahippocampal cortex, as well as right temporal pole, MTG, and posterior STG. Thus, the sentence processing network contains many areas affected by aphasia.

Generally, agrammatism is the term used to describe the syntactic quality of less fluent speech, but it has also been extended to comprehension. Qualities under the umbrella of agrammatism often include the omission or simplification of or difficulty comprehending verbs, function words, morphologically complex words, and syntactically complex sentences. Reliance on canonical word order (e.g., *The boy kicked the ball* as opposed to *The ball was kicked by the boy*) is also characteristic of agrammatism. Generally, paragrammatism is the term used to describe the syntactic quality of more fluent speech, which includes the substitution or blending of syntactic elements while maintaining syntactic complexity. Agrammatism has received a lot more attention and is therefore more well defined than paragrammatism.

Early psycholinguistic accounts proposed a deficit in syntactic knowledge underlying both sentence production and comprehension difficulties (e.g., Trace Deletion Hypothesis, Grodzinsky, 1995; mapping deficit, Schwartz et al., 1980) and relegated these global syntactic deficits to Broca's aphasia, localizing syntactic function to Broca's area. However, dissociations between production and comprehension (Berndt et al., 1996), as well as evidence of intact grammaticality judgment (Linebarger et al., 1983) suggest that there is not a unitary syntactic deficit underlying different sentence processing difficulties. Further, individuals with milder forms of aphasia are less successful at comprehending more complex sentence structures, indicating that a wide variety of aphasia types are affected by sentence comprehension difficulties (e.g., Love & Oster, 2002). Indeed, Caplan (2015) estimates that over 90 percent of individuals with aphasia resulting from left perisylvian damage have difficulties with syntactic comprehension. Thus, difficulties with sentence processing may involve a variety of issues, including difficulties with verb representations (Berndt et al., 1997), speed of processing (Kolk, 1995), verbal working memory (Caplan & Waters, 1999), and/or executive control (Novick et al., 2005).

Matchin and Hickok (2020) recently proposed a new model of syntactic processing that accounts for both production and comprehension difficulties in aphasia. This model assumes the modern lexicalist view[1] that small sections of hierarchical structure (e.g., a noun phrase) are stored and pieced together when needed. They argued that encoding and decoding syntax requires two broad processes – hierarchical structuring and morphosyntactic linearization – and that production and comprehension recruit these processes to different degrees, which results in differential recruitment of clusters of regions for production and comprehension. Specifically, production requires the combination of conceptual information into hierarchical structures, which are then realized as a sequence of morphemes, recruiting pMTG and IFGtri, respectively. Comprehension, on the other hand, requires the decoding of sound sequences into hierarchical structures that are then linked with entity/event knowledge, recruiting pSTG, pMTG, and ATL/AG, respectively. Thus, the authors link damage to IFGtri – as seen in agrammatism – specifically to sentence production difficulties, playing an important role in morphosyntactic sequencing and damage to pMTG – as seen in paragrammatism – to both production and comprehension difficulties, which rely on hierarchical lexical-syntactic processes.

6.2.3.1 Sentence Production

Specific to sentence production, the first psycholinguistic models were serial in nature, positing a top-down process starting with the message and ending with the phonological representation. For example, Garrett (1975) proposed four levels in sentence production, starting with a message level in which the to-be-conveyed thought is formulated. The next level is the functional level – which includes lexical selection and function assignment – where lexical forms are retrieved and assigned thematic roles. This level is where word exchanges – noted in both aphasic and nonaphasic speech – might take place (e.g., *The dog walked the man*). For individuals with aphasia, lexical selection difficulties would have an obvious impact on sentence production. Further, it has been argued that difficulties with verb retrieval may have an additional impact, as verbs are thought to carry additional syntactic information (e.g., regarding thematic role assignment; see Webster & Whitworth, 2012 for a review). In terms of function assignment, difficulties can lead to a mismatch between the lexical item and its role in the action. For example, when asked to provide a sentence starting with "The dog" for an image of a cat biting a dog, a person with aphasia may say "The dog is biting the cat." This task tends to be more difficult when the roles are reversible (i.e., either noun could be the doing or receiving the action). When the roles are not reversible, people with aphasia tend to assign the agent role to the animate noun (e.g., Saffran et al., 1980), and are influenced by plausibility (e.g., Ostrin & Schwartz, 1986). While practical, this may limit conversation to likely events, undermining the human capacity to express impossible or hypothetical events.

Next, at the positional level – which includes constituent assembly and inflection – word order is arranged, function words are inserted, and morphological endings are applied. This level is where stranding errors – noted in both aphasic and nonaphasic speech – might be made (e.g., *The man dogged the walk*). For individuals with aphasia, difficulties at this level may include the omission/substitution of both bound and free morphemes. For example, function words (e.g., the, for) tend to be omitted in agrammatism. Inflectional morphemes, like the suffix -ing, may be added or omitted from a verb in error. This process includes both morphosyntax and morphophonology.[2] In terms of constituent assembly, production of sentences is often simplified to frequent canonical structures, such as active sentences with no embeddings (Caplan & Hanna, 1998).

More recently, models of parallel processing have been developed for sentence production. For example, Chang and colleagues (2006), like Garrett, proposed that there are two systems used for grammatical encoding: a meaning system (akin to the functional level) and a sequencing system (akin to the positional level). However, unlike Garrett, Chang and colleagues proposed a dual-path model in which these systems are separate, but work in parallel and interact. This connectionist model, which uses error-based learning, is an important contribution because it accounts for both language acquisition (e.g., elicited production of a transitive sentence with a novel verb like *glorp*) and language use (e.g., syntactic priming).

6.2.3.2 Sentence Comprehension

For sentences to be comprehended, the incoming words must be recognized and their relationships with one another analyzed. Friederici (2011) described the basic process in three phases. First, word category information helps form the local phrase structure, for example, transitive verbs require two arguments. Then thematic role assignment takes place. The third phase is an optional phase for more difficult sentences in which context and world knowledge assist with interpretation. These phases interact with linguistic prosody, which supports interpretation.

Recent research highlights the importance of the interaction of syntax and semantics and contributions of memory processes for sentence comprehension. Although syntactic and semantic processes can be dissociated in certain patient populations (e.g., syntax is often preserved in semantic dementia; Gorno-Tempini et al., 2011), individuals with stroke-based aphasia have long shown semantic influences in sentence comprehension. For example, when given implausible sentences (e.g., *The ball kicked the boy*), individuals with aphasia are more likely than those without aphasia to interpret the sentence according to the more plausible interpretation, rather than the literal interpretation (e.g., Caramazza & Zurif, 1976). While a global syntactic deficit could explain this behavior, Gibson and colleagues (2016) have proposed that it is more likely (given that this behavior can even occur in highly frequent, canonical, written sentences with low working memory constraints) that individuals with aphasia rely on a Bayesian approach to sentence comprehension, in which they rely heavily on construction frequency and semantic plausibility because there is a high chance that the speaker's intended meaning of a sentence has been compromised by the listener's language difficulties. The authors have coined this the noisy-channel or rational inference approach.

While the notion that memory contributes to sentence comprehension is not new (e.g., Kolk & Van Grunsven, 1985), the exact nature of its contribution is still not fully understood. Early accounts proposed that sentence comprehension difficulties in aphasia were linked with verbal short-term memory (STM) (e.g., Saffran & Martin, 1990) or working memory deficits (e.g., Miyake et al., 1994), with Martin and Romani (1994) arguing for specialized components of working memory dedicated to phonology, semantics, and syntax. In a recent meta-analysis of research studies in aphasia since 1980, Varkanitsa and Caplan (2018) found a clear relationship between performance on verbal STM/WM tasks and sentence comprehension. However, the authors did not distinguish between STM and WM (and proposed that composite scores provide more reliable results), and pointed out their results could not disambiguate between online (e.g., self-paced reading) and post-interpretive effects (e.g., sentence-to-picture matching) of memory, the latter of which was the only effect found by Caplan and colleagues (2013) in a study of sixty-one people with aphasia.

A slightly different take on the role of memory in sentence comprehension is the declarative/procedural (DP) theory (Ullman, 2013), which proposes that language is subserved by the declarative and procedural memory systems in much the same way as other, nonlinguistic cognitive functions. For language, the declarative memory system supports the acquisition and retention of the knowledge of phonology, semantics, and syntax, while the procedural memory system supports the implicit learning, knowledge, and use of rule-governed aspects of phonology and syntax. Specific to aphasia, Ullman (2013) suggests that nonfluent aphasia is related to damage to the procedural memory system (e.g., IFG) while fluent aphasia is related to damage to the declarative memory system (e.g., temporal regions).

One debate in the literature has been whether the memory required for sentence processing was domain general or domain specific. In a recent article examining working memory during naturalistic language comprehension, Shain and colleagues (2021) found that when individuals passively listened to a story, working memory load predictors based on the dependency locality theory (Gibson, 2000; e.g., storage and integration costs) explained activity in the language network (defined in Fedorenko et al., 2011), but not in the multiple demand (i.e., domain general, defined in Duncan, 2010) network. The authors argue that their results support the view that the working memory required for sentence processing is language-domain specific. This study is important for two reasons: First, it uses naturalistic language, which has more ecological validity. Second, if language-domain-specific working memory is required for sentence comprehension and is carried out within the language network, then sentence comprehension difficulties in aphasia can be plausibly linked with disruption to this domain-specific working memory.

6.2.3.3 Neural Correlates of Sentence Production and Comprehension

While semantic and phonological processing generally map onto respective ventral and dorsal streams in dual-route models (Friederici & Gierhan, 2013; Hickok & Poeppel, 2007), ventral and dorsal regions have both been implicated in sentence production and comprehension. For example, in their meta-analysis, Walenski and colleagues (2019) found activation specific to sentence production in left MFG and SMA, bilateral SFG, left temporo-occipital MTG, and bilateral lateral occipital cortex. Activation specific to sentence comprehension was found in the left frontal cortex, including IFG, precentral gyrus, and SMA, left temporal and parietal cortex, including MTG, posterior STG, AG, and fusiform, and right temporal cortex, including temporal pole, MTG, and posterior STG. In addition to the inclusion of both ventral and dorsal stream regions (particularly for comprehension), the authors also noted high overlap between auditory and visual modalities.

In a series of papers using connectome LSM, Fridriksson and colleagues (e.g., den Ouden et al., 2019; Fridriksson et al., 2018; Gleichgerrcht et al., 2021) found that damage to white matter connections within and between the streams underlay both sentence production and comprehension deficits. Moreover, LSM findings have also yielded counterevidence to classic models attributing syntactic processes to Broca's area. For example, Magnusdottir and colleagues (2013) found that noncanonical versus canonical sentence deficits (which reflects agrammatism specifically) were linked to damage to the left ATL rather than IFG. Thothathiri and colleagues (2012) reported that deficits in reversible sentence comprehension were associated with damage to left posterior temporoparietal cortex but not IFG, and other researchers similarly found posterior temporal rather than inferior frontal regions for both sentence production (Henseler et al., 2014) and comprehension (Lwi et al., 2021) impairments. Nonetheless, damage to and disconnection of IFG has been reported in LSM studies of post-stroke aphasia, particularly for deficits of processing syntactically complex sentences (e.g., noncanonical versus canonical) (den Ouden et al., 2019; Gleichgerrcht et al., 2021; Sheppard et al., 2022). This collective evidence speaks to the distributed and complex nature of sentence and syntactic processing that is captured in modern models of sentence processing (e.g., Matchin & Hickok, 2020).

6.3 Caveats and Summary

This chapter summarizes recent advances in the fields of neurolinguistics and psycholinguistics through the lens of aphasia. As such, this chapter does not include recent psycholinguistic and neurolinguistic evidence from the perspective of language acquisition or developmental language disorders. Although post-stroke aphasia has been used as a disease model for investigating the neurobiology of language since Broca and Wernicke first described their classic cases, there are some caveats to the evidence presented in this chapter.

First, this chapter heavily relies on lesion symptom mapping evidence in patients with chronic aphasia. By the chronic post-stroke stage (i.e., $\geq$ six months after stroke), some people with aphasia have made substantial recovery due to functional reorganization of higher-order brain systems. As such, there may be a mismatch in the neural correlates of those processes in neurologically healthy people and the brain regions implicated in persistent deficits in chronic aphasia. LSM methods are also limited by the lesion distributions of the included participants. Based on a recent LSM methods paper (Lorca-Puls et al., 2018), many LSM papers have smaller than desired samples. Another caveat when interpreting lesion findings pertains to the LSM method employed. Region of interest (ROI)-based LSM interrogates the brain at a lower level of granularity than voxel-based methods, and therefore, provides information about brain-behavior relationships with less spatial specificity. Early voxel-based LSM studies used univariate analyses that did not account for the fact that neighboring voxels are likely damaged together and thus should be jointly considered in statistical analyses.

Another important consideration is the variety of measures researchers have used to examine semantics, phonology, and syntax. Use of extensive neuropsychological batteries means linguistic constructs can be assessed more thoroughly and characterization of a deficit in a given area will be less likely to just reflect idiosyncrasies of a single assessment. However, this approach results in a plethora of measures that make analysis unwieldy. Consequently, several LSM studies incorporated principal component analysis (PCA) for data reduction of linguistic measures. One key tradeoff of PCA is that sometimes variables are grouped that have opposing theoretical underpinnings, resulting in lost construct specificity that in turn, has implications for brain-behavior relationships.

Despite these limitations, this chapter illustrates the many developments pertaining to neurolinguistics and psycholinguistics that have transpired in the last twenty years. Knowledge within these fields will continue to grow as neuroimaging and computational modeling continue to evolve in parallel and are applied to understand the complex systems that underlie language behavior and neural processes.

Notes

1 Generally, lexicalist approaches assume that lexical items contain both semantic and syntactic information that guide the argument structure of a sentence while constructivist approaches assume that lexical items contain no syntactic information and that argument structure is guided by syntax (Lohndal, 2014).

2 Morphosyntax is the inflection of a word according to syntactic rules (e.g., the addition of -ed makes a verb past-tense) while morphophonology is the realization of the sound patterns associated with morphological inflection (e.g., while regular verbs like "fix" take the past-tense ending -ed, irregular verbs like "eat" change form) (Kemmerer, 2022).

References

Alyahya, R. S. W., Halai, A. D., Conroy, P., & Lambon Ralph, M. A. (2018). Noun and verb processing in aphasia: Behavioural profiles and neural correlates. *NeuroImage: Clinical, 18*, 215–230. https://doi.org/10.1016/j.nicl.2018.01.023

Alyahya, R. S. W., Halai, A. D., Conroy, P., & Lambon Ralph, M. A. (2020). A unified model of post-stroke language deficits including discourse production and their neural correlates. *Brain, 143*(5), 1541–1554. https://doi.org/10.1093/brain/awaa074

Barsalou, L. W. (2008). Grounded cognition. *Annual Review of Psychology, 59*(1), 617–645. https://doi.org/10.1146/annurev.psych.59.103006.093639

Berndt, R. S., Haendiges, A. N., Mitchum, C. C., & Sandson, J. (1997). Verb retrieval in aphasia. 2. Relationship to sentence processing. *Brain and Language, 56*(1), 107–137. https://doi.org/10.1006/brln.1997.1728

Berndt, R. S., Mitchum, C. C., & Haendiges, A. N. (1996). Comprehension of reversible sentences in "agrammatism": A meta-analysis. *Cognition, 58*(3), 289–308. https://doi.org/10.1016/0010-0277(95)00682-6

Binder, J. R. (2016). Phoneme perception. In *Neurobiology of Language* (pp. 447–461). Elsevier. https://doi.org/10.1016/B978-0-12-407794-2.00037-7

Binder, J. R., & Desai, R. H. (2011). The neurobiology of semantic memory. *Trends in Cognitive Sciences, 15*(11), 527–536. https://doi.org/10.1016/j.tics.2011.10.001

Binder, J. R., Desai, R. H., Graves, W. W., & Conant, L. L. (2009). Where is the semantic system? A critical review and meta-analysis of 120 functional neuroimaging studies. *Cerebral Cortex, 19*(12), 2767–2796. https://doi.org/10.1093/cercor/bhp055

Bird, H., Howard, D., & Franklin, S. (2003). Verbs and nouns: The importance of being imageable. *Journal of Neurolinguistics, 16*(2–3), 113–149. https://doi.org/10.1016/S0911-6044(02)00016-7

Blumstein, S. E. (2016). Psycholinguistic approaches to the study of syndromes and symptoms of aphasia. In *Neurobiology of Language* (pp. 923–933). Elsevier. https://doi.org/10.1016/B978-0-12-407794-2.00074-2

Bolhuis, J. J., Tattersall, I., Chomsky, N., & Berwick, R. C. (2014). How could language have evolved?*PLoS Biology, 12*(8), e1001934. https://doi.org/10.1371/journal.pbio.1001934

Borghi, A. M., Barca, L., Binkofski, F., Castelfranchi, C., Pezzulo, G., & Tummolini, L. (2019). Words as social tools: Language, sociality and inner grounding in abstract concepts. *Physics of Life Reviews, 29*, 120–153. https://doi.org/10.1016/j.plrev.2018.12.001

Breier, J. I., Hasan, K. M., Zhang, W., Men, D., & Papanicolaou, A. C. (2008). Language dysfunction after stroke and damage to white matter tracts evaluated using diffusion tensor imaging. *American Journal of Neuroradiology, 29*(3), 483–487. https://doi.org/10.3174/ajnr.A0846

Breier, J. I., Juranek, J., & Papanicolaou, A. C. (2011). Changes in maps of language function and the integrity of the arcuate fasciculus after therapy for chronic aphasia. *Neurocase, 17*(6), 506–517. https://doi.org/10.1080/13554794.2010.547505

Broca, P. (1861). Nouvelle observation d'aphémie produite par une lésion de la moitié postérieure des deuxième et troisième circonvolution frontales gauches. *Bull Soc Anat Paris, 36*, 398–407.

Bucur, M., & Papagno, C. (2021). An ALE meta-analytical review of the neural correlates of abstract and concrete words. *Scientific Reports,* 11(1), 15727. https://doi.org/10.1038/s41598-021-94506-9

Butler, R. A., Lambon Ralph, M. A., & Woollams, A. M. (2014). Capturing multidimensionality in stroke aphasia: Mapping principal behavioural components to neural structures. *Brain, 137*(12), 3248–3266. https://doi.org/10.1093/brain/awu286

Caplan, D. (1993). Toward a psycholinguistic approach to acquired neurogenic language disorders. *American Journal of Speech-Language Pathology, 2*(1), 59–83. https://doi.org/10.1044/1058-0360.0201.59

Caplan, D. (2015). The neural basis of syntatic processing. In A. E. Hillis (Ed.), *The Handbook of Adult Language Disorders* (2nd ed., pp. 355–374). Psychology Press.

Caplan, D., & Hanna, J. E. (1998). Sentence production by aphasic patients in a constrained task. *Brain and Language, 63*(2), 184–218. https://doi.org/10.1006/brln.1998.1930

Caplan, D., Michaud, J., & Hufford, R. (2013). Short-term memory, working memory, and syntactic comprehension in aphasia. *Cognitive Neuropsychology, 30*(2), 77–109. https://doi.org/10.1080/02643294.2013.803958

Caplan, D., & Waters, G. S. (1999). Verbal working memory and sentence comprehension. *Behavioral and Brain Sciences, 22*(01). https://doi.org/10.1017/S0140525X99001788

Caramazza, A., & Shelton, J. R. (1998). Domain-specific knowledge systems in the brain: The animate-inanimate distinction. *Journal of Cognitive Neuroscience, 10*(1), 1–34. https://doi.org/10.1162/089892998563752

Caramazza, A., & Zurif, E. B. (1976). Dissociation of algorithmic and heuristic processes in language comprehension: Evidence from aphasia. *Brain and Language, 3*(4), 572–582. https://doi.org/10.1016/0093-934X(76)90048-1

Catani, M., & ffytche, D. H. (2005). The rises and falls of disconnection syndromes. *Brain: A Journal of Neurology, 128*(Pt 10), 2224–2239. https://doi.org/10.1093/brain/awh622

Chang, F., Dell, G. S., & Bock, K. (2006). Becoming syntactic. *Psychological Review, 113*(2), 234–272. https://doi.org/10.1037/0033-295X.113.2.234

Chapman, C. A., Hasan, O., Schulz, P. E., & Martin, R. C. (2020). Evaluating the distinction between semantic knowledge and semantic access: Evidence from semantic dementia and comprehension-impaired stroke aphasia. *Psychonomic Bulletin & Review, 27*(4), 607–639. https://doi.org/10.3758/s13423-019-01706-6

Charidimou, A., Kasselimis, D., Varkanitsa, M., Selai, C., Potagas, C., & Evdokimidis, I. (2014). Why is it difficult to predict language impairment and outcome in patients with aphasia after stroke? *Journal of Clinical Neurology, 10*(2), 75. https://doi.org/10.3988/jcn.2014.10.2.75

Cloutman, L., Gottesman, R., Chaudhry, P., Davis, C., Kleinman, J. T., Pawlak, M., Herskovits, E. H., Kannan, V., Lee, A., Newhart, M., Heidler-Gary, J., & Hillis, A. E. (2009). Where (in the brain) do semantic errors come from?*Cortex, 45*(5), 641–649. https://doi.org/10.1016/j.cortex.2008.05.013

Collins, A. M., & Loftus, E. F. (1975). A spreading-activation theory of semantic processing. *Psychological Review, 82*(6), 407–428. https://doi.org/10.1037/0033-295X.82.6.407

Coltheart, M., Rastle, K., Perry, C., Langdon, R., & Ziegler, J. (2001). DRC: A dual route cascaded model of visual word recognition and reading aloud. *Psychological Review, 108*(1), 204–256. https://doi.org/10.1037/0033-295X.108.1.204

Conca, F., Borsa, V. M., Cappa, S. F., & Catricalà, E. (2021). The multidimensionality of abstract concepts: A systematic review. *Neuroscience & Biobehavioral Reviews, 127*, 474–491. https://doi.org/10.1016/j.neubiorev.2021.05.004

DeLeon, J., Gottesman, R. F., Kleinman, J. T., Newhart, M., Davis, C., Heidler-Gary, J., Lee, A., & Hillis, A. E. (2007). Neural regions essential for distinct cognitive processes underlying picture naming. *Brain, 130*(5), 1408–1422. https://doi.org/10.1093/brain/awm011

Dell, G. S., & O'Seaghdha, P. G. (1992). Stages of lexical access in language production. *Cognition, 42*(1–3), 287–314. https://doi.org/10.1016/0010-0277(92)90046-K

Dell, G. S., Schwartz, M. F., Martin, N., Saffran, E. M., & Gagnon, D. A. (1997). Lexical access in aphasic and nonaphasic speakers. *Psychological Review, 104*(4), 801–838.

Dell, G. S., Schwartz, M. F., Nozari, N., Faseyitan, O., & Branch Coslett, H. (2013). Voxel-based lesion-parameter mapping: Identifying the neural correlates of a computational model of word production. *Cognition, 128*(3), 380–396. https://doi.org/10.1016/j.cognition.2013.05.007

den Ouden, D. B., Malyutina, S., Basilakos, A., Bonilha, L., Gleichgerrcht, E., Yourganov, G., Hillis, A. E., Hickok, G., Rorden, C., & Fridriksson, J. (2019). Cortical and structural-connectivity damage correlated with impaired syntactic processing in aphasia. *Human Brain Mapping, 40*(7), 2153–2173. https://doi.org/10.1002/hbm.24514

Desai, R. H., Reilly, M., & van Dam, W. (2018). The multifaceted abstract brain. *Philosophical Transactions of the Royal Society B: Biological Sciences, 373*(1752), 20170122. https://doi.org/10.1098/rstb.2017.0122

Devinsky, O. (2009). Norman Geschwind: Influence on his career and comments on his course on the neurology of behavior. *Epilepsy & Behavior, 15*(4), 413–416. https://doi.org/10.1016/j.yebeh.2009.04.029

Diveica, V., Koldewyn, K., & Binney, R. J. (2021). Establishing a role of the semantic control network in social cognitive processing: A meta-analysis of functional neuroimaging studies. *NeuroImage, 245*, 118702. https://doi.org/10.1016/j.neuroimage.2021.118702

Dronkers, N. F., Plaisant, O., Iba-Zizen, M. T., & Cabanis, E. A. (2007). Paul Broca's historic cases: High resolution MR imaging of the brains of Leborgne and Lelong. *Brain, 130*(5), 1432–1441. https://doi.org/10.1093/brain/awm042

Duncan, J. (2010). The multiple-demand (MD) system of the primate brain: Mental programs for intelligent behaviour. *Trends in Cognitive Sciences, 14*(4), 172–179. https://doi.org/10.1016/j.tics.2010.01.004

Ellis, A. W., & Young, A. W. (1988). *Human Cognitive Neuropsychology*. L. Erlbaum Associates, Publishers.

Fedorenko, E., Behr, M. K., & Kanwisher, N. (2011). Functional specificity for high-level linguistic processing in the human brain. *Proceedings of the National Academy of Sciences, 108*(39), 16428–16433. https://doi.org/10.1073/pnas.1112937108

Fedorenko, E., & Thompson-Schill, S. L. (2014). Reworking the language network. *Trends in Cognitive Sciences, 18*(3), 120–126. https://doi.org/10.1016/j.tics.2013.12.006

Fodor, J. A. (1975). *The Language of Thought* (Digit. repr). Harvard University Press.

Fridriksson, J., Bonilha, L., & Rorden, C. (2007). Severe Broca's aphasia without Broca's area damage. *Behavioural Neurology, 18*(4), 237–238. https://doi.org/10.1155/2007/785280

Fridriksson, J., den Ouden, D.-B., Hillis, A. E., Hickok, G., Rorden, C., Basilakos, A., Yourganov, G., & Bonilha, L. (2018). Anatomy of aphasia revisited. *Brain, 141*(3), 848–862. https://doi.org/10.1093/brain/awx363

Fridriksson, J., Kjartansson, O., Morgan, P. S., Hjaltason, H., Magnusdottir, S., Bonilha, L., & Rorden, C. (2010). Impaired speech repetition and left parietal lobe damage. *The Journal of Neuroscience, 30*(33), 11057–11061. https://doi.org/10.1523/JNEUROSCI.1120-10.2010

Friederici, A. D. (2011). The brain basis of language processing: From structure to function. *Physiological Reviews, 91*(4), 1357–1392. https://doi.org/10.1152/physrev.00006.2011

Friederici, A. D., Chomsky, N., Berwick, R. C., Moro, A., & Bolhuis, J. J. (2017). Language, mind and brain. *Nature Human Behaviour, 1*(10), 713–722. https://doi.org/10.1038/s41562-017-0184-4

Friederici, A. D., & Gierhan, S. M. (2013). The language network. *Current Opinion in Neurobiology, 23*(2), 250–254. https://doi.org/10.1016/j.conb.2012.10.002

Gainotti, G. (2010). The influence of anatomical locus of lesion and of gender-related familiarity factors in category-specific semantic disorders for animals, fruits and vegetables: A review of single-case studies. *Cortex, 46*(9), 1072–1087. https://doi.org/10.1016/j.cortex.2010.04.002

Garrett, M. F. (1975). The analysis of sentence production. In *Psychology of Learning and Motivation* (Vol. 9, pp. 133–177). Elsevier. https://doi.org/10.1016/S0079-7421(08)60270-4

Geschwind, N. (1965a). Disconnexion syndromes in animals and man. Part I. *Brain: A Journal of Neurology, 88*(2), 237–294. https://doi.org/10.1093/brain/88.2.237

Geschwind, N. (1965b). Disconnexion syndromes in animals and man. Part II. *Brain, 88*(3), 585–585. https://doi.org/10.1093/brain/88.3.585

Geschwind, N. (1970). The organization of language and the brain. *Science (New York, N.Y.), 170*(3961), 940–944. https://doi.org/10.1126/science.170.3961.940

Geva, S., Correia, M. M., & Warburton, E. A. (2015). Contributions of bilateral white matter to chronic aphasia symptoms as assessed by diffusion tensor MRI. *Brain and Language, 150,* 117–128. https://doi.org/10.1016/j.bandl.2015.09.001

Gibson, E. (2000). The dependency locality theory: A distance-based theory of linguistic complexity. *Image, Language, Brain: Papers from the First Mind Articulation Project Symposium.,* 94–126.

Gibson, E., Sandberg, C., Fedorenko, E., Bergen, L., & Kiran, S. (2016). A rational inference approach to aphasic language comprehension. *Aphasiology, 30*(11), 1341–1360. https://doi.org/10.1080/02687038.2015.1111994

Gleichgerrcht, E., Roth, R., Fridriksson, J., den Ouden, D., Delgaizo, J., Stark, B., Hickok, G., Rorden, C., Wilmskoetter, J., Hillis, A., & Bonilha, L. (2021). Neural bases of elements of syntax during speech production in patients with aphasia. *Brain and Language, 222,* 105025. https://doi.org/10.1016/j.bandl.2021.105025

Gorno-Tempini, M. L., Hillis, A. E., Weintraub, S., Kertesz, A., Mendez, M., Cappa, S. F., Ogar, J. M., Rohrer, J. D., Black, S., Boeve, B. F., Manes, F., Dronkers, N. F., Vandenberghe, R., Rascovsky, K., Patterson, K., Miller, B. L., Knopman, D. S., Hodges, J. R., Mesulam, M. M., & Grossman, M. (2011). Classification of primary progressive aphasia and its variants. *Neurology, 76*(11), 1006–1014. https://doi.org/10.1212/WNL.0b013e31821103e6

Grodzinsky, Y. (1995). A restrictive theory of agrammatic comprehension. *Brain and Language, 50*(1), 27–51. https://doi.org/10.1006/brln.1995.1039

Hagoort, P. (2013). MUC (memory, unification, control) and beyond. *Frontiers in Psychology, 4.* https://doi.org/10.3389/fpsyg.2013.00416

Halai, A. D., Woollams, A. M., & Lambon Ralph, M. A. (2017). Using principal component analysis to capture individual differences within a unified neuropsychological model of chronic post-stroke aphasia: Revealing the unique neural correlates of speech fluency, phonology and semantics. *Cortex, 86,* 275–289. https://doi.org/10.1016/j.cortex.2016.04.016

Halai, A. D., Woollams, A. M., & Lambon Ralph, M. A. (2018). Predicting the pattern and severity of chronic post-stroke language deficits from functionally-partitioned structural lesions. *NeuroImage: Clinical, 19,* 1–13. https://doi.org/10.1016/j.nicl.2018.03.011

Han, Z., Ma, Y., Gong, G., He, Y., Caramazza, A., & Bi, Y. (2013). White matter structural connectivity underlying semantic processing: Evidence from brain damaged patients. *Brain, 136*(10), 2952–2965. https://doi.org/10.1093/brain/awt205

Henseler, I., Regenbrecht, F., & Obrig, H. (2014). Lesion correlates of patholinguistic profiles in chronic aphasia: Comparisons of syndrome-, modality- and symptom-level assessment. *Brain, 137*(3), 918–930. https://doi.org/10.1093/brain/awt374

Hickok, G., & Poeppel, D. (2004). Dorsal and ventral streams: A framework for understanding aspects of the functional anatomy of language. *Cognition, 92*(1–2), 67–99. https://doi.org/10.1016/j.cognition.2003.10.011

Hickok, G., & Poeppel, D. (2007). The cortical organization of speech processing. *Nature Reviews Neuroscience, 8*(5), 393–402. https://doi.org/10.1038/nrn2113

Hillis, A. E., & Caramazza, A. (1995). Representation of grammatical categories of words in the brain. *Journal of Cognitive Neuroscience, 7*(3), 396–407. https://doi.org/10.1162/jocn.1995.7.3.396

Hillis, A. E., & Heidler, J. (2002). Mechanisms of early aphasia recovery. *Aphasiology, 16*(9), 885–895. https://doi.org/10.1080/0268703

Hillis, A. E., Kane, A., Tuffiash, E., Ulatowski, J. A., Barker, P. B., Beauchamp, N. J., & Wityk, R. J. (2001). Reperfusion of specific brain regions by raising blood pressure restores selective language functions in subacute stroke. *Brain and Language, 79*(3), 495–510. https://doi.org/10.1006/brln.2001.2563

Hillis, A. E., Kleinman, J. T., Newhart, M., Heidler-Gary, J., Gottesman, R., Barker, P. B., Aldrich, E., Llinas, R., Wityk, R., & Chaudhry, P. (2006). Restoring cerebral blood flow reveals neural regions critical for naming. *Journal of Neuroscience, 26*(31), 8069–8073. https://doi.org/10.1523/JNEUROSCI.2088-06.2006

Hillis, A. E., Wityk, R. J., Tuffiash, E., Beauchamp, N. J., Jacobs, M. A., Barker, P. B., & Selnes, O. A. (2001). Hypoperfusion of Wernicke's area predicts severity of semantic deficit in acute stroke. *Annals of Neurology, 50*(5), 561–566. https://doi.org/10.1002/ana.1265

Hodgson, V. J., Lambon Ralph, M. A., & Jackson, R. L. (2021). Multiple dimensions underlying the functional organization of the language network. *NeuroImage, 241*, 118444. https://doi.org/10.1016/j.neuroimage.2021.118444

Hoffman, P., Binney, R. J., & Lambon Ralph, M. A. (2015). Differing contributions of inferior prefrontal and anterior temporal cortex to concrete and abstract conceptual knowledge. *Cortex, 63*, 250–266. https://doi.org/10.1016/j.cortex.2014.09.001

Hoffman, P., & Lambon Ralph, M. A. (2011). Reverse concreteness effects are not a typical feature of semantic dementia: Evidence for the hub-and-spoke model of conceptual representation. *Cerebral Cortex, 21*(9), 2103–2112. https://doi.org/10.1093/cercor/bhq288

Indefrey, P., & Levelt, W. J. M. (2004). The spatial and temporal signatures of word production components. *Cognition, 92*(1–2), 101–144. https://doi.org/10.1016/j.cognition.2002.06.001

Ivanova, M. V., Isaev, D. Yu., Dragoy, O. V., Akinina, Y. S., Petrushevskiy, A. G., Fedina, O. N., Shklovsky, V. M., & Dronkers, N. F. (2016). Diffusion-tensor imaging of major white matter tracts and their role in language processing in aphasia. *Cortex, 85*, 165–181. https://doi.org/10.1016/j.cortex.2016.04.019

Jefferies, E., & Lambon Ralph, M. A. (2006). Semantic impairment in stroke aphasia versus semantic dementia: A case-series comparison. *Brain, 129*(8), 2132–2147. https://doi.org/10.1093/brain/awl153

Kasselimis, D. S., Simos, P. G., Peppas, C., Evdokimidis, I., & Potagas, C. (2017). The unbridged gap between clinical diagnosis and contemporary research on aphasia: A short discussion on the validity and clinical utility of taxonomic categories. *Brain and Language, 164*, 63–67. https://doi.org/10.1016/j.bandl.2016.10.005

Kemmerer, D. (2014). Word classes in the brain: Implications of linguistic typology for cognitive neuroscience. *Cortex, 58*, 27–51. https://doi.org/10.1016/j.cortex.2014.05.004

Kemmerer, D. (2022). *Cognitive Neuroscience of Language* (2nd ed.). Routledge. https://doi.org/10.4324/9781138318427

Kolk, H. (1995). A time-based approach to agrammatic production. *Brain and Language, 50*(3), 282–303. https://doi.org/10.1006/brln.1995.1049

Kolk, H. H. J., & Van Grunsven, M. M. F. (1985). Agrammatism as a variable phenomenon. *Cognitive Neuropsychology, 2*(4), 347–384. https://doi.org/10.1080/02643298508252666

Kummerer, D., Hartwigsen, G., Kellmeyer, P., Glauche, V., Mader, I., Klöppel, S., Suchan, J., Karnath, H.-O., Weiller, C., & Saur, D. (2013). Damage to ventral and dorsal language pathways in acute aphasia. *Brain, 136*(2), 619–629. https://doi.org/10.1093/brain/aws354

Lacey, E. H., Skipper-Kallal, L. M., Xing, S., Fama, M. E., & Turkeltaub, P. E. (2017). Mapping common aphasia assessments to underlying cognitive processes and their neural substrates. *Neurorehabilitation and Neural Repair, 31*(5), 442–450. https://doi.org/10.1177/1545968316688797

Lambon Ralph, M. A., Jefferies, E., Patterson, K., & Rogers, T. T. (2017). The neural and computational bases of semantic cognition. *Nature Reviews Neuroscience, 18*(1), 42–55. https://doi.org/10.1038/nrn.2016.150

Landrigan, J.-F., Zhang, F., & Mirman, D. (2021). A data-driven approach to post-stroke aphasia classification and lesion-based prediction. *Brain*, awab010. https://doi.org/10.1093/brain/awab010

Levelt, W. (2012). *A History of Psycholinguistics: The Pre-Chomskyan Era*. Oxford University Press. https://doi.org/10.1093/acprof:oso/9780199653669.001.0001

Levelt, W. J. M. (1989). *Speaking: From Intention to Articulation* (pp. xiv, 566). The MIT Press.

Lichtheim, L. (1885). On aphasia. *Brain, 7*, 433–484. https://doi.org/10.7551/mitpress/6393.001.0001

Linebarger, M. C., Schwartz, M. F., & Saffran, E. M. (1983). Sensitivity to grammatical structure in so-called agrammatic aphasics. *Cognition, 13*(3), 361–392. https://doi.org/10.1016/0010-0277(83)90015-X

Lohndal, T. (2014). Introduction. In T. Lohndal, *Phrase Structure and Argument Structure* (1st ed., pp. 1–21). Oxford University Press. https://doi.org/10.1093/acprof:oso/9780199677115.003.0001

Lorca-Puls, D. L., Gajardo-Vidal, A., White, J., Seghier, M. L., Leff, A. P., Green, D. W., Crinion, J. T., Ludersdorfer, P., Hope, T. M. H., Bowman, H., & Price, C. J. (2018). The impact of sample size on the reproducibility of voxel-based lesion-deficit mappings. *Neuropsychologia, 115*, 101–111. https://doi.org/10.1016/j.neuropsychologia.2018.03.014

Love, T., & Oster, E. (2002). On the categorization of aphasic typologies: The SOAP (a test of syntactic complexity). *Journal of Psycholinguistic Research, 31*(5), 503–529. https://doi.org/10.1023/A:1021208903394

Lwi, S. J., Herron, T. J., Curran, B. C., Ivanova, M. V., Schendel, K., Dronkers, N. F., & Baldo, J. V. (2021). Auditory comprehension deficits in post-stroke aphasia: Neurologic and demographic correlates of outcome and recovery. *Frontiers in Neurology, 12,* 680248. https://doi.org/10.3389/fneur.2021.680248

Madden, E., Kendall, D., & Riley, E. (2022). Acquired disorders of reading: Modeling, assessment, and treatment. In I. Papathanasiou & P. Coppens (Eds.), *Aphasia and Related Neurogenic Communication Disorders* (3rd ed.). Jones & Bartlett Learning.

Magnusdottir, S., Fillmore, P., den Ouden, D. B., Hjaltason, H., Rorden, C., Kjartansson, O., Bonilha, L., & Fridriksson, J. (2013). Damage to left anterior temporal cortex predicts impairment of complex syntactic processing: A lesion-symptom mapping study. *Human Brain Mapping, 34*(10), 2715–2723. https://doi.org/10.1002/hbm.22096

Mahon, B. Z., & Caramazza, A. (2009). Concepts and categories: A cognitive neuropsychological perspective. *Annual Review of Psychology, 60*(1), 27–51. https://doi.org/10.1146/annurev.psych.60.110707.163532

Mahon, B. Z., & Caramazza, A. (2011). What drives the organization of object knowledge in the brain?*Trends in Cognitive Sciences, 15*(3), 97–103. https://doi.org/10.1016/j.tics.2011.01.004

Martin, N. (2022). Disorders of word production. In I. Papathanasiou & P. Coppens (Eds.), *Aphasia and Related Neurogenic Communication Disorders* (3rd ed.). Jones & Bartlett Learning.

Martin, R. C., & Romani, C. (1994). Verbal working memory and sentence comprehension: A multiple-components view. *Neuropsychology, 8*(4), 506.

Matchin, W., & Hickok, G. (2020). The cortical organization of syntax. *Cerebral Cortex, 30*(3), 1481–1498. https://doi.org/10.1093/cercor/bhz180

McNeil, M. R., & Pratt, S. R. (2001). Defining aphasia: Some theoretical and clinical implications of operating from a formal definition. *Aphasiology, 15*(10–11), 901–911. https://doi.org/10.1080/02687040143000276

Meier, E. L., Sheppard, S. M., Goldberg, E. B., Kelly, C. R., Walker, A., Ubellacker, D. M., Vitti, E., Ruch, K., & Hillis, A. E. (2021). Dysfunctional tissue correlates of unrelated naming errors in acute left hemisphere stroke. *Language, Cognition and Neuroscience, 1–18.* https://doi.org/10.1080/23273798.2021.1980593

Mirman, D., Chen, Q., Zhang, Y., Wang, Z., Faseyitan, O. K., Coslett, H. B., & Schwartz, M. F. (2015). Neural organization of spoken language revealed by lesion-symptom mapping. *Nature Communications, 6,* 6762. https://doi.org/10.1038/ncomms7762

Mirman, D., Landrigan, J.-F., & Britt, A. E. (2017). Taxonomic and thematic semantic systems. *Psychological Bulletin, 143*(5), 499–520. https://doi.org/10.1037/bul0000092

Mirman, D., Zhang, Y., Wang, Z., Coslett, H. B., & Schwartz, M. F. (2015). The ins and outs of meaning: Behavioral and neuroanatomical dissociation of semantically-driven word retrieval and multimodal semantic recognition in aphasia. *Neuropsychologia, 76,* 208–219. https://doi.org/10.1016/j.neuropsychologia.2015.02.014

Miyake, A., Carpenter, P. A., & Just, M. A. (1994). A capacity approach to syntactic comprehension disorders: Making normal adults perform like aphasic patients. *Cognitive Neuropsychology, 11*(6), 671–717. https://doi.org/10.1080/02643299408251989

Mohr, J. P., Pessin, M. S., Finkelstein, S., Funkenstein, H. H., Duncan, G. W., & Davis, K. R. (1978). Broca aphasia: Pathologic and clinical. *Neurology, 28*(4), 311–311. https://doi.org/10.1212/WNL.28.4.311

Noonan, K. A., Jefferies, E., Visser, M., & Lambon Ralph, M. A. (2013). Going beyond inferior prefrontal involvement in semantic control: Evidence for the additional contribution of dorsal angular gyrus and posterior middle temporal cortex. *Journal of Cognitive Neuroscience, 25*(11), 1824–1850. https://doi.org/10.1162/jocn_a_00442

Norris, D., & McQueen, J. M. (2008). Shortlist B: A Bayesian model of continuous speech recognition. *Psychological Review, 115*(2), 357–395. https://doi.org/10.1037/0033-295X.115.2.357

Novick, J. M., Trueswell, J. C., & Thompson-Schill, S. L. (2005). Cognitive control and parsing: Reexamining the role of Broca's area in sentence comprehension. *Cognitive, Affective, & Behavioral Neuroscience, 5*(3), 263–281. https://doi.org/10.3758/CABN.5.3.263

Ostrin, R. K., & Schwartz, M. F. (1986). Reconstructing from a degraded trace: A study of sentence repetition in agrammatism. *Brain and Language, 28*(2), 328–345. https://doi.org/10.1016/0093-934X(86)90109-4

Paivio, A. (1971). Imagery and language. In *Imagery* (pp. 7–32). Elsevier. https://doi.org/10.1016/B978-0-12-635450-8.50008-X

Pillay, S. B., Stengel, B. C., Humphries, C., Book, D. S., & Binder, J. R. (2014). Cerebral localization of impaired phonological retrieval during rhyme judgment. *Annals of Neurology, 76*(5), 738–746. https://doi.org/10.1002/ana.24266

Riccardi, N., Yourganov, G., Rorden, C., Fridriksson, J., & Desai, R. (2020). Degradation of praxis brain networks and impaired comprehension of manipulable nouns in stroke. *Journal of Cognitive Neuroscience, 32*(3), 467–483. https://doi.org/10.1162/jocn_a_01495

Rogalsky, C., Pitz, E., Hillis, A. E., & Hickok, G. (2008). Auditory word comprehension impairment in acute stroke: Relative contribution of phonemic versus semantic factors. *Brain and Language, 107*(2), 167–169. https://doi.org/10.1016/j.bandl.2008.08.003

Rolheiser, T., Stamatakis, E. A., & Tyler, L. K. (2011). Dynamic processing in the human language system: Synergy between the arcuate fascicle and extreme capsule. *Journal of Neuroscience, 31*(47), 16949–16957. https://doi.org/10.1523/JNEUROSCI.2725-11.2011

Saffran, E. M., & Martin, N. (1990). 16. Short-term memory impairment and sentence processing: A case study. *Neuropsychological Impairments of Short-Term Memory, 428*. https://doi.org/10.1017/CBO9780511665547.021

Saffran, E. M., Schwartz, M. F., & Marin, O. S. (1980). The word order problem in agrammatism: II. Production. *Brain and Language, 10*(2), 263–280. https://doi.org/10.1016/0093-934X(80)90056-5

Sandberg, C. W., & Kiran, S. (2014). Analysis of abstract and concrete word processing in persons with aphasia and age-matched neurologically healthy adults using fMRI. *Neurocase, 20*(4), 361–388. https://doi.org/10.1080/13554794.2013.770881

Schumacher, R., Halai, A. D., & Lambon Ralph, M. A. (2019). Assessing and mapping language, attention and executive multidimensional deficits in stroke aphasia. *Brain, 142*(10), 3202–3216. https://doi.org/10.1093/brain/awz258

Schwanenflugel, P. J., & Shoben, E. J. (1983). Differential context effects in the comprehension of abstract and concrete verbal materials. *Journal of Experimental Psychology: Learning, Memory, and Cognition, 9*(1), 82–102. https://doi.org/10.1037/0278-7393.9.1.82

Schwartz, M. F. (1984). What the classical aphasia categories can't do for us, and why. *Brain and Language, 21*(1), 3–8. https://doi.org/10.1016/0093-934X(84)90031-2

Schwartz, M. F., Faseyitan, O., Kim, J., & Coslett, H. B. (2012). The dorsal stream contribution to phonological retrieval in object naming. *Brain, 135*(12), 3799–3814. https://doi.org/10.1093/brain/aws300

Schwartz, M. F., Kimberg, D. Y., Walker, G. M., Faseyitan, O., Brecher, A., Dell, G. S., & Coslett, H. B. (2009). Anterior temporal involvement in semantic word retrieval: Voxel-based lesion-symptom mapping evidence from aphasia. *Brain, 132*(12), 3411–3427. https://doi.org/10.1093/brain/awp284

Schwartz, M. F., Saffran, E. M., & Marin, O. S. M. (1980). The word order problem in agrammatism. *Brain and Language, 10*(2), 249–262. https://doi.org/10.1016/0093-934X(80)90055-3

Shain, C., Blank, I. A., Fedorenko, E., Gibson, E., & Schuler, W. (2021). Robust effects of working memory demand during naturalistic language comprehension in language-selective cortex [Preprint]. *Neuroscience.* https://doi.org/10.1101/2021.09.18.460917

Sheppard, S. M., Meier, E. L., Kim, K. T., Breining, B. L., Keator, L. M., Tang, B., Caffo, B. S., & Hillis, A. E. (2022). Neural correlates of syntactic comprehension: A longitudinal study. *Brain and Language, 225*, 105068. https://doi.org/10.1016/j.bandl.2021.105068

Thompson, C. K. (2003). Unaccusative verb production in agrammatic aphasia: The argument structure complexity hypothesis. *Journal of Neurolinguistics, 16*(2–3), 151–167. https://doi.org/10.1016/S0911-6044(02)00014-3

Thothathiri, M., Kimberg, D. Y., & Schwartz, M. F. (2012). The Neural basis of reversible sentence comprehension: Evidence from voxel-based lesion symptom mapping in aphasia. *Journal of Cognitive Neuroscience, 24*(1), 212–222. https://doi.org/10.1162/jocn_a_00118

Tremblay, P., & Dick, A. S. (2016). Broca and Wernicke are dead, or moving past the classic model of language neurobiology. *Brain and Language, 162*, 60–71. https://doi.org/10.1016/j.bandl.2016.08.004

Troche, J., Crutch, S., & Reilly, J. (2014). Clustering, hierarchical organization, and the topography of abstract and concrete nouns. *Frontiers in Psychology, 5*. https://doi.org/10.3389/fpsyg.2014.00360

Tsapkini, K., Jarema, G., & Kehayia, E. (2002). A morphological processing deficit in verbs but not in nouns: A case study in a highly inflected language. *Journal of Neurolinguistics, 15*(3–5), 265–288. https://doi.org/10.1016/S0911-6044(01)00039-2

Ullman, M. T. (2013). The role of declarative and procedural memory in disorders of language. *Linguistic Variation, 13*(2), 133–154. https://doi.org/10.1075/lv.13.2.01ull

Varkanitsa, M., & Caplan, D. (2018). On the association between memory capacity and sentence comprehension: Insights from a systematic review and meta-analysis of the aphasia literature. *Journal of Neurolinguistics, 48*, 4–25. https://doi.org/10.1016/j.jneuroling.2018.03.003

Walenski, M., Europa, E., Caplan, D., & Thompson, C. K. (2019). Neural networks for sentence comprehension and production: An ALE-based meta-analysis of neuroimaging studies. *Human Brain Mapping, 40*(8), 2275–2304. https://doi.org/10.1002/hbm.24523

Walker, G. M., Schwartz, M. F., Kimberg, D. Y., Faseyitan, O., Brecher, A., Dell, G. S., & Coslett, H. B. (2011). Support for anterior temporal involvement in semantic error production in aphasia: New evidence from VLSM. *Brain and Language, 117*(3), 110–122. https://doi.org/10.1016/j.bandl.2010.09.008

Wang, J., Conder, J. A., Blitzer, D. N., & Shinkareva, S. V. (2010). Neural representation of abstract and concrete concepts: A meta-analysis of neuroimaging studies. *Human Brain Mapping, 31*(10), 1459–1468. https://doi.org/10.1002/hbm.20950

Warrington, E. K. (1975). The selective impairment of semantic memory. *Quarterly Journal of Experimental Psychology, 27*(4), 635–657. https://doi.org/10.1080/14640747508400525

Warrington, E. K., & Mccarthy, R. A. (1987). Categories of knowledge. Further fractionations and an attempted integration. *Brain, 110*(5), 1273–1296. https://doi.org/10.1093/brain/110.5.1273

Webster, J., & Whitworth, A. (2012). Treating verbs in aphasia: Exploring the impact of therapy at the single word and sentence levels: Verb therapy in aphasia. *International Journal of Language & Communication Disorders, 47*(6), 619–636. https://doi.org/10.1111/j.1460-6984.2012.00174.x

Wernicke, C. (1874). *Der Aphasische Symptomencomplex: Eine Psychologische Studie auf Anatomischer Basis*. Cohn & Weigert.

Willmes, K., & Poeck, K. (1993). To what extent can aphasic syndromes be localized?*Brain, 116*(6), 1527–1540. https://doi.org/10.1093/brain/116.6.1527

Woollams, A. M., Halai, A., & Lambon Ralph, M. A. (2018). Mapping the intersection of language and reading: The neural bases of the primary systems hypothesis. *Brain Structure and Function, 223*(8), 3769–3786. https://doi.org/10.1007/s00429-018-1716-z

Yarkoni, T., Poldrack, R. A., Nichols, T. E., Van Essen, D. C., & Wagner, T. D. (2011). NeuroSynth: A new platform for large-scale automated synthesis of human functional neuroimaging data. *Frontiers in Neuroinformatics, 5*. https://doi.org/10.3389/conf.fninf.2011.08.00058

Yourganov, G., Fridriksson, J., Rorden, C., Gleichgerrcht, E., & Bonilha, L. (2016). Multivariate connectome-based symptom mapping in post-stroke patients: Networks supporting language and speech. *Journal of Neuroscience, 36*(25), 6668–6679. https://doi.org/10.1523/JNEUROSCI.4396-15.2016

Yourganov, G., Smith, K. G., Fridriksson, J., & Rorden, C. (2015). Predicting aphasia type from brain damage measured with structural MRI. *Cortex, 73*, 203–215. https://doi.org/10.1016/j.cortex.2015.09.005

Zhao, Y., Halai, A. D., & Lambon Ralph, M. A. (2020). Evaluating the granularity and statistical structure of lesions and behaviour in post-stroke aphasia. *Brain Communications, 2*(2), fcaa062. https://doi.org/10.1093/braincomms/fcaa062

7

The Default Mode Network and Language Impairment in Stroke and Neurodegeneration

E. Susan Duncan, Aswathy Anakkathil Pradeep, Juhi Kidwai, and Adam Buchwald

7.1 Introduction: Organization of the Brain at "Rest"

The brain is a complex organ comprising many interconnected regions. When anatomically distinct regions support a shared function, they are linked through multisynaptic pathways that facilitate their coordinated operation as a network (Fox et al., 2005). These neural networks can be identified using the blood oxygen level-dependent (BOLD) signal detected via functional magnetic resonance imaging (fMRI). The BOLD signal reflects relative increases or decreases in blood flow and oxygenation within the cerebrovascular system, which responds adaptively to local metabolic demands based on the extent of neuronal activity. Brain regions that consistently demonstrate highly correlated signals are considered to be functionally connected; such connectivity is attributed to their tendency to subserve common processes (Biswal et al., 1995). Given the basic principle of Hebbian plasticity – that neurons that fire together, wire together – these functional networks can be observed even in the absence of any task or stimulus and are often examined using resting-state fMRI (rs-fMRI). While it is notable that the coordinated activity of these networks can be observed during explicit tasks that require their engagement, their persistence in the absence of such demands has led to them most commonly being referred to as "resting state networks" (RSNs), although "intrinsic connectivity networks" might be more accurate (Seeley et al., 2007). The question of how RSN properties relate to abilities and deficits in multiple domains, such as cognition and language, is an area of active research.

7.1.1 The Default Mode Network

The earliest-identified and most widely studied RSN is the default mode network (DMN). The DMN has also been labeled the "task-negative" network, as it was first characterized as an assemblage of regions demonstrating suppressed activity during externally cued tasks, in contrast to periods with no task (Fox et al., 2005; Raichle et al., 2001). The widely distributed bilateral brain regions now collectively known as the DMN are robustly active during rs-fMRI; even on an individual basis, the DMN can be identified via multiple techniques that capitalize on finding positive temporal correlations in neural signals. Still, as a functionally defined network, the regions cited as constituting the DMN sometimes vary between studies (Alves et al., 2019). The core areas that are consistently included are the posterior cingulate cortex (PCC), precuneus, medial prefrontal cortices (MPFC; subsumes dorsal and ventral), and the angular gyrus of the posterior inferior parietal lobe (Buckner et al., 2008; see Figure 7.1). Additionally, temporal regions are often reported, including the hippocampal formation of the medial temporal lobes and the anterolateral middle temporal cortices (Dixon et al., 2017; Mars et al., 2012). These areas, while jointly classified as the DMN, are heavily interconnected with other brain regions outside this network. In particular, the PCC, MPFC, and angular gyrus operate as "connector hubs" that support information integration from various functional subsystems across the whole brain (Buckner et al., 2009).

Studies of the DMN, and other RSNs, may report on functional connectivity within a single network (intranetwork) or between two or more networks (internetwork). Both intra- and internetwork functional connectivity can be stronger (more positively correlated) or weaker (less positively or even negatively correlated; Fox et al., 2005). Connectivity differences observed across various tasks, populations, and performance levels suggest

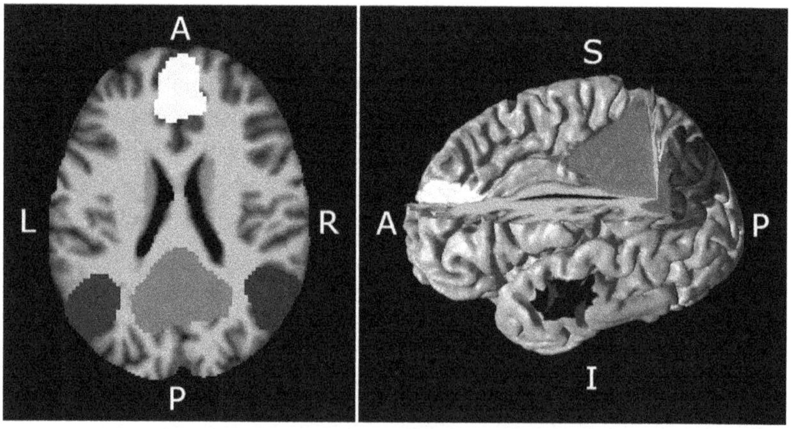

Figure 7.1 The default mode network (DMN). Axial view on left; lateral view (cut out to show medial cortices) on right. Medial prefrontal cortex in white, precuneus/posterior cingulate in light gray, inferior parietal lobe/angular gyrus in dark gray, and lateral temporal cortex in black. A= anterior, P= posterior, L= left, R= right, S= superior, I= inferior.

the importance of appropriately coordinated and complementary RSN activity in healthy brain function.

7.1.2 DMN in the Healthy Brain

The "default" moniker was originally employed to introduce the DMN as a tonic activity pattern (i.e., sustained firing) in the unconstrained brain, which could then be used as a stable baseline condition against which task-evoked responses could be reliably contrasted (Raichle et al., 2001). This was followed by identification of various externally oriented "task-positive" networks, such as the dorsal attention network, salience network, and a network with multiple labels including executive control, multiple demand, and (lateral) frontoparietal. These "task-positive" networks consist of regions that are anticorrelated with the "task-negative" DMN during rest and also positively correlated with regions active during visual attention tasks requiring working memory (Fox et al., 2005; Seeley et al., 2007; Uddin et al., 2019). However, further study has revealed that this task-positive vs. task-negative description is an oversimplification, as the DMN is active not only during the resting state, but across a wide variety of internally and externally oriented goal-directed processes (Elton & Gao, 2015). Additionally, DMN activity flexibly couples with that of other networks as needed to support task demands (Elton & Gao, 2015; Spreng & Schacter, 2012).

7.1.2.1 DMN and Cognition

Cognitive functions for which increased DMN activity has been reported include autobiographical memory (Buckner et al., 2008), future planning (Xu et al., 2016), introspection (Andrews-Hanna, 2012), mental imagery (Pearson, 2019), mind-wandering (Mason et al., 2007), and social cognition (Mars et al., 2012). Notably, these are complex processes requiring abstract thought. In addition to involvement in these functions, there is evidence that intra- and internetwork DMN connectivity during task performance may account for individual differences in wide-ranging cognitive measures (Andrews-Hanna et al., 2010; Elton & Gao, 2015). The DMN comprises higher-order association cortices that are at some distance from the primary brain areas dedicated to vision, hearing, somatosensation, and motor function (Margulies et al., 2016). These association cortices are later to develop within the lifespan, as they were within our evolutionary history (Buckner & Krienen, 2013). Therefore, it has been proposed that the DMN supports higher-level capacities that we consider fundamentally human, such as consciousness, self-awareness, and our sense of identity (Davey et al., 2016), as well the crucial ability to create a holistic understanding of our world by integrating concrete perceptual information within the context of our past experiences, beliefs, and social relationships (Yeshurun et al., 2021). Likely related to the numerous and diverse functions subserved by this network, shifts in DMN connectivity have been tied to significant behavioral effects across multiple clinical conditions (Buckner et al., 2008), including stroke and neurodegeneration.

7.1.2.2 DMN and Language

It has long been recognized that regions typically implicated in language demonstrate significant correlations during rest, resembling the activation and connectivity measured during language tasks (Cordes et al., 2000; Hampson et al., 2002). More recent examination of larger resting state datasets has revealed that these patterns are consistent with a distinct left-lateralized language RSN that is highly reproducible (Tomasi & Volkow, 2012). Identifiable on an individual basis, this network overlaps with regions activated by language tasks for which the DMN does not respond (Braga et al., 2020). Various DMN regions and subnetworks also show dissociable response patterns for language compared to other cognitive tasks (Williams et al., 2022), even when identical stimuli are used to probe the different domains (Seghier & Price, 2012). However, as with other complex human capacities, the precise role of the DMN in language function remains to be established. It has been suggested that the DMN may serve to balance activity within the language circuitry (Schafer & Constable, 2009), although it has also been proposed that language-associated DMN activation is simply an artifact of experimental and naturalistic paradigms that require parallel processing of nonlinguistic semantic and pragmatic information (Campbell & Tyler, 2018).

Given the flexible involvement of its constituent regions across tasks, it has been widely noted that the conception of a singular monolithic DMN is flawed. Instead, recent work has suggested that the larger network often called "the DMN" is in fact a network of networks, some of which preferentially connect with the language network during rs-fMRI (Gordon et al., 2020). Alternatively, some authors have proposed that the DMN and language network are both parts of the same medial frontoparietal macro-network (Uddin et al., 2019). This perspective posits that a single large-scale functional network fractionates when certain highly controlled experimental conditions elicit heterogeneous levels of activity to meet varying task demands. While this is not inconsistent with the claim that the DMN comprises differentiable subnetworks subserving different functions, it directly speaks to a potentially closer coupling between the DMN and language regions than is perhaps standardly conceived. This proposal could provide insight into various conflicting findings, such as the occasional inclusion of regions associated with language in the DMN (e.g., inferior frontal gyri and lateral temporal cortices; Dixon et al., 2017), and the association of both DMN and language nodes with semantic processing (Binder et al., 2009). Several of the functions believed to be facilitated by the DMN, such as future planning and introspection, are challenging to conceive of without a language component, and might even be unattainable without it (Skipper, 2022).

An early study coinciding with the initial introduction of the DMN (Raichle et al., 2001) found these same regions to be more active during rest than during a variety of cued cognitive tasks spanning visuomotor skills, language, mental calculation, and mental imagery (Mazoyer et al., 2001). The researchers also administered a questionnaire to determine the types of mental activities in which participants engaged during the quiet resting period. The majority reported reminiscing on recent or remote events (i.e., autobiographical memory), which included various visual or auditory components such as faces, scenes, conversations, stories, and music. Along with this mental imagery, inner speech was the most reported mental event during the rest period (Mazoyer et al., 2001). This study emphasizes the prevalence of language processes during rest and presents a more nuanced depiction of language than isolated participation in externally directed experimental tasks. The co-occurrence of inner speech and conscious rest provides a rationale for strong connections between language networks and the DMN. This is consistent with findings of strong positive correlations between the time series of a language network hub in the left inferior frontal gyrus ("Broca's area") and parts of the DMN including left PCC and right angular gyrus (Muller & Meyer, 2014).

There is clearly at least one brain network (or subnetwork) that uniquely supports language and is distinct from the DMN. What remains ambiguous is how or whether this requires DMN function; even if the DMN does not independently play a language-specific role, it likely provides domain-general contributions to social and cognitive processes required for typical language use. It has also been proposed that all of the abilities associated with the DMN require engaging in internal discourse, with dependence on language as their common denominator (Gordon et al., 2020; Skipper, 2022). Therefore, the DMN may be materially intertwined with language substrates given its own shared evolutionary history with, and reliance on, language.

7.2 DMN in Aging

As risk of stroke and neurodegenerative disorders increases with age, understanding typical age-associated changes helps contextualize effects of these clinical conditions. Studies investigating gray matter in the aging DMN find volume decreases consistent with atrophy. Compared to younger controls, healthy older individuals present with reduced cortical volumes in posterior DMN regions (Damoiseaux et al., 2008). Longitudinal studies following older participants (~50 to >80 years) also find significant gray matter loss in the DMN over time spans ranging from two to four years, which accelerates with increasing age (Ng et al., 2016; Staffaroni et al., 2018).

Even when controlling for gray matter volumes, reduced resting activation in the DMN (but not other RSNs) is seen when older participants are compared to young adults or younger geriatric subjects (Damoiseaux et al., 2008). In addition to changes during rest, lesser task-related *deactivation* has also been detected; while younger adults consistently suppress DMN activity during a visuospatial planning task, older adults do so to a significantly lesser extent (Spreng & Schacter, 2012). Within the context of one longitudinal study, a cross-sectional analysis revealed no significant aging effects in the cohort on the whole, but older adults showed reduced DMN deactivation when compared to their younger selves (approximately six years earlier; Persson et al., 2014).

Augmenting evidence of change in overall DMN activation, many studies have investigated its intrinsic connectivity, which is stronger in younger adults compared to either children or older adults (Mak et al., 2017; Zhang et al., 2016). In general, increasing age is associated with a decline in intranetwork DMN connectivity, whether assessed longitudinally or between cohorts (Farrás-Permanyer et al., 2019; Ng et al., 2016; Sambataro et al., 2010; Staffaroni et al., 2018). This may serve to divide the network into anterior and poster portions (Mevel et al., 2013), although some studies indicate changes in activation with no significant change in intranetwork connectivity (e.g., Persson et al., 2014). Seemingly paradoxically, *increased* functional connectivity has been found in the oldest participants (80–89; Farrás-Permanyer et al., 2019). This could represent some form of compensatory process but may instead indicate that the relatively few individuals over eighty who participate in research studies are less representative of the broader population than are younger volunteers.

Studies of healthy aging have also documented internetwork changes. Some of these highlight specific relationships with executive control networks, which couple with the DMN selectively in youth depending on task demands (e.g., for autobiographical but not visuospatial planning), but maintain stronger coupling across multiple domains in older participants (Spreng & Schacter, 2012). Malagurski and colleagues (2020) report that network structure becomes more flexible with advancing age, with DMN nodes variably shifting connectivity to join multiple other networks over time. Despite this significant observation, the DMN remains overall more stable than other RSNs, and age-associated changes in network flexibility do not correlate with processing speed or learning (Malagurski et al., 2020). Such flexibility might be a candidate for attempted functional compensation, as *greater* DMN stability has been associated with *poorer* episodic memory in middle-aged and older adults (Ousdal et al., 2020).

Other significant relationships have been identified between the DMN and behavior in aging, especially in domains with the potential to notably impact online language performance. Reduced resting DMN activity correlates with slower processing speed and poorer executive functioning in older, but not younger, adults (Damoiseaux et al., 2008). Decreased intranetwork connectivity is also present in individuals with poorer autobiographical (Mevel et al., 2013), episodic (Staffaroni et al., 2018), recognition (Persson et al., 2014), and working memory (Sambataro et al., 2010). Independent of intrinsic network connectivity, coupling between the executive control network and DMN is positively correlated with processing speed (Ng et al., 2016).

In sum, the brain–behavior relationships of the DMN in healthy aging are complex and could significantly impact language use. There is evidence to suggest that the well-established relationship between age and executive functioning is mediated by functional connectivity within the posterior DMN, and that this relationship is, in turn, mediated by glucose metabolism (Xu et al., 2022).

While these studies differ in precise analysis methods, the overall trajectory seems clear. Despite being a robust and relatively stable network, DMN function becomes less optimal with aging. These changes are associated with deterioration of multiple functions, including memory, executive control, processing speed, and learning, all of which may have implications for real-world language use. The existence of these associations suggests a role for the DMN in pathological conditions, including stroke and neural degeneration discussed later in this chapter.

7.3 DMN in Stroke

Stroke, or cerebrovascular accident (CVA), results in the death of brain tissue due to oxygen deprivation and the toxic effects of excessive neurotransmitters or interstitial blood following cellular or vascular damage. While the physical damage itself is restricted to a specific area, the global organization of the brain is affected immediately. Although RSNs remain identifiable after stroke, there is overall reduced intranetwork integration and compromised internetwork segregation (Siegel et al., 2018). The gross preservation of the brain's functional topology (i.e., large-scale RSNs remain identifiable) can be attributed to a subset of core network hubs that maintain their prominent status and interconnectivity with other core hubs, although local connections within and between RSNs may be markedly reduced. Compared to healthy controls, patients with subacute stroke demonstrate reduced functional connectivity among medial parts of the DMN including the MPFC, PCC, and medial temporal lobes, despite sparing of these and other DMN regions (Tuladhar et al., 2013). Functional disruptions in the DMN and salience network have also been linked to cognitive impairment following subacute ischemic stroke (Miao et al., 2022). These disconnection effects can be observed with a wide variety of anatomically remote lesions that have no direct physical impact on any DMN nodes. For example, acute ischemic brainstem strokes result in reduced intra- and internetwork connectivity of the cortical DMN and executive control network (Geng et al., 2022).

Over the first year of stroke recovery, the modular structure of RSNs begins to return, and internetwork relationships come to more closely resemble those of controls. This restoration of balanced network activity is associated with functional behavioral recovery (Siegel et al., 2018). However, behavioral and neurological changes linger for years, even following minor cerebral events such as transient ischemic attack (TIA). Compared to controls with no neurological history, those who have experienced a minor stroke or TIA show persistently poorer performance on executive function tasks in addition to disrupted functional connectivity between the DMN and frontoparietal networks, with differences detectable at least four years following the event (Nicolas et al., 2021). While we focus here on strokes that result in lasting language impairment (i.e., aphasia), it is important to underscore that network disruptions may accompany a variety of cerebrovascular events and coincide with multiple functional deficits.

7.3.1 DMN in Post-Stroke Aphasia

Aphasia is a communication disorder resulting from damage to brain regions and connections that uniquely facilitate language, most commonly due to left hemisphere stroke. Past research indicates that connectivity between the DMN and other RSNs is disrupted in subacute stages of post-stroke aphasia, which continues into chronic stages. Compared to healthy controls, there is significantly decreased coupling between the DMN and frontoparietal networks at one month (Zhang et al., 2021) and one year (Balaev et al., 2016) following aphasia onset. Geranmayeh and colleagues contrasted internetwork connectivity between speech and rest in a large cohort of fifty-three participants with a four-month history of post-stroke aphasia and a group of age-matched healthy controls. In controls, there was an inverse correlation between activity in the left frontoparietal network and DMN during speech, while the right frontoparietal network was positively correlated with DMN activity during speech. In patients, the directions of both correlations were *reversed*. While this study lacked a post-stroke control group with no aphasia, it is notable that these aberrant patterns were significantly associated with language performance. Patients with larger activation contrasts between the left frontoparietal network and DMN while speaking produced more informational content during an object definition task, while the same pattern between the right frontoparietal network and DMN was tied to poorer performance (Geranmayeh et al., 2016). As language abilities were not associated with the actual quantity of activation within any of the implicated networks, these findings underscore the importance of internetwork balance for functional language skills in aphasia.

Across multiple RSNs, individuals with chronic post-stroke aphasia also demonstrate weaker intranetwork connectivity than healthy controls (Sandberg, 2017). Within the DMN, higher functional connectivity (i.e., more like controls) is positively correlated with higher scores on a language battery (Western Aphasia Battery [WAB]; Kertesz, 2006). Better performance is significantly associated with stronger connections between the left angular gyrus and both MPFC and PCC, connections which are also stronger in healthy controls than those with aphasia (Sandberg, 2017). This functional relationship might be expected given the dual role of the left angular gyrus as both a putative language region and part of the DMN (Buckner & Krienen,

2013), although some models suggest that these represent two adjacent but differentiable nodes from separate networks (Mineroff et al., 2018).

7.3.2 DMN and Language Impairment in Post-Stroke Aphasia

Existing evidence suggests that stroke adversely affects both intra- and inter-network DMN functional connectivity (Siegel et al., 2018). In the case of aphasia, this disruption is accompanied by, and persists along with, language deficits. Further, there appears to be a strong relationship between the extent of DMN disturbance and the severity of language difficulty, which is also found when controlling for related factors such as lesion size (e.g., Balaev et al., 2016; Sandberg, 2017). A naturally ensuing avenue of interest queries how DMN connectivity covaries with aphasia therapy and recovery.

7.3.2.1 DMN Connectivity and Language Recovery

Marcotte and colleagues (2013) were the first to identify the DMN in participants with aphasia by using spatial independent component analysis in the context of a treatment study (Marcotte et al., 2013). While a strength of this work was its inclusion of a group of healthy older controls, the use of task-based fMRI (i.e., overt picture naming rather than rest) may limit comparison with other studies. Both patients and controls had MRI sessions before and after a period of language training. For patients, this training was an evidence-based treatment targeting object naming (Semantic Feature Analysis; Maddy et al., 2014), while controls learned words in a foreign language. The authors examined functional connectivity (integration) within and between two DMN subnetworks (anterior and posterior). The anterior DMN (consisting of frontal regions) did not differ between controls and patients, nor did it change with training. The posterior DMN (including temporoparietal regions) showed consistently higher integration for controls compared to patients. The patient group also demonstrated significantly increased posterior DMN integration following therapy, more closely approximating normative levels. While not reaching statistical significance in this underpowered sample (nine participants with chronic aphasia), there were trends suggesting that higher pre-therapy integration in the anterior DMN – and higher integration between anterior and posterior subnetworks – may have been predictive of post-therapy improvement (Marcotte et al., 2013). This early study suggested a role for the DMN in post-stroke aphasia recovery, both in potential responsiveness and observed response to behavioral intervention, strongly supporting further investigation.

An additional study examined DMN changes following a high-intensity period of constraint-induced aphasia therapy, in which participants are discouraged or prohibited from using non-speech means of communication. This study featured pre- and post-therapy fMRI sessions where participants were tasked with the silent reading of single words without any requested response (Dreyer et al., 2021). The only region showing significant plasticity following this intensive period of therapy was the left precuneus, and its activation change tracked positively with behavioral improvement on a language battery (Aachen Aphasia Test [AAT]; Huber et al., 1980). A follow-up analysis explored whether this positive correlation was present in the rest of the DMN or for the salience network, as a control. While the salience network was not significantly associated with behavioral change, the DMN correlation remained even when the precuneus was removed from the network (Dreyer et al., 2021). Baliki and colleagues (2018) also found a unique contribution of the DMN following high-intensity treatment provided in an Intensive Comprehensive Aphasia Program (ICAP). Following 120 hours of therapy over the course of 4 weeks, patients showed significant improvement across a variety of language and attention tasks. Language changes (WAB Aphasia Quotient; WAB-AQ) were positively correlated with baseline connectivity between the DMN and auditory cortices, while changes on an attention measure correlated with baseline connectivity between the salience network and visual cortices (Baliki et al., 2018). Both these studies suggest a role for the DMN in aphasia recovery that is (1) not shared by other domain-general networks, and (2) differentiable from related cognitive processes such as attention.

7.3.2.2 Conflicting Evidence

Musso and colleagues (2022) conducted an innovative treatment study that entailed training people with aphasia to attend to certain targets embedded in a sequence of words. There was no overt behavioral task; instead, performance accuracy was judged using an EEG-based brain-computer interface, with feedback provided depending on how accurately an individual's brain signals corresponded with a normative response to the target (determined from prior trials with healthy controls performing the same task). Following thirty hours of training over the course of four weeks, ten participants with aphasia demonstrated significant increases across multiple language domains, including naming, self-reported functional communication, and all subtests of the AAT. Cognitive scores did not mirror this pattern. Following intervention, rs-fMRI analysis indicated three interesting findings: (1) lower intrinsic DMN connectivity; (2) higher intrinsic language network connectivity; and (3) lower internetwork connectivity between the language network and DMN (Musso et al., 2022). This contrasts with previous associations between greater DMN activity or integration and better language performance, which may reflect that different interventions can induce unique patterns of neuroplasticity that may still converge on similarly adaptive behavioral outcomes.

7.3.2.3 Potential Avenues for Further Evidence

At present, the limited number of studies explicitly investigating DMN function in aphasia recovery provide evidence that remains more provocative than confirmatory. However, there are several adjacent studies to support further consideration of the DMN in post-stroke language outcomes. Immediate improvement following aphasia therapy has previously been correlated with increased activity in DMN regions including the precuneus and PCC (Menke et al., 2009; van Hees et al., 2014; Vitali et al., 2007; see also Sihvonen et al., 2020). Additional studies have reported on the broader question of RSN interactions in aphasia recovery without singling out any particular network (e.g., Duncan & Small, 2016, 2018), while others report findings that could more directly implicate the DMN but are presented as changes in networks that might overlap or subsume it (e.g., Griffis et al., 2017; Johnson et al., 2020; Sandberg et al., 2015; Zhu et al., 2014).

7.4 DMN in Neurodegeneration

In neurodegenerative disorders, such as Alzheimer's disease (AD), there is an atypical acceleration of neuronal death that leads to brain atrophy and behavioral impairment. While multiple etiologies and pathological processes may result in neurodegeneration, we focus here only on those most associated with declining language skills. Primary progressive aphasia (PPA) is an umbrella term for a variety of clinical symptoms and neurodegenerative pathophysiologies that, by definition, disproportionately affect language, especially in early stages. The three main identified variants of PPA are nonfluent (also known as agrammatic; nfPPA), semantic (also known as semantic dementia; svPPA), and logopenic (lvPPA). The first two are typically associated with frontotemporal lobar degeneration (FTLD), a syndrome which may preferentially impair behavior rather than language (behavioral variant frontotemporal dementia; bvFTD). The latter, lvPPA, is considered an atypical presentation of AD. For further description of the PPA variants, including diagnostic criteria, the interested reader is directed to Gorno-Tempini and colleagues (2011). It should also be noted that, while impairments in nonlanguage domains are prominent in dementia diagnoses excluding PPA, language deficits are often present in amnestic AD and frontotemporal syndromes, even in early stages (Smits et al., 2014).

To differing extents, the degenerative conditions causing language impairment may be associated with excessive accumulation of various proteins in the brain, such as intracellular tau or extracellular beta-amyloid. Therefore, like the well-known amnestic form of AD (in which memory deficits are defining), research on PPA often relies on positron emission tomography (PET), in which a radioactive tracing agent is injected into the body to permit visualization of metabolic processes or the deposition of specific proteins. PET imaging with fluorodeoxyglucose F 18 (FDG) marks glucose uptake (metabolism), while Pittsburgh compound B (PiB) and Flortaucipir (^{18}F) can be used to identify the presence of beta-amyloid and tau, respectively.

The DMN is markedly vulnerable to atrophy and deposits of beta-amyloid proteins that develop into intercellular plaques, and these processes are associated with the onset and progress of neurodegeneration resulting in behavioral impairment, or dementia (Smits et al., 2014). In particular, DMN degeneration has been consistently identified in AD (Buckner et al., 2009). Therefore, the DMN has been investigated far more in lvPPA (as a manifestation of probable AD) than in the other PPA variants.

7.4.1 DMN in PPA

While the contributions of abnormal findings for gray matter volumes, functional connectivity, and protein aggregation may be difficult to disentangle among the various neurodegenerative processes, the DMN has been identified as the site of peak brain atrophy when collapsing across AD subtypes, including patients with PPA (Lehmann et al., 2013b). Putcha and colleagues (2022) investigated the relationship between functional connectivity and tau pathology in a group of nonamnestic AD patients with PET scans indicating the presence of beta-amyloid. While not analyzed separately, 40 percent (n=10) of the sample met clinical criteria for lvPPA. Compared to amyloid-negative control participants with normal cognition, all patient groups had reduced gray matter volumes, reduced intranetwork connectivity, and significant tau accumulation within the posterior DMN, defined as the bilateral PCC, precuneus, and temporoparietal cortices (Putcha et al., 2022). This is consistent with other studies that have found common patterns of gray matter atrophy, hypometabolism, decreased functional connectivity, and amyloid deposition within the same posterior DMN regions across diverse AD presentations including lvPPA (Laforce et al., 2014; Lehmann et al., 2013a; Lehmann et al., 2013b; Ossenkoppele et al., 2015).

Despite the finding of shared DMN involvement, AD subgroups do also show unique patterns of atrophy across other networks consistent with their defining clinical symptoms. For lvPPA, this denotes greater abnormalities in left-lateralized language regions compared to patients with amnestic or other non-amnestic AD classifications. Reductions have been found in gray matter volumes (Lehmann et al., 2013b; Ossenkoppele et al., 2015), metabolism (Bergeron et al., 2020; Laforce et al., 2014; Lehmann et al., 2013a), and functional connectivity (Martersteck et al., 2020; Whitwell et al., 2015) consistent with an impaired language network, although without observed differences in amyloid deposition (Laforce et al., 2014). Yet, independent of syndrome-specific changes, a pattern of common DMN atrophy is evident even in the most intact AD brains; in the most severely affected, the extent of atrophy in the DMN and other networks ultimately converges (Ossenkoppele et al., 2015).

While primary language deficits in AD most commonly present as lvPPA, they can also be associated with other variants. Martersteck and colleagues (2020) contrasted a group of PPA patients with AD pathology (indicated via PET imaging or autopsy) with an amnestic AD group and healthy controls. PPA patients were diagnosed with either lvPPA, nfPPA, or unclassifiable PPA.

In a comparison of DMN, language, and memory networks, both PPA and amnestic AD groups showed reduced intranetwork DMN connectivity with no difference between groups, but, as expected, with significant differences for language and memory networks depending on clinical presentation (Martersteck et al., 2020). There was no comparison reported between the PPA variants or for correlates with language abilities. No patients with svPPA were included in this study, although reduced intrinsic DMN has also been reported in this population (Popal et al., 2020), and the main focal point of atrophy in svPPA (left temporal pole) shows strong functional connectivity with the DMN in the healthy brain (Collins et al., 2017).

Other researchers have found no DMN differences between healthy participants and those with PPA (Bonakdarpour et al., 2017), particularly in ventral or anterior regions (Lehmann et al., 2013a; Lehmann et al., 2015; Whitwell et al., 2015). Some studies emphatically reinforce this dissociation by using the DMN as a control network with which to compare functional language network connectivity changes in nfPPA with FTLD pathology (e.g., Mandelli et al., 2016; Mandelli et al., 2018). It remains difficult to resolve how these seemingly conflicting DMN findings may be attributable to language function vs. underlying disease processes, given the imperfect correspondence between clinical profiles and pathophysiology in PPA (Spinelli et al., 2017). In sum, the strongest evidence for structural and/or functional DMN changes in PPA comes from lvPPA as a manifestation of AD, although significant effects cannot yet be ruled out for other variants and etiologies.

7.4.2 DMN and Language Impairment in PPA

There is limited existing evidence for a relationship between the DMN and language in PPA and other neurodegenerative conditions. Some studies have identified positive correlations between intranetwork DMN connectivity and language-dependent cognitive measures (e.g., Putcha et al., 2022; Weiler et al., 2014; Zhu et al., 2016) or reduced deactivation during a reading-based semantic classification task (Lustig et al., 2003). Gola and colleagues (2015) analyzed gray matter volumes associated with conversational storytelling skills in a group of participants with various neurodegenerative diagnoses including svPPA, bvFTD, and AD, along with healthy controls. The peak regions of atrophy in individuals with deficits in the temporal organization of their narratives overlapped most significantly with regions of the DMN, and DMN atrophy was also implicated in individuals who perseverated inappropriately or showed deficits in attending to social cues in this discourse context (Gola et al., 2015).

Two known studies have directly interrogated changes in the DMN related to post-therapy language abilities in PPA, albeit with apparently conflicting findings (Ficek et al., 2019; Pytel et al., 2021). Each provided three weeks of daily intervention using a putatively excitatory neurostimulation approach. In the first study, Ficek and colleagues (2019) applied anodal transcranial direct current stimulation (tDCS) to stimulate Broca's area in twenty-four participants with nfPPA, lvPPA, or svPPA. Half received active stimulation during therapy sessions targeting oral and written naming, and these improved to a significantly greater extent than those who received sham. The active tDCS group showed an improvement-associated *decrease* in post-therapy language network connectivity while intranetwork DMN connectivity was unchanged. There was reduced internetwork connectivity between language and DMN nodes, but these were not associated with behavioral change (Ficek et al., 2019).

In the second study, Pytel and colleagues (2021) applied high-frequency repetitive transcranial magnetic stimulation (TMS) to stimulate a personalized site (varying, but predominantly located in the left prefrontal cortex) in twenty participants with nfPPA or svPPA. Approximately two-thirds received active-site stimulation only, while the remainder were assigned to a crossover group that also received TMS to a control brain site during a separate three-week interval. There was no concurrent speech-language therapy. Compared to the sham condition, significant behavioral improvement was reported following active-site TMS for several language measures including picture naming, syllable repetition, reading accuracy, and the number of words produced during a storytelling task. PET imaging following active-site TMS indicated multiple brain areas with increased metabolism compared to baseline, including regions consistent with a language network and the DMN, while the control condition showed no significant differences (Pytel et al., 2021).

Current evidence for an association between language and the DMN is lacking in PPA. Additionally, given the nature of the neurodegenerative processes that produce PPA, which ultimately progress to impact other domains, any confirmatory reports might reasonably bear skepticism about the extent to which language can truly be dissociated from more domain-general cognitive processes. However, as with post-stroke aphasia, the number of studies examining this potential relationship is highly limited. Therefore, the equivocal state of the research neither substantiates nor precludes a role for the DMN in PPA symptoms.

7.5 DMN as Therapeutic Target

The connection between the DMN and language skills in stroke and neurodegeneration remains tenuous, but largely due to the absence of evidence rather than the presence of contradictory findings. If further exploration does support this relationship, it opens new avenues for treatment methods that specifically target the DMN to enhance language recovery following stroke or better maintain function in the face of neurodegenerative disease. Given the inherent connections between processes facilitated by the DMN and language, many behavioral approaches likely already incorporate these in clinical practice, such as therapy goals that draw on autobiographical memory, explicit planning of future activities, and introspection during communicative tasks (e.g., self-monitoring).

Perhaps more exciting, however, are prospects for directly targeting DMN nodes via brain stimulation. Techniques such as tDCS and TMS have been steadily growing in popularity as biological means to augment language intervention, usually by stimulating regions in the language network (Duncan et al., 2020). Manipulating the DMN not only offers an additional target, but potentially a better one, since – reported impacts notwithstanding – these regions remain better preserved than the language cortices that are the primary site of damage.

One study targeting the right posterior superior temporal gyrus (Wernicke's homologue) in post-stroke aphasia found improved naming after three days of anodal (excitatory) tDCS in combination with behavioral therapy (Flöel et al., 2011). While not conceived to target the DMN, the electrode placement was centered <2 cm from the angular gyrus. Given the electrode size (5×7 cm), it would be difficult to determine which network may have been more affected without neuroimaging, and tDCS of the right angular gyrus has previously been demonstrated to produce changes in the DMN and other RSNs (Clemens et al., 2014). Several other studies have stimulated the left angular gyrus (CP5; Duncan et al., 2020), but considering the left lateralization of the language network, it would be far overreaching to attribute these changes to DMN effects.

Only one brain stimulation study is known to have intentionally targeted the DMN to treat language impairment. Transcranial photobiomodulation is a noninvasive approach that applies red and near-infrared light wavelengths to the scalp to induce metabolic effects in the underlying brain tissue. While still in nascent stages, it has been used to target DMN nodes in a small number of participants with aphasia. Naesar and colleagues (2020) compared a variety of montages to improve naming and found the best behavioral response with targeting of ipsilesional left hemisphere regions in combination with both midline anterior and posterior DMN nodes (MPFC and precuneus). Stimulating only MPFC and left hemisphere did not result in the same behavioral benefit, possibly suggesting a role for greater DMN integration and upregulation or demonstrating the value of multisite network targeting. This protocol resulted in greater resting state connectivity in the DMN as well as salience and central executive networks (Naeser et al., 2020).

There is some other potentially supporting evidence for DMN stimulation in language impairment, if circumstantial. The effectiveness of tDCS targeting left frontal language regions during an oral and written naming therapy has been tied to baseline gray matter volumes in angular gyrus and PCC of PPA patients (de Aguiar et al., 2020). Additional work has found improved word learning with excitatory stimulation overlying precuneus in healthy young adults and AD patients (Huang et al., 2021; Koch et al., 2018). The current lack of brain stimulation studies targeting the DMN to treat language deficits offers uncharted territory that could provide benefit for individuals with post-stroke or neurodegenerative aphasia.

7.6 Conclusion

There is little evidence for the role of the DMN in language impairment secondary to stroke or neurodegeneration, but, at present, this is primarily due to insufficient investigation. There is sound rationale for suspecting that a greater role might be found with further study. The DMN plays a crucial role in many higher-level abilities that are dependent on language, and it is implicated in multiple capacities on which language depends. It is affected in stroke and neurodegeneration, including when these result in language impairment. What is lacking in the literature, or perhaps simply lacking, is adequate evidence for how DMN disruption may vary when co-occurring with language deficits vs. cases with similar pathophysiology and grossly intact language. Regardless of the role the DMN may play in impairment, its potential to aid functional recovery or maintenance bears further elucidation, as enhancing domain-general skills via DMN stimulation could offer a viable and untapped target for bolstering language.

Notes

This work was supported by the National Institute of Deafness and other Communication Disorders (NIDCD) of the National Institutes of Health (NIH) under Grants R21-DC018364 (PI Duncan) and R01-DC018589 (PI Buchwald).

References

Alves, P., Foulon, C., Karolis, V., Bzdok, D., Margulies, D., Volle, E., & Thiebaut de Schotten, M. (2019). An improved neuroanatomical model of the default-mode network reconciles previous neuroimaging and neuropathological findings. *Communications Biology*, *2*, 370.

Andrews-Hanna, J. (2012). The brains default network and its adaptive role in internal mentation. *The Neuroscientist*, *18*(3), 251–270.

Andrews-Hanna, J., Reidler, J., Huang, C., & Buckner, R. (2010). Evidence for the default networks role in spontaneous cognition. *Journal of Neurophysiology*, *104*(1), 322–335.

Balaev, V., Petrushevsky, A., & Martynova, O. (2016). Changes in functional connectivity of default mode network with auditory and right frontoparietal networks in poststroke aphasia. *Brain Connectivity*, *6*(9), 714–723.

Baliki, M., Babbitt, E., & Cherney, L. (2018). Brain network topology influences response to intensive comprehensive aphasia treatment. *NeuroRehabilitation*, *43*(1), 63–76.

Bergeron, D., Beauregard, J., Soucy, J., Verret, L., Poulin, S., Matias-Guiu, J., Cabrera-Martín, M., Bouchard, R., & Laforce, R. (2020). Posterior cingulate cortex hypometabolism in non-amnestic variants of Alzheimer's disease. *Journal of Alzheimer's Disease, 77*(4), 1569–1577.

Binder, J., Desai, R., Graves, W., & Conant, L. (2009). Where is the semantic system? A critical review and meta-analysis of 120 functional neuroimaging studies. *Cerebral Cortex, 19*(12), 2767–2796.

Biswal, B., Yetkin, F., Haughton, V., & Hyde, J. (1995). Functional connectivity in the motor cortex of resting human brain using echo-planar MRI. *Magnetic Resonance in Medicine, 34*(4).

Bonakdarpour, B., Rogalski, E., Wang, A., Sridhar, J., Mesulam, M., & Hurley, R. (2017). Functional connectivity is reduced in early-stage primary progressive aphasia when atrophy is not prominent. *Alzheimer Disease and Associated Disorders, 31*(2), 101–106.

Braga, R., DiNicola, L., Becker, H., & Buckner, R. (2020). Situating the left-lateralized language network in the broader organization of multiple specialized large-scale distributed networks. *Journal of Neurophysiology, 124*(5), 1415–1448.

Buckner, R., Andrews-Hanna, J., & Schacter, D. (2008.) The brain's default network: Anatomy, function, and relevance to disease. *Annals of the New York Academy of Sciences, 1124*, 1–38.

Buckner, R., & Krienen, F. (2013). The evolution of distributed association networks in the human brain. *Trends in Cognitive Sciences, 17*(12), 648–665.

Buckner, R., Sepulcre, J., Talukdar, T., Krienen, F., Liu, H., Hedden, T., Andrews-Hanna, J., Sperling, R., & Johnson, K. (2009). Cortical hubs revealed by intrinsic functional connectivity: Mapping, assessment of stability, and relation to Alzheimer's disease. *The Journal of Neuroscience, 29*(6), 1860–1873.

Campbell, K., & Tyler, L. (2018). Language-related domain-specific and domain-general systems in the human brain. *Current Opinion in Behavioral Sciences, 21*, 132–137.

Clemens, B., Jung, S., Mingoia, G., Weyer, D., Domahs, F., & Willmes, K. (2014). Influence of anodal transcranial direct current stimulation (tDCS) over the right angular gyrus on brain activity during rest. *PLoS ONE, 9*(4), e95984.

Collins, J., Montal, V., Hochberg, D., Quimby, M., Mandelli, M., Makris, N., Seeley, W., Gorno-Tempini, M., & Dickerson, B. (2017). Focal temporal pole atrophy and network degeneration in semantic variant primary progressive aphasia. *Brain, 140*(2), 457–471.

Cordes, D., Haughton, V. M., Arfanakis, K., Wendt, G. J., Turski, P. A., Moritz, C. H., Quigley, M. A., & Meyerand, M. E. (2000). Mapping functionally related regions of brain with functional connectivity MR imaging. *American Journal of Neuroradiology, 21*(9), 1636–1644.

Damoiseaux, J., Beckmann, C., Arigita, E., Barkhof, F., Scheltens, P., Stam, C., Smith, S., & Rombouts, S. (2008). Reduced resting-state brain activity in the "default network" in normal aging. *Cerebral Cortex, 18*(8), 1856–1864.

Davey, C., Pujol, J., & Harrison, B. (2016). Mapping the self in the brain's default mode network. *NeuroImage, 132*, 390–397.

de Aguiar, V., Zhao, Y., Faria, A., Ficek, B., Webster, K., Wendt, H., Wang, Z., Hillis, A., Onyike, C., Frangakis, C., Caffo, B., & Tsapkini, K. (2020). Brain volumes as predictors of tDCS effects in primary progressive aphasia. *Brain and Language, 200*, 104707.

Dixon, M., Andrews-Hanna, J., Spreng, R., Irving, Z., Mills, C., Girn, M., & Christoff, K. (2017). Interactions between the default network and dorsal attention network vary across default subsystems, time, and cognitive states. *NeuroImage, 147*, 632–649.

Dreyer, F., Doppelbauer, L., Büscher, V., Arndt, V., Stahl, B., Lucchese, G., Hauk, O., Mohr, B., & Pulvermüller, F. (2021). Increased recruitment of domain-general neural networks in language processing following intensive language-action therapy: fMRI evidence from people with chronic aphasia. *American Journal of Speech-Language Pathology, 30*(1S), 455–465.

Duncan, E.S., Anakkathil-Pradeep, A., & Small, S. (2020). A review of biological interventions in chronic aphasia. *Annals of Indian Academy of Neurology, 23*(Suppl 2).

Duncan, E. S., & Small, S. L. (2016). Increased modularity of resting state networks supports improved narrative production in aphasia recovery. *Brain Connectivity, 6*(7), 524–529.

Duncan, E. S., & Small, S. L. (2018). Changes in dynamic resting state network connectivity following aphasia therapy. *Brain Imaging and Behavior, 12*(4), 1141–1149.

Elton, A., & Gao, W. (2015). Task-positive functional connectivity of the default mode network transcends task domain. *Journal of Cognitive Neuroscience, 27*(12), 2369–2381.

Farrás-Permanyer, L., Mancho-Fora, N., Montalà-Flaquer, M., Bartrés-Faz, D., Vaqué-Alcázar, L., Peró-Cebollero, M., & Guàrdia-Olmos, J. (2019). Age-related changes in resting-state functional connectivity in older adults. *Neural Regeneration Research, 14*(9), 1544–1555.

Ficek, B., Wang, Z., Zhao, Y., Webster, K., Desmond, J., Hillis, A., Frangakis, C., Faria, A., Caffo, B., & Tsapkini, K. (2019). The effect of tDCS on functional connectivity in primary progressive aphasia. *NeuroImage: Clinical, 19*, 703–715.

Flöel, A., Meinzer, M., Kirstein, R., Nijhof, S., Deppe, M., Knecht, S., & Breitenstein, C. (2011). Short-term anomia training and electrical brain stimulation. *Stroke, 42*(7), 2065–2067.

Fox, M. D., Snyder, A. Z., Vincent, J. L., Corbetta, M., Van Essen, D. C., & Raichle, M. E. (2005). The human brain is intrinsically organized into dynamic, anticorrelated functional networks. *Proceeings of the National Academy of Sciences, 102*(27), 9673–9678.

Geng, W., Zhang, J., Shang, S., Chen, H., Shi, M., Jiang, L., Yin, X., & Chen, Y. (2022). Reduced functional network connectivity is associated with upper limb dysfunction in acute ischemic brainstem stroke. *Brain Imaging and Behavior, 16*(2), 802–810.

Geranmayeh, F., Leech, R., & Wise, R. (2016). Network dysfunction predicts speech production after left hemisphere stroke. *Neurology, 86*(14), 1296–1305.

Gola, K., Thorne, A., Veldhuisen, L., Felix, C., Hankinson, S., Pham, J., Shany-Ur, T., Schauer, G., Stanley, C., Glenn, S., Miller, B., & Rankin, K. (2015). Neural substrates of spontaneous narrative production in focal neurodegenerative disease. *Neuropsychologia, 79*(Pt A), 158–171.

Gordon, E., Laumann, T., Marek, S., Raut, R., Gratton, C., Newbold, D., Greene, D., Coalson, R., Snyder, A., Schlaggar, B., Petersen, S., Dosenbach, N., & Nelson, S. (2020). Default-mode network streams for coupling to language and control systems. *Proceedings of the National Academy of Sciences, 117*(29), 17308–17319.

Gorno-Tempini, M. L., Hillis, A. E., Weintraub, S., Kertesz, A., Mendez, M., Cappa, S. F., Ogar, J. M., Rohrer, J. D., Black, S., Boeve, B. F., Manes, F., Dronkers, N. F., Vandenberghe, R., Rascovsky, K., Patterson, K., Miller, B. L., Knopman, D. S., Hodges, J. R., Mesulam, M. M., & Grossman, M. (2011). Classification of primary progressive aphasia and its variants. *Neurology, 76*(11), 1006–1014.

Griffis, J., Nenert, R., Allendorfer, J., Vannest, J., Holland, S., Dietz, A., & Szaflarski, J. (2017). The canonical semantic network supports residual language function in chronic post-stroke aphasia. *Human Brain Mapping, 38*(3), 1636–1658.

Hampson, M., Peterson, B. S., Skudlarski, P., Gatenby, J. C., & Gore, J. C. (2002). Detection of functional connectivity using temporal correlations in MR images. *Human Brain Mapping,* 15(4), 247–262.

Huang, Y., Mohan, A., McLeod, S., Luckey, A., Hart, J., & Vanneste, S. (2021). Polarity-specific high-definition transcranial direct current stimulation of the anterior and posterior default mode network improves remote memory retrieval. *Brain Stimulation, 14*(4), 1005–1014.

Huber, W., Weniger, D., Poeck, K., & Willmes, K. (1980.) [The Aachen Aphasia Test rationale and construct validity (author's translation)]. *Der Nervenarzt, 51*(8), 475–482.

Johnson, J., Meier, E., Pan, Y., & Kiran, S. (2020). Pre-treatment graph measures of a functional semantic network are associated with naming therapy outcomes in chronic aphasia. *Brain and Language, 207,* 104809.

Kertesz, A. (2006). Western Aphasia Battery (Revised), PsychCorp, San Antonio, Tx.

Koch, G., Bonnì, S., Pellicciari, M., Casula, E., Mancini, M., Esposito, R., Ponzo, V., Picazio, S., Di Lorenzo, F., Serra, L., Motta, C., Maiella, M., Marra, C., Cercignani, M., Martorana, A., Caltagirone, C., & Bozzali, M. (2018). Transcranial magnetic stimulation of the precuneus enhances memory and neural activity in prodromal Alzheimer's disease. *NeuroImage, 169,* 302–311.

Laforce, R., Tosun, D., Ghosh, P., Lehmann, M., Madison, C., Weiner, M., Miller, B., Jagust, W., & Rabinovici, G. (2014). Parallel ICA of FDG-PET and PiB-PET in three conditions with underlying Alzheimer's pathology. *NeuroImage: Clinical, 4,* 508–516.

Lehmann, M., Ghosh, P., Madison, C., Laforce, R., Corbetta-Rastelli, C., Weiner, M., Greicius, M., Seeley, W., Gorno-Tempini, M., Rosen, H., Miller, B., Jagust, W., & Rabinovici, G. (2013a). Diverging patterns of amyloid deposition and hypometabolism in clinical variants of probable Alzheimer's disease. *Brain, 136*(Pt 3), 844–858.

Lehmann, M., Madison, C., Ghosh, P., Miller, Z., Greicius, M., Kramer, J., Coppola, G., Miller, B., Jagust, W., Gorno-Tempini, M., Seeley, W., & Rabinovici, G. (2015). Loss of functional connectivity is greater outside the default mode network in nonfamilial early-onset Alzheimer's disease variants. *Neurobiology of Aging, 36*(10), 2678–2686.

Lehmann, M., Madison, C., Ghosh, P., Seeley, W., Mormino, E., Greicius, M., Gorno-Tempini, M., Kramer, J., Miller, B., Jagust, W., & Rabinovici, G. (2013b). Intrinsic connectivity networks in healthy subjects explain clinical variability in Alzheimer's disease. *Proceedings of the National Academy of Sciences, 110*(28), 11606–11611.

Lustig, C., Snyder, A., Bhakta, M., OBrien, K., McAvoy, M., Raichle, M., Morris, J., & Buckner, R. (2003). Functional deactivations: Change with age and dementia of the Alzheimer type. *Proceedings of the National Academy of Sciences, 100*(24), 14504–14509.

Maddy, K., Capilouto, G., & McComas, K. (2014). The effectiveness of semantic feature analysis: An evidence-based systematic review. *Annals of Physical and Rehabilitation Medicine, 57*(4), 254–267.

Mak, L., Minuzzi, L., MacQueen, G., Hall, G., Kennedy, S., & Milev, R. (2017). The default mode network in healthy individuals: A systematic review and meta-analysis. *Brain Connectivity, 7*(1), 25–33.

Malagurski, B., Liem, F., Oschwald, J., Mérillat, S., & Jäncke, L. (2020). Longitudinal functional brain network reconfiguration in healthy aging. *Human Brain Mapping, 41*(17), 4829–4845.

Mandelli, M., Vilaplana, E., Brown, J., Hubbard, H., Binney, R., Attygalle, S., Santos-Santos, M., Miller, Z., Pakvasa, M., Henry, M., Rosen, H., Henry, R., Rabinovici, G., Miller, B., Seeley, W., & Gorno-Tempini, M. (2016). Healthy brain connectivity predicts atrophy progression in non-fluent variant of primary progressive aphasia. *Brain, 139*(Pt 10), 2778–2791.

Mandelli, M., Welch, A., Vilaplana, E., Watson, C., Battistella, G., Brown, J., Possin, K., Hubbard, H., Miller, Z., Henry, M., Marx, G., Santos-Santos, M., Bajorek, L., Fortea, J., Boxer, A., Rabinovici, G., Lee, S., Deleon, J., Rosen, H., Miller, B., Seeley, W., & Gorno-Tempini, M. (2018). Altered topology of the functional speech production network in non-fluent/agrammatic variant of PPA. *Cortex, 108*, 252–264.

Marcotte, K., Perlbarg, V., Marrelec, G., Benali, H., & Ansaldo, A. I. (2013) Default-mode network functional connectivity in aphasia: Therapy-induced neuroplasticity. *Brain and Language, 124*(1), 45–55.

Margulies, D., Ghosh, S., Goulas, A., Falkiewicz, M., Huntenburg, J., Langs, G., Bezgin, G., Eickhoff, S., Castellanos, F., Petrides, M., Jefferies, E., & Smallwood, J. (2016). Situating the default-mode network along a principal gradient of macroscale cortical organization. *Proceedings of the National Academy of Sciences, 113*(44), 12574–12579.

Mars, R., Neubert, F., Noonan, M., Sallet, J., Toni, I., & Rushworth, M. (2012). On the relationship between the "default mode network" and the "social brain." *Frontiers in Human Neuroscience, 6*, 189.

Martersteck, A., Sridhar, J., Rader, B., Coventry, C., Parrish, T., Mesulam, M., & Rogalski, E. (2020). Differential neurocognitive network perturbation in amnestic and aphasic Alzheimer disease. *Neurology, 94*(7), e699–e704.

Mason, M., Norton, M., Van Horn, J., Wegner, D., Grafton, S., & Macrae, C. (2007). Wandering minds: The default network and stimulus-independent thought. *Science, 315*(5810), 393–395.

Mazoyer, B., Zago, L., Mellet, E., Bricogne, S., Etard, O., Houdé, O., Crivello, F., Joliot, M., Petit, L., & Tzourio-Mazoyer, N. (2001). Cortical networks for working memory and executive functions sustain the conscious resting state in man. *Brain Research Bulletin, 54*(3), 297–298.

Menke, R., Meinzer, M., Kugel, H., Deppe, M., Baumgärtner, A., Schiffbauer, H., Thomas, M., Kramer, K., Lohmann, H., Flöel, A., Knecht, S., & Breitenstein, C. (2009). Imaging short- and long-term training success in chronic aphasia. *BMC Neuroscience, 10*, 118.

Mevel, K., Landeau, B., Fouquet, M., La Joie, R., Villain, N., Mézenge, F., Perrotin, A., Eustache, F., Desgranges, B., & Chételat, G. (2013). Age effect on the default mode network, inner thoughts, and cognitive abilities. *Neurobiology of Aging, 34*(4), 1292–1301.

Miao, G., Rao, B., Wang, S., Fang, P., Chen, Z., Chen, L., Zhang, X., Zheng, J., Xu, H., & Liao, W. (2022). Decreased functional connectivities of low-degree level rich club organization and caudate in post-stroke cognitive impairment based on resting-state fMRI and radiomics features. *Frontiers in Neuroscience, 15*, 796530.

Mineroff, Z., Blank, I., Mahowald, K., & Fedorenko, E. (2018). A robust dissociation among the language, multiple demand, and default mode networks: Evidence from inter-region correlations in effect size. *Neuropsychologia, 119*, 501–511.

Muller, A. M., & Meyer, M. (2014). Language in the brain at rest: New insights from resting state data and graph theoretical analysis. *Frontiers in Human Neuroscience, 8*, 228.

Musso, M., Hübner, D., Schwarzkopf, S., Bernodusson, M., LeVan, P., Weiller, C., & Tangermann, M. (2022). Aphasia recovery by language training using a brain–computer interface: A proof-of-concept study. *Brain Communications, 4*(1), fcac008.

Naeser, M., Ho, M., Martin, P., Hamblin, M., & Koo, B. (2020). Increased functional connectivity within intrinsic neural networks in chronic stroke following treatment with red/near-infrared transcranial photobiomodulation: Case series with improved naming in aphasia. *Photobiomodulation, Photomedicine, and Laser Surgery, 38*(2), 115–131.

Ng, K., Lo, J., Lim, J., Chee, M., & Zhou, J. (2016). Reduced functional segregation between the default mode network and the executive control network in healthy older adults: A longitudinal study. *NeuroImage, 133*, 312–330.

Nicolas, K., Goodin, P., Visser, M., Michie, P., Bivard, A., Levi, C., Parsons, M., & Karayanidis, F. (2021). Altered functional connectivity and cognition persists 4 years after a transient ischemic attack or minor stroke. *Frontiers in Neurology, 12*, 612177.

Ossenkoppele, R., Cohn-Sheehy, B., La Joie, R., Vogel, J., Möller, C., Lehmann, M., van Berckel, B., Seeley, W., Pijnenburg, Y., Gorno-Tempini, M., Kramer, J., Barkhof, F., Rosen, H., van der Flier, W., Jagust, W., Miller, B., Scheltens, P., & Rabinovici, G. (2015). Atrophy patterns in early clinical stages across distinct phenotypes of Alzheimer's disease. *Human Brain Mapping, 36*(11), 4421–4437.

Ousdal, O., Kaufmann, T., Kolskår, K., Vik, A., Wehling, E., Lundervold, A., Lundervold, A., & Westlye, L. (2020). Longitudinal stability of the brain functional connectome is associated with episodic memory performance in aging. *Human Brain Mapping, 41*(3), 697–709.

Pearson, J. (2019). The human imagination: The cognitive neuroscience of visual mental imagery. *Nature Reviews Neuroscience, 20*(10), 624–634.

Persson, J., Pudas, S., Nilsson, L., & Nyberg, L. (2014). Longitudinal assessment of default-mode brain function in aging. *Neurobiology of Aging, 35*(9), 2107–2117.

Popal, H., Quimby, M., Hochberg, D., Dickerson, B., & Collins, J. (2020). Altered functional connectivity of cortical networks in semantic variant primary progressive aphasia. *NeuroImage: Clinical, 28*, 102494.

Putcha, D., Eckbo, R., Katsumi, Y., Dickerson, B., Touroutoglou, A., & Collins, J. (2022). Tau and the fractionated default mode network in atypical Alzheimer's disease. *Brain Communications, 4*(2), fcac055.

Pytel, V., Cabrera-Martín, M., Delgado-Álvarez, A., Ayala, J., Balugo, P., Delgado-Alonso, C., Yus, M., Carreras, M., Carreras, J., Matías-Guiu, J., & Matías-Guiu, J. (2021). Personalized repetitive transcranial magnetic stimulation for primary progressive aphasia. *Journal of Alzheimer's Disease, 84*(1), 151–167.

Raichle, M., MacLeod, A., Snyder, A., Powers, W., Gusnard, D., & Shulman, G. (2001). A default mode of brain function. *Proceedings of the National Academy of Sciences, 98*(2), 676–682.

Sambataro, F., Murty, V., Callicott, J., Tan, H., Das, S., Weinberger, D., & Mattay, V. (2010). Age-related alterations in default mode network: Impact on working memory performance. *Neurobiology of Aging, 31*(5), 839–852.

Sandberg, C. W. (2017). Hypoconnectivity of resting-state networks in persons with aphasia compared with healthy age-matched adults. *Frontiers in Human Neuroscience*, 11.

Sandberg, C. W., Bohland, J. W., & Kiran, S. (2015). Changes in functional connectivity related to direct training and generalization effects of a word finding treatment in chronic aphasia. *Brain and Language, 150*, 103–116.

Schafer, R., & Constable, T. (2009). Modulation of functional connectivity with the syntactic and semantic demands of a noun phrase formation task: A possible role for the default network. *NeuroImage, 46*(3), 882–890.

Seeley, W., Menon, V., Schatzberg, A., Keller, J., Glover, G., Kenna, H., Reiss, A., & Greicius, M. (2007). Dissociable intrinsic connectivity networks for salience processing and executive control. *The Journal of Neuroscience, 27*(9), 2349–2356.

Seghier, M. L., & Price, C. J. (2012). Functional heterogeneity within the default network during semantic processing and speech production. *Frontiers in Psychology, 3*, 281.

Siegel, J., Seitzman, B., Ramsey, L., Ortega, M., Gordon, E., Dosenbach, N., Petersen, S., Shulman, G., & Corbetta, M. (2018). Re-emergence of modular brain networks in stroke recovery. *Cortex, 101*, 44–59.

Sihvonen, A., Leo, V., Ripollés, P., Lehtovaara, T., Ylönen, A., Rajanaro, P., Laitinen, S., Forsblom, A., Saunavaara, J., Autti, T., Laine, M., Rodríguez-Fornells, A., Tervaniemi, M., Soinila, S., & Särkämö, T. (2020). Vocal music enhances memory and language recovery after stroke: Pooled results from two RCTs. *Annals of Clinical and Translational Neurology, 7*(11), 2272–2287.

Skipper, J. (2022). A voice without a mouth no more: The neurobiology of language and consciousness. *Neuroscience and Biobehavioral Reviews, 140*, 104772.

Smits, M., Jiskoot, L., & Papma, J. (2014). White matter tracts of speech and language. *Seminars in Ultrasound, CT, and MR, 35*(5), 504–516.

Spinelli, E., Mandelli, M., Miller, Z., Santos-Santos, M., Wilson, S., Agosta, F., Grinberg, L., Huang, E., Trojanowski, J., Meyer, M., Henry, M., Comi, G., Rabinovici, G., Rosen, H., Filippi, M., Miller, B., Seeley, W., & Gorno-Tempini, M. (2017). Typical and atypical pathology in primary progressive aphasia variants. *Annals of Neurology, 81*(3), 430–443.

Spreng, R., & Schacter, D. (2012). Default network modulation and large-scale network interactivity in healthy young and old adults. *Cerebral Cortex, 22*(11), 2610–2621.

Staffaroni, A., Brown, J., Casaletto, K., Elahi, F., Deng, J., Neuhaus, J., Cobigo, Y., Mumford, P., Walters, S., Saloner, R., Karydas, A., Coppola, G., Rosen, H., Miller, B., Seeley, W., & Kramer, J. (2018). The longitudinal trajectory of default mode network connectivity in healthy older adults varies as a function of age and is associated with changes in episodic memory and processing speed. *The Journal of Neuroscience, 38*(11), 2809–2817.

Tomasi, D., & Volkow, N. (2012). Resting functional connectivity of language networks: Characterization and reproducibility. *Molecular Psychiatry, 17*(8), 841–854.

Tuladhar, A., Snaphaan, L., Shumskaya, E., Rijpkema, M., Fernandez, G., Norris, D., & de Leeuw, F. (2013). Default mode network connectivity in stroke patients. *PLoS ONE, 8*(6), e66556.

Uddin, L., Yeo, B., & Spreng, R. (2019). Towards a universal taxonomy of macro-scale functional human brain networks. *Brain Topography, 32*(6), 926–942.

van Hees, S., McMahon, K., Angwin, A., de Zubicaray, G., Read, S., & Copland, D. A. (2014). A functional MRI study of the relationship between naming treatment outcomes and resting state functional connectivity in post-stroke aphasia. *Human Brain Mapping, 35*(8), 3919–3931.

Vitali, P., Abutalebi, J., Tettamanti, M., Danna, M., Ansaldo, A., Perani, D., Joanette, Y., & Cappa, S. (2007). Training-induced brain remapping in chronic aphasia: A pilot study. *Neurorehabilitation and Neural Repair, 21*(2), 152–160.

Weiler, M., Fukuda, A., Massabki, L., Lopes, T., Franco, A., Damasceno, B., Cendes, F., & Balthazar, M. (2014). Default mode, executive function, and language functional connectivity networks are compromised in mild Alzheimer's disease. *Current Alzheimer Research, 11*(3), 274–282.

Whitwell, J., Jones, D., Duffy, J., Strand, E., Machulda, M., Przybelski, S., Vemuri, P., Gregg, B., Gunter, J., Senjem, M., Petersen, R., Jack, C., & Josephs, K. (2015). Working memory and language network dysfunctions in logopenic aphasia: A task-free fMRI comparison with Alzheimer's dementia. *Neurobiology of Aging, 36*(3), 1245–1252.

Williams, K., Numssen, O., & Hartwigsen, G. (2022). Task-specific network interactions across key cognitive domains. *Cerebral Cortex, 32*(22), 5050–5071.

Xu, K., Niu, N., Li, X., Chen, Y., Wang, D., Zhang, J., Chen, Y., Li, H., Wei, D., Chen, K., Cui, R., Zhang, Z., & Yao, L. (2022). The characteristics of glucose metabolism and functional connectivity in posterior default network during nondemented aging: Relationship with executive function performance. *Cerebral Cortex, 33*(6), 2901–2911.

Xu, X., Yuan, H., & Lei, X. (2016). Activation and connectivity within the default mode network contribute independently to future-oriented thought. *Scientific Reports, 6*, 21001.

Yeshurun, Y., Nguyen, M., & Hasson, U. (2021). The default mode network: Where the idiosyncratic self meets the shared social world. *Nature Reviews Neuroscience, 22*(3), 181–192.

Zhang, C., Cahill, N. D., Arbabshirani, M. R., White, T., Baum, S. A., & Michael, A. M. (2016). Sex and age effects of functional connectivity in early adulthood. *Brain Connectivity, 6*(9), 700–713.

Zhang, C., Xia, Y., Feng, T., Yu, K., Zhang, H., Sami, M., Xiang, J., & Xu, K. (2021). Disrupted functional connectivity within and between resting-state networks in the subacute stage of post-stroke aphasia. *Frontiers in Neuroscience, 15*, 746264.

Zhu, D., Chang, J., Freeman, S., Tan, Z., Xiao, J., Gao, Y., & Kong, J. (2014). Changes of functional connectivity in the left frontoparietal network following aphasic stroke. *Frontiers in Behavioral Neuroscience, 8*, 167.

Zhu, H., Zhou, P., Alcauter, S., Chen, Y., Cao, H., Tian, M., Ming, D., Qi, H., Wang, X., Zhao, X., He, F., Ni, H., & Gao, W. (2016). Changes of intranetwork and internetwork functional connectivity in Alzheimer's disease and mild cognitive impairment. *Journal of Neural Engineering, 13*(4), 046008.

8

Language and Cognition

Sarah Villard, Lauryn Zipse, and Sofia Vallila-Rohter

8.1 Introduction

The nature of the relationship between language abilities and cognitive abilities, and the extent to which these two skill sets may be intertwined, have long been central topics of discussion in the literature on language processing and its neural underpinnings. This discussion has intensified during recent years, as new research techniques and tools have appeared on the scene, allowing for more precise tracking of brain activation and connectivity during experimental tasks thought to recruit specific cognitive and/or linguistic skills. The current chapter takes a step back from focusing directly on language to summarize our current understanding of four essential cognitive processes: attention, memory, working memory, and executive function. The chapter will delve into each of these cognitive processes in turn, first outlining the current knowledge concerning its function and neural substrates, and then discussing recent findings on its relationship to language processing. Note that the section on memory will encompass various types of short- and long-term memory, while working memory will be discussed separately due to its unique characteristics.

While these four cognitive processes will be discussed in distinct sections for the sake of clarity, it is important to bear in mind that they undeniably share underlying relationships. For example, most of us would probably agree that executive function could not exist in a vacuum without working memory, that the process of encoding information into memory likely requires some amount of attention, and so on. Throughout the chapter, we will discuss several examples of models in which these constructs are conceptualized differently or with more overlap.

Finally, the reader should bear in mind that the literature on cognition is vast, and the selection of models discussed herein has necessarily been curated by the authors of this chapter, based on our own experience, interests, and expertise. The goal in writing this chapter was not to provide an exhaustive overview of the current knowledge on cognition and language, but rather to create a cohesive narrative introduction to the major cognitive processes and some of the ways in which they interact with language processing.

8.2 Attention

8.2.1 Overview of Attention

We will begin with what is perhaps the most fundamental of the cognitive processes: attention. Because attention is closely tied to sensation and perception, it is often considered a necessary precondition for the execution of other, more complex, cognitive operations, including those that will be discussed later in this chapter. If I cannot, for example, pay attention to the auditory stream of a person speaking to me (or to the words printed on the page of a book), it will be virtually impossible for me to encode the message contained therein, analyze it, recall it later, or use that information for any other purpose. Attention is the gate through which information must pass if it is to be stored or manipulated by the brain.

Despite this key role, attention can be notoriously difficult to define; however, it is often conceptualized as a "spotlight" mechanism (Posner et al., 1980) that selectively enhances selected stimuli in the environment for further processing, while other stimuli are ignored or suppressed. In many complex audiovisual environments, the brain's capacity limitations may not allow for simultaneous processing of all the available stimuli (imagine trying to pay simultaneous attention to every sight and sound in a busy restaurant or train station). The spotlight, therefore, serves to direct – or *allocate* – attention to a specific stimulus or task (Kahneman, 1973). While the spotlight analogy was originally developed to describe visual attention, it has been applied to auditory attention as well (Fritz et al., 2007). The characterization of attention as a mechanism allowing certain stimuli to be focused on while others are ignored is known as "selective attention" and is perhaps the most frequently studied. However, many of the foundational twentieth-century clinical and neuropsychological models of attention include multiple additional subtypes of this cognitive construct (Mirsky et al., 1991; Posner & Petersen, 1990; Sohlberg & Mateer, 1987, 2001), such as orienting attention, sustained attention, and/or switching attention. Another key finding of the early foundational attention literature is that attention involves a combination of top-down (goal-directed) and bottom-up (stimulus-driven) processes (Treisman, 1960); this combination of processes accounts for how humans are able to selectively attend to a task but may still respond to unexpected, salient stimuli in the environment (e.g., if one's own name is heard unexpectedly, it is still likely to be attended to). Other work has also highlighted the role of reward and bias in attentional selection, demonstrating that individuals may be more likely to attend to objects in a spatial location that they have been conditioned to favor based on experience (Failing & Theeuwes, 2018).

8.2.2 Neural Basis of Attention

Recent research has built upon these foundational principles of attention, increasing and solidifying our understanding of this construct and, in particular, its neural substrates. Corbetta and Shulman's seminal (2002) paper describes two key attention networks centered in separate neuroanatomical regions but functioning in tandem. According to this framework, top-down control of the attention network operates in the dorsal frontoparietal system, encompassing the bilateral dorsal posterior parietal and frontal lobes, and reflects attention to task-based goals. The complementary bottom-up attention network is centered in the right temporoparietal junction as well as the ventral frontal cortex; this network monitors any unexpected and potentially relevant sensory stimuli, which could include oddball stimuli or stimuli similar to the anticipated target, or to sudden movements or noises that might disrupt the ongoing cognitive task. Later research has provided further support for the flexible interplay between the dorsal and ventral attention systems (e.g., Itti & Koch, 2001; Vossel et al., 2014).

In addition to these general findings, other elements of neural attention processing can differ depending on the modality of the stimuli. The following two sections will outline several major findings within the fields of visual and auditory attention, beginning with visual attention.

8.2.3 Visual Attention

A key principle that has emerged in the study of visual attention is that specific visual stimuli in the environment can be consciously selected for either enhancement or suppression by the viewer, depending on whether they are task-relevant or task-irrelevant. While enhancement vs. suppression cannot be mapped perfectly onto top-down vs. bottom-up attentional control, there is certainly a strong connection between top-down (task-driven) attention and enhancement of task-relevant stimuli. E lectroencephalography (EEG) and magnetoencephalography (MEG) studies have implicated oscillatory power in the alpha frequency range (~8–12 Hz) as a key neural marker during tasks that involve selection and enhancement of some stimuli, and concurrent suppression of other stimuli. There is broad agreement that task-induced increases in alpha oscillatory power are associated with suppression of task-irrelevant stimuli, whereas task-induced decreases in alpha power are associated with enhancement of task-relevant stimuli (Händel et al., 2011; Jensen et al., 2012; Klimesch, 2012; Schneider et al., 2021). Importantly, however, the question of whether the primary emphasis in characterizing alpha should be on suppression or enhancement is still under discussion (e.g., Foster & Awh, 2019; Schneider et al., 2021).

The identification of task-relevant and task-irrelevant stimuli in the visual environment is dependent largely on spatial location. A consistent finding within the alpha literature is that differences in alpha may be lateralized depending on where in space attention is directed, with higher alpha in the hemisphere of the brain that is processing task-irrelevant stimuli but lower alpha in the hemisphere processing task-relevant stimuli (e.g., Kelly et al., 2006; Thut et al., 2006; Worden et al., 2000). Other recent findings have emphasized the ability of the brain to selectively focus attention not just on the right versus left areas of visual space, but on highly precise spatial locations. In a functional magnetic resonance imaging (fMRI) experiment, Datta and DeYoe (2009) found that the blood-oxygen-level dependent (BOLD) signal in occipital cortex could predict the spatial direction of covert attention (i.e., attention not reflected by change in direction of eye gaze) with high accuracy. There is strong evidence that this ability of the brain to direct attention to specific areas in space is driven by past experience and bias related to the perceived probability of reward; for example, if attending to a particular location in space results in a reward, the visual attention system will prioritize that location moving forward (see Failing & Theeuwes, 2018, for a review).

Finally, although a thorough discussion of the processes supporting reading is outside the scope of this chapter, it should be briefly noted that reading is a specific language task that requires substantial visual attentional processes. Recent work on attention deficit/hyperactivity disorder (ADHD) (Stern & Shalev, 2013), as well as on dyslexia (e.g., Bosse et al., 2007; Shaywitz & Shaywitz, 2008; Stein, 2014) has provided evidence that the reading difficulties encountered by individuals with these disorders may be linked to, or (in the case of dyslexia) possibly largely attributed to, impairments in visual attentional mechanisms.

8.2.4 Auditory Attention

Much of the work on selective auditory attention since the 1950s has focused on the "cocktail party problem," or the challenge of attending to target speech when background sounds, or "maskers," are also present (Broadbent, 1952; Cherry, 1953). While some masking is the simple result of spectrotemporal overlap between target and masker energy, maskers consisting of intelligible speech often introduce additional masking known as informational masking. Informational masking is masking that occurs despite acoustic availability of target information and is thought to be mediated by higher-level cognitive skills, particularly attention (Durlach et al., 2003; Kidd & Colburn, 2017; Kidd et al., 2016). Recent work has suggested that successful listening performance in the face of informational masking is supported by a distributed oscillatory network across the frontal, temporal, and parietal lobes (Szalárdy et al., 2019), and that activation in the superior temporal gyrus may also be associated with the ability to suppress informational maskers (Zhang et al., 2021).

An additional line of research within the recent literature on auditory attention has focused on the mechanisms that allow listeners to group incoming sounds together into perceptual auditory objects (Shinn-Cunningham, 2008), so that a given auditory object (e.g., the speech stream of a specific talker, or the sound of a lawnmower outside the window) can be either enhanced or suppressed (Bregman, 1990). A number of recent computational models accounting for this mechanism have been proposed (e.g., Brown et al., 2010; Mesgarani et al., 2010). A key finding thus far is that sounds with temporal coherence are highly likely to be perceptually grouped together (Lu et al., 2017; Shamma et al., 2011; Thakur et al., 2015).

8.2.5 Interplay/Integration between Visual and Auditory Attention

While work on auditory attention and visual attention have unfolded separately in some respects, these two areas of research have also converged in several interesting ways. For example, some experiments employing both visual and auditory stimuli have found that the processing of auditory stimuli is enhanced by the presence of temporally coherent visual stimuli (Atilgan et al., 2018; Maddox et al., 2015). Behavioral work on attention allocation using dual-task experiments has also demonstrated that attentional resources may be flexibly allocated, and reallocated, to both visual and auditory tasks, depending on shifting task demands (e.g., Morey et al., 2011). Additionally, while visual and auditory attention tasks typically activate different modality-specific processing areas (occipital areas during visual tasks and temporal areas during auditory tasks), there is also evidence in support of a supramodal attention processing hub in the parietal cortex (Banerjee et al., 2011; Green et al., 2005) which may also encompass parts of frontal cortex (e.g., Green et al., 2011).

Other work has explored the possibility of utilizing attention-related information from one modality to compensate for deficits in the other modality. An ongoing line of research has focused on developing a visually guided hearing aid (Best et al., 2007; Kidd, 2017; Kidd et al., 2013, 2020) that could monitor an individual's eye gaze direction in order to identify a target auditory source from a spatial array of sources, just as a listener in a real-world situation might direct their gaze toward the target speaker in a room of talkers. After identifying the spatial source of the target auditory stream, this system would apply an acoustic beamforming algorithm to the auditory stimuli in the environment, increasing the level of the stimuli originating from the directed gaze area and attenuating the stimuli originating from other spatial locations. In this way, the system would use a "spotlight" approach integrating visual and auditory information to compensate for the challenges faced by listeners (particularly those with hearing loss) in complex auditory environments.

8.2.6 Attention and Language

Several different lines of research have provided frameworks for understanding the role that attention plays in language processing. One important finding concerns the sharing of attention resources between linguistic and nonlinguistic tasks. Dual-task paradigms examining attention allocation have demonstrated that performance on a word production task can be slowed or disrupted when participants are required to simultaneously allocate some attention to a secondary, nonlinguistic task (see Roelofs & Piai, 2011 for a review). Other studies have examined this question in contexts involving driving simulations (Gherri & Eimer, 2011) as well as pedestrian safety (Nasar et al., 2008) and have found that when individuals are actively engaged in a secondary language task (e.g., talking on the phone), their performance on a nonlinguistic task is decreased and/or slowed. These findings are consistent with the idea that attention is essentially a domain-general resource; that is, the human brain is in possession of a single, finite pool of attention resources that can be flexibly allocated to linguistic as well as nonlinguistic tasks. This finding is important because it frames attention as a prerequisite for language processing.

The literature on aphasia, or neurological language impairment, has also provided evidence in support of attention as a cognitive process providing critical support for language tasks. While the traditional conceptualization of aphasia has been that it is a disorder that affects language while sparing cognition, studies closely examining cognition in individuals with aphasia have found that, in many cases, not only language but also attention is impacted (e.g., Murray, 1999, 2012; Villard & Kiran, 2015). One notable theory regarding the co-occurrence of language and attention impairments in aphasia is that the language deficit may in fact be the result, at least in part, of underlying deficits in attention (Hula & McNeil, 2008; Murray, 1999; Villard & Kiran, 2017). Hula and McNeil (2008) take a particularly strong stance on this issue, framing the attention deficit as "the assumed primary underlying deficit in aphasia" and citing phenomena common in the presentation of aphasia, such as stimulability and moment-to-moment variability in language performance, as support for their position that aphasia is a failure to access linguistic knowledge in the brain, rather than a loss of linguistic knowledge per se. Other researchers have suggested that impaired attention may not be the primary deficit but may strongly contribute to impaired language performance in these patients (Murray, 1999; Villard & Kiran, 2017). Finally, a separate body of research has suggested that impairments in attention may cause particular problems for individuals with aphasia during masked listening tasks where target speech is presented alongside competing distractor speech (Rankin et al., 2014; Villard & Kidd, 2019). These findings are all consistent with the view that attention plays a key role in supporting language processing.

Another, more specific, example of how attention is required for language tasks even in individuals without aphasia comes from the literature on lexical selection during spoken work production tasks. Evidence from picture-word interference experiments, where participants are asked to name a target picture/object in the presence of semantic or phonologic distractors, has suggested that attention resources play a key role in the word retrieval process required to resist interference and correctly name the target (e.g., Mädebach et al., 2011; Oppermann et al., 2014).

8.3 Memory

8.3.1 Overview of Memory

We turn now to memory, another aspect of cognition important for language learning, language formulation and use, and comprehension. Like attention, memory is thought to encompass a number of different subtypes and is described from a variety of perspectives such as sensory, short-term, long-term, implicit, explicit, episodic, semantic and more. In this section, we will discuss the memory systems most relevant to language and communication.

Broadly speaking, memory is described as a process of encoding, storing and retrieving information (Melton, 1963). As individuals navigate the environment, perceived information (physical and mental) enters into the memory system encoded visually, acoustically, or semantically (McDermott & Roediger, 2018; Tulving, 2002). Recoding, or transformation, occurs as lived experience is registered, interpreted, and associated with prior experience to make sense of input. Recoding strategies such as chunking (see Norris & Kalm, 2021), creating imagery (Meneghetti et al., 2011; Miller et al., 2015), or forming associations (Hunt & McDaniel, 1993) have been explored as methods to form longer or distinctive memories, which may be easier or more likely to be remembered.

After it has been recoded, information is maintained, or stored, short term or long term. Short-term memory (STM) has been used to describe a capacity-limited memory store for information immediately available and accessible for a short period of time (Cowan, 2008; Shiffrin & Atkinson, 1969), highly dependent upon the hippocampus (Milner et al., 1998; Squire, 2009a). Long-term memory (LTM) describes neocortical, capacity-unlimited storage of information for long periods of time. Hippocampal STM is understood to be important for long-term memory consolidation, yet the persistence of the role of the hippocampus remains debated. Within standard models of consolidation, information is thought to be encoded in the hippocampus (STM) and neocortex (LTM) simultaneously. The hippocampus serves as an index that trains long-term memories, which eventually become independent of STM (Squire et al., 2015; Squire & Alvarez, 1995). In contrast, multiple memory trace theory (Nadel & Moscovitch, 1997; Nadel et al., 2000) proposes that STM traces of the hippocampus persist particularly for episodic memories for past events and experiences. Importantly, LTM is thought to be highly dynamic (Dudai & Morris, 2013). Memory traces are influenced by a person's preexisting knowledge and interpretations and thus are not direct replicas of reality (Brod et al., 2013). Events, associations, and reconstructions continuously affect LTM as experiences are remembered, sometimes even leading to false remembering (Schacter & Dodson, 2001). Information in LTM is classified as either explicit (declarative) and available to conscious awareness, or implicit (nondeclarative) and below conscious awareness (Poldrack & Gabrieli, 1997; Squire, 1992). Explicit and implicit LTM likely interact and information may shift from one system to another (Packard & McGaugh, 1996).

Finally, information stored in memory can be accessed, or retrieved, when an aspect of the environment such as a sound or image acts as a retrieval cue bringing a memory trace to awareness (for review, see Frankland et al., 2019). During access, memories are revisited and reconstructed within the current context and knowledge (Barry & Maguire, 2019). Retrieval of memories is dependent upon the nature and extent of match of a cue, as well as upon the specificity of earlier encoding (Tulving & Pearlston, 1966). Each retrieval of a memory increases its likelihood for subsequent retrieval. Repeated retrieval, however, can dampen a memory or related traces if they are not retrieved due to inhibition or competition, a process referred to as retrieval-induced forgetting (for a review, see Jonker et al., 2015; Murayama et al., 2014; Storm & Levy, 2012; Verde, 2012). For example, when asked to retrieve a memory about a morning commute two weeks ago, the memory of the morning commute from last month or yesterday provides competition and can lead to retrieval-induced forgetting.

8.3.2 Neural Basis of Memory

Early insights into the neural bases of memory and existence of multiple memory systems came from patient studies in which individuals with bilateral medial temporal lobe amnesia had severely impaired recall and recognition yet still managed to demonstrate skill learning ability (Corkin, 1968; Milner et al., 1968). Patient H.M., for example, an individual who underwent an experimental bilateral hippocampus ablation (extending into parahippocampal cortex) to treat uncontrollable seizures, became unable to form new memories (e.g. could not recognize new faces, or recall completing tasks) yet demonstrated the ability to learn new skills, as demonstrated through decreased response times when repeatedly completing certain tasks like producing a mirror drawing of a star (Annese et al., 2014; Corkin, 1968; Squire, 2009b). Findings such as these set foundations for the two systems now described as explicit (declarative) and implicit (nondeclarative) memory, which frame many of the discussions of the neural bases of memory. Declarative memory refers to memory for facts and concepts (semantic memory), events, episodes, and experiences of daily life (episodic memory), explicitly available to conscious awareness (e.g. Dickerson & Eichenbaum, 2010; Graf and Schacter, 1985; Tulving, 1972). It has been established that the medial temporal lobe (MTL) is critically important for declarative memory (see Bisaz et al., 2014; Reber, 2008), specifically, the hippocampal formation, perirhinal and parahippocampal cortices (Squire & Zola, 1996). The MTL operates with the support of multiple neural regions, including the prefrontal cortex (PFC), thought to modulate activity supporting encoding effort, recollection, and retrieval monitoring (Blumenfeld & Ranganath, 2007; Buckner, 2002; Davachi & Dobbins, 2008; Fernandez, 2017; Javadi & Walsh, 2012;); the lateral temporal cortex, supporting encoding of verbal material; and posterior parietal areas, supporting memory retrieval (Ciaramelli et al., 2020; Ojemann et al., 2009; Wagner et al., 2005). While the hippocampus has often stolen the spotlight during discussions of declarative memory and is important during initial trace formation in STM, research increasingly draws attention to the critical role of the larger widespread and interconnected declarative memory network described above (see Eichenbaum, 2001).

Nondeclarative memory refers to implicit memory processes not dependent upon the MTL that occur largely automatically without conscious or effortful retrieval (see Squire & Zola, 1996). Nondeclarative memory supports us in learning skills and habits or carrying out routine daily activities like navigating a familiar neighborhood. In a review of the neural bases of implicit learning, Reber (2013) astutely describes the challenge of studying a system defined by *absence* (absence of awareness, absence of MTL dependence) and encourages framing implicit learning as a principle rather than a system – an approach that we will adopt here. Implicit memory processes are thought to recruit the cerebellum and basal ganglia as observed in experiments of sequence learning, classification learning, or skill acquisition (Reber, 2013). Additional types of nondeclarative learning such as category-learning or fear conditioning have shown plasticity in visual cortical areas (Reber et al., 2008) and the amygdala (Kim & Cho, 2020; McGaugh et al., 1996), respectively (see LeDoux, 2014 for a discussion on fear involving conscious and unconscious processes). It is important to note that declarative and nondeclarative processes are not fully dissociated from one another and exhibit complex interactions yet to be fully understood (see Cabeza & Moscovitch, 2013; Dew & Cabeza, 2011).

8.3.3 Memory and Language: Semantic Memory

Turning now to memory as it relates to language, fundamental to the act of communication is the ability to assign meaning to what we experience in the world, and this is tied closely to semantic memory. First, perceptual representations must be recognized as objects (Rogers et al., 2003). For example, when we see a cat, visuospatial features evoke perceptual representations. These are thought to be meaning free (Humphreys et al., 1999) or to support the emergence of meaning when additional cat "objects" are encountered and associated with representations (e.g. object recognition, see Humphreys & Forde, 2001; Rogers et al., 2003). Semantic memory, described as a "mental thesaurus" (Tulving, 1972), associates perceptual representations with verbal descriptions and patterns of activity in memory, storing acquired information related to word meanings and their referents. While it is debated whether object recognition and semantic memory are independent or associated, studies of individuals with semantic dementia lend support for the associative view, as impaired semantic memory has been found to degrade visual representations of meaningful objects. Individuals with semantic dementia for example, have been observed to draw additional features on known target items, for example drawing a flamingo with four legs rather than two (Bozeat et al., 2003; Rogers et al., 2003, 2004).

Semantic memory is considered a division of explicit or declarative, long-term memory; specifically, it is memory for facts or knowledge that can be consciously recalled (Tulving, 1972). Semantic memory supports us in knowing that cats are living creatures that are often furry and domesticated or that a peninsula is a projection of land almost entirely surrounded by water. Importantly, when semantic memories are retrieved into conscious awareness from long-term memory, the stored information is free of contextual information in which the event was experienced (see Schendan, 2012). This context-free representation is what makes semantic memory central to conceptual knowledge and differentiates semantic memory from episodic memory, which stores representations related to the spatiotemporal context within which the memory was encoded. For example, in calling up a memory of petting a cat in episodic memory, the retrieval might contain information about when the event occurred, where you were seated, how the cat felt, what it sounded like, who was present, etc. In retrieving this episode, conceptual knowledge of *cat* stored in semantic memory was retrieved and enriched by episodic memory of the specific experience. Whether episodic memory is dependent on semantic memory or vice versa is debated. Some propose that semantic memory can be formed without explicit knowledge of the episodes over which that knowledge was acquired (and thus is not dependent on episodic memory, e.g., Tulving). Others propose that episodic memory is a precursor, serving as an instance memory that contributes to the formation of a semantic memory (see Squire & Zola, 1998). Research in individuals with semantic dementia supports a distinction between episodic and semantic memory, as object identification and concept recognition degrades in the setting of relatively preserved episodic memory (for review see Hodges & Graham, 2001; Hodges and Patterson, 2007; Yee et al., 2013)

8.3.4 Memory and Language: Language Acquisition

Memory is often described as a key process involved in language acquisition during both infancy and adulthood. In infants, implicit memory processes are particularly important. Infants are thought to acquire aspects of phonology, syntax, and semantics through exposure to language, supported by mechanisms of implicit and statistical learning (see Romberg & Saffran, 2010; Saffran & Kirkham, 2018 for review). Implicit learning that occurs at a level beneath the learner's conscious awareness depends upon nondeclarative memory processes. As infants gain exposure to language, an innate ability to implicitly extract properties of regularly occurring patterns (statistical learning) supports them in creating phonemic categories relevant to their language, recognizing syllable boundaries, and learning grammatical regularities. While the acquired information is utilized in the development of language, specific knowledge of the patterns and regularities remains unavailable to conscious access. A large body of research has studied statistical learning in infants, illustrating its role in language acquisition (see Saffran, 2020). While there is evidence to support the development of declarative memory in infants (see Richmond & Nelson, 2007 for review), a relatively meager body of research has been dedicated to exploring contributions of declarative memory to infant language acquisition given the challenge of evaluating conscious awareness or explicit recall of learned information in infants. There does appear to be a relationship between declarative STM and vocabulary development as well as new word learning in a second language (Avons, et al., 1998; Cheung, 1996; de Abreu et al., 2011; Gathercole & Masoura, 2005; Gathercole et al., 1997; Jarrold et al., 2009; Majerus et al., 2006; Mosse & Jarrold, 2008) observed in young and adolescent children. It is thought that in these contexts, temporary representations of word forms stored in STM, specifically verbal STM, support the acquisition and eventual encoding of phonological forms into LTM (Baddley et al., 1998; Gathercole, 2006).

In adults, rather than taking place in immersive contexts through exposure, second language learning often takes place in classroom contexts where instruction is explicit (Brill-Schuetz & Morgan-Short, 2014). Research suggests that the context of learning likely influences second language learning outcomes (Norris & Ortega, 2000) as well as the neural structures supporting the learning (Morgan-Short et al., 2012). Procedural, nondeclarative memory ability may be tied to advanced second language performance depending on the conditions of training and exposure (Ettlinger et al., 2014; Morgan-Short et al., 2014).

8.3.5 A Memory-Based Account of Language

The declarative-procedural model of language (Ullman et al., 1997; Ullman 2001, 2004) places particular emphasis on memory systems, proposing that language learning, storage, and use are highly dependent on declarative and procedural memory. Procedural memory is a type of implicit, nondeclarative memory specifically related to supporting skill development. The declarative-procedural model of language differs from the majority of language theories that focus on language specific functions, development, processing and neural substrates, drawing attention instead to domain-general systems that contribute to language (Ullman, 2016). Within this model, the explicit, declarative system is thought to support the mental lexicon, a repository (or mental dictionary) of knowledge of the phonological, morphological, semantic, and syntactic properties of words. In contrast, procedural memory is proposed to support the acquisition and retention of regularities of language, such as hierarchical relations of sentence structure (for example the fact that a noun phrase such as *the cat* is composed of a noun [cat] and a determiner [the] and that an adjective [black] could be introduced as a modifier: *the black cat*) or rules that constrain morphological endings (e.g., adding *-ed* to form past tense). This mental grammar contains numerous complex rules that are generally unavailable to conscious access (e.g., nondeclarative). The declarative-procedural model has received attention for its proposed explanations of impairments underlying developmental language disorders (Conti-Ramsden et al., 2015; Ullman, 2004), application to language changes observed in healthy aging, and for its explanations of second language learning (Ullman, 2020). The declarative-procedural theory has been at times critiqued or contested (Joanisse & Seidenberg, 1999; also see Gordon & Dell, 2003 and Seidenberg & Plaut, 2014), but remains important to discussions of memory and language.

8.4 Working Memory

8.4.1 Overview of Working Memory

Working memory (WM) involves the temporary storage and manipulation of information, and it is important for performing a wide range of complex, goal-directed cognitive activities. As Miller (2013) put it simply, "Working memory is, essentially, what we think of as thought" (p. 411). The concept of WM arose from the earlier idea of short-term memory (STM), which concerns the temporary storage of a limited amount of information (Atkinson & Shiffrin, 1968). Baddeley and Hitch (1974) expanded upon the idea of short-term storage to add an element of attentional control, as a more active, executive component. This yielded the most influential model of WM to date, which comprises four components: the phonological loop, the visuospatial sketchpad, the central executive, and the episodic buffer (Baddeley, 2000; see Baddeley, 2012, for an excellent historical overview).

The first two components of the model are modality-specific short-term storage systems. The *phonological loop,* likely the most thoroughly explored aspect of the model, facilitates temporary verbal-acoustic storage, and itself includes two subcomponents: the phonological store and articulatory rehearsal (Baddeley, 1992). The phonological store briefly holds a memory trace for auditory information, like a short audio recording, such as when a person hears a phone number spoken aloud and can remember the last few digits without trying. Articulatory rehearsal can be used to maintain verbal information in the phonological store, or to transcode visual information into the phonological store. The *visuospatial sketchpad* supports short-term storage for visual, spatial, and potentially kinesthetic information (Klauer & Zhao, 2004; Smyth & Pendleton, 1990).

The central executive is the most complex component of the working memory model, involved in attentional control and coordination (Baddeley, 2012). Because working memory is a limited-capacity system, there is a need to "choose" between task-relevant and task-irrelevant information as information is retrieved from LTM and environmental stimuli are perceived (Unsworth & Engle, 2007). The central executive serves this function and, in this way, is closely related, and even overlaps with, the cognitive constructs of attention (discussed earlier in this chapter) and executive functions (discussed later). The central executive controls and coordinates the other components of working memory.

The episodic buffer was the last component added to the model, as the need was seen for a system that could take information from the modality-specific short-term stores and unite it with information in LTM (Baddeley, 2000). To fill this need, the episodic buffer was proposed as a component that chunks information into episodes, integrating information from different modalities along with prior knowledge. In this way, the episodic buffer is important both for forming and temporarily maintaining unitary representations (Rudner & Rönnberg, 2008).

8.4.2 Neural Basis of Working Memory

Through neuroimaging studies as well as the study of people with brain lesions, the various components of working memory have been localized to different brain regions. The phonological store has been localized to left temporoparietal regions, in particular the supramarginal gyrus, while articulatory rehearsal is associated with Broca's area (Rogalsky et al., 2008). In contrast, the visuospatial sketchpad is primarily reliant on right-hemisphere regions (Smith & Jonides, 1997). The central executive is a domain-general system, not bound to one particular sensory modality or single function, and it is broadly localized to the frontal lobes (Stuss, 2011; Wager & Smith, 2003). The episodic buffer is associated with middle temporal lobe regions including the entorhinal and perirhinal cortices and hippocampus (Rudner & Rönnberg, 2008; Schmidt-Wilcke et al., 2009). In addition to the approach of localizing model components to brain regions, more recent neural models of working memory have highlighted the importance of networks. In particular, a dopamine-driven network involving the prefrontal cortex, basal ganglia, striatum, and thalamus, but also influencing a broad range of brain regions, is suggested to support the process of linking new experiences to each other and to long-term knowledge, recursively building and updating representations (Miller, 2013; Miller & Cohen, 2001). This process is overseen by the prefrontal cortex, which controls the flow of information throughout various brain networks via oscillatory activity. Different oscillatory rhythms have been implicated in different types of processing (Miller et al., 2018).

8.4.3 Measuring Working Memory Capacity

Many tasks commonly used to measure "working memory ability" in fact rely on the short-term storage aspect of the system, for example, digit span (Murray et al., 2018). More recently, there has been a preference for tasks that emphasize the "working" aspect of working memory, that is, the manipulation of information, and not just maintenance. Such tasks include *n*-back (Kirchner, 1958), updating (Morris & Jones, 1990; Vallat-Azouvi et al., 2007), and complex memory span tasks (Conway et al., 2005). In *n*-back tasks, stimuli are presented sequentially, with the examinee instructed to respond when a stimulus matches the one presented *n* items ago; *n* is a prespecified number, typically 1, 2, or 3. Updating tasks also include sequential presentation of stimuli, but with the prompt to recall the previous *n* items once the sequence has concluded. Sequences vary in length unpredictably, for example, being *n*, *n*+2, *n*+4, or *n*+6 items long. Complex span tasks have received a particular amount of attention in the research literature, in part because it is possible to use parallel versions of these tasks that require different specific skills such as reading or computational abilities, but share a domain general "working memory" demand that can be statistically identified as a latent variable (Kane et al., 2004). Daneman and Carpenter (1980) are recognized as having developed the first complex span task: reading span. In this task, an examinee reads a series of sentences and makes a true/false judgment about each. After completing the series of sentences, the examinee is prompted to recall the last word of each sentence in the set, and the span is the largest number of words recalled. Subsequently developed complex span tasks include operation span, symmetry span, and rotation span (Foster et al., 2014; Kane et al., 2004). What these tasks all have in common is the requirement to recall elements (letters, numbers, objects, spatial locations) whose serial presentation is interleaved with the performance of a task likely to interfere with that recall (Oswald et al., 2015).

The degree to which working memory measures are domain general or domain specific is a point of discussion especially relevant to the relationship between language and other cognitive abilities. Performance on a working memory task correlates with measures of the specific ability required in that task, for example, reading, visuospatial, or arithmetic skills (Oswald et al., 2015). However, different working memory tasks have been found to produce scores that are highly correlated, even when the tasks have different specific processing demands. This suggests these tasks share some common demands that are the essence of working memory (Engle et al., 1999; Kane et al., 2004). Because of concerns about task-specific demands, it has been deemed best to use multiple tasks or paradigms targeting working memory, coupled with statistical techniques such as latent variable analysis, in order to get a valid measure of working memory capacity (Conway et al., 2005; Foster et al., 2014). This is particularly important in clinical populations for whom the effects of task-specific demands may be particularly strong (e.g., Murray et al., 2018).

8.4.4 Importance of Language for Working Memory Ability

The nature of the information held in verbal WM was explored by early studies showing the effects of phonological similarity and word length on immediate serial recall performance, indicating that information is often stored in a phonological form. Similar-sounding words are harder to recall in sequence compared to words with more distinct phonology, and the effect is specific to phonological similarity; immediate recall is not equally affected by semantic similarity (Baddeley, 1966, 2003; Conrad & Hull, 1964). Shorter words are recalled better than longer ones, with a clear advantage for monosyllabic vs. five-syllable words. This latter effect can be abolished with articulatory suppression, that is, blocking articulatory rehearsal by asking the learner to repeat an unrelated word or syllable, which indicates that articulatory rehearsal is responsible for the word-length effect (Baddeley et al., 1975; Klatte et al., 2002). Articulatory rehearsal does not require overt articulation abilities, since people with severe dysarthria also show word length effects (Baddeley & Wilson, 1985). However, the ability to *program* speech motor commands appears to be needed for subvocal rehearsal, as people with apraxia of speech do not show effects of rehearsal (Waters et al., 1992). Collectively, these findings indicate that information in the phonological store is indeed phonological in nature, and motor programming abilities are needed to maintain information in the store through rehearsal.

Making use of the phonological loop, in particular articulatory rehearsal, is a common strategy for maintaining serial order information, even for simple visual items – perhaps even in situations when it is not the best available strategy (Hughes & Marsh, 2017; Peteranderl & Oberauer, 2018; Souza & Oberauer, 2018). Indeed, there is evidence that visual information is often transcoded to be held in the phonological loop when serial order matters, so long as the visual items can be efficiently verbally labeled (Henson et al., 2000). This may be the case because an auditory signal or articulatory action is inherently sequential, in contrast to a visual scene. While verbally mediated strategies are frequently used to maintain serial order information, they are not obligatory. Abstract visual and spatial information (which are not easily verbally labeled) can be recalled sequentially, and current models of memory for serial order include domain-general mechanisms (Hurlstone & Hitch, 2018; Hurlstone et al., 2014; Majerus, 2019).

Language ability and working memory capacity likely have a complex, bidirectional relationship, as suggested by findings in both adult and pediatric clinical populations. At the group level, adults with aphasia exhibit deficits of both verbal and visual working memory, and some studies have shown aphasia severity to correlate with both visual and verbal span tasks, although these associations tend to be higher for memory tasks with more verbal content (DeDe et al., 2014; Lang & Quitz, 2012; Potagas et al., 2011). Associations between *nonverbal* working memory tasks and aphasia severity may be attributable to a domain general working memory deficit in at least some people with aphasia, and/or to verbal strategies being useful even in these supposedly nonverbal tasks, such that people with aphasia remain at a disadvantage due to their language impairment.

8.4.5 Importance of Working Memory Capacity for Language Ability

Working memory capacity comes to bear on the performance of many tasks and abilities, including language abilities. It plays an important role in comprehension of sentences and longer passages (i.e., discourse), where ideas must be linked to one another and information retained as the listener or reader constructs meaning. In fact, the demands of language comprehension epitomize a working memory task: the listener or reader maintains representations of recently encountered words and ideas while computing and updating propositions (essentially, statements that can be true or false), while also retrieving information from long-term memory as needed to make sense of the incoming information (Just & Carpenter, 1992; Kintsch et al., 1999). It is therefore not surprising that reading and auditory comprehension abilities are correlated with verbal working memory (e.g., Cain et al., 2004; Choinski et al., 2020; Peng et al., 2018).

Working memory capacity is important even at the single-sentence level, especially for understanding syntactically complex or ambiguous sentences. Phonological loop capacity is thought to be important for maintaining the first part of the sentence in immediate memory in order to interpret the later part of the sentence (Fallon et al., 2006; Lauro et al., 2010). Baddeley (2003) referred to this function as “a backup to comprehension” (p. 193), which highlights the particular importance of phonological loop capacity in situations where comprehension is not immediate, due to structural properties of the sentence, pragmatic aspects of the communicative situation, or a language impairment. In this way, language impairments often seen in aphasia, such as delayed or otherwise impaired single-word comprehension, may be compounded by the reduced phonological loop capacity often reported in this same disorder. While the importance of the phonological loop has been highlighted in the literature on WM and sentence comprehension, other aspects of WM should not be discounted. Caplan and Waters (1999) proposed a specialized system for WM resources specific to syntactic processing, and more recently, evidence has been presented for the importance of semantic WM for sentence comprehension (Martin, 2021).

Working memory capacity has also been implicated in language learning, especially word learning. Relevant evidence has been reported both for adults learning new words in a non-native language as well as children performing word learning tasks (Gray et al., 2022; Kempe et al., 2010; Martin & Ellis, 2012). Studies in adults have shown that occupying the phonological loop impedes word learning in a foreign language, and that word length and phonological similarity affect word learning (Papagno et al., 1991; Papagno & Vallar, 1992). In children, measures of working memory predict word learning performance, even after accounting for expressive vocabulary and nonverbal IQ (Gray et al., 2022). Children with developmental language disorder (DLD) show lower scores on working memory measures when compared to typically developing peers (Archibald & Gathercole, 2006). In addition, children with DLD consistently show impaired performance on nonword repetition, a task that requires the phonological loop (Estes et al., 2007). As children get older, the relationship between verbal working memory abilities and vocabulary acquisition becomes more reciprocal, with a rich vocabulary supporting nonword repetition (Baddeley, 2003; Munson et al., 2005).

8.5 Executive Function

8.5.1 Overview of Executive Functions

Finally, we turn to executive functions (EFs), which are cognitive functions that direct other cognitive functions. As such, they are complex, multidimensional, and challenging to define. Examples of executive skills include inhibiting, maintaining a goal, shifting among tasks or stimuli, self-monitoring, and (in some models, e.g., Engle, 2002) the allocation of attention. The tripartite model of EFs offered by Miyake and colleagues is arguably the most prominent and analytically well-supported EF model to date (Miyake et al., 2000; Miyake & Friedman, 2012). The model initially included three EFs: updating (monitoring and refreshing WM), shifting (flexibly switching among tasks, stimuli, or mental sets), and inhibition (overriding a habitual or prepotent response) (Miyake et al., 2000). It is noteworthy that within the earlier version of this model, updating was, essentially, synonymous with WM. The relationship between working memory and EFs has been described substantially differently from model to model: while Baddeley incorporated executive attention and skills into his model of WM (the central executive), other EF models often place WM under the umbrella of EFs (Baddeley, 2000; Baddeley & Hitch, 1974; Garon et al., 2008; Miyake et al., 2000). In the tripartite Miyake and Friedman model, the EFs are correlated but separable; the authors refer to this conceptualization as "unity and diversity" (a term that apparently originated with Teuber, 1972; Friedman & Robbins, 2021; Miyake et al., 2000; Miyake & Friedman, 2012). Later confirmatory factor analysis, however, revealed that the three EFs all shared a common variance ("common EF"), which was suggested to reflect maintenance of goals and information needed to achieve them. Inhibition was found to correlate perfectly with common EF and did not contribute any unique variance to the model (Friedman et al., 2008, 2011).

Of course, there are other models of EFs, and while they typically include at least some of the components from the Miyake and Friedman model, they often use different labels, suggesting slightly different conceptions of the exact type of function included, for example, cognitive flexibility instead of shifting, WM instead of updating (Ackerman & Friedman-Krauss, 2017). Alternatively, some models further subdivide components (e.g., separating out inhibition and interference control; Diamond, 2013). Furthermore, many EF models also incorporate different components altogether, such as attentional control, goal setting, and information processing (Ackermann & Friedman-Krauss, 2017; Jones et al., 2016).

Interestingly, models of EF *development* highlight that children younger than about eight years likely have a more integrated executive system than adults, and therefore the tripartite model might not yet apply (Karr et al., 2018; Wiebe et al., 2008). The different components of EF likely emerge sequentially, with a gradual increase in the ability to sustain selective attention supporting the parallel development of inhibitory control and WM. These skills need to be coordinated to accomplish goal-directed tasks, leading to the development of cognitive flexibility and planning (Garon et al., 2008). More recently, the role of motivation and affective factors are increasingly being considered in conceptions of EFs, for example, "hot" and "cool" EFs (Botvinick & Braver, 2015; Chavez-Arana et al., 2018; Liew, 2011; Metcalfe & Mischel, 1999).

8.5.2 Neural Basis of Executive Functions

EFs have long been associated with the frontal lobes, specifically the prefrontal cortex (PFC). In an effort to understand the neural instantiation of the various components of EFs, areas of the PFC have been grouped into subregions or systems that support different aspects of EF. For example, the medial prefrontal cortex (mPFC) has been associated with attentional control and set-shifting, while the orbital prefrontal cortex (OFC) may support response inhibition and aspects of cognitive flexibility (Logue & Gould, 2014). The PFC is connected to many cortical and subcortical brain regions, reflecting its role in influencing neural activity across a range of domains, including visuospatial and verbal (Friedman & Robbins, 2021). Different networks have been identified, through which the PFC is connected to various brain regions to subserve different complex functions, for example, the dorsal and ventral attention networks (Menon & D'Esposito, 2022). In this view, the PFC is the hub that connects the networks involved in cognitive control, but it is more than a passive hub (Friedman & Robbins, 2021). Miller and Cohen (2001) provided an influential account of how the PFC maintains patterns of activation that represent goals and a way to meet them. This activation results in bias signals sent to other brain areas, such that inputs from sensory cortices, information about internal states, and appropriate responses are all linked. This function of the PFC is consistent with the conception of EFs as cognitive activity that directs other cognitive activity, as well as the role of EFs in planning and maintaining goal-directed behavior. Moreover, PFC activation is particularly important when the needed response to a particular stimulus is not well established (e.g., solving a novel problem, or acting contrary to habit) or is changing during a task; notably, these are situations where people with frontal lobe damage often struggle (Levine et al., 2000).

8.5.3 Measuring Executive Functions

Even within the bounds of the tripartite model, EFs are measured in various ways across studies. Tasks used to target inhibition include versions of the Stroop, Simon, Eriksen flanker, antisaccade, and go/no-go tasks (Kaushanskaya et al., 2017, provide examples with references, p. 913). Inhibition has been further divided into *response inhibition*, which involves suppressing a dominant or trained response (e.g., antisaccade, go/no-go, Stroop), and *interference control*, which involves resisting distractor stimuli and requires selective attention (e.g., Flanker task, Simon task) (Pauls & Archibald, 2016). Updating is measured with many of the same tasks used to evaluate WM, since there is broad overlap in these constructs: Corsi blocks, *n*-back, various span tasks, and updating tasks requiring the participant to always remember the last *n* items, where *n* is set at a rather low number, for example, three (Kaushanskaya et al., 2017; Miyake et al., 2000). Cognitive flexibility has been measured by a range of tasks, and may be subdivided into those that require switching between tasks (e.g., Wisconsin Card Sorting Task, trail making tasks) and those that require creative production of unique items (e.g., design generation, verbal fluency) (Karr et al., 2018; Pauls & Archibald, 2016; Tucha et al., 2012).

Because EFs are complex and somewhat abstract, it is impossible to evaluate them in a "pure" form; the tasks used to evaluate them always require other cognitive abilities such as visual attention, language comprehension, or object recognition. It is therefore becoming increasingly preferred to use multiple tasks targeting the same EF, and employ statistical techniques to isolate shared variance and identify it as a latent factor, as has been done with measures of WM (e.g., Friedman et al., 2016; Seer et al., 2021).

8.5.4 Associations between Language and Executive Functioning

There is a complex interplay between language abilities and EFs, with each affecting the other. The relationship between these abilities has been most thoroughly examined in children and in groups with various clinical diagnoses, likely in part because these populations exhibit more variance in both sets of abilities, offering more statistical power to examine associations. The evidence from studies of children suggests that language skills, EFs, and social cognition likely develop in tandem, with intertwined causal relationships (Howland, 2014). Verbal abilities are correlated with EFs in children, and a number of longitudinal studies have demonstrated predictive relationships between language and EF (e.g., Fuhs & Day, 2011; Hughes et al., 2010). For example, low-income preschoolers' language skills at their entry into school predicted EF gains during the school year (Kuhn et al., 2016). Associations between language ability and EFs have been demonstrated in a range of clinical groups including children with DLD (Kapa & Plante, 2015) and adults with aphasia (e.g., El Hachioui et al., 2014). Children with DLD show poorer executive control of attention, with lower scores on measures of sustained selective attention, compared to age-matched peers (Ebert & Kohnert, 2011; Park et al., 2019). They also show lower scores than peers with typical language on tasks requiring working memory, shifting, or inhibition (Kapa & Plante, 2015; Vugs et al., 2014). A meta-analysis of forty-six studies found that children with DLD aged from four to fourteen performed worse than age-matched peers on measures of inhibition and cognitive flexibility, even when age, severity of language impairment, and linguistic demands were accounted for (Pauls & Archibald, 2016).

Adults with aphasia are more likely than those without aphasia to show EF impairments (Mohapatra & Marshall, 2020; Purdy, 2002). In one study of 147 individuals with aphasia, 39 percent showed EF impairment at one year post-stroke, with EF measured as a composite across three tasks not requiring a spoken response (El Hachioui et al., 2014). In another study of thirty-eight individuals with aphasia, 79 percent scored below the range of typical performance on at least one of four nonverbal tasks with EF demands (Olsson et al., 2020). Even adults with relatively mild aphasia showed poorer shifting across a range of tasks compared to an age- and education-matched control sample (Chiou & Kennedy, 2009). However, language ability does not predict nonverbal cognitive ability at the individual level in aphasia, as people with relatively mild aphasia can exhibit notable EF deficits, while others with more severe aphasia can perform very well on nonverbal EF measures (Helm-Estabrooks, 2002). This holds true when simple language tasks, such as naming, word repetition, category judgment, or rhyme judgment, are used to evaluate language ability. In contrast, when these same types of tasks were administered in a more cognitively demanding context (i.e., five-second filled interval between stimulus and response), performance was found to correlate with nonlinguistic measures of EF, in particular inhibition (Obermeyer et al., 2020). This relationship was particularly strong for language tasks requiring lexical semantic processing, perhaps because inhibition is needed to select a lexical item from among semantic competitors (e.g., Martin et al., 2012; Obermeyer et al., 2020; see also Schnur et al., 2006). This highlights that the association between language and EF abilities is context- and task-dependent.

8.5.5 Importance of Language for Executive Functioning

In considering the relationship between language and EFs, another issue that arises is the extent to which executive skills are mediated by, or even dependent on, language ability. Language skills can affect a person's ability to make use of inner speech for self-regulation, and inner speech improves performance on problem solving, strategy planning, and other EF tasks (Howland, 2014; Kirkham et al., 2003; Lidstone et al., 2010; Miyake et al., 2004). For example, use of inner speech or self-talk was associated with better EF performance in preschoolers completing a challenging task, and young children performing a card sorting task did better when instructed to verbally label the target feature prior to sorting each card (Alarcón-Rubio et al., 2014; Kirkham et al., 2003). In fact, self-talk is taught as a strategy for emotional regulation and problem solving for children, as well as for adults with EF impairments secondary to traumatic brain injury (e.g., Fahy, 2014; Howland, 2014; Kennedy & Coelho, 2005). Furthermore, language ability may affect EF task performance by allowing a person to maintain information in verbal short-term memory (Fatzer & Roebers, 2012). Individuals with language impairments such as aphasia may not have facile access to this strategy, limiting their ability to readily maintain and update information in verbal WM for use during complex tasks used to measure EFs.

8.5.6 Importance of Executive Functions in Language Processing

Finally, it is important to note that while language abilities can facilitate executive functioning, the reverse has also been shown to be true: EFs can support language processing and development. Inhibition may be a particularly important EF during language acquisition because children acquiring language need to suppress irrelevant acoustic information to develop phonological representations and need to inhibit semantic competitors in order to subsequently attach labels to concepts. Deficits in inhibitory control could therefore affect word learning (Pauls & Archibald, 2016). Indeed, inhibition predicted word learning outcomes in children with DLD, even after accounting for nonverbal IQ, and receptive vocabulary (Kapa & Erikson, 2020). Inhibition has also been associated with syntactic processing abilities in typically developing school-aged children. Measures of syntactic ability, including following complex instructions and detecting syntactic errors, were correlated with the ability to inhibit information in the context of a nonverbal task (Kaushanskaya et al., 2017). Other EFs are likely also important for language learning, as updating and shifting abilities are needed to flexibly apply linguistic rules to acquire new linguistic forms and content (Howland, 2014). In adults with typical language abilities, there is data supporting relationships between distinct EFs and particular aspects of language processing; for example, the ability to appropriately revise one's interpretation of a garden path sentence is associated with inhibition and shifting (Novick et al., 2005; Peristeri et al., 2020). However, the evidence on these sorts of relationships is somewhat limited, likely due to the challenge of adequately controlling all the relevant factors.

Executive abilities are particularly important at the discourse level of language, that is, beyond single sentences. This includes stories and conversation, where planning, self-monitoring, goal maintenance, and other executive abilities are needed to produce an organized narrative, or to participate appropriately in an extended interaction with a communication partner (Alexander, 2006). Moreover, EF deficits manifest in discourse that is disorganized, illogical, perseverative, or otherwise incoherent; this has been well described in the literature on communication deficits after traumatic brain injury, as well as the schizophrenia literature describing "thought disorder" and related communication difficulties (e.g., Coelho, 2007; Covington et al., 2005; Kuperberg et al., 2010). This role of EFs at the more global level – that is, beyond the mechanics of syntax or the regulation of lexical activation – is also highlighted by evidence regarding the importance of EFs for successful functional communication in PWAs (Fridriksson et al., 2006; Nicholas & Connor, 2017; Olsson et al., 2019). Moreover, EFs are important for PWAs to maximize the linguistic abilities they have, and use compensatory strategies such as gesture, drawing, or circumlocution, as well as to make use of formal augmentative and alternative communication (AAC) devices (Nicholas et al., 2011). This may be especially true for people with severe aphasia, who have an increased need to flexibly switch between modalities and monitor communication partners' responses for understanding that the intended message has been successfully conveyed.

8.5.7 Challenges of Examining Relationships between Executive Functions and Language

Studying the relationship between EFs and language is particularly challenging because both systems are complex and multifaceted. Each construct is difficult to measure in its own right: laboratory tasks to examine language abilities typically place demands on processing speed (which may become an important third variable when examining the relationship between language and EF; Kaushanskaya et al., 2017), and EF tasks can be particularly subject to familiarity or fatigue, lowering reliability (Martella et al., 2011; Müller et al., 2012). The majority of studies exploring the relationship between language and EFs are limited in scope, including one or a few EF measures and one or a few language tasks, limiting the ability of any individual study to capture the complex interplay between language and EFs. A more fundamental problem is the fact that the tasks used to measure EFs typically have language processing demands or can reasonably be thought to benefit from language use (e.g., recalling task instructions or covertly rehearsing strategies). Language is the "medium" in which EFs are typically evaluated, and language demands include reading or listening to instructions, efficiently comprehending verbal stimuli, and making spoken responses. Even when these demands are minimized, many tasks targeting EF likely benefit from verbal strategies or mediation. Language demands are a particularly insidious aspect of the "task impurity problem" when it comes to EFs (Miyake et al., 2000).

8.6 Conclusion

We hope that the individual sections above have provided the reader with an overview of some of the current models on attention, memory, working memory, and executive functions, as well as how these skills are thought to be interconnected with language processing. As discussed in the introduction to this chapter, we chose to present these four constructs individually in service of clarity, while recognizing that this approach allows for relatively little nuance in terms of the overlap between these four constructs. Some models conceptualize these constructs, and/or the boundaries between them, somewhat differently, as has been noted at various points throughout the chapter. Nevertheless, we believe that developing a solid understanding of each of these four constructs separately will serve as a useful base from which to draw one's own conclusions about the best way to conceptualize cognitive and language skills, as well as to design and implement future studies elucidating the nature of each of these constructs.

References

Ackerman, D. J., & Friedman-Krauss, A. H. (2017). Preschoolers' executive function: Importance, contributors, research needs and assessment options. *ETS Research Report Series, 2017*(1), 1–24. https://doi.org/10.1002/ets2.12148

Alarcón-Rubio, D., Sánchez-Medina, J. A., & Prieto-García, J. R. (2014). Executive function and verbal self-regulation in childhood: Developmental linkages between partially internalized private speech and cognitive flexibility. *Early Childhood Research Quarterly, 29*(2), 95–105. https://doi.org/10.1016/j.ecresq.2013.11.002

Alexander, M. P. (2006). Impairments of procedures for implementing complex language are due to disruption of frontal attention processes. *Journal of the International Neuropsychological Society, 12*(2), 236–247. https://doi.org/10.1017/s1355617706060309

Annese, J., Schenker-Ahmed, N. M., Bartsch, H., Maechler, P., Sheh, C., Thomas, N., ... & Corkin, S. (2014). Postmortem examination of patient HM's brain based on histological sectioning and digital 3D reconstruction. *Nature Communications*, 5(1), 3122. https://doi.org/10.1038/ncomms4122

Archibald, L. M., & Gathercole, S. E. (2006). Short-term and working memory in specific language impairment. *International Journal of Language & Communication Disorders, 41*(6), 675–693. https://doi.org/10.1080/13682820500442602

Atilgan, H., Town, S. M., Wood, K. C., Jones, G. P., Maddox, R. K., Lee, A. K., & Bizley, J. K. (2018). Integration of visual information in auditory cortex promotes auditory scene analysis through multisensory binding. *Neuron, 97*(3), 640–655. https://doi.org/10.1016/j.neuron.2017.12.034

Atkinson, R. C., & Shiffrin, R. M. (1968). Human memory: A proposed system and its control processes. *Psychology of Learning and Motivation, 89–195.* https://doi.org/10.1016/s0079-7421(08)60422-3

Avons, S. E., Wragg, C. A., Cupples, W. L., & Lovegrove, W. J. (1998). Measures of phonological short-term memory and their relationship to vocabulary development. *Applied Psycholinguistics, 19*(4), 583–601. https://doi.org/10.1017/s0142716400010377

Baddeley, A. (1992). Working memory. *Science, 255*(5044), 556–559. https://doi.org/10.1126/science.1736359

Baddeley, A. (2000). The episodic buffer: A new component of working memory? *Trends in Cognitive Sciences, 4(*11), 417–423. https://doi.org/10.1016/s1364-6613(00)01538-2

Baddeley, A. (2003). Working memory and language: An overview. *Journal of Communication Disorders, 36*(3), 189–208. https://doi.org/10.1016/s0021-9924(03)00019-4

Baddeley, A. (2012). Working memory: Theories, models, and controversies. *Annual Review of Psychology, 63*(1), 1–29. https://doi.org/10.1146/annurev-psych-120710-100422

Baddeley, A. D. (1966). Short-term memory for word sequences as a function of acoustic, semantic and formal similarity. *Quarterly Journal of Experimental Psychology, 18*(4), 362–365. https://doi.org/10.1080/14640746608400055

Baddeley A. D., & Hitch, G. J. (1974). Working memory. In G. A. Bower (Ed.), *The Psychology of Learning and Motivation: Advances in Research and Theory* (Vol. 8, pp. 47–89). Academic Press. http://doi.org/10.1016/s0079-7421(08)60452-1

Baddeley, A. D., Thomson, N., & Buchanan, M. (1975). Word length and the structure of short-term memory. *Journal of Verbal Learning and Verbal Behavior, 14*(6), 575–589. https://doi.org/10.1016/s0022-5371(75)80045-4

Baddeley, A., & Wilson, B. (1985). Phonological coding and short-term memory in patients without speech. *Journal of Memory and Language, 24*(4), 490–502. https://doi.org/10.1016/0749-596x(85)90041-5

Baddeley, A., Gathercole, S., & Papagno, C. (1998). The phonological loop as a language learning device. *Psychological Review, 105*(1), 158–173. https://doi.org/10.1037/0033-295x.105.1.158

Banerjee, S., Snyder, A. C., Molholm, S., & Foxe, J. J. (2011). Oscillatory alpha-band mechanisms and the deployment of spatial attention to anticipated auditory and visual target locations: Supramodal or sensory-specific control mechanisms? *Journal of Neuroscience, 31*(27), 9923–9932. https://doi.org/10.1523/jneurosci.4660-10.2011

Barry, D. N., & Maguire, E. A. (2019). Remote memory and the hippocampus: A constructive critique. *Trends in Cognitive Sciences, 23*(2), 128–142. https://doi.org/10.1016/j.tics.2018.11.005

Best, V., Ozmeral, E. J., & Shinn-Cunningham, B. G. (2007). Visually-guided attention enhances target identification in a complex auditory scene. *Journal for the Association for Research in Otolaryngology, 8*(2), 294–304. https://doi.org/10.1007/s10162-007-0073-z

Bisaz, R., Travaglia, A., & Alberini, C. M. (2014). The neurobiological bases of memory formation: From physiological conditions to psychopathology. *Psychopathology, 47*(6), 347–356. https://doi.org/10.1159/000363702

Blumenfeld, R. S., & Ranganath, C. (2007). Prefrontal cortex and long-term memory encoding: An integrative review of findings from neuropsychology and neuroimaging. *The Neuroscientist, 13*(3), 280–291. https://doi.org/10.1177/1073858407299290

Bosse, M.-L., Tainturier, M. J., & Valdois, S. (2007). Developmental dyslexia: The visual attention span deficit hypothesis. *Cognition, 104*(2), 198–230. https://doi.org/10.1016/j.cognition.2006.05.009

Botvinick, M., & Braver, T. (2015). Motivation and cognitive control: From behavior to neural mechanism. *Annual Review of Psychology, 66,* 83–113. https://doi.org/10.1146/annurev-psych-010814-015044

Bozeat, S., Lambon Ralph, M. A., Graham, K. S., Patterson, K., Wilkin, H., Rowland, J., Rogers, T. T., & Hodges, J. R. (2003). A duck with four legs:

Investigating the structure of conceptual knowledge using picture drawing in semantic dementia. *Cognitive neuropsychology, 20*(1), 27–47. https://doi.org/10.1080/02643290244000176

Bregman, A. S. (1990). *Auditory Scene Analysis: The Perceptual Organization of Sound*. The MIT Press. https://doi.org/10.7551/mitpress/1486.001.0001

Brill-Schuetz, K., & Morgan-Short, K. (2014). The role of procedural memory in adult second langauge acquisition. *Proceedings of the Annual Meeting of the Cognitive Science Society, 36*(36). https://doi.org/10.1037/e502412013-500

Broadbent, D. E. (1952). Failures of attention in selective listening. *Journal of Experimental Psychology, 44*(6), 428–433. https://doi.org/10.1037/h0057163

Brod, G., Werkle-Bergner, M., & Shing, Y. L. (2013). The influence of prior knowledge on memory: A developmental cognitive neuroscience perspective. *Frontiers in Behavioral Neuroscience, 7*, 139. https://doi.org/10.3389/fnbeh.2013.00139

Brown, G. J., Ferry, R. T., & Meddis, R. (2010). A computer model of auditory efferent suppression: Implications for the recognition of speech in noise. *The Journal of the Acoustical Society of America, 127*(2), 943–954. https://doi.org/10.1121/1.3273893

Buckner, R. L. (2002). Frontally mediated control processes contribute to source memory retrieval. *Neuron, 35*(5), 817–818. https://doi.org/10.1016/s0896-6273(02)00866-8

Cabeza, R., & Moscovitch, M. (2013). Memory systems, processing modes, and components: Functional neuroimaging evidence. *Perspectives on Psychological Science, 8*(1), 49–55. https://doi.org/10.1177/1745691612469033

Cain, K., Oakhill, J., & Bryant, P. (2004). Children's reading comprehension ability: Concurrent prediction by working memory, verbal ability, and component skills. *Journal of Educational Psychology, 96*(1), 31–42. https://doi.org/10.1037/0022-0663.96.1.31

Caplan, D., & Waters, G. S. (1999). Verbal working memory and sentence comprehension. *Behavioral and Brain Sciences, 22*, 77–126. https://doi.org/10.1017/S0140525X99001788

Chavez-Arana, C., Catroppa, C., Carranza-Escarcega, E., Godfrey, C., Yanez-Tellez, G., Prieto-Corona, B., ... Anderson, V. (2018). A systematic review of interventions for hot and cold executive functions in children and adolescents with acquired brain injury. *Journal of Pediatric Psychology, 43*(8), 928–942. https://doi.org/10.1093/jpepsy/jsy013

Cherry, E.C. (1953) Some experiments on the recognition of speech, with one and with two ears. *Journal of the Acoustical Society of America, 25*, 974–979. http://dx.doi.org/10.1121/1.1907229

Cheung, H. (1996). Nonword span as a unique predictor or second-language vocabulary language. *Developmental Psychology, 32*(5), 867. https://doi.org/10.1037/0012-1649.32.5.867

Chiou, H. S., & Kennedy, M. R. T. (2009). Switching in adults with aphasia. *Aphasiology, 23*(7–8), 1065–1075. https://doi.org/10.1080/02687030802642028

Choinski, M., Szelag, E., Wolak, T., & Szymaszek, A. (2020). Working memory in aphasia: The role of temporal information processing. *Frontiers in Human Neuroscience, 14.* https://doi.org/10.3389/fnhum.2020.589802

Ciaramelli, E., Burioanová, H., Vallesi, A., Cabeza, R., & Moscovitch, M. (2020). Funcitonal interplay between posterior parietal cortex and hippocampus durign detection of memory targets and non-targets. *Frontiers in Neuroscience, 14.* https://doi.org/10.3389/fnins.2020.563768

Coelho, C. A. (2007). Management of discourse deficits following traumatic brain injury: Progress, caveats, and needs. *Seminars in Speech and Language, 28*(2), 122–135. https://doi.org/10.1055/s-2007-970570

Conrad, R., & Hull, A. J. (1964). Information, acoustic confusion and memory span. *British Journal of Psychology, 55*(4), 429–432. https://doi.org/10.1111/j.2044-8295.1964.tb00928.x

Conti-Ramsden, G., Ullman, M. T., & Lum, J. A. (2015). The relation between receptive grammar and procedural, declarative, and working memory in specific language impairment. *Frontiers in Psychology, 6*, 1090. https://doi.org/10.3389/fpsyg.2015.01090

Conway, A. R. A., Kane, M. J., Bunting, M. F., Hambrick, D. Z., Wilhelm, O., & Engle, R. W. (2005). Working memory span tasks: A methodological review and user's guide. *Psychonomic Bulletin & Review, 12*(5), 769–786. https://doi.org/10.3758/bf03196772

Corbetta, M., & Shulman, G. L. (2002). Control of goal-directed and stimulus-driven attention in the brain. *Nature Reviews Neuroscience, 3*(3), 201–215. https://doi.org/10.1038/nrn755

Corkin, S. (1968). Acquisition of motor skill after bilateral medial temporal-lobe excision. *Neuropsychologia. 6*(3), 255–265.

Covington, M. A., He, C., Brown, C., Naçi, L., McClain, J. T., Fjordbak, B. S., Semple, J., & Brown, J. (2005). Schizophrenia and the structure of language: The linguist's view. *Schizophrenia Research, 77*(1), 85–98. https://doi.org/10.1016/j.schres.2005.01.016

Cowan, N. (2008). What are the differences between long-term, short-term, and working memory?. In W. S. Sossin, J.-C. Lacaille, V. F. Castellucci, & S. Belleville (Eds.), *Progress in Brain Research* (Vol. 169, pp. 323–338). Elsevier. https://doi.org/10.1016/S0079-6123(07)00020-9

Daneman, M., & Carpenter, P. A. (1980). Individual differences in working memory and reading. *Journal of Verbal Learning & Verbal Behavior, 19*(4), 450–466. https://doi.org/10.1016/S0022-5371(80)90312-6

Datta, R., & DeYoe, E. A. (2009). I know where you are secretly attending! The topography of human visual attention revealed with fMRI. *Vision Research, 49*(10), 1037–1044. https://10.1016/j.visres.2009.01.014

Davachi, L., & Dobbins, I. G. (2008). Declarative memory. *Current Directions in Psychological Science, 17*(2), 112–118. https://doi.org/10.1111/j.1467-8721.2008.00559.x

de Abreu, P. M. J. E., Gathercole, S. E., & Martin, R. (2011). Disentangling the relationship between working memory and language: The roles of short-term storage and cognitive control. Learning and Individual Differences, 21(5), 569–574. https://doi.org/10.1016/j.lindif.2011.06.002

DeDe, G., Ricca, M., Knilans, J., & Trubl, B. (2014). Construct validity and reliability of working memory tasks for people with aphasia. *Aphasiology, 28*(6), 692–712. https://doi.org/10.1080/02687038.2014.895973

Dew, I. T. Z., & Cabeza, R. (2011). The porous boundaries between explicit and implicit memory: Behavioral and neural evidence. *Annals of the New York Academy of Sciences, 1224*(1), 174–190. https://doi.org/10.1111/j.1749-6632.2010.05946.x

Diamond, A. (2013). Executive functions. *Annual Review of Psychology, 64*(1), 135–168. https://doi.org/10.1146/annurev-psych-113011-143750

Dickerson, B. C., & Eichenbaum, H. (2010). The episodic memory system: Neurocircuitry and disorders. *Neuropsychopharmacology, 35*(1), 86–104. https://doi.org/10.1038/npp.2009.126

Dudai, Y., & Morris, R. G. (2013). Memorable trends. *Neuron, 80*(3), 742–750. https://doi.org/10.1016/j.neuron.2013.09.039

Durlach, N. I., Mason, C. R., Kidd, G. Jr,, Arbogast, T. L., Colburn, H. S., & Shinn-Cunningham, B. G. (2003). Note on informational masking (L). *The Journal of the Acoustical Society of America, 113*(6), 2984–2987. https://doi.org/10.1121/1.1570435

Ebert, K. D., & Kohnert, K. (2011). Sustained attention in children with primary language impairment: A meta-analysis. *Journal of Speech, Language, and Hearing Research, 54*(5), 1372–1384. https://doi.org/10.1044/1092-4388(2011/10-0231)

Eichenbaum, H. (2001). The hippocampus and declarative memory: Cognitive mechanisms and neural codes. *Behavioral Brain Research, 127*(1–2), 199–207. https://doi.org/10.1016/s0166-4328(01)00365-5

El Hachioui, H., Visch-Brink, E. G., Lingsma, H. F., van de Sandt-Koenderman, M. W., Dippel, D. W., Koudstaal, P. J., & Middelkoop, H. A. (2014). Nonlinguistic cognitive impairment in poststroke aphasia. *Neurorehabilitation and Neural Repair, 28*(3), 273–281. https://doi.org/10.1177/1545968313508467

Engle, R. W. (2002). Working memory capacity as executive attention. *Current Directions in Psychological Science, 11*(1), 19–23. https://doi.org/10.1111/1467-8721.00160

Engle, R. W., Tuholski, S. W., Laughlin, J. E., & Conway, A. R. (1999). Working memory, short-term memory, and general fluid intelligence: A latent-variable approach. *Journal of Experimental Psychology: General, 128*(3), 309–331. https://doi.org/10.1037/0096-3445.128.3.309

Estes, K. G., Evans, J. L., & Else-Quest, N. M. (2007). Differences in the nonword repetition performance of children with and without specific language impairment: A meta-analysis. *Journal of Speech, Language, and Hearing Research, 50*, 177–195. https://doi.org/10.1044/1092-4388(2007/015)

Ettlinger, M., Bradlow, A. R., & Wong, P. C. (2014). Variability in the learning of complex morphophonology. *Applied Psycholinguistics, 35*(4), 807–831. https://doi.org/10.1017/S0142716412000586

Fahy, J. K. (2014). Language and executive functions: Self-talk for self-regulation. *Perspectives on Language Learning and Education, 21*(2), 61–71. https://doi.org/10.1044/lle21.2.61

Failing, M., & Theeuwes, J. (2018). Selection history: How reward modulates selectivity of visual attention. *Psychonomic Bulletin & Review, 25*(2), 514–538. https://doi.org/10.3758/s13423-017-1380-y

Fallon, M., Peelle, J. E., & Wingfield, A. (2006). Spoken sentence processing in young and older adults modulated by task demands: Evidence from self-paced listening. *The Journals of Gerontology. Series B, Psychological Sciences and Social Sciences, 61,* P10–7. https://doi.org/10.1093/Geronb/61.1.P10

Fatzer, S. T., & Roebers, C. M. (2012). Language and executive functions: The effect of articulatory suppression on executive functioning in children. *Journal of Cognition and Development, 13*(4), 454–472. https://doi.org/10.1080/15248372.2011.608322

Fernández, G. (2017). The medial prefrontal cortex is a critical hub in the declarative memory system. In N. Axmacher & B. Rasch (Eds.), *Cognitive Neuroscience of Memory Consolidation* (pp. 45–56). Springer. https://doi.org/10.1007/978-3-319-45066-7_3

Foster, J. J., & Awh, E. (2019). The role of alpha oscillations in spatial attention: Limited evidence for a suppression account. *Current Opinion in Psychology, 29,* 34–40. https://doi.org/10.1016/j.copsyc.2018.11.001

Foster, J. L., Shipstead, Z., Harrison, T. L., Hicks, K. L., Redick, T. S., & Engle, R. W. (2014). Shortened complex span tasks can reliably measure working memory capacity. *Memory & Cognition, 43*(2), 226–236. https://doi.org/10.3758/s13421-014-0461-7

Frankland, P. W., Josselyn, S. A., & Köhler, S. (2019). The neurobiological foundation of memory retrieval. *Nature neuroscience, 22*(10), 1576–1585. https://doi.org/10.1038/s41593-019-0493-1

Fridriksson, J., Nettles, C., Davis, M., Morrow, L., & Montgomery, A. (2006). Functional communication and executive function in aphasia. *Clinical Linguistics & Phonetics, 20*(6), 401–410. https://doi.org/10.1080/02699200500075781

Friedman, N. P., & Robbins, T. W. (2021). The role of the prefrontal cortex in cognitive control and executive function. *Neuropsychopharmacology, 47*(1), 72–89. https://doi.org/10.1038/s41386-021-01132-0

Friedman, N. P., Miyake, A., Altamirano, L. J., Corley, R. P., Young, S. E., Rhea, S. A., & Hewitt, J. K. (2016). Stability and change in executive function abilities from late adolescence to early adulthood: A longitudinal twin study. *Developmental Psychology, 52*(2), 326–340. https://doi.org/10.1037/dev0000075

Friedman, N. P., Miyake, A., Robinson, J. L., & Hewitt, J. K. (2011). Developmental trajectories in toddlers' self-restraint predict individual differences in executive functions 14 years later: A behavioral genetic analysis. *Developmental Psychology, 47*(5), 1410–1430. https://doi.org/10.1037/a0023750

Friedman, N. P., Miyake, A., Young, S. E., DeFries, J. C., Corley, R. P., & Hewitt, J. K. (2008). Individual differences in executive functions are almost entirely genetic in origin. *Journal of Experimental Psychology: General, 137*(2), 201–225. https://doi.org/10.1037/0096-3445.137.2.201

Fritz, J. B., Elhilali, M., David, S. V., & Shamma, S. A. (2007). Auditory attention: Focusing the searchlight on sound. *Current Opinion in Neurobiology, 17*(4), 437–455. https://doi.org/10.1016/j.conb.2007.07.011

Fuhs, M. W., & Day, J. D. (2011). Verbal ability and executive functioning development in preschoolers at head start. *Developmental Psychology, 47*(2), 404–416. https://doi.org/10.1037/a0021065

Garon, N., Bryson, S. E., & Smith, I. M. (2008). Executive function in preschoolers: A review using an integrative framework. *Psychological Bulletin, 134*(1), 31–60. https://doi.org/10.1037/0033-2909.134.1.31

Gathercole, S. E. (2006). Complexities and constraints in nonword repetition and word learning. Applied Psycholinguistics, 27(4), 599–613. https://doi.org/10.1017/S014271640606053X

Gathercole, S. E., Alloway, T. P., Willis, C., & Adams, A.-M. (2006). Working memory in children with reading disabilities. *Journal of Experimental Child Psychology, 93*(3), 265–281. https://doi.org/10.1016/j.jecp.2005.08.003

Gathercole, S. E., Hitch, G. J., Service, E., & Martin, A. J. (1997). Phonological short-term memory and new word learning in children. *Developmental Psychology, 33*(6), 966–979. https://doi.org/10.1037/0012-1649.33.6.966

Gathercole, S. E., & Masoura, E. V. (2005). Contrasting contributions of phonological short-term memory and long-term knowledge to vocabulary learning in a foreign language. *Memory, 13*(3–4), 422–429. https://doi.org/10.1080/09658210344000323

Gherri, E., & Eimer, M. (2011). Active listening impairs visual perception and selectivity: An ERP study of auditory dual-task costs on visual attention. *Journal of Cognitive Neuroscience, 23*(4), 832–844. https://doi.org/10.1162/jocn.2010.21468

Gordon, J. K., & Dell, G. S. (2003). Learning to divide the labor: An account of deficits in light and heavy verb production. *Cognitive Science, 27*(1), 1–40. https://doi.org/10.1207/s15516709cog2701_1

Graf Estes, K., Evans, J. L., & Else-Quest, N. M. (2007). Differences in the nonword repetition performance of children with and without specific language impairment: A meta-analysis. *Journal of Speech, Language, and Hearing Research, 50*(1), 177–195. https://doi.org/10.1044/1092-4388(2007/015)

Graf, P., & Schacter, D. L. (1985). Implicit and explicit memory for new associations in normal and amnesic subjects. *Journal of Experimental Psychology: Learning, Memory, and Cognition, 11*(3), 501–518. https://doi.org/10.1037/0278-7393.11.3.501

Gray, S. I., Levy, R., Alt, M., Hogan, T. P., & Cowan, N. (2022). Working memory predicts new word learning over and above existing vocabulary and nonverbal IQ. *Journal of Speech, Language, and Hearing Research, 65*(3), 1044–1069. https://doi.org/10.1044/2021_jslhr-21-00397

Green, J. J., Doesburg, S. M., Ward, L. M., & McDonald, J. J. (2011). Electrical neuroimaging of voluntary audio spatial attention: Evidence for a supramodal attention control network. *Journal of Neuroscience, 31*(10), 3560–3564. https://doi.org/10.1523/jneurosci.5758-10.2011

Green, J. J., Teder-Sälejärvi, W. A., & McDonald, J. J. (2005). Control mechanisms mediating shifts of attention in auditory and visual space: A spatio-temporal ERP analysis. *Experimental Brain Research, 166*(3), 358–369. https://doi.org/10.1007/s00221-005-2377-8

Händel, B. F., Haarmeier, T., & Jensen, O. (2011). Alpha oscillations correlate with the successful inhibition of unattended stimuli. *Journal of Cognitive Neuroscience, 23*(9), 2494-2502. https://doi.org/10.1162/jocn.2010.21557

Helm-Estabrooks, N. (2002). Cognition and aphasia: A discussion and a study. *Journal of Communication Disorders, 35*(2), 171–186. https://doi.org/10.1016/s0021-9924(02)00063-1

Henson, R. N. A., Burgess, N., & Frith, C. D. (2000). Recoding, storage, rehearsal and grouping in verbal short-term memory: An fMRI study. *Neuropsychologia, 38*(4), 426–440. https://doi.org/10.1016/s0028-3932(99)00098-6

Hodges, J. R., & Graham, K. S. (2001). Episodic memory: Insights from semantic dementia. *Philosophical Transactions of the Royal Society of London. Series B: Biological Sciences, 356*(1413), 1423–1434. https://doi.org/10.1098/rstb.2001.0943

Hodges, J. R., & Patterson, K. (2007). The neuropsychology of frontotemporal dementia. *Frontotemporal Dementia Syndromes, 35*(102–133), 20–36. https://doi.org/10.1017/cbo9781316135457.006

Howland, K. (2014). Developing executive control skills in preschool children with language impairment. *Perspectives on Language Learning and Education, 21*(2), 51–60. https://doi.org/10.1044/lle21.2.51

Hughes, C., Ensor, R., Wilson, A., & Graham, A. (2010). Tracking executive function across the transition to school: A latent variable approach. *Developmental Neuropsychology, 35*(1), 20–36. https://doi.org/10.1080/87565640903325691

Hughes, R. W., & Marsh, J. E. (2017). The functional determinants of short-term memory: Evidence from perceptual-motor interference in verbal serial recall. *Journal of Experimental Psychology: Learning, Memory, and Cognition, 43*(4), 537–551. https://doi.org/10.1037/xlm0000325

Hula, W. D., & McNeil, M. R. (2008). Models of attention and dual-task performance as explanatory constructs in aphasia. *Seminars in Speech and Language, 29*(3), 169–187. https://doi.org/10.1055/s-0028-1082882

Humphreys, G. W., & Forde, E. M. (2001). Hierarchies, similarity, and interactivity in object recognition: "Category-specific" neuropsychological deficits. *Behavioral and Brain Sciences, 24*(3), 453–509. https://doi.org/10.1017/s0140525x01004150

Humphreys, G. W., Price, C. J., & Riddoch, M. J. (1999). From objects to names: A cognitive neuroscience approach. *Psychological Research, 62*(2–3), 118–130. https://doi.org/10.1007/s004260050046

Hunt, R. R., & McDaniel, M. A. (1993). The enigma of organization and distinctiveness. *Journal of Memory and Language, 32*(4), 421–445. https://doi.org/10.1006/jmla.1993.1023

Hurlstone, M. J., & Hitch, G. J. (2018). How is the serial order of a visual sequence represented? Insights from transposition latencies. *Journal of Experimental Psychology: Learning, Memory, and Cognition, 44*(2), 167–192. https://doi.org/10.1037/xlm0000440

Hurlstone, M. J., Hitch, G. J., & Baddeley, A. D. (2014). Memory for serial order across domains: An overview of the literature and directions for future research. *Psychological Bulletin, 140*(2), 339–373. https://doi.org/10.1037/a0034221

Itti, L., & Koch, C. (2001). Computational modeling of visual attention. *Nature Reviews Neuroscience, 2*(3), 194–203. https://doi.org/10.1038/35058500

Jarrold, C., Thorn, A. S. C., & Stephens, E. (2009). The relationships among verbal short-term memory, phonological awareness, and new word learning: Evidence from typical development and Down syndrome. *Journal of Experimental Child Psychology, 102*(2), 196–218. https://doi.org/10.1016/j.jecp.2008.07.001

Javadi, A. H., & Walsh, V. (2012). Transcranial direct current stimulation (tDCS) of the left dorsolateral the prefrontal cortex modulates declarative memory. *Brain Stimulation, 5*(3), 231–241. https://doi.org/10.1016/j.brs.2011.06.007

Jensen, O., Bonnefond, M., & VanRullen, R. (2012). An oscillatory mechanism for prioritizing salient unattended stimuli. *Trends in Cognitive Sciences, 16*(4), 200–206. https://doi.org/10.1016/j.tics.2012.03.002

Joanisse, M. F., & Seidenberg, M. S. (1999). Impairments in verb morphology after brain injury: A connectionist model. *Proceedings of the National Academy of Sciences, 96*(13), 7592–7597. https://doi.org/10.1073/pnas.96.13.7592

Jones, S. M., Bailey, R., Barnes, S. P., & Partee, A. (2016). Executive function mapping project: Untangling the terms and skills related to executive function and self-regulation in early childhood. OPRE Report # 2016-88, Office of Planning, Research and Evaluation, Administration for Children and Families, U.S. Department of Health and Human Services. www.acf.hhs.gov/sites/default/files/documents/opre/efmapping_report_101416_final_508.pdf

Jonker, T. R., Seli, P., & MacLeod, C. M. (2015). Retrieval-induced forgetting and context. *Current Directions in Psychological Science, 24*(4), 273–278. https://doi.org/10.1177/0963721415573203

Just, M. A., & Carpenter, P. A. (1992). A capacity theory of comprehension: Individual differences in working memory. *Psychological Review, 99*(1), 122–149. https://doi.org/10.1037/0033-295x.99.1.122

Kahneman, D. (1973). *Attention and Effort* (Vol. 1063). Prentice-Hall.

Kane, M. J., Hambrick, D. Z., Tuholski, S. W., Wilhelm, O., Payne, T. W., & Engle, R. W. (2004). The generality of working memory capacity: A latent-variable approach to verbal and visuospatial memory span and reasoning. *Journal of Experimental Psychology: General, 133*(2), 189–217. https://doi.org/10.1037/0096-3445.133.2.189

Kapa, L. L., & Erikson, J. A. (2020). The relationship between word learning and executive function in preschoolers with and without developmental language disorder. *Journal of Speech, Language, and Hearing Research, 63*(7), 2293–2307. https://doi.org/10.1044/2020_jslhr-19-00342

Kapa, L. L., & Plante, E. (2015). Executive function in SLI: Recent advances and future directions. *Current Developmental Disorders Reports, 2*(3), 245–252. https://doi.org/10.1007/s40474-015-0050-x

Karr, J. E., Areshenkoff, C. N., Rast, P., Hofer, S. M., Iverson, G. L., & Garcia-Barrera, M. A. (2018). The unity and diversity of executive functions: A systematic review and re-analysis of latent variable studies. *Psychological Bulletin, 144*(11), 1147–1185. https://doi.org/10.1037/bul0000160

Kaushanskaya, M., Park, J. S., Gangopadhyay, I., Davidson, M. M., & Weismer, S. E. (2017). The relationship between executive functions and language abilities in children: A latent variable approach. *Journal of Speech, Language, and Hearing Research, 60*(4), 912–923. https://doi.org/10.1044/2016_jslhr-l-15-0310

Kelly, S. P., Lalor, E. C., Reilly, R. B., & Foxe, J. J. (2006). Increases in alpha oscillatory power reflect an active retinotopic mechanism for distracter suppression during sustained visuospatial attention. *Journal of Neurophysiology, 95*(6), 3844–3851. https://doi.org/10.1152/jn.01234.2005

Kempe, V., Brooks, P. J., & Kharkhurin, A. (2010). Cognitive predictors of generalization of Russian grammatical gender categories. *Language Learning, 60*(1), 127–153. https://doi.org/10.1111/j.1467-9922.2009.00553.x

Kennedy, M. R. T., & Coelho, C. (2005). Self-regulation after traumatic brain injury: A framework for intervention of memory and problem solving. *Seminars in Speech and Language, 26*(04), 242–255. https://doi.org/10.1055/s-2005-922103

Kidd, G. Jr., (2017). Enhancing auditory selective attention using a visually guided hearing aid. *Journal of Speech, Language, and Hearing Research, 60*(10), 3027–3038. https://doi.org/10.1044/2017_jslhr-h-17-0071

Kidd, G. Jr., & Colburn, H. S. (2017). Informational masking in speech recognition. In J. C. Middlebrooks, J. Z. Simon, A. N. Popper, & R. R. Fay (Eds.), *The Auditory System at the Cocktail Party* (pp. 75–109). Springer.

Kidd, G. Jr., Favrot, S., Desloge, J. G., Streeter, T. M., & Mason, C. R. (2013). Design and preliminary testing of a visually guided hearing aid. *The Journal of the Acoustical Society of America, 133*(3), EL202–EL207. https://doi.org/10.1121/1.4791710

Kidd, G. Jr., Jennings, T. R., & Byrne, A. J. (2020). Enhancing the perceptual segregation and localization of sound sources with a triple beamformer. *The Journal of the Acoustical Society of America, 148*(6), 3598–3611. https://doi.org/10.1121/10.0002779

Kidd, G. Jr., Mason, C. R., Swaminathan, J., Roverud, E., Clayton, K. K., & Best, V. (2016). Determining the energetic and informational components of speech-on-speech masking. *The Journal of the Acoustical Society of America, 140*(1), 132-144. https://doi.org/10.1121/1.4954748

Kim, W. B., & Cho, J.-H. (2020). Encoding of contextual fear memory in hippocampal–amygdala circuit. *Nature Communications, 11*(1), 1–22. https://doi.org/10.1038/s41467-020-15121-2

Kintsch, W., Patel, V. L., & Ericsson, K. A. (1999). The role of long-term working memory in text comprehension. *Psychologia, 42*(4), 186–198.

Kirchner, W. K. (1958). Age differences in short-term retention of rapidly changing information. *Journal of Experimental Psychology, 55*(4), 352–358. https://doi.org/10.1037/h0043688

Kirkham, N. Z., Cruess, L., & Diamond, A. (2003). Helping children apply their knowledge to their behavior on a dimension-switching task. *Developmental Science, 6*(5), 449–467. https://doi.org/10.1111/1467-7687.00300

Klatte, M., Lee, N., & Hellbruck, J. (2002). Effects of irrelevant speech and articulatory suppression on serial recall of heard and read materials. *Psychologische Beiträge, 44*(2), 166–186.

Klauer, K. C., & Zhao, Z. (2004). Double dissociations in visual and spatial short-term memory. *Journal of Experimental Psychology: General, 133*(3), 355–381. https://doi.org/10.1037/0096-3445.133.3.355

Klimesch, W. (2012). Alpha-band oscillations, attention, and controlled access to stored information. *Trends in Cognitive Sciences, 16*(12), 606–617. https://doi.org/10.1016/j.tics.2012.10.007

Kuhn, L. J., Willoughby, M. T., Vernon-Feagans, L., Blair, C. B., & The Family Life Project Key Investigators. (2016). The contribution of children's time-specific and longitudinal expressive language skills on developmental trajectories of executive function. *Journal of Experimental Child Psychology, 148*, 20–34. https://doi.org/10.1016/j.jecp.2016.03.008

Kuperberg, G. R., Kreher, D. A., & Ditman, T. (2010). What can event-related potentials tell us about language, and perhaps even thought, in schizophrenia? *International Journal of Psychophysiology, 75*(2), 66–76. https://doi.org/10.1016/j.ijpsycho.2009.09.005

Lang, C. J., & Quitz, A. (2012). Verbal and nonverbal memory impairment in aphasia. *Journal of Neurology, 259*(8), 1655–1661. https://doi.org/10.1007/s00415-011-6394-1

LeDoux, J. E. (2014). Coming to terms with fear. *Proceedings of the National Academy of Sciences, 111*(8), 2871–2878. https://doi.org/10.1073/pnas.1400335111

Levine, B., Dawson, D., Boutet, I., Schwartz, M. L., & Stuss, D. T. (2000). Assessment of strategic self-regulation in traumatic brain injury: Its relationship to injury severity and psychosocial outcome. *Neuropsychology, 14*(4), 491–500. https://doi.org/10.1037/0894-4105.14.4.491

Lidstone, J. S. M., Meins, E., & Fernyhough, C. (2010). The roles of private speech and inner speech in planning during middle childhood: Evidence from a dual task paradigm. *Journal of Experimental Child Psychology, 107*(4), 438–451. https://doi.org/10.1016/j.jecp.2010.06.002

Liew, J. (2011). Effortful control, executive functions, and education: Bringing self-regulatory and social-emotional competencies to the table. *Child Development Perspectives, 6*(2), 105–111. https://doi.org/10.1111/j.1750-8606.2011.00196.x

Logue, S. F., & Gould, T. J. (2014). The neural and genetic basis of executive function: Attention, cognitive flexibility, and response inhibition. *Pharmacology Biochemistry and Behavior, 123*, 45–54. https://doi.org/10.1016/j.pbb.2013.08.007

Lu, K., Xu, Y., Yin, P., Oxenham, A. J., Fritz, J. B., & Shamma, S. A. (2017). Temporal coherence structure rapidly shapes neuronal interactions. *Nature Communications, 8*(1), 1–12. https://doi.org/10.1038/ncomms13900

Maddox, R. K., Atilgan, H., Bizley, J. K., & Lee, A. K. (2015). Auditory selective attention is enhanced by a task-irrelevant temporally coherent visual stimulus in human listeners. *eLife, 4*. https://doi.org/10.7554/elife.04995

Mädebach, A., Jescheniak, J. D., Oppermann, F., & Schriefers, H. (2011). Ease of processing constrains the activation flow in the conceptual-lexical system during speech planning. *Journal of Experimental Psychology: Learning, Memory, and Cognition, 37*(3), 649–660. https://doi.org/10.1037/a0022330

Majerus, S. (2019). Verbal working memory and the phonological buffer: The question of serial order. *Cortex, 112*, 122–133. https://doi.org/10.1016/j.cortex.2018.04.016

Majerus, S., Poncelet, M., Grefee, C., & Van der Linden, M. (2006). Relations between vocabulary development and verbal short-term memory for serial order and item information. *Journal of Experimental Child Psychology, 93*(2), 95–119. https://doi.org/10.1016/j.jecp.2005.07.005

Martella, D., Casagrande, M., & Lupiáñez, J. (2011). Alerting, orienting and executive control: The effects of sleep deprivation on attentional networks. *Experimental Brain Research, 210*(1), 81–89. https://doi.org/10.1007/s00221-011-2605-3

Martin, K. I., & Ellis, N. C. (2012). The roles of phonological short-term memory and working memory in L2 grammar and vocabulary learning. *Studies in Second Language Acquisition, 34*(3), 379–413. https://doi.org/10.1017/S0272263112000125

Martin, N., Kohen, F., Kalinyak-Fliszar, M., Soveri, A., & Laine, M. (2012). Effects of working memory load on processing of sounds and meanings of words in aphasia. *Aphasiology, 26*(3–4), 462–493. https://doi.org/10.1080/02687038.2011.619516

Martin, R. C. (2021). The critical role of semantic working memory in language comprehension and production. *Current Directions in Psychological Science, 30*(4), 283–291. https://doi.org/10.1177/0963721421995178

McDermott, K. B., & Roediger, H. L. (2018). Memory (encoding, storage, retrieval). In R. Biswas-Diener & E. Diener (Eds.), *General Psychology FA18* (pp. 117–140). Noba Project.

McGaugh, J. L., Cahill, L., & Roozendaal, B. (1996). Involvement of the amygdala in memory storage: Interaction with other brain systems. *Proceedings of the National Academy of Sciences, 93*(24), 13508–13514. https://doi.org/10.1073/pnas.93.24.13508

Melton, A. W. (1963). Implications of short-term memory for a general theory of memory. *Journal of Verbal Learning and Verbal Behavior, 2*(1), 1–21. https://doi.org/10.1016/s0022-5371(63)80063-8

Meneghetti, C., De Beni, R., Pazzaglia, F., & Gyselinck, V. (2011). The role of visuo-spatial abilities in recall of spatial descriptions: A mediation model. *Learning and Individual Differences, 21*(6), 719–723. https://doi.org/10.1016/j.lindif.2011.07.015

Menon, V., & D'Esposito, M. (2022). The role of PFC networks in cognitive control and executive function. *Neuropsychopharmacology*, *47*(1), 90–103. https://doi.org/10.1038/s41386-021-01152-w

Mesgarani, N., Fritz, J., & Shamma, S. (2010). A computational model of rapid task-related plasticity of auditory cortical receptive fields. *Journal of Computational Neuroscience*, *28*(1), 19–27. https://doi.org/10.1007/s10827-009-0181-3

Metcalfe, J., & Mischel, W. (1999). A hot/cool-system analysis of delay of gratification: Dynamics of willpower. *Psychological Review*, *106*(1), 3–19. https://doi.org/10.1037/0033-295X.106.1.3

Miller, E. K. (2013). The "working" of working memory. *Dialogues in Clinical Neuroscience*, *15*(4), 411–418. https://doi.org/10.31887/dcns.2013.15.4/emiller

Miller, E. K., & Cohen, J. D. (2001). An integrative theory of prefrontal cortex function. *Annual Review of Neuroscience*, *24*(1), 167–202. https://doi.org/10.1146/annurev.neuro.24.1.167

Miller, E. K., Lundqvist, M., & Bastos, A. M. (2018). Working memory 2.0. *Neuron*, *100*(2), 463–475. https://doi.org/10.1016/j.neuron.2018.09.023

Miller, S., McCulloch, S., & Jarrold, C. (2015). The development of memory maintenance strategies: Training cumulative rehearsal and interactive imagery in children aged between 5 and 9. *Frontiers in Psychology*, *6*, 1–10. https://doi.org/10.3389/fpsyg.2015.00524

Milner, B., Corkin, S., & Teuber, H. L. (1968). Further analysis of the hippocampal amnesic syndrome: A 14-year follow-up study of H.M. *Neuropsychologia*, *6*(3), 215–234. https://doi.org/10.1016/0028-3932(68)90021-3

Milner, B., Squire, L. R., & Kandel, E. R. (1998). Cognitive neuroscience and the study of memory. *Neuron*, *20*(3), 445–468. https://doi.org/10.1016/s0896–6273(00)80987-3

Mirsky, A. F., Anthony, B. J., Duncan, C. C., Ahearn, M. B., & Kellam, S. G. (1991). Analysis of the elements of attention: A neuropsychological approach. *Neuropsychology Review*, *2*(2), 109–145. https://doi.org/10.1007/bf01109051

Miyake, A., & Friedman, N. P. (2012). The nature and organization of individual differences in executive functions. *Current Directions in Psychological Science*, *21*(1), 8–14. https://doi.org/10.1177/0963721411429458

Miyake, A., Emerson, M. J., Padilla, F., & Ahn, J. (2004). Inner speech as a retrieval aid for task goals: The effects of cue type and articulatory suppression in the random task cueing paradigm. *Acta Psychologica*, *115*, 123–142. https://doi.org/10.1016/j.actpsy.2003.12.004

Miyake, A., Friedman, N. P., Emerson, M. J., Witzki, A. H., Howerter, A., & Wager, T. D. (2000). The unity and diversity of executive functions and their contributions to complex "frontal lobe" tasks: A latent variable analysis. *Cognitive Psychology*, *41*(1), 49–100. https://doi.org/10.1006/cogp.1999.0734

Mohapatra, B., & Marshall, R. S. (2019). Performance differences between aphasia and healthy aging on an executive function test battery. *International Journal of Speech-Language Pathology*, *22*(4), 487–497. https://doi.org/10.1080/17549507.2019.1691262

Morey, C. C., Cowan, N., Morey, R. D., & Rouder, J. N. (2011). Flexible attention allocation to visual and auditory working memory tasks: Manipulating reward induces a trade-off. *Attention, Perception, & Psychophysics, 73*(2), 458–472. https://doi.org/10.3758/s13414-010-0031-4

Morgan-Short, K., Faretta-Stutenberg, M., Brill-Schuetz, K. A., Carpenter, H., & Wong, P. C. (2014). Declarative and procedural memory as individual differences in second language acquisition. *Bilingualism: Language and Cognition, 17*(1), 56–72. https://doi.org/10.1017/S1366728912000715

Morgan-Short, K., Steinhauer, K., Sanz, C., & Ullman, M. T. (2012). Explicit and implicit langauge training differentially affect the achievement of native-like brain activation patterns. *Journal of Cognitive Neuroscience, 24*(4), 933–947. https://doi.org/10.1162/jocn_a_00119

Morris, N., & Jones, D. M. (1990). Memory updating in working memory: The role of the central executive. *British Journal of Psychology, 81*(2), 111–121. https://doi.org/10.1111/j.2044-8295.1990.tb02349.x

Mosse, E. K., & Jarrold, C. (2008). Short article: Hebb learning, verbal short-term memory, and the acquisition of phonological forms in children. *Quarterly Journal of Experimental Psychology, 61*(4), 505–514. https://doi.org/10.1080/17470210701680779

Müller, U., Kerns, K. A., & Konkin, K. (2012). Test-retest reliability and practice effects of executive function tasks in preschool children. *The Clinical Neuropsychologist, 26*(2), 271–287. https://doi.org/10.1080/13854046.2011.645558

Munson, B., Kurtz, B. A., & Windsor, J. (2005). The influence of vocabulary size, phonotactic probability, and word likeness on nonword repetitions of children with and without specific language impairment. *Journal of Speech, Language, and Hearing Research, 48*(5), 1033–1047. https://doi.org/10.1044/1092-4388(2005/072)

Murayama, K., Miyatsu, T., Buchli, D., & Storm, B. C. (2014). Forgetting as a consequence of retrieval: A meta-analytic review of retrieval-induced forgetting. *Psychological Bulletin, 140*(5), 1383–1409. https://doi.org/10.1037/a0037505

Murray, L. L. (1999). Review attention and aphasia: Theory, research and clinical implications. *Aphasiology, 13*(2), 91–111. https://doi.org/10.1080/026870399402226

Murray, L. L. (2012). Attention and other cognitive deficits in aphasia: Presence and relation to language and communication measures. *American Journal of Speech-Language Pathology, 21*(2), 51–64. https://doi.org/10.1044/1058-0360(2012/11-0067)

Murray, L., Salis, C., Martin, N., & Dralle, J. (2018). The use of standardized short-term and working memory tests in aphasia research: A systematic review. *Neuropsychological Rehabilitation, 28*(3), 309–351. https://doi.org/10.1080/09602011.2016.1174718

Nadel, L., & Moscovitch, M. (1997). Memory consolidation, retrograde amnesia and the hippocampal complex. *Current Opinion in Neurobiology, 7*(2), 217–227. https://doi.org/10.1016/s0959-4388(97)80010-4

Nadel, L., Samsonovich, A., Ryan, L., & Moscovitch, M. (2000). Multiple trace theory of human memory: Computational, neuroimaging, and neuropsychological results. *Hippocampus, 10*(4), 352–368. https://doi.org/10.1002/1098-1063(2000)10:4<352::aid-hipo2>3.0.co;2-d

Nasar, J., Hecht, P., & Wener, R. (2008). Mobile telephones, distracted attention, and pedestrian safety. *Accident Analysis & Prevention, 40*(1), 69–75. https://doi.org/10.1016/j.aap.2007.04.005

Nicholas, M., & Connor, L. T. (2017) People with aphasia using AAC: Are executive functions important? *Aphasiology, 31*(7), 819–836. https://doi.org/10.1080/02687038.2016.1258539

Nicholas, M., Sinotte, M. P., & Helm-Estabrooks, N. (2011). C-speak aphasia alternative communication program for people with severe aphasia: Importance of executive functioning and semantic knowledge. *Neuropsychological Rehabilitation, 21*(3), 322–366. https://doi.org/10.1080/09602011.2011.559051

Norris, D., & Kalm, K. (2021). Chunking and data compression in verbal short-term memory. *Cognition, 208,* 104534. https://doi.org/10.1016/j.cognition.2020.104534

Norris, J., & Ortega, L. (2000). Effectiveness of L2 instruction: A research synthesis and quantitative meta-analysis. *Language Learning, 50*(3), 417–528. https://doi.org/10.1111/0023-8333.00136

Novick, J. M., Trueswell, J. C., & Thompson-Schill, S. L. (2005). Cognitive control and parsing: Reexamining the role of Broca's area in sentence comprehension. *Cognitive, Affective, & Behavioral Neuroscience, 5*(3), 263–281. https://doi.org/10.3758/cabn.5.3.263

Obermeyer, J., Schlesinger, J., & Martin, N. (2020). Evaluating the contribution of executive functions to language tasks in cognitively demanding contexts. *American Journal of Speech-Language Pathology, 29*(1S), 463–473. https://doi.org/10.1044/2019_ajslp-cac48-18-0216

Ojemann, G. A., Schoenfield-McNeill, J., & Corina, D. (2009). The roles of human lateral temporal cortical neuronal activity in recent verbal memory encoding. *Cerebral Cortex, 19*(1), 197–205. https://doi.org/10.1093/cercor/bhn071

Olsson, C., Arvidsson, P., & Blom Johansson, M. (2019). Relations between executive function, language, and functional communication in severe aphasia. *Aphasiology, 33*(7), 821–845. https://doi.org/10.1080/02687038.2019.1602813

Olsson, C., Arvidsson, P., & Blom Johansson, M. (2020). Measuring executive function in people with severe aphasia: Comparing neuropsychological tests and informant ratings. *NeuroRehabilitation, 46*(3), 299–310. https://doi.org/10.3233/nre-192998

Oppermann, F., Jescheniak, J. D., & Görges, F. (2014). Resolving competition when naming an object in a multiple-object display. *Psychonomic Bulletin & Review, 21*(1), 78–84. https://doi.org/10.3758/s13423-013-0465-5

Oswald, F. L., McAbee, S. T., Redick, T. S., & Hambrick, D. Z. (2015). The development of a short domain-general measure of working memory capacity. *Behavior Research Methods, 47*(4), 1343–1355. https://doi.org/10.3758/s13428-014-0543-2

Packard, M. G., and McGaugh, J. L. (1996). Inactivation of hippocampus or caudate nucleus with lidocaine differentially affects expression of place and response learning. *Neurobiology of Learning and Memory, 65*(1), 65–72. https://doi.org/10.1006/nlme.1996.0007

Papagno, C., & Vallar, G. (1992). Phonological short-term memory and learning of novel words: The effect of phonological similarity and item length. *The Quarterly Journal of Experimental Psychology Section A, 44(*1), 47–67. https://doi.org/10.1080/14640749208401283

Papagno, C., Valentine, T., & Baddeley, A. (1991). Phonological short-term memory and foreign-language vocabulary learning. *Journal of Memory & Language, 30*(3), 331–347. https://doi.org/10.1016/0749-596x(91)90040-q

Park, J., Miller, C. A., Sanjeevan, T., van Hell, J. G., Weiss, D. J., & Mainela-Arnold, E. (2019). Bilingualism and attention in typically developing children and children with developmental language disorder. *Journal of Speech, Language, and Hearing Research, 62*(11), 4105–4118. https://doi.org/10.1044/2019_jslhr-l-18-0341

Pauls, L. J., & Archibald, L. M. (2016). Executive functions in children with specific language impairment: A meta-analysis. *Journal of Speech, Language, and Hearing Research, 59*(5), 1074–1086. https://doi.org/10.1044/2016_jslhr-l-15-0174

Peng, P., Barnes, M., Wang, C., Wang, W., Li, S., Swanson, H. L., Dardick, W., & Tao, S. (2018). A meta-analysis on the relation between reading and working memory. *Psychological Bulletin, 144*(1), 48–76. https://doi.org/10.1037/bul0000124

Peristeri, E., Tsimpli, I. M., Dardiotis, E., & Tsapkini, K. (2020). Effects of executive attention on sentence processing in aphasia. *Aphasiology, 34*(8), 943–969. https://doi.org/10.1080/02687038.2019.1622647

Peteranderl, S., & Oberauer, K. (2018). Serial recall of colors: Two models of memory for serial order applied to continuous visual stimuli. *Memory & Cognition, 46*(1), 1–16. https://doi.org/10.3758/s13421-017-0741-0

Poldrack, R. A., & Gabrieli, J. D. (1997). Functional anatomy of long-term memory. *Journal of Clinical Neurophysiology, 14*(4), 294–310. https://doi.org/10.1097/00004691-199707000-00003

Posner, M. I., & Petersen, S. E. (1990). The attention system of the human brain. *Annual Review of Neuroscience, 13*(1), 25–42. https://doi.org/10.1146/annurev.ne.13.030190.000325

Posner, M. I., Snyder, C. R., & Davidson, B. J. (1980). Attention and the detection of signals. *Journal of Experimental Psychology: General, 109*(2), 160–174. https://doi.org/10.1037/0096-3445.109.2.160

Potagas, C., Kasselimis, D., & Evdokimidis, I. (2011). Short-term and working memory impairments in aphasia. *Neuropsychologia, 49*(10), 2874–2878. https://doi.org/10.1016/j.neuropsychologia.2011.06.013

Purdy, M. (2002). Executive function ability in persons with aphasia. *Aphasiology, 16*(4–6), 549–557. https://doi.org/10.1080/02687030244000176

Rankin, E., Newton, C., Parker, A., & Bruce, C. (2014). Hearing loss and auditory processing ability in people with aphasia. *Aphasiology, 28*(5), 576–595. https://doi.org/10.1080/02687038.2013.878452

Reber, P. J. (2008). Cognitive neuroscience of declarative and nondeclarative memory. In M. Guadagnoli, A. S. Benjamin, J. S. de Belle, B. Etnyre, & T. A. Polk (Eds.), *Human Learning: Biology, Brain and Neuroscience* (Vol. 139, pp. 113–123). Elsevier.

Reber, P. J. (2013). The neural basis of implicit learning and memory: A review of neuropsychological and neuroimaging research. *Neuropsychologia, 51*(10), 2026–2042. https:/doi.org/10.1016/j.neuropsychologia.2013.06.019

Richmond, J., & Nelson, C. A. (2007). Accounting for change in declarative memory: A cognitive neuroscience perspective. *Developmental Review, 27*(3), 349–373. https://doi.org/10.1016/j.dr.2007.04.002

Roelofs, A., & Piai, V. (2011). Attention demands of spoken word planning: A review. *Frontiers in Psychology, 2.* https://doi.org/10.3389/fpsyg.2011.00307

Rogalsky, C., Matchin, W., & Hickok, G. (2008). Broca's area, sentence comprehension, and working memory: An fMRI study. *Frontiers in Human Neuroscience, 2.* https://doi.org/10.3389/neuro.09.014.2008

Rogers, T., Hodges, J., Patterson, K., & Lambon Ralph, M. (2003). Object recognition under semantic impairment: The effects of conceptual regularities on perceptual decisions. *Language and Cognitive Processes, 18*(5–6), 625–662. https://doi.org/10.1080/01690960344000053

Rogers, T. T., Lambon Ralph, M. A., Garrard, P., Bozeat, S., McClelland, J. L., Hodges, J. R., & Patterson, K. (2004). Structure and deterioration of semantic memory: A neuropsychological and computational investigation. *Psychological Review, 111*(1), 205–235. https://doi.org/10.1037/0033-295x.111.1.205

Romberg, A. R., & Saffran, J. R. (2010). Statistical learning and language acquisition. *WIREs Cognitive Science, 1*(6), 906–914. https://doi.org/10.1002/wcs.78

Romero Lauro, L. J., Reis, J., Cohen, L. G., Cecchetto, C., & Papagno, C. (2010). A case for the involvement of phonological loop in sentence comprehension. *Neuropsychologia, 48*(14), 4003–4011. https://doi.org/10.1016/j.neuropsychologia.2010.10.019

Rudner, M., & Rönnberg, J. (2008). The role of the episodic buffer in working memory for language processing. *Cognitive Processing, 9*(1), 19–28. https://doi.org/10.1007/s10339-007-0183-x

Saffran, J. R. (2020). Statistical language learning in infancy. *Child Development Perspectives, 14*(1), 49–54. https://doi.org/10.1111/cdep.12355

Saffran, J. R., & Kirkham, N. Z. (2018). Infant statistical learning. *Annual Review of Psychology, 69*(1), 181–203. https://doi.org/10.1146/annurev-psych-122216-011805

Schacter, D. L., & Dodson, C. S. (2001). Misattribution, false recognition and the sins of memory. *Philosophical Transactions of the Royal Society of London. Series B: Biological Sciences, 356*(1413), 1385–1393. https://doi.org/10.1098/rstb.2001.0938

Schendan, H. E. (2012). Semantic memory. In V. S. Ramachandran (Ed.), *Encyclopedia of Human Behavior* (pp. 350–358). Elsevier. https://doi.org/10.1016/B978-0-12-375000-6.00315-3

Schmidt-Wilcke, T., Poljansky, S., Hierlmeier, S., Hausner, J., & Ibach, B. (2009). Memory performance correlates with gray matter density in the ento-/perirhinal cortex and posterior hippocampus in patients with mild cognitive impairment and healthy controls: A voxel based morphometry study. *NeuroImage*, *47*(4), 1914–1920. https://doi.org/10.1016/j.neuroimage.2009.04.092

Schneider, D., Herbst, S. K., Klatt, L.-I., & Wöstmann, M. (2021). Target enhancement or distractor suppression? Functionally distinct alpha oscillations form the basis of attention. *European Journal of Neuroscience*, *55*(11–12), 3256–3265. https://doi.org/10.1111/ejn.15309

Schnur, T. T., Schwartz, M. F., Brecher, A., & Hodgson, C. (2006). Semantic interference during blocked-cyclic naming: Evidence from aphasia. *Journal of Memory and Language*, *54*(2),199–227. https://doi.org/10.1016/j.jml.2005.10.002

Seer, C., Sidlauskaite, J., Lange, F., Rodríguez-Nieto, G., & Swinnen, S. P. (2021). Cognition and action: A latent variable approach to study contributions of executive functions to motor control in older adults. *Aging*, *13*(12), 15942–15963. https://doi.org/10.18632/aging.203239

Seidenberg, M. S., & Plaut, D. C. (2014). Quasiregularity and its discontents: The legacy of the past tense debate. *Cognitive Science*, *38*(6), 1190–1228. https://doi.org/10.1111/cogs.12147

Shamma, S. A., Elhilali, M., & Micheyl, C. (2011). Temporal coherence and attention in auditory scene analysis. *Trends in Neurosciences*, *34*(3), 114–123. https://doi.org/10.1016/j.tins.2010.11.002

Shaywitz, S. E., & Shaywitz, B. A. (2008). Paying attention to reading: The neurobiology of reading and dyslexia. *Development and Psychopathology*, *20*(4), 1329–1349. https://doi.org/10.1017/s0954579408000631

Shiffrin, R. M., & Atkinson, R. C. (1969). Storage and retrieval processes in long-term memory. *Psychological Review*, *76*(2), 179–193. https://doi.org/10.1037/h0027277

Shinn-Cunningham, B. G. (2008). Object-based auditory and visual attention. *Trends in Cognitive Sciences*, *12*(5), 182–186. https://doi.org/10.1016/j.tics.2008.02.003

Smith, E. E., & Jonides, J. (1997). Working memory: A view from neuroimaging. *Cognitive Psychology*, *33*(1), 5–42. https://doi.org/10.1006/cogp.1997.0658

Smyth, M. M., & Pendleton, L. R. (1990). Space and movement in working memory. *The Quarterly Journal of Experimental Psychology Section A*, *42*(2), 291–304. https://doi.org/10.1080/14640749008401223

Sohlberg, M. M., & Mateer, C. A. (1987). Effectiveness of an attention-training program. *Journal of Clinical and Experimental Neuropsychology*, *9*(2), 117–130. https://doi.org/10.1080/01688638708405352

Sohlberg, M. M., & Mateer, C. A. (2001). *Cognitive Rehabilitation: An Integrative Neuropsychological Approach*. Guilford Press.

Souza, A. S., & Oberauer, K. (2018). Does articulatory rehearsal help immediate serial recall? *Cognitive Psychology*, *107*, 1–21. https://doi.org/10.1016/j.cogpsych.2018.09.002

Squire, L. R. (1992). Declarative and nondeclarative memory: Multiple brain systems supporting learning and memory. *Journal of Cognitive Neuroscience*, *4*(3), 232–243. https://doi.org/10.1162/jocn.1992.4.3.232

Squire, L. R. (2009a). Memory and brain systems: 1969–2009. *Journal of Neuroscience, 29*(41), 12711–12716. https://doi.org/10.1523/jneurosci.3575-09.2009

Squire, L. R. (2009b). The legacy of patient HM for neuroscience. Neuron, 61(1), 6–9. https://doi.org/10.1016/j.neuron.2008.12.023

Squire, L. R., & Alvarez, P. (1995). Retrograde amnesia and memory consolidation: A neurobiological perspective. *Current Opinion in Neurobiology, 5*(2), 169–177. https://doi.org/10.1016/0959-4388(95)80023-9

Squire, L. R., Genzel, L., Wixted, J. T., & Morris, R. G. (2015). Memory consolidation. *Cold Spring Harbor Perspectives in Biology, 7*(8), a021766. https://doi.org/10.1101/cshperspect.a021766

Squire, L. R., & Zola, S. M. (1996). Structure and function of declarative and nondeclarative memory systems. *Proceedings of the National Academy of Sciences, 93*(24), 13515–13522. https://doi.org/10.1073/pnas.93.24.13515

Squire, L. R., & Zola, S. M. (1998). Episodic memory, semantic memory, and amnesia. *Hippocampus, 8*(3), 205–211. https://doi.org/10.1002/(sici)1098-1063(1998)8:3<205::aid-hipo3>3.0.co;2-i

Stein, J. (2014). Dyslexia: The role of vision and visual attention. *Current Developmental Disorders Reports, 1*(4), 267–280. https://doi.org/10.1007/s40474-014-0030-6

Stern, P., & Shalev, L. (2013). The role of sustained attention and display medium in reading comprehension among adolescents with ADHD and without it. *Research in Developmental Disabilities, 34*(1), 431–439. https://doi.org/10.1016/j.ridd.2012.08.021

Storm, B. C., & Levy, B. J. (2012). A progress report on the inhibitory account of retrieval-induced forgetting. *Memory & Cognition, 40*(6), 827–843. https://doi.org/10.3758/s13421-012-0211-7

Stuss, D. T. (2011). Functions of the frontal lobes: Relation to executive functions. *Journal of the International Neuropsychological Society, 17*(5), 759–765. https://doi.org/10.1017/S1355617711000695

Szalárdy, O., Tóth, B., Farkas, D., György, E., & Winkler, I. (2019). Neuronal correlates of informational and energetic masking in the human brain in a multi-talker situation. *Frontiers in Psychology, 10*. https://doi.org/10.3389/fpsyg.2019.00786

Teuber, H. L. (1972). Unity and diversity of frontal lobe functions. *Acta Neurobiologiae Experimentalis, 32*(2), 615–656.

Thakur, C. S., Wang, R. M., Afshar, S., Hamilton, T. J., Tapson, J. C., Shamma, S. A., & van Schaik, A. (2015). Sound stream segregation: A neuromorphic approach to solve the "cocktail party problem" in real-time. *Frontiers in Neuroscience, 9*. https://doi.org/10.3389/fnins.2015.00309

Thut, G., Nietzel, A., Brandt, S. A., & Pascual-Leone, A. (2006). Alpha-band electroencephalographic activity over occipital cortex indexes visuospatial attention bias and predicts visual target detection. *Journal of Neuroscience, 26*(37), 9494–9502. https://doi.org/10.1523/jneurosci.0875-06.2006

Treisman, A. M. (1960). Contextual cues in selective listening. *Quarterly Journal of Experimental Psychology, 12*(4), 242–248. https://doi.org/10.1080/17470216008416732

Tucha, L., Aschenbrenner, S., Koerts, J., & Lange, K. W. (2012). The five-point test: Reliability, validity, and normative data for children and adults. *PLoS ONE, 7*(9), e46080. https://doi.org/10.1371/journal.pone.0046080

Tulving E. (1972). Episodic and semantic memory. In E. Tulving & W. Donaldson (Eds.), *Organization of Memory* (pp. 381–403). Academic Press.

Tulving, E. (2002). Does memory encoding exist? In M. Naveh-Benjamin, M. Moscovitch, & H. L. Roediger, III (Eds.), *Perspectives on Human Memory and Cognitive Aging: Essays in Honor of Fergus Craik* (pp. 6–27). Psychology Press.

Tulving, E., & Pearlstone, Z. (1966). Availability versus accessibility of information in memory for words. *Journal of Verbal Learning and Verbal Behavior, 5*(4), 381–391. https://doi.org/10.1016/s0022-5371(66)80048-8

Ullman, M. T. (2001). A neurocognitive perspective on language: The declarative/procedural model. *Nature Reviews Neuroscience, 2*(10), 717–726. https://doi.org/10.1038/35094573

Ullman M.T. (2004). Contributions of memory circuits to language: The declarative/procedural model. *Cognition, 92*(1–2), 231–270. https://doi.org/10.1016/j.cognition.2003.10.008

Ullman, M. T. (2016). The declarative/procedural model: A neurobiological model of language learning, knowledge, and use. In *Neurobiology of Language* (pp. 953–968). Academic Press.

Ullman, M. T. (2020). The declarative/procedural model: A neurobiologically motivated theory of first and second language 1. In *Theories in Second Language Acquisition* (pp. 128–161). Routledge.

Ullman, M. T., Corkin, S., Coppola, M., Hickok, G., Growdon, J. H., Koroshetz, W. J., & Pinker, S. (1997). A neural dissociation within language: Evidence that the mental dictionary is part of declarative memory, and that grammatical rules are processed by the procedural system. *Journal of Cognitive Neuroscience, 9*(2), 266–276. https://doi.org/10.1162/jocn.1997.9.2.266

Unsworth, N., & Engle, R. W. (2007). The nature of individual differences in working memory capacity: Active maintenance in primary memory and controlled search from secondary memory. *Psychological Review, 114*(1), 104–132. https://doi.org/10.1037/0033-295x.114.1.104

Vallat-Azouvi, C., Weber, T., Legrand, L., & Azuovi, P. (2007). Working memory after severe traumatic brain injury. *Journal of the International Neuropsychological Society, 13*(05), 770–780. https://doi.org/10.1017/s1355617707070993

Verde, M. F. (2012). Retrieval-induced forgetting and inhibition: A critical review. In B. H. Ross (Ed.), *Psychology of Learning and Motivation,* (Vol. 56, pp. 47–80). Academic Press. https://doi.org/10.1016/b978-0-12-394393-4.00002-9

Villard, S., & Kidd, G., Jr. (2019). Effects of acquired aphasia on the recognition of speech under energetic and informational masking conditions. *Trends in Hearing, 23,* 1–22. https://doi.org/10.1177/2331216519884480

Villard, S., & Kiran, S. (2015). Between-session intra-individual variability in sustained, selective, and integrational non-linguistic attention in aphasia. *Neuropsychologia, 66*, 204–212. https://doi.org/10.1016/j.neuropsychologia.2014.11.026

Villard, S., & Kiran, S. (2017). To what extent does attention underlie language in aphasia? *Aphasiology, 31*(10), 1226–1245. https://doi.org/10.1080/02687038.2016.1242711

Vossel, S., Geng, J. J., & Fink, G. R. (2014). Dorsal and ventral attention systems: Distinct neural circuits but collaborative roles. *The Neuroscientist, 20*(2), 150–159. https://doi.org/10.1177/1073858413494269

Vugs, B., Hendriks, M., Cuperus, J., & Verhoeven, L. (2014). Working memory performance and executive function behaviors in young children with SLI. *Research in Developmental Disabilities, 35*(1), 62–74. https://doi.org/10.1016/j.ridd.2013.10.022

Wager, T. D., & Smith, E. E. (2003). Neuroimaging studies of working memory: A meta-analysis. *Cognitive, Affective, & Behavioral Neuroscience, 3*(4), 255–274. https://doi.org/10.3758/cabn.3.4.255

Wagner, A. D., Shannon, B. J., Kahn, I., & Buckner, R. L. (2005). Parietal lobe contributions to episodic memory retrieval. *Trends in Cognitive Sciences, 9*(9), 445–453. https://doi.org/10.1016/j.tics.2005.07.001

Waters, G. S., Rochon, E., & Caplan, D. (1992). The role of high-level speech planning in rehearsal: Evidence from patients with apraxia of speech. *Journal of Memory and Language, 31*(1), 54–73. https://doi.org/10.1016/0749-596X(92)90005-I

Wiebe, S. A., Espy, K. A., & Charak, D. (2008). Using confirmatory factor analysis to understand executive control in preschool children: I. *Latent structure. Developmental Psychology, 44*(2), 575–587. https://doi.org/10.1037/0012-1649.44.2.575

Winkielman, P., Schwarz, N., Fazendeiro, T. A., & Reber, R. (2003). The hedonic marking of processing fluency: Implications for evaluative judgment. In J. Musch & K. C. Klauer (Eds.), *The Psychology of Evaluation: Affective Processes in Cognition and Emotion* (pp. 189–217). Lawrence Erlbaum Associates Publishers.

Worden, M. S., Foxe, J. J., Wang, N., & Simpson, G. V. (2000). Anticipatory biasing of visuospatial attention indexed by retinotopically specific alpha-bank electroencephalography increases over occipital cortex. *Journal of Neuroscience, 20*(6), RC63. https://doi.org/10.1523/jneurosci.20-06-j0002.2000

Yee, E., Chrysikou, E. G., Hoffman, E., & Thompson-Schill, S. L. (2013). Manual experience shapes object representations. *Psychological Science, 24*(6), 909–919. https://doi.org/10.1177/0956797612464658

Zhang, M., Alamatsaz, N., & Ihlefeld, A. (2021). Hemodynamic responses link individual differences in informational masking to the vicinity of superior temporal gyrus. *Frontiers in Neuroscience, 15*. https://doi.org/10.3389/fnins.2021.675326

9

Prediction in Language Processing

Some Ideas about How It's Done

Matthew J. Traxler and Tamara Y. Swaab

When we listen to speech or read text, we can use the sensory input to activate stored linguistic and nonlinguistic conceptual information to assign or derive the overall meaning for linguistic expressions (Kintsch, 1988; Kintsch & van Dijk, 1978; Traxler, 2023). Hypothetically, a clean linguistic signal provides all the information necessary for the comprehender to understand the speaker's intended message and communicative intentions (absent some kind of production error; Levelt, 1992). Thus, a purely bottom-up processing model could, in principle, explain how comprehenders turn linguistic input into outputs that capture overall meaning and communicative intentions. However, the processing burden of real-time language production and comprehension is high for several reasons. Both language production and comprehension proceed at incredible speed – the average speaker of English can produce 110–150 words per minute and comprehend 3 words per second. Given that adult language users have knowledge of at least 50,000 words, this is an incredible feat. In addition, the linguistic input is often complex or ambiguous (e.g., lexical or syntactic ambiguities) and comprehension may be hampered due to noisy situations (caused either by external environmental noise or internal neural noise). Furthermore, there are many contextual factors and additional sources of information, not necessarily contained in the linguistic signal itself, that are essential to comprehension of the speaker's intention and for communication more generally (e.g., pragmatic information about the intention of the speaker). Thus, many have argued that purely bottom-up processing solutions are insufficient (Elman & McClelland, 1988; Gibson et al., 2013; Marslen-Wilson, 1987; McClelland & Rumelhart, 1981). How does the human language processing system solve this problem? One answer relies on top-down processing, using our knowledge of, and prior experience with, language to resolve complexities and "fill in the holes" in noisy or degraded signals. On these accounts, the representation of the signal itself derives both from bottom-up signal processing and access to previously stored information that becomes active during language processing.

Rooted in century-old ideas of Kant (Swanson, 2016), recent theories of perception, cognition, language, and neuroscience have taken these top-down accounts a step further, proposing that active generation of top-down *predictions* of imminent sensory input guide our perceptual experiences. These predictions can be based on long-term knowledge representations stored in memory, experience-based probabilistic constraints and immediate contextual constraints[1] (Clark, 2013; Friston, 2018; Lupyan & Clark, 2015; Rao & Ballard, 1999; see Ryskin & Nieuwland, 2023 for a review). Several prominent contemporary approaches to language interpretation assign significant impact to prediction or predictive processing during language interpretation (Altmann & Kamide, 1999; Gibson et al., 2013; Hale, 2001, 2003, 2006; Heilbron et al., 2022; Kuperberg, 2021; Kutas et al., 2011; Levy, 2008; Ryskin & Nieuwland, 2023), and some accounts suggest that the production system is actively involved in prediction during listening and reading comprehension (Pickering & Gambi, 2018; Pickering & Garrod, 2013). According to these accounts, comprehenders do not wait until they encounter processing difficulties due to ambiguous input, significant noise, or degradation to engage in top-down processing. Instead, they reduce the computational processing load by tonically and proactively engaging in a set of processes by which they can predict and preactivate imminent bits of the linguistic input. Many of these models focus on the prediction of *words*[2] as an easily identifiable component of the signal (e.g., Hale, 2001, 2006; Levy, 2008). However, other prominent accounts of prediction during language processing (e.g., Kuperberg, 2021; Pickering & Gambi, 2018; Wang et al., 2023) assume that prediction generation occurs at multiple linguistic levels (form: phonology/orthography; and semantics: lexical and event-semantics). Furthermore, several studies show that imminent syntactic categories and structures can be predicted as well, constrained by frequency information as well as knowledge of grammar and syntactic rules (Steedman & Altmann, 1989; Bohnage et al., 2015; Brothers & Traxler, 2016; Hale, 2001; Levy, 2008; Lau et al., 2006; MacDonald et al., 1994; see Traxler, 2014 for a review). These accounts fall within the category of *incremental* processing accounts (Just & Carpenter, 1980; Pickering & Traxler, 1998), as they assume that predictions are updated as each word is processed and understood. They propose that at each word in an expression, comprehenders use the prior context to calculate how likely different words are to appear next (see also Elman, 2004). Making successful predictions is claimed to produce faster processing, and in fact, words that are predictable in context are processed faster than the same words in contexts that make them less predictable (Balota et al., 1985; Drieghe et al., 2005). Processing benefits of predictability during language processing have been inferred from reduced reaction times in behavioral studies, greater skipping rates and reduced reading times in eyetracking studies, reduced electrophysiological responses in ERP/EEG studies and reduced BOLD responses in fMRI studies. More recently, machine learning approaches have been used to distinguish between brain responses to predicted words and words that are semantically related but not predicted (e.g., Trammel et al., 2023). Making inaccurate predictions is claimed to contribute to language learning in children (e.g., Fazekas et al., 2020) and adults (e.g., Fitz & Chang, 2019), as feedback from inaccurate predictions affects the system in which likelihood estimates are computed (see also Rescorla & Wagner, 1972). Prediction and predictability may also play a role in shaping form information in language, including the relationship between word length and frequency (Zipf, 1935/1965).

9.1 Prediction in Language Processing: Classic and Contemporary Approaches

Some accounts of prediction specify in detail how probability estimates are calculated (Hale, 2001, 2006; Levy, 2008). These accounts suggest that linguistic expressions have two properties that influence processing load: *entropy* and *surprisal.* Entropy refers to the degree of uncertainty about future input at any given point. Expressions that allow for more different possible continuations have more entropy than expressions that have fewer different possible continuations. Surprisal refers to the degree to which the actual input matches the predicted input. In a low entropy expression, such as (1),

(1) *At home, I park my car inside my* ______________.

The word *garage* has very low surprisal, but *barn* has high surprisal (even though, in the real world, it would be possible to park a car in a barn). Entropy and surprisal are, in principle, different dimensions, but in practice interact, because, for example, it is not possible to have both high entropy and low surprisal. By contrast, low entropy allows for both low and high surprisal, depending on how an expression continues. On these accounts, classic sentence processing experiments may be viewed as manipulating surprisal. For example, Kutas and Hillyard's *I like my coffee with cream and sugar/dog* ERP experiment produced significant between-condition differences, with *sugar* eliciting a reduced *N400* relative to *cream* (Kutas & Hillyard, 1980). Since this original finding, many studies showed that this N400 effect is graded by the fit of an incoming word to its preceding context, such that N400 amplitudes are reduced to highly predictable words (e.g., Wlotko & Federmeier, 2012), as in Example 2 from Brothers and colleagues (2019):

(2) High Cloze: *"Driving in the countryside, I could only hear static on my radio."*
Low Cloze: *"Unfortunately I was only able to find crummy stuff on my radio."*

This N400 effect results from facilitated processing of words that are predictable in the context, which could be interpreted in terms of surprisal.

These contemporary approaches align with an older account by Paula Schwanenflugel and colleagues (Schwanenflugel, 1991; Schwanenflugel & Lacount, 1988; Schwanenflugel & Shoben, 1985). According to Schwanenflugel's account, sentences differ in terms of their degree of *constraint* (equivalent to entropy). Words in sentences differ in terms of whether they are *expected* or *unexpected*. Schwanenflugel conducted a series of cross-modal lexical decision experiments, where critical target words were shown at the end of spoken sentences that varied in degree of constraint – some were highly constraining, and some did not place much constraint on possible sentence-ending words. She found that low-constraint sentences facilitated lexical decision responses to a range of words, but high-constraint sentences facilitated lexical decisions only to the most highly predictable sentence-ending word. These experiments did not show differences in the degree of facilitation between low-entropy/low-surprisal words and modest entropy/moderately surprising words, but the experiments were not designed to evaluate possible differences of that kind. Also, subsequent cross-modal naming and lexical decision experiments suggested that the precise pattern obtained in the original experiments could have reflected a timing artefact (Traxler et al., 2000). Constraint-based models have also been applied to syntactic processing. MacDonald and colleagues (1994), for example, suggested that multiple syntactic, lexical, and semantic constraints were weighted during parsing, and that lexical and semantic constraints could lead the parser to immediately assign an unpreferred syntactic structure.

One persistent challenge in psycholinguistic research, especially research involving predictability, is that multiple potentially relevant factors need to be considered. Some of those factors correlate in natural language and hence may be very difficult to isolate in the lab. Case in point: Many studies investigating sentence and word processing attribute effects to predictability and preactivation of information about imminent words, but they do not make much of an effort to deconfound predictability from other potentially correlated variables, like semantic plausibility, which cannot affect processing until at least part of the input has been processed (i.e., integrated). For example, in highly constraining (low entropy) expressions, any continuation other than the most expected one will be both higher on surprisal and lower on plausibility. If we assume, as many language processing models do, that semantic implausibility has effects on processing time, then there is no clear justification for attributing effects to prediction that could arise just as easily from (nonpredictive) semantic integration processes (see Ferreira & Chantavarin, 2018 for a discussion of potentially complementary roles of prediction, anticipation and integration in processing of new and redundant information in discourse).

Several studies have attempted to address this problem. Federmeier and colleagues (2007) did dissociate effects of expectancy/predictability and plausibility/integration by manipulating both cloze probability and contextual constraint as in the following example from their study:

(3a) High Constraint – High Cloze/ Low Cloze *"The children went outside to play/look."*

(3b) Low Constraint – Moderate Cloze/ Low Cloze *"Joy was too frightened to move/look."*

Typical N400 effects of expectancy (cloze probability) were found in both low and high constraint conditions, and this effect was greater in the high constraint condition – that is, a greater reduction in the amplitude of the N400 as a function of cloze probability in the high than low constraint condition. In contrast, the amplitude of the N400 to identical low cloze sentence final words ("look" in the example above) did not differ as a function of contextual constraint. The cost of processing low cloze words in a highly constraining relative to a low constraining context was instead found on the anterior post N400 positivity. This pattern of results suggests that effects of prediction and contextual integration may indeed be distinguishable.

To distinguish between effects of prediction, integration, or metacognitive processes that assess the quality of the interpretation, studies have examined anticipatory effects of prediction on words preceding the critical predictable words. An early effort along those lines used the *visual world* paradigm to identify when participants activated semantic information associated with likely upcoming words (Altmann & Kamide, 1999; 2007; Kamide et al., 2003; Tanenhaus et al., 1995; see, e.g., Perdomo & Kaan, 2021, for a bilingual version of the experiment). In these experiments, participants typically look at a visual display that contained four contrasting objects. In Altmann and Kamide's study (1999), one of the objects was edible (*cake*), while the others were not (*train, tricycle, motorcycle*). Participants listened to spoken sentences containing a verb whose selection restrictions were compatible with only one of the pictured objects (e.g., *eat*; *The boy will eat the ... cake*) or sentences containing a verb that was compatible with multiple objects (e.g., *move*; *The boy will move the ... cake*). Participants' eye-movements were monitored while they listened to the sentences. Altmann and Kamide were able to show that participants fixated the *cake* object before the word *cake* was presented in the *eat* condition, but not in the *move* condition. This type of experiment demonstrates that, at least when the scope of reference is restricted to four possible referents, participants make rapid use of verb selection restriction information to anticipate future mention of a pictured object. Asking a similar question, but using different methods, Delong and colleagues also found evidence that comprehenders were anticipating future input (Delong et al., 2005; Wicha et al., 2004; but see Nieuwland et al., 2020; Poulton & Nieuwland, 2022). In an EEG-ERP study, Delong and colleagues generated highly constraining sentence contexts, such as (4a) and (4b), and manipulated whether the determiner directly preceding the predictable noun was consistent with the expectation generated by the context:

(4a) It was a windy day outside, so the boy went outside to fly *a ... kite*. (expected)

(4b) It was a windy day outside, so the boy went outside to fly *an ... airplane*. (unexpected)

The data showed a difference in the ERP waves at the word *a* (versus *an*) as well as at the subsequent *kite/airplane*. Taken at face value, these results would indicate that not only did participants anticipate an upcoming concept (*kite* versus *airplane*), but that they anticipated the phonological form of the preceding article (*a* versus *an*), hence that comprehenders: (1) definitely engaged in predictive processing; (2) anticipated likely upcoming concepts; and (3) also anticipated the phonological form of the critical word (*kite*) and the preceding article *a*/*an*. (But see: Nieuwland et al., 2018).

Effects of anticipation on words prior to predictable nouns have also been observed in other studies. Several studies have investigated languages with more syntactic gender marking than English. These studies demonstrated that the syntactic gender of upcoming predictable nouns affected processing of preceding adjectives. Specifically, facilitated processing occurred when adjectives matched in gender with the imminent predictable noun, demonstrating that this information was activated well before the processing of that noun (Van Berkum et al., 2005; Szewczyk & Schriefers, 2013). Similar effects have been observed in studies of bilingual language users (e.g., Dussias et al., 2013; Morales et al., 2016). In addition, in our lab, Boudewyn and colleagues showed that congruency of semantic features of adjectives preceding predictable nouns in short discourse passages facilitated processing of those adjectives (Boudewyn et al., 2015). These studies give some indication of the content of the preactivated information, but other studies using representational similarity analyses or machine learning approaches to classify EEG signals are likely to contribute to our understanding of whether and when certain information is preactivated (e.g., animacy, Heikel et al., 2018; Wang et al., 2020); lexical vs phonological information, Gwilliams et al., 2023).

In a series of other studies from our lab, we aimed to isolate predictive processing outcomes from other effects, like the consequences of semantic or contextual integration, by manipulating task, the proportion of sentences with predictable critical words, or the reliability of the speaker. In the first study of its type, we used participants' responses to back-sort data on a trial-by-trial basis (Brothers et al., 2015). In this study, the task of the subjects was to indicate, with a button press, whether they had predicted the final word of two sentence passages. The implementation of this task ensured that participants would engage in predictive processing for many of the stimuli in the experiment. Furthermore, it allowed us to sort trials in the experiment according to the participants' responses – "no"-responses meant that participants had not predicted the identical word, "yes"-responses meant that they had predicted the actual sentence final word. In this study, we began by creating stimuli that produced roughly 50 percent likelihood of any given passage final word. For example, the word *novel* occurs about 50 percent of the time in a cloze norming task at the end of (5a):

(5a) *The author is writing another chapter about the fictional detective. To date, he thinks it will be his most popular novel.*

The word *book* is also about 50 percent likely to appear. Hence, after the word *popular*, participants have about a 50 percent chance of successfully predicting the actual word that turns up. In (5b), the second sentence is identical to (5a), but manipulating the previous context sentence means that the critical word "*novel*" is almost impossible to predict, and it is relatively less plausible (because *chefs* are more known for *food* than for *novels*).

(5b) *Everyone congratulated the chef on all his hard work. To date, he thinks it will be his most popular novel.*

Even though "novel' is not predictable in 5b, the overall contextual constraint was the same as in 3a (moderately constraining for "dish" or "dessert" in the example).

By back sorting the ERP data from trials involving texts like (5a), Brothers and colleagues were able to estimate when prediction "success" versus "failure" occurred. Similarly, by comparing the "failed" prediction trials (participants predict *book* but see *novel*) to trials involving mini stories like (5b), *novel* is unpredicted and more difficult to integrate, they could assess when integration processes impacted the ERP signal.

First, consider processing of "novel" when it is successfully predicted versus unpredicted. The ERP results showed a divergence between these two conditions starting about 250 ms post-stimulus. This early effect had a more frontal distribution than the canonical N400 effect, suggesting that it might be generated by only partically or nonoverlapping neural sources. We identify this as an effect of prediction. Next, consider processing of "novel" in the unsuccessful condition from (5a), where the word is easily integrateable, versus in the condition from (5b), where the word is less easy to integrate. Note that both conditions entail prediction failure (i.e., participants did not predict the critical word in either case). In the contrast of (5a) (prediction failure, easy integration) and (5b) (prediction failure, hard integration), the ERP waves diverged later, in the typical N400 time window (300–500 ms post-stimulus). We identify this second effect as reflecting how easy it was to integrate the presented word (*novel*). As is often the case in ERP experiments, easier integration of *novel* in (5a) led to a smaller N400 effect compared to the same sentence in (5b).

We can draw some conclusions from these results that affect the way we think about prediction. First, the data suggest that prediction and integration are separate mental processes and that word processing outcomes depend on both prediction and integration. If there were a single, unified response to words in mini-story contexts, we would not see a temporal and topographic dissociation between prediction success versus failure (which is observed sooner, by about 250 ms following the critical word) and integration difficulty (which is observed later, about 300–500 ms following the critical word).

The Brothers paradigm has proven useful in investigating other issues relating to predictive processing in language. We know very little, for example, about the extent to which predictive processes transpire uniformly across different individuals. Most research on prediction (and most research in psycholinguistics more generally) involves healthy young adult college students (but see Federmeier, 2007; Kutas & Iragui, 1998; Wlotko & Federmeier, 2012).[3] This population has relatively high processing speed compared to older individuals and tends to read more than non-college-age mates. These are just two factors that may influence the specific predictions that comprehenders draw while processing language, as well as their general ability and propensity to engage in predictive processing (Huettig, 2015). So, what happens when older people like the authors of this chapter process language? Do we engage in predictive processing to the same degree as younger individuals (perhaps because greater experience allows for more accurate assessment of probabilities)? Or does slower processing speed or other age-related factors get in the way?

We addressed those questions in a study of older and younger comprehenders using the Brothers paradigm (Dave et al., 2018). As in the prior study, participants read sentences that had either a moderately predictable or an unpredictable final word (5a vs. 5b). They were asked on each trial whether they had successfully predicted the final word of the mini story. As in our 2015 study, back sorting the data yielded three conditions: successful prediction/plausible; unsuccessful prediction plausible; unsuccessful prediction/less plausible. Overall, the older folks showed smaller effects of prediction success versus failure (comparing 5a successful versus unsuccessful prediction trials), as well as smaller differences between more easily integrateable material (5a) and less easily integrateable material (5b), consistent with previous findings from ERP studies from other labs. However, both younger and older adults produced prediction effects earlier post-stimulus than they produced integration/plausibility effects (for similar results from a study using eye tracking, see Choi et al., 2017).

We take these results as indicating that, while the processing substrate in the older adults may be a bit noisier (leading to reduced differences between conditions), their ability to predict future input remains. More work needs to be done to identify the specific neurophysiological and cognitive factors that produce different outcomes in older versus younger adults. In one of our studies, we showed that neural noise is a likely contributing factor (Dave et al., 2018). Different trajectories of verbal knowledge, processing speed, and neural noise may be fruitful areas for future inquiry.

9.2 How Are Predictions Derived?

Various predictive processing accounts attribute predictions to different cognitive processes and systems. For example, Pickering and Gambi (2018; Pickering & Garrod, 2007) suggest that covert production processes may provide the means by which predictions are generated. On this account, readers and listeners engage in covert production processes using written and spoken material as input. In essence, comprehenders are using their own production processes to anticipate how the writer's or speaker's utterance is likely to continue. This process could come in handy in cases where the speaker suffers a dysfluency (i.e., the listener supplies her own input via the production system rather than waiting around for the speaker to get unstuck). The process could also come in handy as a comprehension *stop rule.* That is, comprehenders stop conducting further interpretive operations when their internally generated predictions line up with the actual input. Other accounts attribute predictions to the tracking of n-grams (sequences of n successive syllables or adjacent phonemes across one or more words; Brennan et al., 2020; Hale, 2003, 2016). According to these models, comprehenders keep track of word strings as they are encountered, similar to the Minerva style model of memory proposed by Hintzman (Hintzman, 1986). As new strings are encountered, comprehenders use stored knowledge of prior strings to calculate what words can appear in a given context and how likely or probable different words are to appear.

Hale's account formalizes these computations as follows (2001, pp. 1–2): "(The model uses) conditional probabilities of the form P(Wn = wn|W1 = w1 ... Wn-1 = wn -1). The probability that the word will actually be Wn, given that the words leading up to then have been w1, w2 ... wn-1. Given some finite lexicon, the probability of each possible outcome for Wn can be estimated using that outcome's relative frequency in a sample." Prior context can make a given word either easier or harder to process than baseline, depending on whether the presence of a given word occurs at greater or lesser frequency in combination with some set of words that has appeared immediately prior to the word in question.

In this framework, as noted above, processing difficulty depends on surprisal chiefly. Hypothetically, more surprising continuations require greater cognitive effort, although the reasons why that is the case are somewhat obscure. Perhaps, as is formalized in models like LOGOGEN, some part of the processing system requires more bottom-up evidence to accumulate before it is willing to accept the presence of a surprising constituent (Morton, 1969).

Kuperberg and colleagues' account suggests that comprehenders track probabilities in multiple, linked systems simultaneously (Kuperberg, 2021; Wang et al., 2023). This account includes words as an important constituent, but it includes more than words. Comprehenders can use knowledge derived from prior exposure to make predictions about events (i.e., the occurrence of certain events will be made more probable by the introduction of other events, see, for example, Metusalem et al., 2012), about specific words (similar to other word prediction models), and about specific orthographic or phonological forms (in light of results like those in Delong et al., 2005, and Wicha et al., 2004). This account likely would be able to accommodate prediction of upcoming syntactic structure as well (see, e.g., Brothers & Traxler, 2016; Lau et al., 2006).

Let us take a few moments to focus on the notion that n-grams serve as the basis for prediction. We could adopt a "weak" version of the n-gram account (as Kuperberg and colleagues do) and propose that conditional probabilities calculated on the basis of n-grams contribute to, but do not entirely govern, processing difficulty for specific words in a discourse. We could, by contrast, adopt a "strong" version of the n-gram account and propose that surprisal (qua conditional probability of a word in context) is the sole or dominant contributor to processing difficulty. Let us focus here on evaluating some evidence relating to the strong version.[4]

One line of evidence relates to the consistency of prediction outcomes across different processing environments. If conditional probabilities govern processing times for words, we would expect to see consistent effects of those probabilities so long as n-gram information is held constant. In fact, experimental manipulations of at least two types cast doubt on the idea that n-gram effects are consistent across different processing contexts. The first involves a manipulation of syntactic information while keeping word-to-word associations constant (Morris, 1994; see also Hess et al., 1995; Traxler et al., 2000). In Morris' experiment, participants read sentences while having their eye movements measured. She held lexical associations constant across conditions while manipulating syntactically assigned thematic roles. Locally, (6a and b) and (6c and d) had the same n-grams, hence the entropy associated with the critical word *mustache* should be the same. However, (6a) and (6c) are more plausible overall, because the syntax of the sentence assigns normal agents and themes to the described action (i.e., *barbers* are more likely to trim mustaches than *gardeners* are).

(6a) The gardener watched as the barber trimmed the mustache.

(6b) The gardener who watched the barber trimmed the mustache.

(6c) The barber who watched the gardener trimmed the mustache.

(6d) The barber watched as the gardener trimmed the mustache.

Note that any n-gram account of surprisal differences across 6a–d would have to encompass more than three prior words. Even so, the n-gram account might still struggle to explain why both *the gardener trimmed the mustache* and *the barber trimmed the mustache* yield comparable processing outcomes.[5]

Experiments from our own lab also support the conclusion that local n-gram configurations do not explain the full range of observed effects. In an ERP experiment, we held local n-gram configurations constant while manipulating aspects of the greater global context (Brothers et al., 2017; see also Delaney-Busch et al., 2017). In the first experiment, mini-story stimuli were taken from the previous prediction versus integration study (Brothers et al., 2015). We used ERPs to assess comprehenders' responses to the stimuli under two task conditions. In the first, participants were instructed to read for comprehension. In the second, participants were instructed to try to predict the final word of each passage. The stimuli were identical across conditions, so we could isolate the effects of task instructions. The "predict" instruction condition produced larger N400 effects and a larger anterior post-N400 positivity (PNP) than the "comprehend" instruction condition. Minimally, these results indicate that processing of the critical word depends on more than just conditional probability based on lexical co-occurrence.

In the second experiment, we presented the same critical stimuli in two different global contexts and measured participants' self-paced reading times. Half of the critical stimuli had high constraint/low entropy (on average, one final word was 96 percent likely), as in (7a):

(7a) The web had been spun by the large *spider* on the porch. (*spider* is the critical word)

(7b) Alex said he wanted to watch the large *spider* on the porch.

Stimuli like (7b) had low constraint/high entropy at the critical word (<1 percent cloze probability). We expected to see large differences in reading time on the critical words between the predictable (7a) and unpredictable (7b) conditions. However, and this is the crucial bit, we created three different conditions by manipulating characteristics of the filler sentences. We used highly constraining filler sentences across three *global validity* conditions. In the *high validity* condition, 100 percent of the high-constraint filler sentences were completed with a predictable word, such as (8a):

(8a) The volleyball shot barely made it over the *net*. (*net* is 95 percent likely on average)

In the *medium validity* condition, only 50 percent of sentences such as (8a) ended with the predictable word. The other filler sentences ended with a highly unpredictable word (<1 percent cloze), as in (8b):

(8b) The volleyball shot barely made it over the *car*.

In the *low validity* condition, all the filler sentences (100 percent) had unpredictable final words (like 8b). Given that all the same critical sentences appeared in the different validity conditions, calculations based on n-grams yield the same values, and so we would not expect any differences across the different global validity conditions. If instead participants are sensitive to the utility of prediction in a given processing environment, then we would expect to see different outcomes across different global validity conditions. In fact, the benefits of predictability scaled more or less linearly with global validity. The effects on local predictability (reading time for *spider* in 7a versus 7b) were bigger in the high validity condition, smaller in the medium validity condition, and absent in the low validity condition. The difference was driven largely by changes across validity conditions in reading time for the predictable words. Unpredictable words (spider in 7b) were read at about the same speed across the different global validity conditions. We conclude from these results that readers do not automatically, reflexively, and uniformly react to locally predictability information. Instead, they appear to adapt their use of local predictability information based on its utility in a given (global) processing environment. Subsequent studies have also found evidence for *rational adaptation* of this kind (Ness & Meltzer-Asscher, 2021a,b).[6] Many open questions remain about how comprehenders dynamically adapt to local predictability information, but these results would suggest that such reliance is at least partly under strategic control and that comprehenders rely on local predictability information more when doing so is beneficial.

The n-gram approach to prediction pays scant attention to potential differences between different speakers. However, it is well established that listeners rapidly adapt to speaker characteristics. For example, different speakers produce widely divergent acoustic patterns for the same expressions; and the same speaker can produce widely divergent acoustic patterns for the same expressions across different utterances, depending on numerous internal factors (Johnson & Sjerps, 2021). N-gram approaches focus on language-wide patterns, as expressed in large-scale corpora. However, an individual speaker may or may not follow those patterns. In fact, there is likely considerable between-individual variation based on multiple factors (education level, developmental stage, differential exposure to different genres, differential exposure to written versus spoken input, etc.). Might lexical prediction, like speaker normalization processes, lead comprehenders to respond differently to predictability information coming from different speakers? We evaluated this possibility in a series of EEG-ERP experiments.

Our approach followed from our prior studies of cue validity (Brothers et al., 2017, 2019). In the first ERP experiment, our critical stimuli were like those in (7a) and (7b), presenting the same target words in contexts that either supported prediction (7a) or did not. We would expect to see differences between (7a) and (7b) at the word *spider* in the N250, N400, and later windows, with less predictable words showing greater frontal positivity. As in the prior self-paced reading study of cue validity, we manipulated characteristics of the filler sentences to create *high validity* and *low validity* conditions. However, instead of a between-participants manipulation of cue validity, we recorded two different speakers to create a within-participant manipulation of cue validity. One speaker – the *reliable* speaker – always produced filler sentences similar to (8a), and so had highly predictable endings to their sentences. The other speaker – the *unreliable* speaker – always produced filler sentences similar to (8b), and so had highly unpredictable endings to their sentences. Half of the critical test items were presented in the reliable speaker's voice; and the other half of the critical items were presented in the unreliable speaker's voice. If n-grams alone serve as the basis of lexical predictions, then it should not matter what voice participants heard. In fact, it did matter to participants whether the voice they were hearing was from a reliable speaker or an unreliable speaker. We observed greater differences in the ERP waves between sentences such as (8a) and (8b) for the reliable speaker than the unreliable speaker. Hence, as in the manipulation of global context in the self-paced reading study, participants appeared to make greater use of predictability information when that information was likely to be useful. Note that we tested the same participants in the same sessions with the same critical test sentences. Hence, we observed greater versus lesser response to predictability information in the same participants during the same test session.

We followed up on this initial study with another assessment of how aging might influence listeners' ability to flexibly adapt to speaker characteristics (Dave et al., 2021). In this study, we again used EEG-ERP methods to assess how older and younger adults respond to predictability information and how they respond when that information comes from unreliable versus reliable sources. The experimental manipulation was the same as the prior study of speaker reliability. We had critical words that were either predictable or not; and we had speakers who either were reliable or not. We added a further between-participants assessment of age effects (older versus younger). In addition, we evaluated individual differences in executive control using an automated STROOP task (Stroop, 1935; Stoet, 2017). As in the prior study of speaker reliability, we found that the younger participants showed larger between-condition effects (predictable versus unpredictable words) when the source was a reliable speaker as opposed to an unreliable speaker, replicating prior speaker adaptation effects. In contrast, the older participants showed a main effect of predictability that was just as large when the speaker was unreliable as when the speaker was reliable. Older adults, unlike younger adults appeared to rely more on local (sentence-level) predictability information and did not change their response for different sources of information (different speakers). However, both older and younger adults' response to speaker reliability differences scaled with their STROOP performance. In both cases, participants who exhibited greater degrees of executive control (as assessed by STROOP) showed greater response to speaker reliability information (see also Ryskin et al., 2020). We conclude from these results that the ability to combine global processing context information with locally sourced predictability information changes across the lifespan – older folks are less able to adapt to aspects of the global task environment. Alternatively, older folks may simply weight n-gram information more heavily than other potentially useful sources. With all that in mind, we have tantalizing clues that the change in response across the lifespan may be related to changes in executive function.

Our results show that listeners are sensitive to predictability information, but that they are not merely reflexively reacting to that information as they build interpretations of linguistic expressions and infer the speaker or writer's intentions. These results align with a viewpoint and a variety of findings in the developmental literature suggesting that language learners similarly are not reflexive pattern learners (Buac et al., 2019; Clement et al., 2004; Jaswal & Neely, 2006; Kinzler et al., 2007; Kinzler et al., 2011; Koenig et al., 2004; Koenig & Harris, 2005; Koenig & Woodward, 2010; Sabbagh & Baldwin, 2001; Scofield & Behrend, 2008). For example, in the word learning literature, young children have been shown to acquire word meanings faster or slower depending on many source characteristics, including (but possibly not limited to): speakers' individual histories of accuracy in expressing word meanings; a speaker's degree of expressed uncertainty; a speaker's age; speaker accent; a speaker's physical appearance; other social cues; and bilingual status of children themselves (bilingual children are more "trusting" of speakers who express uncertainty than are monolingual children). Similarly, children who are exposed to noisy or inaccurate language models do not merely parrot the patterns they observe. Instead, they infer underlying regularities and mold their own production to those inferred rules (Singleton & Newport, 2004). As Koenig and Harris (2005, p. 1261) noted, "Preschoolers mistrust ignorant and inaccurate speakers." We would not expect to see such mistrust if children were merely learning how to predict words in context.

9.3 Conclusions

Recent research in our lab and others has helped fill in some pieces of the puzzle regarding factors that affect real-time processing and learning from language exposure. We believe our results can help narrow the range of possible models of prediction and its effects on language processing and interpretation. First, our results support a framework in which anticipatory and integration processes both matter in the sense that both kinds of operations produce separate, observable effects in language processing tasks. Hence, to understand and build adequate descriptions of language processing and interpretation, we need to pay attention to more than just prediction. Fortunately, we have a variety of existing accounts to choose from to gain insight into linguistic processing at different levels (phonological, lexical, syntactic, discourse, etc.). Second, our results suggest that n-grams are not the ultimate source or basis of predictions, or at least not the sole source. If they were, we would not expect to see effects of n-grams wax and wane as other aspects of the language processing environment change. In our view, the information sources that feed into predictive processing and the precise nature of the calculations that generate predictions remain to be identified.

Returning to Kintsch's directive to contemplate how stored knowledge enables comprehenders to understand messages, we would like to encourage readers to view the research on prediction in the wider context of language understanding and language learning. Prediction, when successful, may facilitate aspects of language processing (many accounts focus on increasing the speed of word recognition and activation of stored lexical information), but it does not in and of itself offer a satisfactory explanation of how comprehenders assign meanings and infer communicative intentions.

Theories of language processing and interpretation should accommodate the fact that humans are complex creatures. Instead of reacting reflexively to pattern information, they appear to evaluate characteristics of information sources during real-time processing and distill what they learn from processing language based on those characteristics. These types of findings are more surprising if we focus narrowly on prediction, but less so in light of more nuanced descriptions of the human language processing system.

Notes

1 Please note that prediction during language processing at the computational level does not necessarily imply *predictive coding*, which suggests that the brain actively anticipates imminent sensory input.
2 Without necessarily defining what, exactly, is captured by the term "word." Given the complex nature of words and word representations, and cross-linguistic differences in the way "words" are constructed, this lack of specificity may not be entirely benign.
3 As well as crafting conditions that are most highly conducive to supporting successful predictions.
4 Evaluating the more nuanced version would require a much larger expenditure of ink. Interested readers should consult recent papers from the Kuperberg lab (e.g., Brothers & Kuperberg, 2021; Eddine et al., 2022; Kuperberg & Jaeger, 2016).
5 Pure associationist accounts of semantics might also struggle, but that is beyond the scope of this chapter (Landauer & Dumais, 1997; Lund & Burgess, 1996).
6 There has been some recent conversation about whether rational adaptation is an artefact of blocked presentation and may not always occur, or may occur during later stages of processing (Kroczek & Gunter, 2021; Van Wonderen & Nieuwland, 2023).

References

Altmann, G. T., & Kamide, Y. (1999). Incremental interpretation at verbs: Restricting the domain of subsequent reference. *Cognition, 73*, 247–264.

Altmann, G. T., & Kamide, Y. (2007). The real-time mediation of visual attention by language and world knowledge: Linking anticipatory (and other) eye movements to linguistic processing. *Journal of Memory and Language, 57*, 502–518.

Balota, D. A., Pollatsek, A., & Rayner, K. (1985). The interaction of contextual constraints and parafoveal visual information in reading. *Cognitive Psychology, 17*, 364–390.

Bonhage, C. E., Mueller, J. L., Friederici, A. D., & Fiebach, C. J. (2015). Combined eye tracking and fMRI reveals neural basis of linguistic predictions during sentence comprehension. *Cortex*, 68, 33–47. https://doi.org/10.1016/j.cortex.2015.03.014

Boudewyn, M. A., Long, D. L., & Swaab, T. Y. (2015). Graded expectations: Predictive processing and the adjustment of expectations during spoken language comprehension. *Cognitive, Affective, & Behavioral Neuroscience, 15*(3), 607–624. https://doi.org/10.3758/s13415-015-0340-0

Brennan, J. R., Dyer, C., Kuncoro, A., & Hale, J. T. (2020). Localizing syntactic predictions using recurrent neural network grammars. *Neuropsychologia, 146*, 107479.

Brothers, T., & Kuperberg, G. R. (2021). Word predictability effects are linear, not logarithmic: Implications for probabilistic models of sentence comprehension. *Journal of Memory and Language, 116*, 104174.

Brothers, T., & Traxler, M. J. (2016). Anticipating syntax during reading: Evidence from the boundary change paradigm. *Journal of Experimental Psychology: Learning, Memory, and Cognition, 42*(12), 1894.

Brothers, T., Dave, S., Hoversten, L. J., Traxler, M. J., & Swaab, T. Y. (2019). Flexible predictions during listening comprehension: Speaker reliability affects anticipatory processes. *Neuropsychologia, 135*, 107225.

Brothers, T., Swaab, T. Y., & Traxler, M. J. (2015). Effects of prediction and contextual support on lexical processing: Prediction takes precedence. *Cognition, 136*, 135–149.

Brothers, T., Swaab, T. Y., & Traxler, M. J. (2017). Goals and strategies influence lexical prediction during sentence comprehension. *Journal of Memory and Language, 93*, 203–216.

Buac, M., Tauzin-Larché, A., Weisberg, E., & Kaushanskaya, M. (2019). Effect of speaker certainty on novel word learning in monolingual and bilingual children. *Bilingualism: Language and Cognition, 22*(4), 883–895.

Choi, W., Lowder, M. W., Ferreira, F., Swaab, T. Y., & Henderson, J. M. (2017). Effects of word predictability and preview lexicality on eye movements during reading: A comparison between young and older adults. *Psychology and Aging, 32*, 232–242. https://doi.org/10.1037/pag0000160

Clark, A. (2013). Expecting the world: Perception, prediction, and the origins of human knowledge. *The Journal of Philosophy, 110*(9), 469–496. https://doi.org/10.5840/jphil20131109/2

Clement, F, Koenig, M, & Harris, P. (2004). The ontogenesis of trust. *Mind & Language, 19*, 360–379.

Dave, S., Brothers, T. A., & Swaab, T. Y. (2018). 1/f neural noise and electrophysiological indices of contextual prediction in aging. *Brain Research, 1691*, 34–43.

Dave, S., Brothers, T. A., Traxler, M. J., Ferreira, F., Henderson, J. M., & Swaab, T. Y. (2018). Electrophysiological evidence for preserved primacy of lexical prediction in aging. *Neuropsychologia, 117*, 135–147.

Dave, S., Brothers, T., Hoversten, L. J., Traxler, M. J., & Swaab, T. Y. (2021). Cognitive control mediates age-related changes in flexible anticipatory processing during listening comprehension. *Brain Research, 1768*, 147573.

Delaney-Busch, N., Morgan, E., Lau, E. F., & Kuperberg, G. R. (2017). Comprehenders rationally adapt semantic predictions to the statistics of the local environment: A Bayesian model of trial-by-trial N400 amplitudes. In Proceedings of the 39th Annual Meeting of the Cognitive Science Society, London.

DeLong, K. A., Urbach, T. P., & Kutas, M. (2005). Probabilistic word preactivation during language comprehension inferred from electrical brain activity. *Nature Neuroscience, 8*, 1117–1121.

Drieghe, D., Rayner, K., & Pollatsek, A. (2005). Eye movements and word skipping during reading revisited. *Journal of Experimental Psychology: Human Perception and Performance, 31*, 954–969.

Dussias, P. E., Kroff, J. R. V., Tamargo, R. E. G., & Gerfen, C. (2013). When gender and looking go hand in hand: Grammatical gender processing in L2 Spanish. *Studies in Second Language Acquisition, 35*(2), 353–387. https://doi.org/10.1017/S0272263112000915

Eddine, S. N., Brothers, T., & Kuperberg, G. R. (2022). The N400 in silico: A review of computational models. *Psychology of Learning and Motivation, 76*, 123–206.

Elman, J. L. (2004). An alternative view of the mental lexicon. *Trends in Cognitive Sciences, 8*, 301–306.

Elman, J. L., & McClelland, J. L. (1988). Cognitive penetration of the mechanisms of perception: Compensation for coarticulation of lexically restored phonemes. *Journal of Memory and Language, 27*, 143–165.

Fazekas, J., Jessop, A., Pine, J., & Rowland, C. (2020). Do children learn from their prediction mistakes? A registered report evaluating error-based theories of language acquisition. *Royal Society Open Science, 7*(11), 180877. https://doi.org/10.1098/rsos.180877

Federmeier, K. D. (2007). Thinking ahead: The role and roots of prediction in language comprehension. *Psychophysiology, 44*(4), 491–505.

Federmeier, K. D., Wlotko, E. W., De Ochoa-Dewald, E., & Kutas, M. (2007). Multiple effects of sentential constraint on word processing. *Brain Research, 1146*, 75–84.

Ferreira, F., & Chantavarin, S. (2018). Integration and prediction in language processing: A synthesis of old and new. *Current Directions in Psychological Science, 27*(6), 443–448. https://doi.org/10.1177/0963721418794491

Fitz, H., & Chang, F. (2019). Language ERPs reflect learning through prediction error propagation. *Cognitive Psychology, 111*, 15–52. https://doi.org/10.1016/j.cogpsych.2019.03.002

Friston, K. (2018). Does predictive coding have a future?*Nature Neuroscience, 21*(8), Article 8. https://doi.org/10.1038/s41593-018-0200-7

Gibson, E., Bergen, L., & Piantadosi, S. T. (2013). Rational integration of noisy evidence and prior semantic expectations in sentence interpretation. *Proceedings of the National Academy of Sciences, 110*, 8051–8056.

Gwilliams, L., Marantz, A., Poeppel, D., & King, J.-R. (2023). Top-down information shapes lexical processing when listening to continuous speech. *Language, Cognition and Neuroscience, 39*, 1045–1058.

Hale, J. (2001). A probabilistic Earley parser as a psycholinguistic model. In Second meeting of the North American chapter of the Association for Computational Linguistics.

Hale, J. (2003). The information conveyed by words in sentences. *Journal of Psycholinguistic Research, 32*, 101–123.

Hale, J. (2006). Uncertainty about the rest of the sentence. *Cognitive Science, 30*, 643–672.

Hale, J. (2016). Information-theoretical complexity metrics. *Language and Linguistics Compass, 10*(9), 397–412.

Heikel, E., Sassenhagen, J., & Fiebach, C. J. (2018). Decoding semantic predictions from EEG prior to word onset (p. 393066). bioRxiv. https://doi.org/10.1101/393066

Heilbron, M., Armeni, K., Schoffelen, J.-M., Hagoort, P., & de Lange, F. P. (2022). A hierarchy of linguistic predictions during natural language comprehension. *Proceedings of the National Academy of Sciences, 119*(32), e2201968119. https://doi.org/10.1073/pnas.2201968119

Hess, D. J., Foss, D. J., & Carroll, P. (1995). Effects of global and local context on lexical processing during language comprehension. *Journal of Experimental Psychology: General, 124,* 62.

Hintzman, D. L. (1986). "Schema abstraction" in a multiple-trace memory model. *Psychological Review, 93,* 411–428.

Huang, K. J., Arehalli, S., Kugemoto, M., Muxica, C., Prasad, G., Dillon, B., & Linzen, T. (2023). Surprisal does not explain syntactic disambiguation difficulty: Evidence from a large-scale benchmark. Preprint on PsyArXiv.

Huettig, F. (2015). Four central questions about prediction in language processing. *Brain Research, 1626,* 118–135. https://doi.org/10.1016/j.brainres.2015.02.014

Jaswal, V. K., & Neely, L. A. (2006). Adults don't always know best: Preschoolers use past reliability over age when learning new words. *Psychological Science, 17*(9), 757–758.

Johnson, K., & Sjerps, M. J. (2021). Speaker normalization in speech perception. In J. S. Pardo, L. C. Nygaard, R. E. Remez, & D. B. Pisoni, (Eds.), *The Handbook of Speech Perception,* 145–176. John Wiley & Sons.

Just, M. A., & Carpenter, P. A. (1980). A theory of reading: From eye fixations to comprehension. *Psychological Review, 87,* 329–354.

Kamide, Y., Altmann, G. T., & Haywood, S. L. (2003). The time-course of prediction in incremental sentence processing: Evidence from anticipatory eye movements. *Journal of Memory and Language, 49,* 133–156.

Kintsch, W. (1988). The role of knowledge in discourse comprehension: A construction-integration model. *Psychological Review, 95,* 163.

Kintsch, W., & Van Dijk, T. A. (1978). Toward a model of text comprehension and production. *Psychological Review, 85,* 363.

Kinzler, K. D., Corriveau, K. H., & Harris, P. L. (2011). Children's selective trust in native-accented speakers. *Developmental Science, 14*(1), 106–111.

Kinzler, K. D., Dupoux, E., & Spelke, E. S. (2007). The native language of social cognition. *Proceedings of the National Academy of Sciences, 104*(30), 12577–12580.

Koenig, M. A., & Harris, P. L. (2005). Preschoolers mistrust ignorant and inaccurate speakers. *Child Development, 76,* 1261–1277.

Koenig, M. A., & Woodward, A. L. (2010). Sensitivity of 24-month-olds to the prior inaccuracy of the source: Possible mechanisms. *Developmental Psychology, 46*(4), 815.

Koenig, M. A., Clément, F., & Harris, P. L. (2004). Trust in testimony: Children's use of true and false statements. *Psychological Science, 15*(10), 694–698.

Kroczek, L. O., & Gunter, T. C. (2021). The time course of speaker-specific language processing. *Cortex, 141*, 311–321.

Kuperberg, G. R. (2021). Tea with milk? A hierarchical generative framework of sequential event comprehension. *Topics in Cognitive Science, 13*, 256–298.

Kuperberg, G. R., & Jaeger, T. F. (2016). What do we mean by prediction in language comprehension? *Language, Cognition and Neuroscience, 31*, 32–59.

Kutas, M., DeLong, K. A., & Smith, N. J. (2011). A look around at what lies ahead: Prediction and predictability in language processing. *Predictions in the Brain: Using Our Past to Generate a Future, 190207*(10.1093).

Kutas, M., & Hillyard, S. A. (1980). Reading senseless sentences: Brain potentials reflect semantic incongruity. *Science, 207*(4427), 203–205. https://doi.org/10.1126/science.7350657

Kutas, M., & Iragui, V. (1998). The N400 in a semantic categorization task across 6 decades. *Electroencephalography and Clinical Neurophysiology/Evoked Potentials Section, 108*(5), 456–471.

Kutas, M., Van Petten, C. K., & Kluender, R. (2006). Psycholinguistics electrified II (1994–2005). In *Handbook of Psycholinguistics* (pp. 659–724). Academic Press.

Lau, E., Stroud, C., Plesch, S., & Phillips, C. (2006). The role of structural prediction in rapid syntactic analysis. *Brain and Language, 98*(1), 74–88.

Landauer, T. K., & Dumais, S. T. (1997). A solution to Plato's problem: The latent semantic analysis theory of acquisition, induction, and representation of knowledge. *Psychological Review, 104*(2), 211.

Levelt, W. J. (1992). Accessing words in speech production: Stages, processes and representations. *Cognition, 42*, 1–22.

Levy, R. (2008). Expectation-based syntactic comprehension. *Cognition, 106*, 1126–1177.

Lund, K., & Burgess, C. (1996). Producing high-dimensional semantic spaces from lexical co-occurrence. *Behavior Research Methods, Instruments, & Computers, 28*(2), 203–208.

Lupyan, G., & Clark, A. (2015). Words and the World: Predictive Coding and the Language-Perception-Cognition Interface. *Current Directions in Psychological Science, 24*(4), 279–284. https://doi.org/10.1177/0963721415570732

MacDonald, M. C., Pearlmutter, N. J., & Seidenberg, M. S. (1994). The lexical nature of syntactic ambiguity resolution. *Psychological Review*, 101(4), 676.

Marslen-Wilson, W. D. (1987). Functional parallelism in spoken word-recognition. *Cognition, 25*(1–2), 71–102.

McClelland, J. L., & Rumelhart, D. E. (1981). An interactive activation model of context effects in letter perception: I. An account of basic findings. *Psychological Review, 88*, 375.

Metusalem, R., Kutas, M., Urbach, T. P., Hare, M., McRae, K., & Elman, J. L. (2012). Generalized event knowledge activation during online sentence comprehension. *Journal of Memory and Language, 66*(4), 545–567.

Morales, L., Paolieri, D., Dussias, P. E., Kroff, J. R. V., Gerfen, C., & Bajo, M. T. (2016). The gender congruency effect during bilingual spoken-word recognition. *Bilingualism: Language and Cognition, 19*(2), 294–310. https://doi.org/10.1017/S1366728915000176

Morris, R. K. (1994). Lexical and message-level sentence context effects on fixation times in reading. *Journal of Experimental Psychology: Learning, Memory, & Cognition, 20,* 92–103.

Morton, J. (1969). Interaction of information in word recognition. *Psychological Review, 76,* 165–178.

Ness, T., & Meltzer-Asscher, A. (2021a). From pre-activation to pre-updating: A threshold mechanism for commitment to strong predictions. *Psychophysiology, 58*(5), e13797.

Ness, T., & Meltzer-Asscher, A. (2021b). Rational adaptation in lexical prediction: The influence of prediction strength. *Frontiers in Psychology, 12,* 622873.

Nieuwland, M. S., Arkhipova, Y., & Rodríguez-Gómez, P. (2020). Anticipating words during spoken discourse comprehension: A large-scale, pre-registered replication study using brain potentials. *Cortex, 133,* 1–36.

Nieuwland, M. S., Politzer-Ahles, S., Heyselaar, E., Segaert, K., Darley, E., Kazanina, N., Von Grebmer Zu Wolfsthurn, S., Bartolozzi, F., Kogan, V., & Ito, A. (2018). Large-scale replication study reveals a limit on probabilistic prediction in language comprehension. *ELife, 7,* e33468.

Perdomo, M., & Kaan, E. (2021). Prosodic cues in second-language speech processing: A visual world eye-tracking study. *Second Language Research,* **37**, 349–375.

Pickering, M. J., & Gambi, C. (2018). Predicting while comprehending language: A theory and review. *Psychological Bulletin, 144*(10), 1002. https://doi.org/10.1037/bul0000158

Pickering, M. J., & Garrod, S. (2007). Do people use language production to make predictions during comprehension?*Trends in Cognitive Sciences, 11,* 105–110.

Pickering, M. J., & Garrod, S. (2013). An integrated theory of language production and comprehension. *Behavioral and Brain Sciences, 36*(4), 329–392. https://doi.org/10.1017/S0140525X12003238

Pickering, M. J., & Traxler, M. J. (1998). Plausibility and recovery from garden paths: An eye-tracking study. *Journal of Experimental Psychology: Learning, Memory, and Cognition, 24,* 940–961.

Poulton, V. R., & Nieuwland, M. S. (2022). Can you hear what's coming? Failure to replicate ERP evidence for phonological prediction. *Neurobiology of Language, 3,* 556–574.

Rao, R. P. N., & Ballard, D. H. (1999). Predictive coding in the visual cortex: A functional interpretation of some extra-classical receptive-field effects. *Nature Neuroscience, 2*(1), Article 1. https://doi.org/10.1038/4580

Rescorla, R. A., & Wagner, A. R. (1972). A theory of Pavlovian conditioning: Variations in the effectiveness of reinforcement and nonreinforcement. In A. H. Black & W. F. Prokasy (Eds.), *Classical Conditioning II: Current Research and Theory* (Vol. 2, pp. 64–99). Appleton-Century-Crofts.

Ryskin, R., Levy, R. P., & Fedorenko, E. (2020). Do domain-general executive resources play a role in linguistic prediction? Re-evaluation of the evidence and a path forward. *Neuropsychologia, 136,* 107258.

Ryskin, R., & Nieuwland, M. S. (2023). Prediction during language comprehension: What is next?*Trends in Cognitive Sciences, 27*(11). https://doi.org/10.1016/j.tics.2023.08.003

Sabbagh, M. A., & Baldwin, D. A. (2001). Learning words from knowledgeable versus ignorant speakers: Links between preschoolers' theory of mind and semantic development. *Child Development, 72*(4), 1054–1070.

Schwanenflugel, P. J. (1991). Contextual constraint and lexical processing. In *Advances in Psychology* (Vol. 77, pp. 23–45). North-Holland.

Schwanenflugel, P. J., & LaCount, K. L. (1988). Semantic relatedness and the scope of facilitation for upcoming words in sentences. *Journal of Experimental Psychology: Learning, Memory, and Cognition, 14,* 344–354.

Schwanenflugel, P. J., & Shoben, E. J. (1985). The influence of sentence constraint on the scope of facilitation for upcoming words. *Journal of Memory and Language, 24,* 232–252.

Scofield, J., & Behrend, D. A. (2008). Learning words from reliable and unreliable speakers. *Cognitive Development, 23*(2), 278–290.

Singleton, J. L., & Newport, E. L. (2004). When learners surpass their models: The acquisition of American Sign Language from inconsistent input. *Cognitive Psychology, 49*(4), 370–407.

Steedman, M., & Altmann, G. (1989). Ambiguity in context: A reply. *Language and Cognitive Processes, 4*(3–4), SI105–SI122. https://doi.org/10.1080/01690968908406365

Stoet, G., (2017). PsyToolkit: A novel web-based method for running online questionnaires and reaction-time experiments. *Teaching of Psychology, 44,* 24–31.

Stroop, J. R., (1935). Studies of interference in serial verbal reactions. *Journal of Experimental Psychology, 18,* 643.

Swanson, L. R. (2016). The predictive processing paradigm has roots in Kant. *Frontiers in Systems Neuroscience, 10.* www.frontiersin.org/articles/10.3389/fnsys.2016.00079

Szewczyk, J. M., & Schriefers, H. (2013). Prediction in language comprehension beyond specific words: An ERP study on sentence comprehension in Polish. *Journal of Memory and Language, 68*(4), 297–314.

Tanenhaus, M. K., Spivey-Knowlton, M. J., Eberhard, K. M., & Sedivy, J. C. (1995). Integration of visual and linguistic information in spoken language comprehension. *Science, 268*(5217), 1632–1634.

Trammel, T., Khodayari, N., Luck, S. J., Traxler, M. J., & Swaab, T. Y. (2023). Decoding semantic relatedness and prediction from EEG: A classification method comparison. *NeuroImage, 277,* 120268.

Traxler, M. J. (2014). Trends in syntactic parsing: Anticipation, Bayesian estimation, and good-enough parsing. *Trends in Cognitive Sciences, 18,* 605–611. https://doi.org/10.1016/j.tics.2014.08.001

Traxler, M. J. (2023). *Introduction to Psycholinguistics: Understanding Language Science* (2nd ed.). Wiley-Blackwell.

Traxler, M. J., Foss, D. J., Seely, R. E., Kaup, B., & Morris, R. K. (2000). Priming in sentence processing: Intralexical spreading activation, schemas, and situation models. *Journal of Psycholinguistic Research, 29,* 581–595.

Van Berkum, J. J., Brown, C. M., Zwitserlood, P., Kooijman, V., & Hagoort, P. (2005). Anticipating upcoming words in discourse: Evidence from ERPs and reading times. *Journal of Experimental Psychology: Learning, Memory, and Cognition, 31*(3), 443.

Van Wonderen, E., & Nieuwland, M. S. (2023). Lexical prediction does not rationally adapt to prediction error: ERP evidence from pre-nominal articles. *Journal of Memory and Language, 132,* 104435.

Wang, L., Schoot, L., Brothers, T., Alexander, E., Warnke, L., Kim, M., ... & Kuperberg, G. R. (2023). Predictive coding across the left fronto-temporal hierarchy during language comprehension. *Cerebral Cortex, 33,* 4478–4497.

Wang, L., Wlotko, E., Alexander, E., Schoot, L., Kim, M., Warnke, L., & Kuperberg, G. R. (2020). Neural evidence for the prediction of animacy features during language comprehension: Evidence from MEG and EEG representational similarity analysis. *Journal of Neuroscience, 40*(16), 3278–3291. https://doi.org/10.1523/JNEUROSCI.1733-19.2020

Wicha, N. Y., Moreno, E. M., & Kutas, M. (2004). Anticipating words and their gender: An event-related brain potential study of semantic integration, gender expectancy, and gender agreement in Spanish sentence reading. *Journal of Cognitive Neuroscience, 16*(7), 1272–1288.

Wlotko, E. W., & Federmeier, K. D. (2012). So that's what you meant! Event-related potentials reveal multiple aspects of context use during construction of message-level meaning. *NeuroImage, 62*(1), 356–366.

Zipf, G. K. (1935). (reprinted 1965). *The Psycho-Biology of Language.* MIT Press.

Part IV

Variations of Language Representations in the Brain

Part IVA

Building Cognitive Brain Reserve and the Importance of Proficiency

10

Bilingualism as a Dynamic Experience and Its Effects on Brain Structure, Function, and Metabolism

Christos Pliatsikas

10.1 Bilingualism, Cognition and the Brain: Arguments and Controversies, Current Directions

This chapter reviews contemporary evidence about the effects that the cognitively demanding experience of bilingualism has on the brain, in terms of its structure and its function, focusing on effects not related to particular cognitive or linguistic tasks. Research on bilingual cognition has decisively shown that all languages are active in a bilingual's mind at any given time, and they "compete" for production in different environments (Kroll & Bialystok, 2013). This reality means that bilinguals have the constant need to select one of their available languages, while suppressing the other, in order to communicate in a given context, and also the need to efficiently switch between the languages they speak when the necessity arises. Crucially, these needs become more relevant to individuals immersed in contexts where both languages are spoken regularly (*active* bilinguals). Moreover, these needs apply not only to individuals exposed to more than one language from a very early age (*simultaneous* or *early* bilinguals), but also to adult *sequential* learners of additional languages, as long as they are proficient and active users of these languages, and immersed in such environments (Abutalebi & Green, 2016). It then follows that bilingualism constitutes a constant, long-term (and sometimes life-long) cognitive challenge that does not have its equivalent in monolinguals. In order for this challenge to be resolved, bilinguals are required to heavily recruit their executive functions, a set of cognitive skills that allow us to direct attention, switch between tasks, select between competing alternatives and inhibit irrelevant responses (Alrwaita et al., 2022). This sustained taxing of executive functions is thought to "train" the implicated cognitive skills and brain networks

(Bialystok et al., 2012) in that it makes them more efficient in language switching and control. Crucially, this "training" has been suggested to directly and fundamentally change these domain-general processes across the board, leading to more efficient performance reported in bilinguals compared to monolinguals even in nonlinguistic tasks involving executive control (Bialystok, 2017). However, these suggestions remain controversial, owing to the fact that the relevant findings have not been consistently replicated, especially in younger adults (Lehtonen et al., 2018).

Nevertheless, and despite the relative challenges in replicating these effects, considerable evidence remains that bilingualism affects domain general cognition. This may also suggest that it affects the underlying brain structures and mechanisms that support the affected cognitive processes – it has been suggested that any changes in cognition that follow the acquisition of a new skill will also yield commensurate changes in the structure and function of those brain regions that support this skill (Wenger et al., 2017). Indeed, substantial, and relatively more consistent, evidence has recently been provided suggesting that bilingualism directly affects the brain, both in terms of its structure and function. Adaptations resulting from cognitively challenging experiences is a well-documented finding in other domains in the field of experience-based neuroplasticity (Wenger et al., 2017). In the case of bilingualism, such adaptations have been observed in a network of cortical and subcortical brain regions that support language acquisition, processing, and control, and also in the functional and structural connectivity between these regions (Pliatsikas, 2019; Pliatsikas & Luk, 2016). The following paragraphs revisit these findings in the healthy brain, including the developing brain, followed by some evidence on how the reported effects may have an effect on brain decline, including both in healthy aging and in progressive neurodegenerative diseases. This chapter will primarily focus on newer evidence and related theoretical interpretations, to not only provide an up-to-date account of our knowledge in the field, but also to assist in designing future research. In doing so, it will focus on established bilinguals with varied experiences (so evidence from language training studies (or *emerging* bilinguals) will not be reviewed; for an overview, see Korenar & Pliatsikas, 2023), and on designs that did not include specific cognitive tasks, in order to avoid discussing task-specific effects. Finally, the focus will mainly be on studies using bilingual participants, which constitute a substantial body of evidence, whereas evidence from multilinguals remains limited to date (for a detailed review, see Yee et al., 2023).

10.2 Bilingualism and the Healthy Brain

10.2.1 Structure

Most of the evidence to date on how bilingualism affects the brain comes from studies looking at brain structure. This has included studies that have used various metrics of *gray matter* availability, both cortical (thickness, volume, surface area extent), and subcortical (volume, shape). Moreover, several metrics of *white matter* integrity have been used, which typically focus on how freely water molecules move (or *diffuse*) in different types of brain tissue. This movement is affected by the presence of myelin, in that the higher the myelin availability (i.e., the better the white matter integrity), the more restricted the diffusion is. The most commonly used such metrics have been Mean Diffusivity (MD), a measure of diffusion rate that decreases when myelin availability is high, and Fractional Anisotropy (FA), a measure of directionality of diffusion which increases with myelin availability; in brief, greater white matter integrity is thought to be expressed as high FA and low MD (see Soares et al., 2013, for a detailed description of these measures). This literature has provided rich evidence of structural adaptations in the bilingual brain, in regions involved in language acquisition, processing and control, as well as the white matter tracts that provide connectivity between these regions. This has included findings in cortical regions such as, among others, the supramarginal gyrus (Mechelli et al., 2004), the inferior and middle frontal gyri (Kaiser et al., 2015), the middle and superior temporal gyri (Olulade et al., 2016), the anterior cingulate cortex (Felton et al., 2017), and the hippocampus (Li et al., 2017); subcortical regions, including the caudate nucleus (Zou et al., 2012), putamen (Abutalebi et al., 2013), globus pallidus (Burgaleta et al., 2016), and thalamus (Pliatsikas et al., 2017); and the cerebellum (Pliatsikas et al., 2014) (for a comprehensive review of subcortical effects, see Yee et al., 2024). In terms of white matter effects, several key tracts were shown to be affected in respect of their integrity, usually expressed as higher FA and/or lower MD. These include the inferior fronto-occipital, superior longitudinal and uncincate fasciculi (Pliatsikas et al., 2015), the arcuate fasciculus (Hämäläinen et al., 2017), and the corpus callosum (Felton et al., 2017). These studies usually employed between-group comparisons, while longitudinal evidence for bilingualism-induced adaptations remains scarce (e.g., DeLuca et al., 2019) (for a comprehensive review of these earlier findings, and a discussion on the suggested roles of the affected regions and tracts, see Pliatsikas, 2019).

However, these earlier studies suffered from low replicability in terms of the location, direction, and magnitude of the reported effects – for example, while some studies reported volumetric increases in cortical gray matter regions for bilinguals vs. monolinguals, others reported decreases, while others reported subcortical effects or effects in white matter only. Moreover, some of these studies reported differences between multilinguals and bilinguals, and others between simultaneous vs. early vs. sequential bilinguals (Berken, Gracco, et al., 2016; Claussenius-Kalman et al., 2020a; Klein et al., 2014; Tu et al., 2021), further blurring the overall picture. Interestingly, some of those earlier studies revealed that such effects may directly relate to sustained usage and switching between languages, for example, via residence (immersion) in a bilingual environment (e.g., Pliatsikas et al., 2017). In the light of this, it has been suggested that the variability in the earlier studies can be attributed to the variable (and not controlled for) characteristics of the bilingual samples that were tested, for example, differences in their respective language backgrounds, current linguistic environments, opportunities for language switching, etc. (Pliatsikas, DeLuca, et al., 2020; Titone & Tiv, 2022). On the basis of this argument, a recent wave of studies has started treating bilingualism as a spectrum of experiences, instead of a binary value (bilingual/monolingual), and utilized measures of the bilingual experience as predictors of structural adaptations. For example, DeLuca and colleagues (2019) showed a range of adaptations in the shape of subcortical structures, including the caudate nucleus, putamen and the thalamus, as well as the integrity of the corpus callosum, and these effects were predicted in different ways by factors such as the age of acquisition of the second language (L2), L2 use in home and social settings, as measured by a language background questionnaire (Anderson et al., 2018). Other studies have reported significant adaptations on both gray and white matter structure as predicted by a variety of continuous metrics that have been used as proxies of bilingual experiences, including L2 proficiency (Nichols & Joanisse, 2016; Wang et al., 2020), measures of balanced proficiency between the two languages (Claussenius-Kalman et al., 2020b), L2 age of acquisition (Luk et al., 2020), composite scores of L2 exposure and/or use (Del Maschio et al., 2020; Fedeli et al., 2021) and composite scores accounting for L2 proficiency, age of acquisition and daily exposure (Gallo et al., 2021).

Notwithstanding their value in shifting the field and renewing the approach to bilingualism as a multivariate, multifaceted experience, the studies using these continuous approaches did not succeed to explain all the original variability, and they also created some additional variability themselves; for example, the effects of L2 age of acquisition on FA in the corpus callosum are in opposite directions in Pliatsikas and colleagues (2015) and Luk and colleagues (2020). One potential reason for this is that all these studies assumed that the effects of bilingualism on brain structure are linear, and they do not fluctuate as a function of the respective predictors. However, this may not be the case – for example, how does a linear approach reconcile the reported subcortical *increases* in bilinguals (e.g., Pliatsikas et al., 2017) and *decreases* in multilingual interpreters (Elmer et al., 2014) when compared to controls?

Inspired by models describing nonlinear long-term trajectories of experience-based neuroplasticity during skill acquisition (Lövdén et al., 2013), and by previous suggestions that structural effects in the bilingual brain cannot be viewed independently of the actual bilingual experiences (Abutalebi & Green, 2016; Li et al., 2014) and vary over time (Grundy et al., 2017), Pliatsikas (2020) argued that bilingualism should be viewed as a long-term dynamic experience with commensurate adaptations at the structural level. In his Dynamic Restructuring Model (DRM), Pliatsikas described three stages in the bilingual experience, when specific adaptations can be expected: in brief, during the *initial exposure* stage, when individuals are first exposed to language learning, local increases in gray matter tissue should be expected in cortical regions involved in the acquisition of vocabulary, semantics, and phonology, including frontal, parietal and temporal regions, and the hippocampus. These increases are typically interpreted as signifying generation of new local neuronal connections which are required to support the newly acquired skills of additional language acquisition. These effects can be followed by effects in regions related to grammatical acquisition, like the cerebellum, but most crucially those related to language control, like the caudate nucleus. The latter is because acquisition of additional lexical and phonological alternatives would also lead to increased competition between them during production, so the relevant regions will need to be reorganized in order to accommodate this newly required skill. This becomes more prominent in the subsequent, *consolidation* stage, which comes with increased experience. At this stage, significant increases are expected mainly in subcortical structures, including not only the caudate nucleus but also regions related to articulatory control, such as the putamen and the globus pallidus, as fluency in the second language increases. Moreover, at that stage there are likely to be changes in the thalamus, a structure that is key in relaying information between different brain regions, as well as increases in the integrity of white matter tracks which provide intra- and interhemispheric connectivity. Both these effects signify the increased needs for efficient communication between brain regions, as the availability of competing language alternatives increases. Notably, the gray matter regions that increased in volume during the initial exposure stage would now start to return to baseline volumes, or *renormalize*. This is because these regions are thought to undergo a process of *pruning*: once the most efficient networks have been identified among those that were newly created at the initial exposure stage, the superfluous ones get pruned leading to the return to baseline tissue volume which however is now efficient in dealing with the newly learnt skills. At a third stage, that of *peak efficiency*, which applies to the most-experienced bilinguals, including groups such as professional interpreters, it is the turn of the subcortical structures that grew at the previous stage to start returning to their baseline volume, after having themselves become efficient in language control; at the same time, effects in white matter tracts appear to subside from anterior regions and appear in more posterior regions (see also Grundy et al., 2017). In all, and according to the DRM, the trajectory of bilingualism-induced adaptations in brain structure is nonlinear and is a direct effect of increasing experience in using, and switching between, two languages.

This approach has fed into the current shift in the field toward treating bilingualism as a spectrum of quantifiable experiences instead of a binary variable (Leivada et al., 2021; Pliatsikas, DeLuca, et al., 2020; Surrain & Luk, 2019; Titone & Tiv, 2022). With respect to structural effects, several studies have used nonlinear statistical approaches to show that quantified bilingual experiences, usually measured via detailed language background questionnaires, are significant predictors of nonlinear adaptations in the volumes of cortical (Marin-Marin et al., 2022) and subcortical (Korenar et al., 2023) gray matter structures, including the hippocampus (Voits, Robson, et al., 2022). The patterns of these effects have corroborated the suggestions of the DRM by showing the predicted volumetric increases at initial stages of language experience, which tend to reach a plateau and then revert toward baseline volumes as experience increases. This is a promising avenue which needs to be further explored in future studies.

10.2.2 Function

A rapidly emerging body of literature has also provided evidence to suggest that bilingualism also leads to long-term adaptations in brain *function* at rest, that is, in the absence of a specific task, and that these adaptations are not independent from the reported structural effects. The bulk of this evidence comes from studies using resting state functional Magnetic Resonance Imaging (rsfMRI), to study functional connectivity in the brain, that is, systematic correlations in spontaneous activity between different (networks of) brain regions when no task is administered to individuals (for a review of the relevant methods, see Fox & Raichle, 2007). Notably, most of the studies studying functional connectivity, some key ones being reviewed in this paragraph, compared bilingual groups that differed in their language acquisition, proficiency, or use background, further suggesting that such factors are key in understanding how bilingualism affects the brain (for a comprehensive review, see Jafari et al., 2021); for example, simultaneous life-long bilinguals have shown stronger functional connectivity than sequential bilinguals between the left and right inferior frontal gyrus, regions involved in language and cognitive control, and the right inferior parietal lobule, a region involved in vocabulary acquisition and retrieval (Berken, Chai, et al., 2016). According to the authors of that study, these findings suggest that long-term bilinguals are better adapted to manage competing language systems. Looking at large-scale brain networks, simultaneous bilinguals have also demonstrated stronger anticorrelations compared to sequential bilinguals between the default mode network (mostly active when the brain is at rest) and the frontoparietal network (involved in cognitive control), a relationship that was interpreted as evidence for robust interference suppression, and better cognitive control, for the former group (Kousaie et al., 2017). More evidence for the effects of L2 AoA on resting state functional connectivity was provided by Liu and colleagues (2021), while Sun and colleagues (2019) also reported differences between bilingual groups of high and low L2 proficiency.

Following up on the recent direction of treating bilingualism as a spectrum, a handful of studies have started using measures of quantified bilingualism as predictors of functional connectivity in the brain, in ways similar to those used in structural studies. For example, Gullifer and colleagues (2018) used the newly utilized measure of entropy, a metric of social diversity in bilingual language use (Gullifer & Titone, 2020), to show that bilingual experiences positively predict connectivity between regions subserving switching and conflict monitoring & L2 speech articulation, suggesting that the greater the experiences the more efficient these processes become (see also Li et al., 2021). Other studies have used more common metrics to report significant modulations on resting-state brain connectivity, including L2 proficiency (Wang et al., 2020), L2 AoA and exposure (Dash et al., 2022), or the combined effects of such factors (Sulpizio et al., 2020).

Some evidence on how bilingualism can affect brain functioning in adults also comes from a handful of recent studies using resting-state Electroencephalogram (rsEEG), a method that provides fine-grained temporal resolution, and as such is complementary to rsfMRI (Luk, Pliatsikas, et al., 2020). Specifically, rsEEG focuses on synchronized firing of neurons, which is broken down to distinct frequency bands, each of them underlying specific processes (Luk, Pliatsikas, et al., 2020). Studying the characteristics of these bands (power, coherence, etc.) provides information relating to the brain's intrinsic dynamics and connectivity, and it has been used to look at the effects of cognitively demanding experiences on brain function (Bice et al., 2020). With respect to bilingualism, Bice and colleagues (2020) reported not only that resting-state activity significantly differs between bilinguals and monolinguals, but also that it can be predicted by quantified bilingual experiences, such as L2 frequency of use, age of acquisition, and proficiency. More recently, Pereira Soares and colleagues (2021) showed that the earlier the L2 is acquired the higher is the power in the high beta and gamma bands across several brain regions, and that some bands can be modulated differently by different linguistic contexts; for example, the amount of everyday L2 use in social settings positively correlated with alpha and theta coherence between middle frontal and frontotemporal regions. Notably, all the affected bands in this study appear to have an important role in language acquisition and control – for example, combined beta and gamma power has been shown to have a positive effect on the rates of L2 learning (Prat et al., 2016), whereas beta relates to cortical-basal ganglia loops which underlie the regulation of language use in bilinguals (Stocco & Prat, 2014), in that the higher the beta power and coherence is, the more efficient the language control. Moreover, coherence in theta and alpha bands has been suggested to underlie long-distance cortico-cortical interactions and top-down processing, and help integrate multiple simultaneous cognitive processes, meaning that any bilingualism-induced modulations to them may have knock-on effects on domain-general top-down processing (Pereira Soares et al., 2021).

10.2.3 Metabolism

Another way of studying the dynamic effects of bilingualism on the brain has been at the level of *local metabolism*. The task of dealing with increased cognitive load, and the structural and functional adaptations that result from it, are energy-demanding processes, which are underlined by measurable local metabolic activity. One method to study metabolic changes in the brain is *proton magnetic resonance spectroscopy* (^{1}H-MRS) (Soares & Law, 2009), a method that targets the local concentrations of neurochemicals (or *metabolites*) involved in the processes of skill acquisition and neuroplasticity. Like the previously described indices of neuroplasticity, local metabolite concentrations can be quantified and examined as functions of the intensity of an ongoing experience, including the bilingual experience. Indeed, initial evidence has shown significant differences in the concentration of several key metabolites between bilinguals and monolinguals in the anterior cingulate cortex, a region that is crucial for domain-general cognitive control (Weekes et al., 2018). Following up on that study, Pliatsikas, Pereira Soares and colleagues (2021) used ^{1}H-MRS on bilinguals and monolinguals of a large age-range (19–83) to study metabolite concentrations in the basal ganglia, a region of particular importance to language control, which is susceptible to bilingualism-induced structural and functional neuroplasticity (see previous sections). After controlling for the effects of age, Pliatsikas and colleagues reported that bilinguals had higher concentration of *myoinositol* and lower concentration of *N-acetylaspartate* than monolinguals, with both effects being potential markers of synaptic pruning, that is, the process that is suggested to take place as the most efficient networks are identified, and the redundant ones are disposed of. Crucially, both effects were significantly predicted by a continuous measure of bilingual experiences. The authors interpreted this pattern of findings as indicative of the biological basis of neuroplasticity in the bilingual brain, now observed at the microscopic level, but also as supportive of models such as the DRM which suggest that such effects are dynamic, depend on the amount and/or intensity of the bilingual experiences, and can be observed throughout the lifespan.

10.3 Bilingualism and the Developing Brain

If bilingualism can have such a variety of effects on the adult brain, an interesting and important question is to investigate whether these effects also apply to the *developing* bilingual brain, and whether they interact with typical patterns of brain development. The literature on this topic remains somewhat limited, yet informative. An early study showed that proficiency in ten-year old multilingual children positively correlated with the volume of the left inferior parietal lobule, a region involved in language acquisition (Della Rosa et al., 2013). A pair of studies by Mohades and colleagues also showed that simultaneous bilingual children aged eight to eleven had greater integrity in the left inferior longitudinal fasciculus than both sequential bilingual and monolingual age-matched children (Mohades et al., 2012); however, sequential bilingual children subsequently manifested faster increase in integrity in the same white matter tract than the other two groups when scanned again after two years (Mohades et al., 2015), an effect that was attributed to their increasing bilingual experiences over time. Some more evidence is available regarding cortical differences between bilingual and monolingual children and adolescents. For example Archila-Suerte and colleagues (2018) reported *thinner* cortex for nine-year-old bilingual children with balanced proficiency compared to those with unbalanced ones, in cortical regions including the left inferior and middle frontal gyri, and larger putamen for the same group. Although the authors did not provide an interpretation for this pattern of effects, it is reminiscent of the shift from cortical to subcortical increases in tissue that are proposed by the DRM to be a result of increasing experience in using two languages (see also Thieba et al., 2019, for similar patterns in younger multilingual children, which however did not reach statistical significance). In a different study, Brito and Noble (2018) failed to find any major differences in thickness or surface area of the same regions between bilinguals and monolinguals in a large cohort of individuals aged three to twenty-one, and only reported greater surface area for the anterior cingulate cortex for a subgroup of adolescent bilinguals compared to age-matched monolinguals. This study was followed up by Pliatsikas, Meteyard and colleagues (2020) who, inspired by the suggestions of the DRM, used a nonlinear approach to compare the developmental trajectories of the same groups of bilinguals and monolinguals, by using different metrics (volume, thickness, surface area) of cortical and subcortical regions covering the entire brain. Apart from the previously documented overall patterns that suggest progressive reduction of gray matter tissue of these regions over development, Pliatsikas and colleagues also reported that these reductions took place at a slower rate for the bilingual group in a range of frontal, temporal, and parietal regions, all of which have specific roles related to language acquisition, processing, and control. This finding can be interpreted as indicative of the suggested local volumetric increases in bilinguals as described by the DRM – however, that dataset did not include detailed information about the bilinguals' language experiences. A more recent study, also looking at a large cohort of children, reported a different pattern of results; specifically, Vaughn and colleagues (2021) reported thinner cortex for nine-year old bilinguals vs. monolinguals in a range of regions across the entire brain, including several of those reported in Pliatsikas and colleagues. This seeming contradiction may be related to the characteristics of the group in Vaughn and colleagues, who were better described as being active users of their two languages. Nevertheless, it is worth nothing that the pattern was similar to that reported in children of a similar age in Pliatsikas, Meteyard, and colleagues (2020). Taken together, these results reaffirm that bilingualism has effects on brain development which may not be linear and may differ at different ages and with different experiences.

The relative scarcity of evidence from the developing brain does not allow for firm conclusions about the expected effects, but some theoretical suggestions have already been put forward. In their recent theoretical paper, Claussenius-Kalman and colleagues (2021) use a neuroemergentist approach to suggest that any effects of bilingualism on brain development (including its structure and function) should be viewed as the product of the interaction between the genetically determined abilities of the individual and the linguistic environment (or ecosystem) that individuals find themselves in, which is not necessarily stable itself. In other words, this interaction creates a multidimensional developmental space which is probably unique for each individual (see also Hernandez et al., 2019). This variability needs to be captured by carefully designed longitudinal studies measuring all aspects of variance in individuals and their linguistic environments (e.g., Goksan et al., 2020), which in turn may help in formulating more advanced models of bilingual brain development.

10.4 Bilingualism and the Declining Brain

The literature on bilingualism-induced brain adaptations in brain structure, function, and metabolism is definitely of theoretical importance, if bilingualism is to be treated as a test case for the general effects of experience-dependent neuroplasticity. However, it is also important to the extent that these effects may interact with aging of the brain, in ways similar to how they appear to interact with development, for example, influence processes and trajectories of tissue loss (Pliatsikas, Ansaldo, et al., 2021). Indeed, most of the ongoing discussion about whether the experience of bilingualism can be beneficial to general cognition has focused on whether, and how, these effects are expressed in older age and in the declining brain, and the changes in brain structure, function, and metabolism that underlie those effects (Bialystok, 2021; Gallo et al., 2022). Specifically, bilingualism has been suggested to contribute to two different types of *reserves* (Gallo et al., 2020; Gold, 2016; Voits, DeLuca, et al., 2022), which can be defined as follows: cognitive reserve, defined as "a property of the brain allows for cognitive performance that is better than expected given the degree of life-course related brain changes and brain injury or disease," and brain reserve, defined as reflecting "the neurobiological status of the brain (numbers of neurons, synapses, etc.) at any point in time" (Stern et al., 2023), meaning that greater brain reserves help the brain cope better with age- or disease-related decline. In the case of bilingualism, brain reserve is usually reported in healthy aging bilinguals when compared to age-matched monolinguals, and it is usually expressed as better preserved structure in the former (for a detailed review, see Pliatsikas, 2023). This has been reported in a range of white matter tracts (Anderson, Grundy, et al., 2018; DeLuca & Voits, 2022; Luk et al., 2011; Olsen et al., 2015) and, quite crucially, in a range of cortical regions related to language acquisition, processing, and control, which have also been shown to be "reinforced" at initial stages of L2 acquisition (Pliatsikas, 2020), including the left anterior cingulate cortex (Abutalebi et al., 2015; Del Maschio et al., 2018), the left inferior temporal gyrus (Abutalebi et al., 2014), the left inferior frontal gyrus (Heim et al., 2019) and the hippocampus (Voits, Robson, et al., 2022). Moreover, Del Maschio, Fedeli, and colleagues (2019) reported better preserved gyrification patterns in the right cingulate and entorhinal cortices for bilinguals vs. monolinguals aged eighteen to seventy-five, suggesting a slower rate of typical age-related gyrification reductions for the former group.

On the other hand, the concept of cognitive reserve more appropriately applies to bilingual samples facing brain decline due to neurodegeneration or brain damage, although suggestions for cognitive reserves have also been based on evidence from healthy samples; for example, Gold and colleagues (2013) reported equal cognitive abilities in healthy aging bilinguals and monolinguals but with significantly reduced white matter integrity in the former group, suggesting efficient utilization of the tissue that was spared from the aging processes; moreover, Berkes and colleagues (2021) matched older healthy bilinguals and monolinguals on measures of white matter integrity, and reported poorer cognitive performance by the latter group, suggesting more efficient utilization of existing brain tissue even before any clinical symptoms appear. Some further evidence for stronger functional connectivity at rest for healthy aging bilinguals vs. monolinguals can also be considered as corroborating evidence for cognitive reserve (de Frutos-Lucas et al., 2020; Grady et al., 2015; Luk et al., 2011). Nevertheless, several studies have reported effects that suggest bilingualism-induced cognitive reserve in patients, that is, equal cognitive abilities between bilinguals and monolinguals on the face of more severe brain degeneration in the former group. Some of this evidence comes from patients diagnosed with mild cognitive impairment (MCI), a condition that usually precedes, and signifies, a subsequent diagnosis of Alzheimer's disease (AD). For example, Costumero and colleagues (2020) reported less global volume of the brain parenchyma (effectively, the combined volumes of gray and white matter) for bilingual vs. monolingual MCI patients of similar cognitive abilities, while Marin-Marin and colleagues (2021) reported higher resting state functional connectivity between posterior brain regions for bilingual vs. monolingual MCI patients. However, evidence from MCI patients remains limited and inconsistent – for example, Duncan and colleagues (2018) reported greater thickness for bilinguals in several frontal, temporal and parietal cortical regions involved in language control, a pattern suggesting a brain reserve. Moreover, Rosselli and colleagues (2019) reported no differences in hippocampal volumes between bilingual and monolingual MCI patients, despite the former outperforming the latter on various memory measures. When treating bilingualism as a spectrum of experiences, Voits and colleagues (2023) reported a nonlinear effect of bilingualism on hippocampal volumes in patients with MCI, which was not accompanied by comparable effects in their performance in tasks tapping episodic memory, echoing findings in healthy aging bilinguals (Voits, Robson, et al., 2022). In terms of white matter, Marin-Marin and colleagues (2020) reported both increased and decreased integrity in different tracts when comparing bilinguals to monolinguals. This variability in patterns is reminiscent of the variability reported in the earlier studies looking at neuroplasticity in the healthy brain, and may be attributed to different methods used (García-Pentón et al., 2016) or different diagnostic, demographic, or language background characteristics of the examined groups (Surrain & Luk, 2019) – nevertheless, the available evidence warrants for further investigation, especially in the light of evidence that bilinguals are diagnosed with MCI later than monolinguals, but they also convert to AD at a faster rate (Berkes et al., 2020).

Some relatively consistent evidence for bilingualism-related cognitive reserve also comes from studies looking at AD patients, and also at the level of brain structure, function, and metabolism. For example, perhaps the earliest study in this subfield reported greater atrophy of the temporal lobe for bilingual vs. age- and cognition-matched monolinguals, an effect clearly indicating a cognitive reserve for the former group (Schweizer et al., 2012). Similarly, Smirnov and colleagues (2019) compared brain structure in bilingual AD patients to healthy bilingual controls, and reported significant atrophy in the former group in a range of frontal temporal and parietal regions, as well as the left rostral anterior cingulate cortex, but not in its caudal portion or in its right homologue. This pattern was interpreted as evidence for cognitive reserve, in the sense that the spared parts of the bilateral anterior cingulate cortex received additional cognitive burden as a result of deterioration of the surrounding regions, which helped those parts resist decline. Corroborating evidence for bilingualism-induced cognitive reserves was presented by Perani and colleagues (2017) who used positron emission tomography (PET) and reported more severe brain hypometabolism in bilinguals with AD compared to monolinguals, which was accompanied by increased connectivity in the executive control and default mode networks but no between-group differences in attentional control (for similar results see also Kowoll et al., 2016; Sala et al., 2022). Interestingly, some evidence also exists pointing toward neural reserves even in AD patients: for example, Raji and colleagues (2020) reported larger volumes for bilingual vs. monolingual AD patients in the thalamus, putamen, ventral diencephalon, and brain stem.

Following up from the reported effects from patients with MCI and AD, a handful of studies have shifted their focus on potential effects of bilingualism on other neurodegenerative diseases (for a detailed review, see Voits et al., 2020). The literature remains limited, but it also appears promising – for example, when it comes to Huntington's disease patients, Martínez-Horta and colleagues (2019) reported that active bilingualism positively predicted gray matter volume in the right inferior frontal gyrus; moreover, the same group of participants also exhibited a positive correlation between active bilingualism and glucose metabolism (measured with PET) in a range of frontal and temporal regions, including those related to cognitive control. The authors interpreted those effects as better preserved brain function and structure of regions that are known to be targeted by Huntington's disease, suggesting that active bilingualism has assisted in accruing brain reserve in these regions. Some evidence also comes from a single study on patients with multiple sclerosis; specifically, Ehling and colleagues (2019) administered L2 training to patients and health controls, and scanned both groups pre- and post-training. They reported structural changes in both groups which however differed in terms of their location, with the patient group demonstrating volumetric increases in a range of structures in the right hemisphere, including the hippocampus and the putamen, whereas the healthy group showed increases in the left insula only. According to the authors, this pattern suggested increased involvement of the right hemisphere in language learning, which in itself could simplify a compensating mechanism in the face of neurodegeneration, that is, a cognitive reserve. These findings, along with some evidence for potential beneficial cognitive effects of bilingualism in multiple sclerosis (Aveledo et al., 2021), signify that bilingualism should be further investigated as a factor bringing about cognitive reserve in a range of conditions that affect the brain, in similar ways that are afforded for factors such as nutrition or exercise (Voits, DeLuca, et al., 2022).

10.5 Conclusion: Synthesizing and Moving Forward

This chapter has showcased that bilingualism may have important, quantifiable, and predictable effects on the brain, at the levels of structure, function, and metabolism, and these effects may underlie the more commonly observed effects on behavior. These effects should be expected to be observed across the lifespan, to be dynamic and depend on the actual bilingual experiences, but also to interact with typical developmental patterns, healthy aging and progressive neurodegeneration. The field is only at its beginnings, and is in need of theoretical models unifying the results from different approaches, indices, and samples (e.g. DeLuca et al., 2020), but also new, well-designed, and powered studies that will help unveil the full array of mechanisms that are in play, and help provide a full account on the effects of bilingualism on the brain.

References

Abutalebi, J., Canini, M., Della Rosa, P. A., Sheung, L. P., Green, D. W., & Weekes, B. S. (2014). Bilingualism protects anterior temporal lobe integrity in aging. *Neurobiology of Aging*, *35*(9), 2126–2133. https://doi.org/10.1016/j.neurobiolaging.2014.03.010

Abutalebi, J., & Green, D. W. (2016). Neuroimaging of language control in bilinguals: Neural adaptation and reserve. *Bilingualism: Language and Cognition*, *19*(04), 1–10. https://doi.org/10.1017/S1366728916000225

Abutalebi, J., Guidi, L., Borsa, V., Canini, M., Della Rosa, P. A., Parris, B. A., & Weekes, B. S. (2015). Bilingualism provides a neural reserve for aging populations. *Neuropsychologia, 69*, 201–210. https://doi.org/10.1016/j.neuropsychologia.2015.01.040

Abutalebi, J., Rosa, P. A. D., Castro Gonzaga, A. K., Keim, R., Costa, A., & Perani, D. (2013). The role of the left putamen in multilingual language production. *Brain and Language, 125*(3), 307–315. https://doi.org/10.1016/j.bandl.2012.03.009

Alrwaita, N., Houston-Price, C., & Pliatsikas, C. (2022). The effects of using two varieties of one language on cognition: Evidence from bidialectalism and diglossia. *Linguistic Approaches to Bilingualism, 13*(6), 830–853. https://doi.org/10.1075/lab.21044.alr

Anderson, J. A. E., Grundy, J. G., De Frutos, J., Barker, R. M., Grady, C., & Bialystok, E. (2018). Effects of bilingualism on white matter integrity in older adults. *NeuroImage, 167*, 143–150. https://doi.org/10.1016/j.neuroimage.2017.11.038

Anderson, J. A. E., Mak, L., Keyvani Chahi, A., & Bialystok, E. (2018). The language and social background questionnaire: Assessing degree of bilingualism in a diverse population. *Behavior Research Methods, 50*(1), 250–263. https://doi.org/10.3758/s13428-017-0867-9

Archila-Suerte, P., Woods, E. A., Chiarello, C., & Hernandez, A. E. (2018). Neuroanatomical profiles of bilingual children. *Developmental Science, 21*(5), e12654. https://doi.org/10.1111/desc.12654

Aveledo, F., Higueras, Y., Marinis, T., Bose, A., Pliatsikas, C., Meldaña-Rivera, A., Martínez-Ginés, M. L., García-Domínguez, J. M., Lozano-Ros, A., Cuello, J. P., & Goicochea-Briceño, H. (2021). Multiple sclerosis and bilingualism: Some initial findings. *Linguistic Approaches to Bilingualism, 11*(4), 551–577. https://doi.org/10.1075/lab.18037.ave

Berken, J. A., Chai, X. J., Chen, J.-K., Gracco, V. L., & Klein, D. (2016). Effects of early and late bilingualism on resting-state functional connectivity. *Journal of Neuroscience, 36*(4), 1165–1172. https://doi.org/10.1523/JNEUROSCI.1960-15.2016

Berken, J. A., Gracco, V. L., Chen, J.-K., & Klein, D. (2016). The timing of language learning shapes brain structure associated with articulation. *Brain Structure and Function, 221*(7), 3591–3600. https://doi.org/10.1007/s00429-015-1121-9

Berkes, M., Bialystok, E., Craik, F. I. M., Troyer, A., & Freedman, M. (2020). Conversion of mild cognitive impairment to Alzheimer disease in monolingual and bilingual patients. *Alzheimer Disease & Associated Disorders, 34*(3), 225–230. https://doi.org/10.1097/WAD.0000000000000373

Berkes, M., Calvo, N., Anderson, J. A. E., & Bialystok, E. (2021). Poorer clinical outcomes for older adult monolinguals when matched to bilinguals on brain health. *Brain Structure and Function, 226*(2), 415–424. https://doi.org/10.1007/s00429-020-02185-5

Bialystok, E. (2017). The bilingual adaptation: How minds accommodate experience. *Psychological Bulletin, 143*(3), 233–262. https://doi.org/10.1037/bul0000099

Bialystok, E. (2021). Bilingualism: Pathway to cognitive reserve. *Trends in Cognitive Sciences, 25*(5), 355–364. https://doi.org/10.1016/j.tics.2021.02.003

Bialystok, E., Craik, F. I. M., & Luk, G. (2012). Bilingualism: Consequences for mind and brain. *Trends in Cognitive Sciences, 16*(4), 240–250. https://doi.org/10.1016/j.tics.2012.03.001

Bice, K., Yamasaki, B. L., & Prat, C. S. (2020). Bilingual language experience shapes resting-state brain rhythms. *Neurobiology of Language, 1*(3), 288–318. https://doi.org/10.1162/nol_a_00014

Brito, N. H., & Noble, K. G. (2018). The independent and interacting effects of socioeconomic status and dual-language use on brain structure and cognition. *Developmental Science, 21*(6), e12688. https://doi.org/10.1111/desc.12688

Burgaleta, M., Sanjuán, A., Ventura-Campos, N., Sebastian-Galles, N., & Ávila, C. (2016). Bilingualism at the core of the brain. Structural differences between bilinguals and monolinguals revealed by subcortical shape analysis. *NeuroImage, 125*, 437–445. https://doi.org/10.1016/j.neuroimage.2015.09.073

Claussenius-Kalman, H., Hernandez, A. E., & Li, P. (2021). Expertise, ecosystem, and emergentism: Dynamic developmental bilingualism. *Brain and Language, 222*, 105013. https://doi.org/10.1016/j.bandl.2021.105013

Claussenius-Kalman, H. L., Vaughn, K. A., Archila-Suerte, P., & Hernandez, A. E. (2020a). Age of acquisition impacts the brain differently depending on neuroanatomical metric. *Human Brain Mapping, 41*(2), 484–502. https://doi.org/10.1002/hbm.24817

Claussenius-Kalman, H. L., Vaughn, K. A., Archila-Suerte, P., & Hernandez, A. E. (2020b). Highly proficient, balanced bilingualism is related to thinner cortex in two cognitive control regions. *Annals of the New York Academy of Sciences*, nyas.14491. https://doi.org/10.1111/nyas.14491

Costumero, V., Marin-Marin, L., Calabria, M., Belloch, V., Escudero, J., Baquero, M., Hernandez, M., Ruiz de Miras, J., Costa, A., Parcet, M. A., & Ávila, C. (2020). A cross-sectional and longitudinal study on the protective effect of bilingualism against dementia using brain atrophy and cognitive measures. *Alzheimer's Research & Therapy, 12*(1), 11. https://doi.org/10.1186/s13195-020-0581-1

Dash, T., Joanette, Y., & Ansaldo, A. I. (2022). Exploring attention in the bilingualism continuum: A resting-state functional connectivity study. *Brain and Language, 224*, 105048. https://doi.org/10.1016/j.bandl.2021.105048

de Frutos-Lucas, J., López-Sanz, D., Cuesta, P., Bruña, R., de la Fuente, S., Serrano, N., López, M. E., Delgado-Losada, M. L., López-Higes, R., Marcos, A., & Maestú, F. (2020). Enhancement of posterior brain functional networks in bilingual older adults. *Bilingualism: Language and Cognition, 23*(2), 387–400. https://doi.org/10.1017/S1366728919000178

Del Maschio, N., Fedeli, D., Sulpizio, S., & Abutalebi, J. (2019). The relationship between bilingual experience and gyrification in adulthood: A cross-sectional surface-based morphometry study. *Brain and Language, 198*, 104680. https://doi.org/10.1016/j.bandl.2019.104680

Del Maschio, N., Sulpizio, S., Gallo, F., Fedeli, D., Weekes, B. S., & Abutalebi, J. (2018). Neuroplasticity across the lifespan and aging effects in bilinguals and monolinguals. *Brain and Cognition, 125,* 118–126. https://doi.org/10.1016/j.bandc.2018.06.007

Del Maschio, N., Sulpizio, S., Toti, M., Caprioglio, C., Del Mauro, G., Fedeli, D., & Abutalebi, J. (2020). Second language use rather than second language knowledge relates to changes in white matter microstructure. *Journal of Cultural Cognitive Science, 4(2),* 165–175. https://doi.org/10.1007/s41809-019-00039-z

Della Rosa, P. A., Videsott, G., Borsa, V. M., Canini, M., Weekes, B. S., Franceschini, R., & Abutalebi, J. (2013). A neural interactive location for multilingual talent. *Cortex, 49*(2), 605–608. https://doi.org/10.1016/j.cortex.2012.12.001

DeLuca, V., Rothman, J., Bialystok, E., & Pliatsikas, C. (2019). Redefining bilingualism as a spectrum of experiences that differentially affects brain structure and function. *Proceedings of the National Academy of Sciences, 116*(15), 7565–7574. https://doi.org/10.1073/pnas.1811513116

DeLuca, V., Rothman, J., & Pliatsikas, C. (2019). Linguistic immersion and structural effects on the bilingual brain: A longitudinal study. *Bilingualism: Language and Cognition, 22*(5), 1160–1175. https://doi.org/10.1017/S1366728918000883

DeLuca, V., Segaert, K., Mazaheri, A., & Krott, A. (2020). Understanding bilingual brain function and structure changes? U bet! A unified bilingual experience trajectory model. *Journal of Neurolinguistics, 56,* 100930. https://doi.org/10.1016/j.jneuroling.2020.100930

DeLuca, V., & Voits, T. (2022). Bilingual experience affects white matter integrity across the lifespan. *Neuropsychologia, 169,* 108191. https://doi.org/10.1016/j.neuropsychologia.2022.108191

Duncan, H. D., Nikelski, J., Pilon, R., Steffener, J., Chertkow, H., & Phillips, N. A. (2018). Structural brain differences between monolingual and multilingual patients with mild cognitive impairment and Alzheimer disease: Evidence for cognitive reserve. *Neuropsychologia, 109,* 270–282. https://doi.org/10.1016/j.neuropsychologia.2017.12.036

Ehling, R., Amprosi, M., Kremmel, B., Bsteh, G., Eberharter, K., Zehentner, M., Steiger, R., Tuovinen, N., Gizewski, E. R., Benke, T., Berger, T., Spöttl, C., Brenneis, C., & Scherfler, C. (2019). Second language learning induces grey matter volume increase in people with multiple sclerosis. *PLoS ONE, 14*(12), e0226525. https://doi.org/10.1371/journal.pone.0226525

Elmer, S., Hänggi, J., & Jäncke, L. (2014). Processing demands upon cognitive, linguistic, and articulatory functions promote grey matter plasticity in the adult multilingual brain: Insights from simultaneous interpreters. *Cortex, 54,* 179–189. https://doi.org/10.1016/j.cortex.2014.02.014

Fedeli, D., Del Maschio, N., Sulpizio, S., Rothman, J., & Abutalebi, J. (2021). The bilingual structural connectome: Dual-language experiential factors modulate distinct cerebral networks. *Brain and Language, 220,* 104978. https://doi.org/10.1016/j.bandl.2021.104978

Felton, A., Vazquez, D., Ramos-Nunez, A. I., Greene, M. R., Macbeth, A., Hernandez, A. E., & Chiarello, C. (2017). Bilingualism influences structural indices of interhemispheric organization. *Journal of Neurolinguistics, 42,* 1–11. https://doi.org/10.1016/j.jneuroling.2016.10.004

Fox, M. D., & Raichle, M. E. (2007). Spontaneous fluctuations in brain activity observed with functional magnetic resonance imaging. *Nature Reviews Neuroscience, 8*(9), 700–711. https://doi.org/10.1038/nrn2201

Gallo, F., DeLuca, V., Prystauka, Y., Voits, T., Rothman, J., & Abutalebi, J. (2022). Bilingualism and aging: Implications for (delaying) neurocognitive decline. *Frontiers in Human Neuroscience, 16,* 819105. https://doi.org/10.3389/fnhum.2022.819105

Gallo, F., Myachykov, A., Shtyrov, Y., & Abutalebi, J. (2020). Cognitive and brain reserve in bilinguals: Field overview and explanatory mechanisms. *Journal of Cultural Cognitive Science, 4*(2), 127–143. https://doi.org/10.1007/s41809-020-00058-1

Gallo, F., Novitskiy, N., Myachykov, A., & Shtyrov, Y. (2021). Individual differences in bilingual experience modulate executive control network and performance: Behavioral and structural neuroimaging evidence. *Bilingualism: Language and Cognition, 24*(2), 293–304. https://doi.org/10.1017/S1366728920000486

García-Pentón, L., Fernández García, Y., Costello, B., Duñabeitia, J. A., & Carreiras, M. (2016). The neuroanatomy of bilingualism: How to turn a hazy view into the full picture. *Language, Cognition and Neuroscience, 31*(3), 303–327. https://doi.org/10.1080/23273798.2015.1068944

Goksan, S., Argyri, F., Clayden, J. D., Liegeois, F., & Wei, L. (2020). Early childhood bilingualism: Effects on brain structure and function. *F1000Research, 9,* 370. https://doi.org/10.12688/f1000research.23216.1

Gold, B. T. (2016). Lifelong bilingualism, cognitive reserve and Alzheimer's disease: A review of findings. *Linguistic Approaches to Bilingualism, 6*(1–2), 171–189. https://doi.org/10.1075/lab.14028.gol

Gold, B. T., Johnson, N. F., & Powell, D. K. (2013). Lifelong bilingualism contributes to cognitive reserve against white matter integrity declines in aging. *Neuropsychologia, 51*(13), 2841–2846. https://doi.org/10.1016/j.neuropsychologia.2013.09.037

Grady, C. L., Luk, G., Craik, F. I. M., & Bialystok, E. (2015). Brain network activity in monolingual and bilingual older adults. *Neuropsychologia, 66,* 170–181. https://doi.org/10.1016/j.neuropsychologia.2014.10.042

Grundy, J. G., Anderson, J. A. E., & Bialystok, E. (2017). Neural correlates of cognitive processing in monolinguals and bilinguals: Neural correlates of bilingualism. *Annals of the New York Academy of Sciences, 1396*(1), 183–201. https://doi.org/10.1111/nyas.13333

Gullifer, J. W., Chai, X. J., Whitford, V., Pivneva, I., Baum, S., Klein, D., & Titone, D. (2018). Bilingual experience and resting-state brain connectivity: Impacts of L2 age of acquisition and social diversity of language use on control networks. *Neuropsychologia, 117,* 123–134. https://doi.org/10.1016/j.neuropsychologia.2018.04.037

Gullifer, J. W., & Titone, D. (2020). Characterizing the social diversity of bilingualism using language entropy. *Bilingualism: Language and Cognition, 23*(2), 283–294. https://doi.org/10.1017/S1366728919000026

Hämäläinen, S., Sairanen, V., Leminen, A., & Lehtonen, M. (2017). Bilingualism modulates the white matter structure of language-related pathways. *NeuroImage, 152*, 249–257. https://doi.org/10.1016/j.neuroimage.2017.02.081

Heim, S., Stumme, J., Bittner, N., Jockwitz, C., Amunts, K., & Caspers, S. (2019). Bilingualism and "brain reserve": A matter of age. *Neurobiology of Aging, 81*, 157–165. https://doi.org/10.1016/j.neurobiolaging.2019.05.021

Hernandez, A. E., Claussenius-Kalman, H. L., Ronderos, J., Castilla-Earls, A. P., Sun, L., Weiss, S. D., & Young, D. R. (2019). Neuroemergentism: A framework for studying cognition and the brain. *Journal of Neurolinguistics, 49*, 214–223. https://doi.org/10.1016/j.jneuroling.2017.12.010

Jafari, Z., Perani, D., Kolb, B. E., & Mohajerani, M. H. (2021). Bilingual experience and intrinsic functional connectivity in adults, aging, and Alzheimer's disease. *Annals of the New York Academy of Sciences, 1*, nyas.14666. https://doi.org/10.1111/nyas.14666

Kaiser, A., Eppenberger, L. S., Smieskova, R., Borgwardt, S., Kuenzli, E., Radue, E.-W., Nitsch, C., & Bendfeldt, K. (2015). Age of second language acquisition in multilinguals has an impact on gray matter volume in language-associated brain areas. *Frontiers in Psychology, 6*. https://doi.org/10.3389/fpsyg.2015.00638

Klein, D., Mok, K., Chen, J.-K., & Watkins, K. E. (2014). Age of language learning shapes brain structure: A cortical thickness study of bilingual and monolingual individuals. *Brain and Language, 131*, 20–24. https://doi.org/10.1016/j.bandl.2013.05.014

Korenar, M., & Pliatsikas, C. (2023). L2 acquisition and neuroplasticity: Insights from the dynamic restructuring model. In *The Routledge Handbook of Second Language Acquisition and Neurolinguistics*. Routledge.

Korenar, M., Treffers-Daller, J., & Pliatsikas, C. (2023). Dynamic effects of bilingualism on brain structure map onto general principles of experience-based neuroplasticity. *Scientific Reports, 13*, 3428. https://doi.org/10.1038/s41598-023-30326-3

Kousaie, S., Chai, X. J., Sander, K. M., & Klein, D. (2017). Simultaneous learning of two languages from birth positively impacts intrinsic functional connectivity and cognitive control. *Brain and Cognition, 117*, 49–56. https://doi.org/10.1016/j.bandc.2017.06.003

Kowoll, M. E., Degen, C., Gorenc, L., Küntzelmann, A., Fellhauer, I., Giesel, F., Haberkorn, U., & Schröder, J. (2016). Bilingualism as a contributor to cognitive reserve? Evidence from cerebral glucose metabolism in mild cognitive impairment and Alzheimer's disease. *Frontiers in Psychiatry, 7*(APR), 1–6. https://doi.org/10.3389/fpsyt.2016.00062

Kroll, J. F., & Bialystok, E. (2013). Understanding the consequences of bilingualism for language processing and cognition. *Journal of Cognitive Psychology, 25*(5), 497–514. https://doi.org/10.1080/20445911.2013.799170

Lehtonen, M., Soveri, A., Laine, A., Järvenpää, J., de Bruin, A., & Antfolk, J. (2018). Is bilingualism associated with enhanced executive functioning in adults? A meta-analytic review. *Psychological Bulletin, 144*(4), 394–425. https://doi.org/10.1037/bul0000142

Leivada, E., Duñabeitia, J. A., Westergaard, M., & Rothman, J. (2021). On the phantom-like appearance of bilingualism effects on cognition: (How) should we proceed? *Bilingualism: Language and Cognition, 24*(1), 197–210. https://doi.org/10.1017/S1366728920000358

Li, L., Abutalebi, J., Emmorey, K., Gong, G., Yan, X., Feng, X., Zou, L., & Ding, G. (2017). How bilingualism protects the brain from aging: Insights from bimodal bilinguals: Bimodal bilingualism prevents brain aging. *Human Brain Mapping, 38*(8), 4109–4124. https://doi.org/10.1002/hbm.23652

Li, P., Legault, J., & Litcofsky, K. A. (2014). Neuroplasticity as a function of second language learning: Anatomical changes in the human brain. *Cortex, 58*, 301–324. https://doi.org/10.1016/j.cortex.2014.05.001

Li, X., Ng, K. K., Wong, J. J. Y., Lee, J. W., Zhou, J. H., & Yow, W. Q. (2021). Bilingual language entropy influences executive functions through functional connectivity and signal variability. *Brain and Language, 222*, 105026. https://doi.org/10.1016/j.bandl.2021.105026

Liu, X., Tu, L., Chen, X., Wang, J., Li, M., Lu, Z., & Huang, R. (2021). Effect of AoA-L2 on L1 and L2 networks in early and late bilinguals. *International Journal of Bilingualism, 25*(6), 136700692110330. https://doi.org/10.1177/13670069211033026

Lövdén, M., Wenger, E., Mårtensson, J., Lindenberger, U., & Bäckman, L. (2013). Structural brain plasticity in adult learning and development. *Neuroscience & Biobehavioral Reviews, 37*(9), 2296–2310. https://doi.org/10.1016/j.neubiorev.2013.02.014

Luk, G., Bialystok, E., Craik, F. I. M., & Grady, C. L. (2011). Lifelong bilingualism maintains white matter integrity in older adults. *The Journal of Neuroscience, 31*(46), 16808–16813. https://doi.org/10.1523/JNEUROSCI.4563-11.2011

Luk, G., Mesite, L., & Leon Guerrero, S. (2020). Onset age of second language acquisition and fractional anisotropy variation in multilingual young adults. *Journal of Neurolinguistics, 56*, 100937. https://doi.org/10.1016/j.jneuroling.2020.100937

Luk, G., Pliatsikas, C., & Rossi, E. (2020). Brain changes associated with language development and learning: A primer on methodology and applications. *System, 89*, 102209. https://doi.org/10.1016/j.system.2020.102209

Marin-Marin, L., Costumero, V., Ávila, C., & Pliatsikas, C. (2022). Dynamic effects of immersive bilingualism on cortical and subcortical grey matter volumes. *Frontiers in Psychology, 13*. www.frontiersin.org/articles/10.3389/fpsyg.2022.886222

Marin-Marin, L., Costumero, V., Belloch, V., Escudero, J., Baquero, M., Parcet, M. -A., & Ávila, C. (2020). Effects of bilingualism on white matter atrophy in mild cognitive impairment: A diffusion tensor imaging study. *European Journal of Neurology, 27*(4), 603–608. https://doi.org/10.1111/ene.14135

Marin-Marin, L., Palomar-García, M.-Á., Miró-Padilla, A., Adrián-Ventura, J., Aguirre, N., Villar-Rodríguez, E., & Costumero, V. (2021). Bilingualism's effects on resting-state functional connectivity in mild cognitive impairment. *Brain Connectivity, 11*(1), 30–37. https://doi.org/10.1089/brain.2020.0877

Martínez-Horta, S., Moreu, A., Perez-Perez, J., Sampedro, F., Horta-Barba, A., Pagonabarraga, J., Gomez-Anson, B., Lozano-Martinez, G. A., Lopez-Mora, D. A., Camacho, V., Fernández-León, A., Carrió, I., & Kulisevsky, J. (2019). The impact of bilingualism on brain structure and function in Huntington's disease. *Parkinsonism & Related Disorders, 60,* 92–97. https://doi.org/10.1016/j.parkreldis.2018.09.017

Mechelli, A., Crinion, J. T., Noppeney, U., O'Doherty, J., Ashburner, J., Frackowiak, R. S., & Price, C. J. (2004). Structural plasticity in the bilingual brain. *Nature, 431*(7010), 757. https://doi.org/10.1038/431757a

Mohades, S. G., Struys, E., Van Schuerbeek, P., Mondt, K., Van De Craen, P., & Luypaert, R. (2012). DTI reveals structural differences in white matter tracts between bilingual and monolingual children. *Brain Research, 1435,* 72–80. https://doi.org/10.1016/j.brainres.2011.12.005

Mohades, S. G., Van Schuerbeek, P., Rosseel, Y., Van De Craen, P., Luypaert, R., & Baeken, C. (2015). White-matter development is different in bilingual and monolingual children: A longitudinal DTI study. *PLoS ONE, 10*(2), e0117968. https://doi.org/10.1371/journal.pone.0117968

Nichols, E. S., & Joanisse, M. F. (2016). Functional activity and white matter microstructure reveal the independent effects of age of acquisition and proficiency on second-language learning. *NeuroImage, 143,* 15–25. https://doi.org/10.1016/j.neuroimage.2016.08.053

Olsen, R. K., Pangelinan, M. M., Bogulski, C., Chakravarty, M. M., Luk, G., Grady, C. L., & Bialystok, E. (2015). The effect of lifelong bilingualism on regional grey and white matter volume. *Brain Research, 1612,* 128–139. https://doi.org/10.1016/j.brainres.2015.02.034

Olulade, O. A., Jamal, N. I., Koo, D. S., Perfetti, C. A., LaSasso, C., & Eden, G. F. (2016). Neuroanatomical evidence in support of the bilingual advantage theory. *Cerebral Cortex, 26*(7), 3196–3204. https://doi.org/10.1093/cercor/bhv152

Perani, D., Farsad, M., Ballarini, T., Lubian, F., Malpetti, M., Fracchetti, A., Magnani, G., March, A., & Abutalebi, J. (2017). The impact of bilingualism on brain reserve and metabolic connectivity in Alzheimer's dementia. *Proceedings of the National Academy of Sciences, 114*(7), 1690–1695. https://doi.org/10.1073/pnas.1610909114

Pereira Soares, S. M., Kubota, M., Rossi, E., & Rothman, J. (2021). Determinants of bilingualism predict dynamic changes in resting state EEG oscillations. *Brain and Language, 223,* 105030. https://doi.org/10.1016/j.bandl.2021.105030

Pliatsikas, C. (2019). Multilingualism and brain plasticity. In J. W. Schwieter & M. Paradis (Eds.), *The Handbook of the Neuroscience of Multilingualism* (1st ed., pp. 230–251). Wiley. https://doi.org/10.1002/9781119387725.ch11

Pliatsikas, C. (2020). Understanding structural plasticity in the bilingual brain: The dynamic restructuring model. *Bilingualism: Language and Cognition, 23*(2), 459–471. https://doi.org/10.1017/S1366728919000130

Pliatsikas, C. (2023): Bilingualism and brain structure: insights from healthy ageing and progressive neurodegenerative diseases. In G. Luk et al. (Eds.), *Understanding Language and Cognition through Bilingualism: In Honor of Ellen Bialystok* (pp. 301–317). John Benjamins.

Pliatsikas, C., Ansaldo, A. I., & Voits, T. (2021). Bilingualism and the declining brain. *Linguistic Approaches to Bilingualism, 11*(4), 453–458. https://doi.org/10.1075/lab.00031.int

Pliatsikas, C., DeLuca, V., Moschopoulou, E., & Saddy, J. D. (2017). Immersive bilingualism reshapes the core of the brain. *Brain Structure and Function, 222*(4), 1785–1795. https://doi.org/10.1007/s00429-016-1307-9

Pliatsikas, C., DeLuca, V., & Voits, T. (2020). The many shades of bilingualism: Language experiences modulate adaptations in brain structure. *Language Learning, 70*(S2), 133–149. https://doi.org/10.1111/lang.12386

Pliatsikas, C., Johnstone, T., & Marinis, T. (2014). Grey matter volume in the cerebellum is related to the processing of grammatical rules in a second language: A structural voxel-based morphometry study. *The Cerebellum, 13*(1), 55–63. https://doi.org/10.1007/s12311-013-0515-6

Pliatsikas, C., & Luk, G. (2016). Executive control in bilinguals: A concise review on fMRI studies. *Bilingualism: Language and Cognition, 19*(4), 699–705. https://doi.org/10.1017/S1366728916000249

Pliatsikas, C., Meteyard, L., Veríssimo, J., DeLuca, V., Shattuck, K., & Ullman, M. T. (2020). The effect of bilingualism on brain development from early childhood to young adulthood. *Brain Structure and Function, 225*(7), 2131–2152. https://doi.org/10.1007/s00429-020-02115-5

Pliatsikas, C., Moschopoulou, E., & Saddy, J. D. (2015). The effects of bilingualism on the white matter structure of the brain. *Proceedings of the National Academy of Sciences, 112*(5), 1334–1337. https://doi.org/10.1073/pnas.1414183112

Pliatsikas, C., Pereira Soares, S. M., Voits, T., Deluca, V., & Rothman, J. (2021). Bilingualism is a long-term cognitively challenging experience that modulates metabolite concentrations in the healthy brain. *Scientific Reports, 11*(1), 1–12. https://doi.org/10.1038/s41598-021-86443-4

Prat, C. S., Yamasaki, B. L., Kluender, R. A., & Stocco, A. (2016). Resting-state qEEG predicts rate of second language learning in adults. *Brain and Language, 157–158*, 44–50. https://doi.org/10.1016/j.bandl.2016.04.007

Raji, C. A., Meysami, S., Merrill, D. A., Porter, V. R., & Mendez, M. F. (2020). Brain structure in bilingual compared to monolingual individuals with Alzheimer's disease: Proof of concept. *Journal of Alzheimer's Disease, 76*(1), 275–280. https://doi.org/10.3233/JAD-200200

Rosselli, M., Loewenstein, D. A., Curiel, R. E., Penate, A., Torres, V. L., Lang, M., Greig, M. T., Barker, W. W., & Duara, R. (2019). Effects of bilingualism on verbal and nonverbal memory measures in mild cognitive impairment. *Journal of the International Neuropsychological Society, 25*(1), 15–28. https://doi.org/10.1017/S135561771800070X

Sala, A., Malpetti, M., Farsad, M., Lubian, F., Magnani, G., Frasca Polara, G., Epiney, J., Abutalebi, J., Assal, F., Garibotto, V., & Perani, D. (2022). Lifelong bilingualism and mechanisms of neuroprotection in Alzheimer dementia. *Human Brain Mapping, 43*(2), 581–592. https://doi.org/10.1002/hbm.25605

Schweizer, T. A., Ware, J., Fischer, C. E., Craik, F. I. M., & Bialystok, E. (2012). Bilingualism as a contributor to cognitive reserve: Evidence from brain atrophy in Alzheimer's disease. *Cortex, 48*(8), 991–996. https://doi.org/10.1016/j.cortex.2011.04.009

Smirnov, D. S., Stasenko, A., Salmon, D. P., Galasko, D., Brewer, J. B., & Gollan, T. H. (2019). Distinct structural correlates of the dominant and nondominant languages in bilinguals with Alzheimer's disease (AD). *Neuropsychologia, 132*, 107131. https://doi.org/10.1016/j.neuropsychologia.2019.107131

Soares, D. P., & Law, M. (2009). Magnetic resonance spectroscopy of the brain: Review of metabolites and clinical applications. *Clinical Radiology, 64*(1), 12–21. https://doi.org/10.1016/j.crad.2008.07.002

Soares, J. M., Marques, P., Alves, V., & Sousa, N. (2013). A hitchhiker's guide to diffusion tensor imaging. *Frontiers in Neuroscience, 7*, 1–14. https://doi.org/10.3389/fnins.2013.00031

Stern, Y., Albert, M., Barnes, C. A., Cabeza, R., Pascual-Leone, A., & Rapp, P. R. (2023). A framework for concepts of reserve and resilience in aging. Neurobiology of Aging, 124, 100–103. https://doi.org/10.1016/j.neurobiolaging.2022.10.015

Stocco, A., & Prat, C. S. (2014). Bilingualism trains specific brain circuits involved in flexible rule selection and application. *Brain and Language, 137*, 50–61. https://doi.org/10.1016/j.bandl.2014.07.005

Sulpizio, S., Del Maschio, N., Del Mauro, G., Fedeli, D., & Abutalebi, J. (2020). Bilingualism as a gradient measure modulates functional connectivity of language and control networks. *NeuroImage, 205*, 116306. https://doi.org/10.1016/j.neuroimage.2019.116306

Sun, X., Li, L., Ding, G., Wang, R., & Li, P. (2019). Effects of language proficiency on cognitive control: Evidence from resting-state functional connectivity. *Neuropsychologia, 129*, 263–275. https://doi.org/10.1016/j.neuropsychologia.2019.03.020

Surrain, S., & Luk, G. (2019). Describing bilinguals: A systematic review of labels and descriptions used in the literature between 2005–2015. *Bilingualism: Language and Cognition, 22*(2), 401–415. https://doi.org/10.1017/S1366728917000682

Thieba, C., Long, X., Dewey, D., & Lebel, C. (2019). Young children in different linguistic environments: A multimodal neuroimaging study of the inferior frontal gyrus. *Brain and Cognition, 134*, 71–79. https://doi.org/10.1016/j.bandc.2018.05.009

Titone, D. A., & Tiv, M. (2022). Rethinking multilingual experience through a systems framework of bilingualism. *Bilingualism: Language and Cognition, 26*(1), 1–16. https://doi.org/10.1017/S1366728921001127

Tu, L., Niu, M., Pan, X., Hanakawa, T., Liu, X., Lu, Z., Gao, W., Ouyang, D., Zhang, M., Li, S., Wang, J., Jiang, B., & Huang, R. (2021). Age of acquisition of Mandarin modulates cortical thickness in high-proficient Cantonese–Mandarin bidialectals. *Journal of Psycholinguistic Research, 50*(4), 723–736. https://doi.org/10.1007/s10936-020-09716-5

Vaughn, K. A., Nguyen, M. V. H., Ronderos, J., & Hernandez, A. E. (2021). Cortical thickness in bilingual and monolingual children: Relationships to language use and language skill. *NeuroImage, 243*, 118560. https://doi.org/10.1016/j.neuroimage.2021.118560

Voits, T., DeLuca, V., & Abutalebi, J. (2022). The nuance of bilingualism as a reserve contributor: Conveying research to the broader neuroscience community. *Frontiers in Psychology, 13*. www.frontiersin.org/articles/10.3389/fpsyg.2022.909266

Voits, T., Pliatsikas, C., Robson, H., & Rothman, J. (2020). Beyond Alzheimer's disease: Can bilingualism be a more generalized protective factor in neurodegeneration? *Neuropsychologia, 147*, 107593. https://doi.org/10.1016/j.neuropsychologia.2020.107593

Voits, T., Robson, H., Rothman, J., & Pliatsikas, C. (2022). The effects of bilingualism on hippocampal volume in ageing bilinguals. *Brain Structure and Function, 227*(3), 979–994. https://doi.org/10.1007/s00429-021-02436-z

Voits, T., Rothman, J., Aguirre, N., Cattaneo, G., Calabria, M., Costumero, V., Hernández, M., Juncadella Puig, M., Marín-Marín, L., Robson, H., Suades, A., Costa, A., & Pliatsikas, C. (2023). Hippocampal adaptations in mild cognitive impairment patients are modulated by bilingual language experiences. *Bilingualism: Language and Cognition, 27*(2), 263–273. https://doi.org/10.1017/S1366728923000354

Wang, R., Ke, S., Zhang, Q., Zhou, K., Li, P., & Yang, J. (2020). Functional and structural neuroplasticity associated with second language proficiency: An MRI study of Chinese-English bilinguals. *Journal of Neurolinguistics, 56*, 100940. https://doi.org/10.1016/j.jneuroling.2020.100940

Weekes, B. S., Abutalebi, J., Mak, H. K., Borsa, V. M., Soares, S. M. P., Chiu, P.-W., & Zhang, L. (2018). Effect of monolingualism and bilingualism in the anterior cingulate cortex: A proton magnetic resonance spectroscopy study in two centers. *Letras de Hoje, 53*(1), 5–12. https://doi.org/dx.doi.org/10.15448/1984-7726.2018.1.30954

Wenger, E., Brozzoli, C., Lindenberger, U., & Lövdén, M. (2017). Expansion and renormalization of human brain structure during skill acquisition. *Trends in Cognitive Sciences, 21*(12), 930–939. https://doi.org/10.1016/j.tics.2017.09.008

Yee, J., Deluca, V., & Pliatsikas, C. (2023): The effects of multilingualism on brain structure, language control and language processing: Insights from MRI. In J. Cabrelli et al. (Eds.), *The Cambridge Handbook of Third Language Acquisition and Processing* (pp. 577–605). Cambridge University Press.

Yee, J., Kořenář, M., Sheehan, A., & Pliatsikas, C. (2024). Subcortical malleability as a result of cognitively challenging experiences: the case of bi-/multilingualism. *Current Opinion in Behavioral Sciences, 59*, 101438. https://doi.org/1016/j.cobeha.2024.101438

Zou, L., Ding, G., Abutalebi, J., Shu, H., & Peng, D. (2012). Structural plasticity of the left caudate in bimodal bilinguals. *Cortex, 48*(9), 1197–1206. https://doi.org/10.1016/j.cortex.2011.05.022

11

Factors in Variable Outcomes in Second Language Acquisition

David Birdsong

11.1 Introduction

In the study of variable outcomes in second language acquisition (L2A), an early research tradition focused on "The Good Language Learner," in order to identify learning strategies that differentiate more successful learners from less successful learners, typically in instructional contexts (Naiman et al., 1978; Rubin, 1975; Stevick, 1989). A different approach looks at the behaviors and neurocognitive makeup of truly exceptional learners, particularly those who distinguish themselves by learning dozens of new languages to high levels of proficiency (Biedroń & Birdsong, 2019; Hyltenstam, 2016; Schneiderman & Desmarais, 1988). Another perspective is situated within a more general research program where differences relating to an extensive suite of factors – some learner-related, some learner-independent – are examined for their conditioning effects across the full spectrum of observed L2A outcomes. This approach is represented in the present chapter.

Well-known conditioning factors in L2A outcomes include the age at which L2 learning begins in earnest, often referred to as age of acquisition (AoA);[1] levels of education and literacy; contexts of learning such as nonimmersion vs. immersion; amount and quality of interaction with native speakers of the L2; psychosocial orientations such as the desire to assimilate with native speakers of the L2; family support; sex; music experience; processing ability in the L1; linguistic factors such as the relatedness of the first language (L1) to the L2; cognitive factors such as learning strategies, aptitude, working memory, perceptual acuity, cognitive aging, and cognitive style; and neurobiological factors such as genetically mediated dopamine and testosterone levels and volumetric and connectivity characteristics of relevant brain structures. (For further information on predictors of L2A outcomes, and for comparisons with L1A, see Birdsong, 2012, 2018, 2021; Bowden et al., 2005; Dabrowska, 2018, 2019; Ettlinger et al., 2014; Hartshorne et al., 2018; Indefrey, 2006; Li et al., 2022; Newman et al., 2012; Pfenninger & Singleton, 2019; Reiterer, 2019; Robinson,

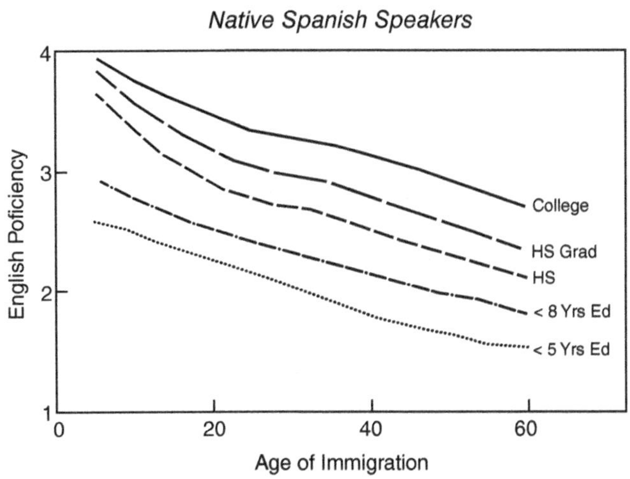

Figure 11.1 Loess fits for self-assessed English proficiency by age of immigration and amount of education among 2+ million Spanish-speaking immigrants to the United States.
From Hakuta, K., Bialystok, E., & Wiley, E. (2003). Critical evidence: A test of the critical-period hypothesis for second-language acquisition. *Psychological Science*, 14(1), 31–38.

2002; Skehan, 1998, 2016; Sparks, 2012; Sparks et al., 2017; Qi et al., 2017; 2017; Ullman et al., 2007; van Hell, 2023; van Hell & Abdollahi, 2017.)

As an illustration the role of such factors in L2A outcomes, consider Figure 11.1 (Hakuta et al., 2003). Here the joint effects of age of immigration and amount of education are revealed in self-reported levels of English L2 proficiency on 1990 United States census forms by 2,016,317 native speakers of Spanish who had immigrated to the United States.

Over all respondents, later AoA is predictive of lower levels of L2 English proficiency. Concurrently, respondents with more years of education attained higher L2 proficiency than those with fewer years of education.

A systematic review of the many factors at play in L2A outcome variability would extend beyond the bounds of a handbook chapter. Instead, the intent here is to identify and discuss a select handful of factors that are particularly instructive in terms of understanding variable outcomes in L2A. The chosen factors are represented by a substantial body of prior work and, at the same time, lend themselves to multiple avenues of future research.

With these intentions, Sections 2–5 examine, respectively, the conditioning roles in L2A outcomes of individuals' working memory capacity, attitudes toward learning, music background, and genetic makeup. Section 6 looks at the relationship of AoA to two distinct dimensions of outcome variability: differences between AoA cohorts in terms of outcome levels and dispersion of results within cohorts representing early vs. late AoA. Section 7 illustrates the instability of outcomes in L2A (and L1A) with examples of changes in L1–L2 dominance relationships in bilingualism. Along with examination of results of pertinent studies, each section includes suggestions for ways to advance the domain of inquiry.

11.2 Working Memory

Since the groundbreaking work of Baddeley and Hitch (1974), who characterized working memory (WM) as consisting of the phonological loop, the visuospatial sketchpad, and the central executive, WM has emerged as perhaps the most extensively studied cognitive factor associated with L2A outcomes. As one indication of WM's prevalence in the literature, Wen (2016, updated 2018) produced a 189-page "Working memory in first and second language: A comprehensive bibliography" that includes around 3500 references. The diversity of studies of WM in language is represented in the forty chapters (800 pages) of Schwieter and Wen (2022). Contributing to the sheer volume of research in this area is the fact that WM has been studied in terms of its individual subcomponents, along both broad and restrictive operational dimensions, with respect to specific domains of application, and under multiple theoretical frameworks; see Logie and colleagues (2021) for approaches to theory integration.

Working memory "refers to the cognitive system(s) responsible for the control, regulation, and active maintenance of information" (Linck et al., 2014: 862), typically in the context of noise or distraction. As with other dimensions of cognitive functioning, people differ in "our cognitive ability to simultaneously maintain and process a small amount of information in our head temporarily so as to complete some mental tasks" (Wen, 2021: 279). See Mashburn and colleagues (2021) for elaboration on individual differences in maintenance of information in working memory and disengagement from irrelevant information.

To get a sense of how variable outcomes are tied to WM differences, consider a simplified generic description of the workings of WM in language learning (L1A and L2A), loosely based on Just and Carpenter (1992), Caplan and Waters (1999), Farmer and colleagues (2012) and Baddeley and colleagues (2021). Language processing occurs across sequentially and hierarchically organized strings of linguistic input (phones, morphemes, words, collocations, phrases, prosodically-marked information structure, etc.) of varying complexities and lengths. Here at the front end of learning, limitations on online memory condition the accuracy and breadth of tracking of formal regularities, transitional probabilities, sound-meaning mappings, and other features of the linguistic input. Information derived from processing is then stored and rehearsed; this information constitutes the epistemological basis for recognizing, inducing, or deducing language structure and for the development of language categories, representations, and generalizations. Interindividual differences in memory capacity at the processing stage, and in the ability to order and manage information at the storage stage – along with efficiency (speed and accuracy) of eventual retrieval of information – contribute to heterogeneity of outcomes in language learning. (For elaboration see Baddeley, 2012, 2015; Baddeley et al., 2021; Caplan & Waters, 1999;

Cunnings, 2017; Farmer et al., 2012; Hopp, 2014; Kane et al., 2007; Kidd et al., 2018; Kim et al., 2018; Kim & Christianson, 2013; Kousaie et al., 2021; Schwering & MacDonald, 2020; Swets et al., 2007.)

Moreover, WM is generally viewed as a component of aptitude for learning an L2. Thus, individuals who score higher on simple and complex WM span tasks generally outperform lower WM-scorers on various metrics of L2 proficiency and processing. For example, Zhou and colleagues (2017) tested fifty adult (M = 23 yoa) Chinese native speakers who had started learning English in middle school at age twelve on their ability to judge grammaticality in subject-extraction English sentences (e.g., *Who do you think loved the comedian with all his heart?*) and object-extraction sentences (e.g., *Who do you think the comedian loved with all his heart?*). Participants were divided into equal-number groups on the basis of their scores on the operation span task (OSPAN; Conway et al., 2005), a measure of WM capacity that has been shown to correlate with other WM measures (Unsworth et al., 2005). Judgment accuracy rates were similar between the high- and low-working memory groups. However, in the critical sentence region for parsing (i.e., after "think"), the reading latencies for the low WM group were significantly longer than those for the high WM group. Importantly, for the sequence of words "Who do you think" leading up to this critical region, the reading latencies between the low- and high-WM groups did not differ significantly, suggesting that the researchers identified the precise locus of the advantage of high WM capacity.

Some studies have shown that the processing of complex L2 sentences by some high-WM L2 learners is comparable to that of low-WM native speakers (e.g., Havik et al., 2009; Hopp, 2014; Indefrey, 2006). Juffs and Harrington (2011) note that WM effects are absent in some L2 processing studies (e.g., Felser & Roberts, 2007), and argue that WM effects in L2 may vary depending on the linguistic domains, tasks, and ages of participants. Linck and colleagues (2014) conducted a meta-analysis over seventy-nine published experiments that employed various operationalizations of WM and L2 outcome measures. Effect sizes in all but two of the experiments were positive, with a range of -0.15 to 0.61 and a population value over aggregated data of 0.25 in the random effects model.

Numerous studies have reported WM declines over age starting in early adulthood (e.g., Craik & Salthouse, 2000; Daselaar & Cabeza, 2013; McNab et al., 2015; Nyberg et al., 2012; Pliatsikas et al., 2019). In Section 6, we look at whether these post-maturational declines map onto declines in L2A outcomes. We go on to examine whether outcome declines are accompanied by greater WM variability amongst individuals within a given age-of-L2-learning cohort, and whether such variability could be linked to greater dispersion of L2 learning and performance. We note as well that individual variations in working memory are subject to genetic control, a consideration addressed in Section 5.

11.3 Attitudinal Factors

In the present contribution, attitude is thought of as an umbrella term for a variety of self-originated orientations toward L2 learning. These orientations include several well-studied types of motivation, an individual attitudinal variable that has attracted the most extensive study in the L2A literature, and to which we devote a subsection below to illustrate its relationship to variability in L2A outcomes. Among the other attitudes that have been explored in the context of L2A outcomes are: degree of identification with and acculturation toward the L2 language and culture (e.g., Lybeck, 2002; Rubenfeld et al., 2006; Schmidt, 1983); self-regulation of learning (e.g., Zhang & Papi, 2021; Zhang & Zou, 2022; Zheng et al., 2018); self-determination (e.g., Al-Hoorie et al., 2022; Noels et al., 2019); and desire to sound like or be taken for a native speaker (e.g., Magnusson & Stroud, 2012; Piller, 2002). For examples of interactions and strengths of attitudinal types in L2A outcomes, see Csizér and Dörnyei (2005); Dörnyei and Mentzelopoulos (2022); Kormos and Csizér (2014); Papi and Teimouri (2012); Zhang and Papi (2021).

Attitudes toward L2 learning also include powerful conative factors such as choice and will (e.g., Magnusson & Stroud, 2012; Piller, 2002; Young & Perkins, 1995). For example, L2 learners may choose to engage socially with natives as they strive to improve their speaking skills. As a conspicuous illustration of the deterministic role of purposeful choice in L2A outcomes, Walters (2011) describes the non-nativelike pronunciation of French /r/ by Tunisian adult males who speak L2 French. Unlike Tunisian women who, just like French natives, pronounce their L2 French /r/ with the uvular variant [ʁ], Tunisian men speaking L2 French produce the apical variant [r]. As their L1 (Tunisian Arabic) phonology includes the distinctive phonemes /r/ and /ʁ/, the issue among Tunisian men is not an inability to master the pronunciation of [ʁ]. Rather, as a matter of choice, and with the purpose of asserting their linguistic, cultural, and gender identity, Tunisian men conspicuously distinguish themselves from native French speakers and from Tunisian women who speak L2 French. See Harvey (2017); Henry (2017); Norton (2000); and Setter and Jenkins (2005) for further discussion of the relationship of choice and identity to L2 outcomes.

11.3.1 Motivation

Since the pioneering work of Gardner and Lambert (1972), scores of studies of the role of motivation in L2A outcomes have been published. The breadth of work in this area, as well as methodological and conceptual advances, are exposed in Dörnyei (2005, 2009, 2019); Dörnyei and Mentzelopoulos (2022); Dörnyei and Ushioda (2021); Dörnyei and colleagues (2015); Masgoret and Gardner (2003); Papi and Hiver (2020); Ushioda (2009, 2016, 2019a); Waninge and colleagues (2014); You and Dörnyei (2016). For a comprehensive treatment of motivation in language learning, consult the thirty-two chapters (~700 pages) in Lamb and colleagues (2019).

De Bot and colleagues (2007: 14) consider motivation a compensatory resource in L2A: at the level of the individual "a low aptitude may be compensated by high motivation." The strength of the motivation factor in determining L2A outcomes varies across studies. As a classic example, among native Chinese and Korean learners of L2 English, Johnson and Newport (1989) found that motivation – defined as a composite of scalar responses to two questions: "Is it important to you to speak English well?" and "Do you plan on staying in the United States?" – correlated significantly ($r = .39$, $p < .01$) with accuracy scores on a test of knowledge of English morphosyntax. Contrastingly, using the same test materials and administration methods as Johnson and Newport (1989) and sampling Spanish native speakers, Birdsong and Molis (2001) found that scalar (1–5) responses to the question "In the first years of learning English, how motivated to learn were you?" did not significantly correlate with English morphosyntax knowledge among late arrivals, early arrivals, and across all participants. Responses to the question "How motivated are you (presently) to improve your English?" did not correlate significantly with the test scores of late arrivals and across all participants. At the same time, early arriving participants' responses to this question were mildly correlated negatively with test scores ($r = -.409$, $p < .1$). Taking the disparate results of these two studies as a simple test case, we can ask if the strength of the motivation factor in L2A outcomes varies by the formulation of the relevant questions, by the L1 of participants, and by participants' AoA. Transcending these factors, effects on L2 outcomes could be due to particular choices or identificational orientations toward L2 attainment associated with different groups of learners; see above and below.

In a related vein, motivation researchers point out that not all motivation types contribute equally (or positively) to L2A outcomes. A notable example of this disparity is the divergent L2A outcomes that are associated with ideal-self motivation versus ought-to-self motivation, two dimensions of Dorynei's L2 motivational self system (Dornyei, 2009). Among 152 Chinese learners of L2 English, Zhao and colleagues (2022: 3) found that assessed ideal-self ("an ideal image the L2 learner aspires to become in the future, which encompasses a wide range of components such as cultural interest, integrativeness, instrumentality, promotion and attitude to the L2 community") was positively correlated with English proficiency, similarly to findings of prior studies of L2 Iranian, Japanese, and Korean. In contrast, the learners' assessed ought-to-self ("attributes that the learners think they ought to possess as a result of a series of duties, obligations, or responsibilities," such as meeting expectations of superiors or avoiding negative consequences) negatively predicted L2 English proficiency. This finding is situated among a mixed bag of results from prior studies (some positive, some negative, some neutral) regarding the association of ought-to self with L2A outcomes (see also: Feng & Papi, 2020; Papi & Teimouri, 2012).

We note that motivation, in its multiple varieties and domains of influence, is an unstable attribute of individuals (e.g., Waninge et al., 2014), and therefore an unstable predictor of L2A outcomes. For example, motivation to make changes of behaviors may decline over age (see overview in Carstensen & Hartel, 2006). In the individual L2A context, motivation levels may be enhanced by positive experiences with L2 learning, suggesting a reciprocity or bidirectionality of influence (learning <-> motivation). Motivation levels are known to fluctuate over short timespans, notably in the aftermath of study abroad, and have been monitored on a per-second timescale (MacIntyre & Serroul, 2014).

The takeaway from the situational, up-and-down, character of motivation – which sets it apart from stable factors such as genetic makeup and the relatedness of the L1 and L2 – is the need for sensitive methodology and circumspect interpretation in assessing the conditioning role of motivation in L2A outcomes. For a review of appropriate methods in all areas of L2A motivation research, see Ushioda (2019b).

11.4 Music Background

Music and language have much in common. They are both constrained, hierarchically organized combinatorial systems; they share design features characterized by levels of structure of varying degrees of complexity; and they are subserved by some of the same neural structures and associated acoustic and cognitive mechanisms. The acoustic signals of spoken languages and music have tonal, rhythmic, melodic, and intensity properties. Production of music and language involves neuromuscular coordination, while processing in both involves segmentation and signal-from-noise extraction. Neural components of music processing are involved in "pitch perception, syntax and harmonic structural processing, semantics, timbre and speaker identification, attending in auditory scenes, and rhythm" (Ogg & Slevc, 2019: 907). Production and processing of music and language require development of skills relating to memory, executive control, and attention. Statistical learning is observed for both music and language (see: Asano, 2015; Besson & Schön, 2001; Fenk-Oczlon, 2022; Fenk-Oczlon & Fenk, 2009; Fritz et al., 2013; Jänke, 2012; McMullen & Saffran, 2004; Rebuschat et al., 2012; Slevc & Okada, 2015).

It has been shown that music experience and training can enhance general auditory skills (e.g., Kraus & Chandrasekaran, 2010; Pesnot Lerousseau et al., 2020); verbal memory (e.g., Chan et al., 1998; Ho et al., 2003; Franklin et al., 2008); auditory working memory (e.g., Nie et al., 2022); Parbery-Clark et al., 2011a,b); auditory and visual updating (Slevc et al., 2016); classification of subcortical speech representations (Weiss & Bidelman, 2015); speech segmentation (e.g., François et al., 2012), and speech processing (e.g., Patel, 2011, 2014). Compared to non-musicians, musicians display more distributed neural activity when listening to speech sounds (e.g., Jantzen et al., 2014; Parbery-Clark et al., 2012). Musicianship has been shown to slow aging-associated loss in white matter integrity, specifically in the bilateral superior longitudinal fasciculus and bilateral uncinate fasciculus tracts (Andrews et al., 2021); see Novén and colleagues (2021) for correlations of the morphological characteristics of these subcortical areas with language learning aptitudes. Bidelman and colleagues (2010, 2013), Choi (2021), White and colleagues (2013), and Zhang and colleagues (2020) examine bidirectional transfer of skills between musical and language features; see Chang and colleagues (2016) for mixed evidence and Peretz and colleagues (2011) for counterevidence.

Against this backdrop of the relationship of music experience to skills that are known to underlie language learning and processing, the emphasis in this section is on how variations in musical experience may be related to variations in L2A outcomes. For example, Shook and colleagues (2013) found that highly trained musicians outperformed lesser-trained musicians on a statistical language-learning task involving novel Morse code-based words. The authors suggest that musical experience, by enhancing the fidelity of auditory encoding, improves the processing of statistical information in both artificial and natural languages. In another statistical learning experiment, this one involving sequences of tones that varied in level of grammaticality (high grammatical, low grammatical, ungrammatical), Van Hedger and colleagues (2022) showed that training on tones produced by familiar instruments (piano, violin) resulted in reduced accuracy as compared to training on artificially produced tones. In addition, participants (roughly half of whom were musically trained) were biased toward judging tone sequences as grammatical when the sequences approximated intervals common to Western music.

Turning to natural languages, Marques and colleagues (2007) examined the perceptual acuity of musician and nonmusician French native speakers listening to Portuguese, a language they were all unfamiliar with. Musicians (who had at least fourteen years of instrumental music training and practiced regularly) were better than nonmusicians at perceiving subtle pitch variations corresponding to weak prosodic incongruities. Further, compared to the nonmusicians, the musicians displayed shorter evoked response potential (ERP) response latencies in processing strong F0 incongruities. In a similarly designed study involving French natives listening to unfamiliar Mandarin, Marie and colleagues (2011) found that musicians relative to nonmusicians were more accurate in detecting tonal and segmental variations and demonstrated earlier N2/N3 increases in response to segmental variations. In line with behavioral data, musicians relative to nonmusicians showed shorter P3b latencies in categorizing tonal variations. Moreover, larger P3b amplitudes among the musicians suggested greater confidence in their decisions.

Lee and Hung (2008) tested musician and nonmusician speakers of English for their ability to identify Mandarin tones, with which neither group had familiarity. The musicians were university music majors and played at least two instruments. When listening to tones produced by thirty-two different speakers of Mandarin, the musicians performed above chance in identifying intact tones (68 percent correct) and silent-center tones (54 percent correct), while nonmusicians' performance was below chance on both intact tones (44 percent correct) and silent-center tones (36 percent correct). In identifying onset-only tones, the musicians' accuracy (31 percent) exceeded that of nonmusicians (28 percent), with the difference approaching significance ($p = 0.06$). For similar studies involving Mandarin tones, see Chen and colleagues (2021) and Gottfried and colleagues (2004, 2005).

In further conceptualizing the relationship of music experience to L2A, particularly with respect to appropriate levels and foci of analysis, one can think of musical backgrounds in terms of phenotypes, each with potentially different contributions to L2A. For example, one can anticipate distinct types and magnitudes of L2A advantages between those individuals who have some basic music training and those who have developed musical virtuosity over many years of rehearsal and performance. The same goes for those who read music (involving visual analysis of structure, pitch, timing, etc.) and those who don't; those who are or are not accomplished in music theory (involving harmonies, tonal systems, intervals, durations, rhythms, dissonance, counterpoint, ornamentation, improvisation, etc.); singers vs. instrumentalists; percussion players vs. string players vs. horn players vs. keyboard players; and so on. In principle, the various skills (neuromuscular, auditory, visual, vocal, analytical, synthetic, attentional, etc.) associated with these distinct musical backgrounds or phenotypic characteristics should underlie different L2 processes and behaviors and should align with different L2 domains. A percussionist might outperform any nonmusician in recognizing and producing L2 rhythmic characteristics, such as phrasal lengths, pauses, weights, and durations. An opera singer, but not necessarily a percussionist, should perform well on tasks involving both phonetic discrimination and phonetic production in the L2, and be adept at learning tone languages. Thus, in research involving the relationship of music experience to L2A, the desired focus in developing research questions, in designing studies, tasks and measures, and in interpreting results is on connecting well-delineated backgrounds in music – and, importantly, the particular set of *abilities* associated with each background – with apposite performance in particular areas of L2 learning and processing.

In this vein, Slevc and Miyake (2006) studied fifty Japanese native speakers who had arrived in Colorado after eleven years of age and had lived there for at least six months. The researchers looked at how various features of the participants' L2 English proficiency relate to circumscribed musical abilities, as opposed to less-granularly defined factors of prior music experience or training. Included in the design were language experience variables such as L2 use and L2 AoA, cognitive variables such as nonverbal intelligence and phonological short-term memory, and motivation to use the L2. Specific music abilities were assessed on three subtests from Wing (1968): analysis of the number of notes played in a chord; pitch change in comparing two chords; a receptive tonal memory test where participants indicated the position of a change of one note between two short tunes (three to ten notes in length). In addition, participants performed a productive tonal memory task where they were instructed to sing a short tune (three to seven notes in length) from immediate memory. Defined in these ways, and controlling for other factors in the design, musical ability predicted English pronunciation accuracy at the level of words, sentences, and passages, as well as phonological perception relating to discrimination of minimal pairs at the word and sentence levels and detection of mispronunciations while listening to a recording of a short story. In contrast, musical ability was not predictive of knowledge of English syntax (in this case, grammaticality judgment items from Johnson & Newport, 1989) or English vocabulary knowledge.

The Slevc and Miyake (2006) study illustrates the value of aligning well-defined and properly tested music abilities with specific outcomes of L2A. In this case, the evidence supports the reasonable connection of a set of attributes of musical "ear" with productive and receptive L2 phonological skills, but not with L2 syntactic knowledge or lexical knowledge. We note as well that the focus of Slevc and Miyake is on proficiency outcomes resulting from considerable amounts of L2 immersion and use, as opposed to the results of training studies, where periods of treatment are comparatively limited.

To look in more detail at the relationship of music background to L2A variable outcomes, it would be of interest to compare the perception of sounds in novel languages between nonmusicians and (targeted types of) musicians who have been matched for domain-general auditory capacities. Matching for preexistent auditory capacities is the key design feature, in light of Mankel and Bidelman's (2018) finding that some nonmusicians are endowed with musician-like auditory neurobiological function.

An additional research angle is suggested by Parbery-Clark and colleagues (2012), who report that musicians differ from nonmusicians by virtue of their enhanced subcortical discrimination of closely related speech sounds (e.g., [ba] vs [ga]), along with the general finding that they are better at extracting signal from noise (see, however, discussion of mixed results for signal-in-noise performance by Zhang et al., 2019). Evidence of this nature prompts further exploration of the mapping of particular experiential variables, cognitive abilities, and neurofunction onto L2 behavioral and brain-based evidence.

For a review of the contributions of music experience and expertise to L2A, with emphasis on correlates of brain activity with music training, see Chobert and Besson (2013). On behavioral and neurological relationships between music skills and language learning aptitude, see Turker and Reiterer (2021).

11.5 Genetic Makeup

Differences in genetic makeup between individuals account for 50 percent to 70 percent of the variation in cognitive abilities such as processing speed, memory, and reasoning (Bouchard & McGue, 1981; Haworth et al., 2010;

Tucker-Drob et al., 2013). Substantial contributions of genetics to learning and educational attainment have also been established (Bueno, 2019; Cessarini & Visscher, 2017; Kovas et al., 2013; Okbay et al., 2016; Rimfield et al., 2018).

For both L1A and L2A, specific genes and their variants (single-nucleotide polymorphisms (SNP)) are associated with learning abilities. In the L1A context, much of the research over the past twenty-five years has focused on links between individuals' genetic makeup and non-normal learning outcomes among individuals with specific language impairment, stuttering, speech-sound disorder, dyslexia, and developmental verbal dyspraxia (e.g., Graham & Fisher, 2013; Lai et al., 2001; Mountford et al., 2022; Newbury & Monaco, 2010; Smith et al., 2015; Stromswold, 1998; Van der Lely & Stollwerck, 1996).

For L2A, researchers have made significant advances in the study of genetic traits that are associated with differing levels of attainment among nonimpaired individuals. Some studies have looked at genetic variation with respect to cognitive capacities that condition L2A outcomes. For example, the study of monozygotic and dizygotic twins by Ando and colleagues (2001) revealed that a common genetic factor accounted for 26 percent of the variance in verbal working memory and 64 percent of the variance in spatial working memory. (On the genetic contributions to WM, see also Karlsgodt et al., 2010, 2011; Knowles et al., 2014; Kremen et al., 2007; Takahashi et al., 2021; Thevenon et al., 2016; van Leeuwen et al., 2009; Wright et al., 2001.)

Wong and colleagues (2012) showed how individual genetic differences relating to the encoding of dopaminergic (DA) binding and transmission can modulate learning of procedural features of L2 grammar. (DA is essential in working memory, executive function, and inhibitory control. Procedural learning, which is carried out in the prefrontal cortex and basal ganglia, is thought to enable representations of constituent order in syntax.) The genes that Wong and colleagues identified as being implicated with procedural learning and other cognitive aspects of language learning include *DAN1, COMT, DRD1*, and two SNP of *DRD2*.

Mamiya and colleagues (2016) explored the contribution of fractional anisotropy (FA) values and variants of the *COMT* gene to L2 learning. (FA measures white matter fiber tract connectivity, and FA values have been related to skill learning. Different polymorphisms of *COMT* and their genotypes (Met/Met, Val/Val, and Val/Met) have been associated with dopamine levels and with differential activity of enzymes that degrade catecholamines.) Among twenty-five Chinese students enrolled in university language classes, Mamiya and colleagues analyzed the relationship of the participants' FA values and their *COMT* genotypes (11 Val/Val, 11 Met/Val, 3 Met/Met) to course grades. The researchers found that FA values in the superior longitudinal fasciculus accounted for 18.6 percent of the variance, and that including *COMT* genotype as additional predictor increased the explained variance to 46.2 percent.

A large-scale study (n = 940) of adult native Cantonese speakers of Han Chinese descent by Wong and colleagues (2022) looked simultaneously at genetic and nongenetic contributions to variable learning outcomes in L2A (English) and L3A (various languages). A particular issue addressed in the study is whether the same set of genetic variants contributes to the learning of different languages, and if the same variants are involved in both native and L2/L3 learning. The genetic factors examined were twenty-eight SNP of fifteen distinct genes, which included *COMT, CNTNAP2, FOXP2, DCDC2,* and *DRD2*. The nongenetic factors selected for study included Gender, Musical Training, Nonverbal IQ, Family SES, L2 age of learning, and internal/external motivation (in the L3 analysis). For L1 proficiency, significant predictors under bidirectional stepwise regression were Gender, Family SES, one SNP of the *DRD2* gene and two SNPs of *DCDC2*, with combined genetic factors explaining more of the variance (3.7 percent) than Gender (2.2 percent) and Family SES (1.4 percent). By contrast, significant predictors for L2 English proficiency under the model were Family SES (5 percent), Music Training (2.6 percent), along with two SNP of the *FOXP2* gene and one SNP of *CNTNAP2*, which combined explained 3.5 percent of the variance. The significant predictors for L3 proficiency were Internal Motivation (5.4 percent) and a variant of *S100B* (~1 percent). Separate analyses under structural equation modeling generally confirmed the respective roles in L1, L2, and L3 proficiencies of the same genetic and nongenetic factors found under regression. Further analyses led the authors to suggest that the association of genetic makeup with language learning outcomes varies from language to language.

Future research might be directed to the role of genetically regulated acetylcholine (ACh) transmission and reception in L2A variable outcomes. The neurotransmitter ACh participates in neuronal excitability and synaptic plasticity (Dani & Bertrand, 2007); underlies neuromodulatory changes in brain circuits, thereby potentiating behavioral adaptations to environmental stimuli (Ballinger et al., 2016; Picciotto et al., 2012); undergirds attentional processes by enhancing focus and reducing the impact of distractors (Berry et al., 2015; English et al., 2009); interacts with DA and glutamate in age-related brain function (Freeman & Gibson, 1988); and bidirectionally regulates learning and memory (Huang et al., 2022; Leaderbrand et al., 2016). Hasselmo (2006) describes the function of ACh in neural encoding in the entorhinal and perirhinal cortices, as well as in the hippocampus, and the involvement of ACh in the formation and maintenance of episodic memories (see, e.g., Hamrick et al., 2019; Jiang & Forster, 2001; Zhang et al., 2021 for how individual differences in episodic memory ability predict L2 lexical knowledge). Newman and colleagues (2012) review how cholinergic manipulations affect performance in tasks relevant to L2 learning that involve attention, working memory, episodic memory, and spatial memory.

The gene groups associated with cholinergic transportation and reception in humans are cholinergic receptors nicotine subunits (CHRN), and cholinergic receptors muscarinic (CHRM) (see, www.genenames.org/data/genegroup/#!/group/175 (retrieved August 13, 2022) and https://medlineplus.gov/genetics/gene/chrna4/ (retrieved August 13, 2022)).

11.6 Age of Acquisition, Magnitude Differences, and Heterogeneity of L2A Outcomes

In the L2A context, age of acquisition (AoA) is understood to be a proxy for the initial state of L2 learning. This moment in the lifespan of individuals, which coincides with their current degree of L1 development and entrenchment, is characterized by a complex bundle of interacting or confounded variables. In this sense, AoA is properly characterized as a "meta-variable" (Flege et al., 1999) that reflects "the sum of an individual's cognitive, neurological and linguistic development, along with motivational, identificational, attitudinal and experiential characteristics" (Birdsong, 2018: 2).

As seen in Figure 11.1, L2 attainment levels typically decline with increasing AoA. An extensive body of literature considers the schematic geometry of the AoA – L2 attainment function – steep vs. shallow declines; roughly linear decline vs. stretched-7 shape (plateau at ceiling, followed by linear decline) vs. stretched-L shape (decline from ceiling, followed by levelling off) – alongside the timing of the onset of decline, that is, in infancy, during childhood, around adolescence, post-adolescence, adulthood. The literature also considers sources of the decline, which include relative lack of L2 interaction at later AoA, biological maturation, entrenchment of the L1, bilingualism effects, socialization and motivational factors, age-related declines in cognitive functions that are involved in L2A, etc. (see Bialystok & Hakuta, 1999; Birdsong, 2005, 2006, 2009, 2018; Birdsong & Gertken, 2013; DeKeyser, 2012; DeKeyser et al., 2010; Flege & Bohn, 2021; Hernandez et al., 2021; Johnson & Newport, 1989; Long, 1990; MacWhinney, 2001; Saito, 2024; Werker & Hensch, 2015).

The temporal and geometric features of the decline, as well as potential sources of the decline, are tied to theories and controversies around the critical period hypothesis for L2A, which are beyond the scope of the present contribution (see Birdsong, 1999, 2005, 2014b, 2018, 2022; Birdsong & Molis, 2001; Birdsong & VanHove, 2016; DeLuca et al., 2019; Hakuta et al., 2003; Hartshorne, 2022; Hartshorne et al., 2018; Hernandez et al., 2021; Hyltenstam & Abrahamsson, 2000; Mayberry & Kluender, 2018; Newport et al., 2001; Singleton & Leśniewska, 2021; Vanhove, 2013).

In this main section, we consider AoA-related variability in L2A outcomes from two distinct perspectives. The first perspective, which we have taken up to now, concerns AoA as a conditioning factor in levels of L2A attainment. From this angle on variable outcomes, we are looking at differences between AoA cohorts in terms of *magnitude*, for example, levels of accuracy on tasks that relate to morphosyntactic knowledge or to phonetic discrimination.

From the second perspective on variability, we examine the relationship of AoA to group-level *dispersion*. Outcomes of L2A are typically more heterogeneous among cohorts who start L2 learning late in life, compared to those who start early (although the relationship between AoA and L2 variability may not be linear). The greater outcome dispersion within cohorts with late AoA is observed in multiple domains, most plainly in studies of L2 morphosyntactic knowledge and L2 pronunciation (e.g., Birdsong, 2018; Birdsong & Molis, 2001; DeKeyser et al., 2010; Flege et al., 1999; Hartshorne et al., 2018; Johnson & Newport, 1989).

These two perspectives are illustrated in Figure 11.2, where we see performance by 240 Korean native speakers and 24 native speakers of English on a 144-item binary grammaticality judgment test of knowledge of English L2 morphosyntax (Flege et al., 1999, to be discussed further below). As suggested by comparing the two bracketed areas (brackets added here to enhance visualization), the results of the later-AoA cohort appear both lower in magnitude (accuracy level) and more dispersed than those of the early-AoA cohort.

As another illustration of the distinction in question, consider Figure 11.3 (Hartshorne et al., 2018). In this online 132-item multiple-choice test of knowledge of L2 English syntax taken by 669,498 respondents from dozens of native-language backgrounds, immersion learners outperformed nonimmersion learners, and early learners outperformed later learners in both learning contexts. Along with these differences in magnitude, at late AoA

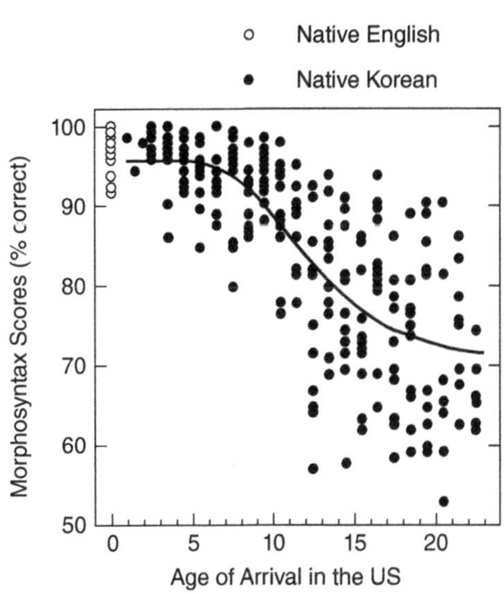

Figure 11.2 The relationship of AoA to accuracies in binary grammaticality judgments for 144 English sentences by 240 native Korean speakers and 24 native English speakers. Gompertz-Makeham line indicates contour of accuracy decrement over AoA; brackets (added) highlight greater dispersion of outcomes among later vs. earlier AoA cohorts.

From Flege, J. E., Yeni-Komshian, G. H., & Liu, S. (1999). Age constraints on second-language acquisition. *Journal of Memory and Language, 41*(1).

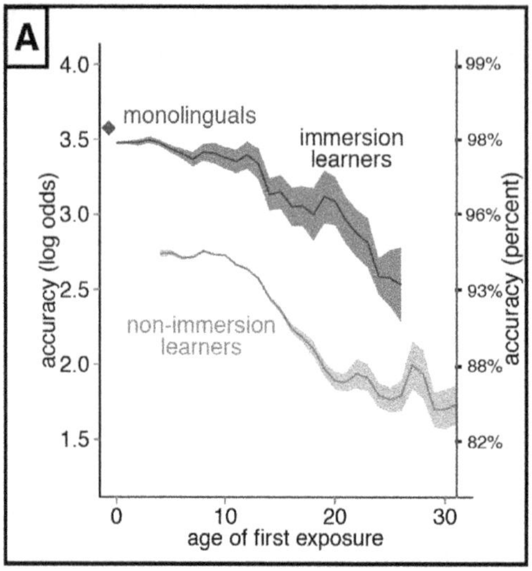

Figure 11.3 Results of an online 132-item multiple choice test of English syntax taken by 2/3 million respondents from diverse language backgrounds.
From Hartshorne, J. K., Tenenbaum, J. B., & Pinker, S. (2018). A critical period for second language acquisition: Evidence from 2/3 million English speakers. *Cognition, 177.*

the dispersions of performance are greater in both the immersion and the nonimmersion group.

Figures 11.2 and 11.3 capture a fundamental two-pronged generalization around L2 learner outcomes: with later AoA, one observes lower L2A attainment levels, and these magnitude decrements are accompanied by greater dispersion of outcomes. For outcome magnitudes, the basic points of reference are group means, whereas for dispersion of outcomes, the relevant data are variances about the group means, expressed as standard deviations or coefficients of variation.[2]

In the subsections that follow, we examine further the relationship between AoA and the magnitude and dispersion of outcomes in L2A. Sections 11.6.1 and 11.6.2 consider possible range effects on dispersion of L2A outcomes over increasing AoA. Section 11.6.3 looks at the possible mapping onto L2A magnitude and dispersion of age-related variations in working memory, which we previously identified (Section 11.4) as conditioning L2A outcomes. The final section considers results from four studies that illustrate how L2 outcomes in experimental contexts may be affected by a learner-external factor, the grammatical status of test items.

11.6.1 Range Effects with Age-Dependent Experiences

Individual-level variability in experiential factors relating to L2 learning[3] increases with age, with the effect of expanding the possible range of L2A outcomes. For example, among individual L2 learners with AoA = 12, the nominal range of years of schooling (a known predictor of L2A outcomes; see

Figure 11.1, also Hartshorne et al., 2018) is 0 to around 7; among those whose AoA is 18, the education range is 0–13 years; and among those whose AoA is 25, the potential range of schooling is 0–20 years. With increasing degrees of freedom for years of schooling between AoA 7 and AoA 25, the predictable conditioning effects of AoA on L2A outcomes become more diverse. The same logic – dispersion of L2A outcomes follows dispersion of L2A-relevant experience – applies to other conditioning experiential factors such as cumulative exposure to and interaction with the L2, the ranges of which are greater with increasing age.

11.6.2 Statistical Range Effects

Consider the relationship of dispersion to departures from ceiling. At early AoA, observed mean L2 values (accuracy scores, accent ratings compared to monolingual controls, etc.) are at or near ceiling, leaving little range for individual values above the mean. As means decline off ceiling with later AoA, the possible range of outcomes around the mean increases concomitantly. Thus, as a result of an expansion of the potential range of values as performance departs from ceiling, dispersion around the mean potentially increases as well. On this notion, at least some of the greater L2 outcome heterogeneity at later AoA could be ascribed to a range effect that creates greater room for dispersion. Note that, by the same logic, increases of dispersion could likewise be anticipated as values range from floor upward; see discussion to follow.

11.6.3 AoA and Working Memory, Magnitude, and Dispersion

Here we examine WM as a test case to illustrate the possibility that graded variations in WM, an aptitude factor in L2 learning (see Section 11.2), are congruent with L2A outcomes. Specifically, we should expect associations of WM direct scores with L2A outcome levels (magnitude) as well as associations of WM variability with L2A outcome heterogeneity (dispersion). As we will see, these predictions are partially supported by the evidence, and depend in part on how WM is tested and whether dispersion is measured in terms of standard deviation (SD) or coefficient of variation (CV).

11.6.3.1 Age-Related WM Direct Scores (Magnitudes) and L2A Attainment Levels

Direct scores on various measures of WM are known to improve from the earliest testing ages, with peaks at between fifteen and twenty-five years of age. These peaks are followed by declines, creating something of an outspread "inverted-U" shape (see Bauer et al., 2017; Buttelmann et al., 2020; Cowan, 2016; Daselaar & Cabeza, 2013; Gathercole et al., 2004; McNab et al., 2015; Michalczyk et al., 2013; Park, 1999; Pliatsikas et al., 2019; Swanson, 2017; and discussion below).

In contrast, there is no evidence that increasing AoA up to the late teens is associated with improved L2 attainment, thus suggesting a mismatch or nonisomorphism of WM direct scores and L2A outcome levels with AoA $\leq$ 15–20 years. However, studies that find L2A magnitude declines starting around AoA's of late teens to mid twenties, and continuing indefinitely thereafter, are consistent with declining WM performance over that age range. Now comparing the geometries of WM magnitudes and L2 attainment levels over the lifespan, we see no match of any of the potential AoA-L2 attainment functions (linear, stretched-7, stretched-L) with the inverted-U-shaped geometry of WM scores over the full range of ages. In sum, WM magnitudes do appear to map onto L2 attainment levels at later AoA, but not at early AoA or over all AoA.

11.6.3.2 Age-Related WM Dispersion and L2 Attainment Dispersion

To investigate a possible relationship of WM score variability over age to dispersion of L2A outcomes over increasing AoA, we turn first to the evidence on WM dispersion over age from four studies: Muñoz-Pradas and colleagues (2021), Klencklen and colleagues (2017), Troutman and Diaz (2020), and Wild-Wall and colleagues (2011). A key consideration in these studies is whether WM dispersion is represented in terms of absolute variability (SD) or as relative variability (CV).

Among 258 participants aged six–twenty-six years, Muñoz-Pradas and colleagues administered nine tests of working memory relating to phonological loop, visuo-spatial sketchpad and central executive. Positive relationships of testing age with direct scores on each of the three WM components were observed. Muñoz-Pradas and colleagues found that, as the magnitude of direct scores increased, so did the dispersion of scores as measured by SD (absolute variability). However, the authors found that relative variability, as measured by CV, remained constant across testing ages. Summarizing their findings, Muñoz-Pradas and colleagues (2021: 6) observe: "there is an increase in the inter-group absolute variability of WM with age, probably due to an indirect effect of magnitude increase of direct scores with age, while there is a maintenance of the relative variability, once the effects of the numerical magnitude of direct scores are eliminated. The constancy of relative variability across ages would guarantee the same effective variability in each age group with the same level of interindividual competence in the different age groups."

A different set of results relating to absolute vs. relative variability in WM scores over age is obtained by Klencklen and colleagues (2017), who looked at Weschler Digit Span scores for thirty-four adults between twenty and thirty years and thirty-five adults aged between sixty-five and seventy-five. As is typical of WM tests of participants older than twenty, that is, older than most of those in Muñoz-Pradas and colleagues (2021), the digit span scores for the twenty to thirty year-old group ($M = 28.21$) were higher than those of the sixy-five to seventy-five year-old group ($M = 23.91$). As the digit span score magnitudes declined with age, so did standard deviations, with the younger group SD (6.67) being larger than the older group SD (4.57). In contrast with Muñoz-Pradas and colleagues, the relative variability between the two groups in Klencklen and colleagues (2017) did not remain constant with the change in direct scores, as the younger group CV (.24) exceeded that of the older group (CV = .19). We note, however, that the between-group CV difference (11 percent) is smaller relative to the between-group SD difference (31 percent). (The CV analyses of the Klencken and colleagues' data are mine.)

In the context of their study of white matter connectivity, Troutman and Diaz (2020) presented data on WM for a younger age cohort (M = 23.7 years; N = 18) and an older age cohort (M = 67 years; N = 19). On four WM tests involving reaction times (nonverbal WM, simple speed of processing, complex speed of processing, digit symbol), the older group reaction times (RT) were longer and their SD were higher than for the younger group. The corresponding CV, with the exception of digit symbol, were also higher in the older group. On forward digit span, the older group span was slightly shorter, and the corresponding SD and CV slightly higher, than that of the older group. On backward digit span the older group span was shorter, the SD slightly smaller, than the younger group, while the CV of the two groups were the same (cf. results of Klencklen et al., 2017).

For dorsal, ventral, and frontostriatal RD, the older group values were significantly higher than in the younger group, reflecting lower fiber integrity. In addition, the corresponding SD and CV of the older group were higher. For dorsal, ventral, and frontostriatal FA, the values in the older group were significantly lower, indicating poorer fiber coherence, while the corresponding SD and CV were higher. (All CV calculations are mine.)

In sum, in the Troutman and Diaz (2020) comparisons of cohorts more than forty years apart in mean age, WM scores for the older group were generally inferior to and more dispersed than for the younger group. At the same time, across all RD and FA observations, the older group showed lower white matter integrity, with greater dispersion of values, than the younger group. In this regard we note that Halliday and colleagues (2019) find a relationship between lower white matter integrity among normally aging older (fifty-nine to eighty-nine years old) adults and greater dispersion of performance in tests of memory, executive function, general cognition, and language.[4]

A fourth set of results presents a more complex picture of age and WM dispersion. Wild-Wall and colleagues (2011) examined the performance on a high-working memory load task (2-back task) by a group of middle-aged male participants (M = 51 years; N =46) and a group of younger males (M = 22 years; N = 44). The relevant measures for targets and nontargets were RT, error rate and inverse efficiencies (IE), defined as RT divided by rate of correct responses. Relevant results are displayed in Table 11.1 (calculations of CV are mine).

Adapted from: Wild-Wall N., Falkenstein M., & Gajewski, P. D. (2011). Age-related differences in working memory performance in a 2-back task. *Frontiers in Psychology, 2*, 186.

Table 11.1 *Performance data for middle-aged and younger groups on the 2-back task in Wild-Wall et al. (2011)*

	Middle-aged group			Younger group		
		SD	*CV*		*SD*	*CV*
RT – targets (ms)	494	*83.5*	*.17*	416	*76.9*	*.18*
RT – non-targets (ms)	433	*73.9*	*.17*	358	*67.7*	*.19*
% error – targets	32.1	*17*	*.53*	29.9	*13.5*	*.45*
% error – non-targets	2.9	*2*	*.69*	2.4	*2.2*	*.91*
IE – targets (ms)	785	*290*	*.37*	623	*199.5*	*.32*
IE – non-targets (ms)	446	*78.1*	*.18*	367	*69.6*	*.19*

For all six measures, the raw WM performance of the middle-aged group was inferior (i.e., slower and with more errors) to that of the younger group. With one exception (percentage error for nontargets), dispersions as represented by SD (absolute variability) for the middle-aged group were higher, corresponding to the larger magnitudes of the raw scores. When looking at dispersion in terms of CV (relative variability), however, the relationship to raw performances of the two groups is not systematic. On only two of the six measures (percentage error - targets and IE - targets) are the CV figures for the middle-aged group higher than those of the younger group. In sum, when represented by SD, dispersion is strongly and positively correlated with raw performance: the performance of middle-aged respondents is slower, more error-prone, and more variable in absolute terms than that of younger participants. On the other hand, when dispersion is expressed in relative terms by CV, a strict relationship of variability to raw performance is not maintained; consequently, the middle-aged and younger groups are not distinguishable in this respect.

For the same two groups of participants, Wild-Wall and colleagues (2011) reported performance on a two-choice task, which involved a low WM load. Results are shown in Table 11.2 (calculations of CV are mine).

Adapted from: Wild-Wall N., Falkenstein M., & Gajewski, P. D. (2011). Age-related differences in working memory performance in a 2-back task. *Frontiers in Psychology*, *2*, 186.

On this less-demanding task, the raw RT and IE values for the younger group were lower (i.e., indicating better performance) than those of the middle-aged group. The corresponding dispersion values, expressed as SD and as CV, were also lower, the lone exception being IE – targets. In contrast, the error rates of the younger group on both targets (12.6 percent) and non-targets (1 percent) were around twice those of the middle-aged group (5.8 percent and 0.4 percent). With respect to dispersion, these error rates are particularly revealing. Specifically, we see that the higher error rates for the younger group are accompanied by higher SD (absolute variability). However, in terms of CV (relative variability) the error rates for the younger group are actually less dispersed than for the middle-aged group.

Table 11.2 *Performance data for middle-aged and younger groups on the two-choice task in Wild-Wall et al. (2011).*

	Middle-aged group			Younger group		
		SD	*CV*		*SD*	*CV*
RT – targets (ms)	420	*52.9*	*.13*	376	*51.7*	*.14*
RT – non-targets (ms)	360	*43.2*	*.12*	312	*35.6*	*.11*
% error – targets	5.8	*4.9*	*.84*	12.6	*7.5*	*.63*
% error – non-targets	0.4	*0.6*	*1.5*	1.0	*1.2*	*1.2*
IE – targets (ms)	447	*57.8*	*.13*	436	*87.4*	*.20*

Distilling the results of Wild-Wall and colleagues (2011), in most instances we observe magnitude differences between younger and middle-aged groups, with the former outperforming the latter. However, the relationship between WM raw performance and dispersion of results depends on the task (2-back vs. two-choice), on the measure (RT, error rate, IE) and on whether dispersion is represented in absolute terms (SD) or in relative terms (CV).

11.6.3.3 Summary: L2A, WM, and AoA

With respect to magnitude differences and dispersion differences over AoA, we have taken a close look at the relationship between working memory and L2A outcomes, previously exposed in Section 11.2. The studies reviewed here suggest that the relationship is nuanced.

To sum up our observations with respect to magnitude, WM direct scores appear to map roughly onto L2A magnitudes at AoA $\geq$ 20, assuming after this point a linear decline in L2 attainment. However, WM direct scores do not align with L2 attainment levels at earlier AoA or across the full AoA range.

With respect to dispersion, the results of Muñoz-Pradas and colleagues (2021), Klencklen and colleagues (2017), Troutman and Diaz (2020), and Wild-Wall and colleagues (2011) revealed that as WM magnitudes increased, SD (absolute variability) also tended to increase. Importantly, this association (high magnitude, high SD; low magnitude, low SD) generally holds when greater magnitudes reflect superior performance (e.g., Weschler Digit Span scores) as well as when larger magnitudes reflect inferior performance (e.g., longer RT, more errors). This insight leads to the understanding that when WM direct score increases reflect inferior performance (e.g., longer RT), the concomitant increase in WM SD maps onto increased L2 attainment dispersion. As we have seen, such increased attainment dispersion goes hand in hand with lower L2 attainment levels at later AoA. We emphasize as well that the positive correlation between the magnitude of WM direct scores and degree of dispersion holds if dispersion is defined as absolute variability (SD). Conversely, when dispersion is understood as relative variability (CV), the relationship between magnitude and dispersion is inconsistent. As CV and SD are equally informative, but in different ways, both may be potentially useful in conceptualizing and representing dispersion. In statistical comparisons of dispersion, however, CV is the measure of choice.

One clear take-away from the evidence and observations around magnitudes and dispersion is the need for precision in characterizing the conditioning role of WM in L2 outcomes. In addition, to the point of this test case, future inquiry along these lines might be applied selectively to other conditioning factors in L2A outcomes besides WM to determine the extent to which their magnitudes and dispersions map onto those of L2 attainment over AoA.

11.6.4 Grammatical Status of Test Items as a Non-learner Factor in L2 Attainment Magnitudes and Dispersion

In this section we look at four studies – Flege and colleagues (1999), Meulman and colleagues (2015), Bolibaugh and Foster (2021), and Birdsong (2021) – that illustrate the role of the grammatical status of experimental test items as an external (non-learner) factor in predicting both magnitude differences and dispersion in L2A.

Flege and colleagues (1999) found that the grammatical status of items interacts with AoA to produce distinct differences in both magnitude and dispersion. As seen in composite results in Figure 11.2, Korean participants of AoA ≤ 7 years performed close to the range of the twenty-four English native controls, whereas at later AoA, the L2 learners' scores are more both lower and more dispersed. Inspection of the scatterplots for seventy-two grammatical items (Figure 11.4, Panel a) and seventy-two ungrammatical items (Figure 11.4, Panel b) reveals that accuracy scores decline over both types of

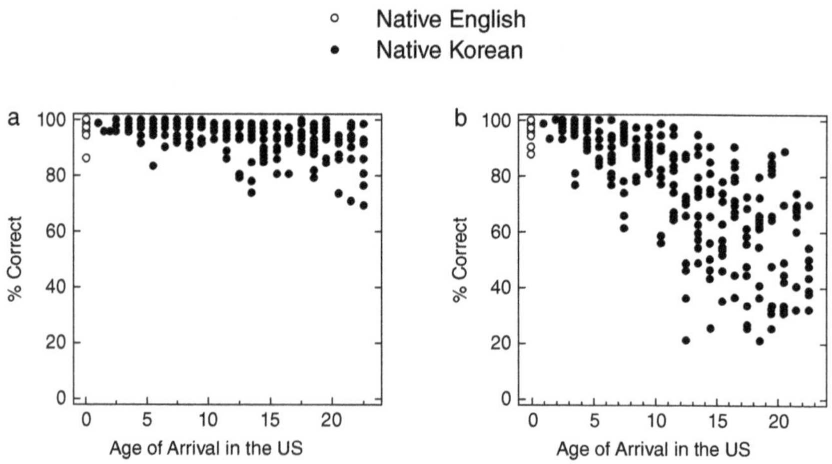

Figure 11.4 Mean accuracies for English binary grammaticality judgments by twenty-four native English speakers and 240 Korean native speakers. Panel A: grammatical items; Panel B: ungrammatical items.

From Flege, J. E., Yeni-Komshian, G. H., & Liu, S. (1999). Age constraints on second-language acquisition. *Journal of Memory and Language, 41*(1).

items, but there is a steeper slope of decline in accuracy scores over AoA for ungrammatical items than grammatical items.

This magnitude differential is accompanied by an overall greater dispersion of accuracy scores for ungrammatical items. The interaction of AoA and grammaticality status also plays out in the degree of dispersion of accuracy scores at later AoA (>10), which is conspicuously greater for ungrammatical items than for grammatical items. Note that, in terms of both dispersion and magnitude, there are no visible effects for grammaticality status among the native controls; however see Meulman and colleagues (2015) and Birdsong (2021) below.[5]

Meulman and colleagues (2015) analyzed binary grammaticality judgments (behavioral data) and simultaneous ERP responses in the processing of non-finite verbs and grammatical gender agreement in German. Participants were twenty-nine native German speakers and sixty-six Slavic L1 speakers with advanced proficiency in German L2. Over all items, the natives outperformed the learners, while in both groups, significantly lower accuracies were observed for ungrammatical gender items compared to grammatical gender items. Among the learners, the smallest AoA effects on grammaticality judgments were observed for verb items that were grammatical, and the largest AoA effects were found for ungrammatical gender items. For example, a learner with AoA of seven years had a probability of 0.98 for accurately answering the grammatical verb items, whereas for a learner with AoA of thirty-six, the probability went down to 0.8. A look at ungrammatical gender items shows accuracy probabilities of 0.92 for an early learner and 0.03 for a late learner.

Of particular interest in Meulman and colleagues (2015) are the dispersions of accuracy as grammaticality interacts with structure (i.e., verb vs. gender). In Figure 11.5 we see that the greatest dispersion of judgment accuracy among learners is observed for ungrammatical gender items, followed by ungrammatical verb items. (Among natives, the greatest dispersion is seen in the ungrammatical gender items, with no clear differences between the other three conditions.) For learners, we note that with gender items, the greater dispersion on ungrammatical vs. grammatical items could at least partially be artifactual, as scores off ceiling allow for a greater range of results. However, with verb items, greater dispersion on ungrammatical vs. grammatical items is observed even though the median accuracies are near ceiling and roughly identical.

Meulman and colleagues (2015) also examined the effects of grammaticality, structure and AoA on ERP amplitudes in the 500–1200 ms time window. Plotting the grammaticality effect (defined as the difference in ERP signal magnitude between grammatical and ungrammatical items) over AoA, the researchers found that the correlation between AoA and the magnitude of the grammaticality effect was not significant in the verb condition, $p = .825$. However, the same correlation was found to be significant for gender items, $p = .009$.

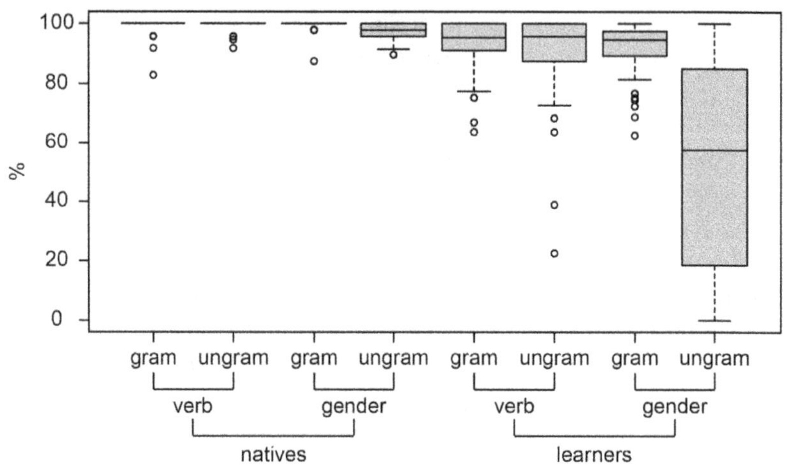

Figure 11.5 Performance on grammaticality judgments made during EEG session. Percentages of accurate responses to German items from twenty-nine native German speakers and sixty-six native Slavic speakers, plotted by group (learners, natives), structure (verb, gender), and grammaticality status of items (grammatical, ungrammatical).

From Meulman, N., Wieling, M., Sprenger, S. A., Stowe, L. A., & Schmid, M. S. (2015). Age effects in L2 grammar processing as revealed by ERPs and how (not) to study them. *PLoS ONE, 10*(12).

Bolibaugh and Foster (2021) investigated the relationship between AoA and item grammaticality on judgment accuracy for two groups of Polish native L2 English learners, one residing in the UK, the other residing in Poland. The mean AoA for the UK cohort was eighteen years, the mean age of testing was fifty-six years, the mean length of residence in the UK was thirty-eight years, and their daily use of English was ~ 50 percent. The corresponding values for the Poland cohort were twelve years, twenty-nine years, sixteen years, and ~ 25 percent. On a 110-item binary grammaticality judgment test, AoA for the UK cohort had a strong negative effect on accuracy for ungrammatical items, and a much weaker negative effect on grammatical items. For the group residing in Poland, increasing AoA was associated with much higher inaccuracy on both grammatical and ungrammatical items. Overall, both learner groups (as well as native controls) were less accurate and exhibited greater dispersion of scores on ungrammatical items than on grammatical items.

With a focus on techniques for analyzing both magnitude and dispersion in acceptability judgments, Birdsong (2021) examined published data from Birdsong (1992) relating to twenty French native speakers' (FNS) and twenty English native speakers' (ENS) scalar acceptability judgments for seventy-six items representing seven syntactic features of French. The ENS were late learners of L2 French (mean AoA in France 28.5 years) who had been living in the Paris area for three to thirty-six years and who exhibited a wide range of French proficiencies. The FNS and ENS were matched for current age, place of residence, education level, and linguistic sophistication. Over all test items, the two groups were statistically indistinguishable on ratings ($p = .55$). In addition to this magnitude similarity, the ENS and FNS did not differ in terms of ratings dispersion, as represented by their respective coefficients of variation ($p = .54$). When broken out by grammaticality status, comparisons of ratings and CV for the two groups were likewise similar for both grammatical items and ungrammatical items. Importantly, however – and in contrast with the results of Flege and colleagues (1999), above – for *both ENS and FNS*, CV values for ungrammatical items were found to be significantly larger than CV's for grammatical items (both groups' p's $< .001$). At the same time, under item-level correlational analysis, the native and learner groups did not converge on ungrammatical item ratings proper ($p = .13$) and, notably, they diverged on ungrammatical item CV as well ($p = .54$). Conversely, for grammatical item ratings the two groups correlated significantly with respect to item ratings ($p < .001$) and item CV ($p < .001$). See Birdsong (2021) for comparisons of ENS and FNS on subsets of items representing high vs. low dispersions of FNS ratings.

Note that no claim is made for the generalizability of these findings. Rather, along with the results of the studies of Flege and colleagues (1999), Meulman and colleagues (2015), and Bolibaugh and Foster (2021), the analyses illustrated in Birdsong (2021) reinforce the value of looking at the grammaticality status of test items as a predictor of both magnitude and dispersion in L2A outcomes among natives and non-natives.

11.7 Instability of L1A and L2A Outcomes: Language Dominance Relationships in Bilingualism

At the level of the individual bilingual, the dominant language is typically the language in which proficiency and fluency are higher and/or processing is easier; this is referred to as dimension-based or skill-based dominance. The dominant language may also be the language that is used more often or in more contexts; this is domain-based dominance.[6] Dominance is a continuous and relative construct and is assessed accordingly. That is, a given bilingual is not simply Language X-dominant, but is dominant in Language X to a measurable degree and by comparison with Language Y. In this section, we look at how dominance relations between the L1 and the L2 of bilinguals can change and how, in turn, these unstable outcomes can contribute to a second level of instability in both L2A and L1A.

Over the course of a bilingual's lifetime, dominance switches or reversals sometimes occur, whereby one language overtakes the other as the dominant language. Dominance shifts, where the degree to which one language is dominant increases or declines without a full-on dominance reversal, are common features of the dynamic nature of bilingualism. The fact that dominance relationships are not fixed demonstrates the transitory nature of outcomes in the two languages of bilinguals.

Changes in domain-based dominance can occur throughout adulthood. For example, life circumstances (e.g., relocation or immigration to a different language community; changes in job or educational setting; changes in parenting, caregiving, and social circle membership) can modulate the relative frequency of contact with and use of the bilingual's languages (for further discussion, see Birdsong, 2014a, b; 2016).

Greater frequency of use in one language over time (a change in domain-based dominance) can promote shifts in dimension-based dominance features such as processing facility. Changes in dimension-based dominance are typically (but not always; see Grosjean, 2010; also Birdsong, 2014a) less likely or less pronounced at later L2 AoA and with increasing chronological age. This is because linguistic skills and proficiencies are conditioned by the degree of entrenchment of the L1, which may in turn reflect the developmental/representational stage of the L1, cumulative use of the L1, and state of neurobiological development.

Moving beyond the factors that underlie the dynamic character of dominance relations, consider now the second level of variable outcomes, namely the effects on the L1 and the L2 of a bilingual when their dominance relations change. Dominance shifts and switches have been observed and studied in several domains of language (see Köpke & Genevska-Hanke, 2018 for an overview and empirical study). For the present purposes it is in the domain of bilingual phonetics that the outcomes of dominance shifts and switches are most instructively revealed. For example, phonetic drift – most often, a variation in acoustic values such as voice onset time (VOT) and formant frequencies (particularly F1 and F2) – is observed with changes in L1–L2 dominance relationships. In some reported cases of dominance shifts, the acoustic drift is assimilatory in nature. Notably, in Chang's (2012) study of adult L1 English students learning Korean L2, the F1 values of the learners' L1 English vowels moved toward Korean-like F1 values after only limited immersive study abroad in Korea – a remarkably small but swift shift in dominance. Dominance shifts can also result in dissimilation effects. Thus, in Flege and Eefting (1987a), L1 Dutch learners of L2 English at high proficiency levels, compared to those at lower English proficiencies, generated greater VOT differences (dissimilation) between their L1 Dutch [t] and their L2 English [t] by virtue of producing lower VOT in their Dutch [t] and higher VOT in their English [t]. Similarly, Flege and Eefting (1987b) observed that relatively shorter Spanish VOT for stop consonants were produced by L1 Spanish speakers learning L2 English than by monolingual Spanish speakers, thereby exceeding the nominal VOT differences between Spanish and English stops in the bilingual group.

In instances of dominance switches, both assimilatory drifts and dissimilatory drifts have been observed. For example, in Kornder and Mennen's (2021) study of Arnold Schwarzenegger, who had lived in the US for more than fifty years since immigrating in his early twenties and had become strongly L2 English dominant, Schwarzenegger's pronunciation of German vowels /i/ and /ɔ/ drifted toward English values for F1 and F2. At the same time, the F1 and F2 for Schwarzenegger's other German vowels became more dissimilar from corresponding English values. In Schwarzenegger's pronunciation of consonants, his German plosive VOTs drifted toward long lags typical of English plosives. Effects of Schwarzenegger's dominance switch were not readily identified for the VOTs of his English plosives, however, as values were highly variable. In short, in Schwarzenegger's case the switch to L2 dominance resulted in local, not systemic, changes in the phonetic characteristics of both his newly dominant L2 English and his partially maintained native German.

To summarize, the preceding examples illuminate two key dimensions of the transitory nature of outcomes in both native and non-native language acquisition. Not only are L1–L2 dominance relationships subject to change in bilingualism, but these very instabilities contribute to outcome impermanence in both languages.[7]

Particularly in the case of the L1, but also in the L2, the impermanence in question is often thought of as attrition. In the literature, attrition is variously understood as a nonpathological neurocognitive destabilization or adjustment in language processing, or as a reconfiguration or repurposing of the L1 and L2 in adapting to new linguistic-experiential situations, or as any discernable influence of either language system on the other (e.g., Gallo et al., 2021; Hernandez et al., 2019; Köpke & Schmid, 2004; Steinhauer & Kasparian, 2020). Attrition is known to be temporary and reversible (Köpke & Genevska-Hanke, 2018) once domain-based dominance returns to initial values. In terms of breadth, depth, and speed, attrition is typically less pronounced at later chronological ages (Birdsong, 2014a). With these observations in mind, one may ask if a change in dominance relationships should be properly conceptualized as a catalyst (among possibly other catalysts) of attrition, or perhaps as a necessary condition for attrition. In a similar vein, proof-of-concept work might be directed toward the question of whether attrition – in at least some areas of language, particularly in subtle phonetic features (see above) – is an insuppressible reflex of dominance changes. These considerations are reminders once again that L2 and L1 outcomes should not be thought of as final states. Their instability reflects the lifespan dynamics of bilingualism, which is characterized by ongoing permeability and reciprocal influences of language systems.

11.8 Conclusion

Consistent with conventional practices in disciplinary handbooks, this chapter is oriented around conceptual, empirical, and methodological axes, with utilitarian emphases on instructiveness and future research. In this spirit, we examined a small set of individual factors in L2A – genetics, music background, AoA, Working Memory, attitudes – that underlie the axiom "different learners, different outcomes." For each case we reviewed key studies that illustrate the workings of the factor, that yield foundational insights into the impact of the factor in L2A outcomes, and that set the stage for further inquiry. We also looked at grammaticality, a learner-independent factor, in terms of its roles in variable L2A outcomes. Along with AoA, the grammatical status of items is associated not only with different levels of outcomes, but also with differences in dispersion of outcomes. We explored the possibility that WM levels and dispersion over the course of the lifespan map congruently onto the magnitudes and the dispersions of L2A outcomes over AoA. On this notion, the evidence generally suggests a relationship of WM declines and greater WM absolute variability (SD) with declining levels and increased dispersion of L2A outcomes. Similar explorations involving other individual variables and relative variability (CV) is called for. With respect to outcomes in bilingualism, we showed that dominance relationships are inherently unstable. In turn, changes in dominance can destabilize features of L1 and L2 production and processing. The dynamics of dominance relationships and the permeability of language systems demonstrate the transient character of language learning outcomes.

Notes

1 The term Age of Acquisition refers to the age at which persistent L2 learning begins (i.e., as opposed to the age at which occasional classroom instruction, if any, begins), most often in immersion contexts such as immigration. For this individual learner factor, various studies employ equivalent terms such as Age of Arrival, Age of Immersion, Age of Immigration, Age of Learning, Age of Onset, and Age of Exposure. In this chapter we use Age of Acquisition (AoA) as a generic cover term.

2 The coefficient of variation (CV), also known as relative variability, is the ratio of absolute variability (SD) to its corresponding mean (CV = SD/M). When comparing variability across sets of data, CV is used instead of SD as a standardized measure of dispersion. The use of CV in L2 studies is illustrated in Birdsong (2021), Harrington (2006) and Hulstijn and colleagues (2009). See Section 11.6.3.2 and 11.6.3.3 in the current chapter for examples of SD vs. CV in studies of working memory performance at various ages.

As a clarification, note that magnitude differences between monolinguals and learners are not strictly AoA-related. For example, Hartshorne and colleagues (2018: 270) and Li and Hartshorne (2022) observe that the performance of simultaneous bilinguals is significantly lower than that of monolinguals. Since the AoA of the two groups is the same (0 years), we must look elsewhere for possible sources of the difference. One consideration relates to the distribution of opportunity to learn. Learners of two languages have less time and fewer occasions for interaction in each language than do individuals who are exposed to a single language. Another source of monolingual–bilingual (simultaneous and sequential) differences derives from the nature of bilingualism, wherein both languages of bilinguals are activated at all times (Dijkstra & Van Heuven, 2002; Kroll et al., 2012; Shook & Marian, 2013; van Hell & Dijkstra, 2002; van Hell & Tanner, 2012; Schwartz & Kroll, 2006). Coactivation and use of two languages result in bidirectional (L2 <–> L1) influences that make it impossible for the L2 *or the L1* of a bilingual to resemble the language of a monolingual (Balcom, 2003; Flege et al., 2003; Grosjean, 1989; Han et al., 2023; Laufer, 2003). The inevitable neurophysiological and behavioral reflexes of bilingualism, which occur no matter what the age of L2 learning, are not the result of lost language learning ability. Thus, in comparing bilinguals to monolinguals, we understand that magnitude differences are not uniquely keyed to AoA, nor should they be thought of as deficits (Birdsong, 2023; van Hell, 2023).

3 It is reasonable to assume that certain nonexperiential factors associated with L2 learning outcomes (L1 processing ability, perceptual acuity, genetic and neurocognitive makeup, L2-L1 relatedness, etc.; see Introduction) are distributed among learners equally across all AoA, and thus are not candidate sources of outcome dispersion differences between early and late L2 learners.

4 See Madden and colleagues (2012) for a review of studies that relate white matter integrity to WM in younger and older adults. See Salthouse (2011) for a review of research relating white matter hyperintensities (WMH), which are reflective of white matter damage, to WM performance. See Prins and Scheltens (2015) for discussion of WMH and cognitive impairment. The meta-analysis of twenty-three studies by Anderson and colleagues (2023) reveals a tract-specific effect for bilingualism in maintaining WM integrity. See Barbeau and colleagues (2023) on the relationship of the dorsal branch of the arcuate fasciculus to phonological working memory.

5 Flege and colleagues (1999) analyzed the native speaker responses and found no effect for grammatical status on magnitude (accuracy) in this group. No analyses were performed to quantify the interaction of AoA and grammaticality status on dispersion of learner scores.

6 A distinction is made between individual-level language dominance, our concern here, and language dominance at the level of the bilingual speech community. The language of community-level dominance is typically spoken by the greatest numbers of residents, and is often the language of official communications and instruction in schools. The dominant community language may also be the prestige language or the lingua franca. At the level of the individual bilingual, the dominant language is associated with more frequent use, or with greater communicative skill or ease of processing. The day-to-day prevalence of the community-level dominant language may naturally feed into in individual-level dominance. Note as well that the matters in question are not to be confused with hemispheric dominance or neurofunctional lateralization.

7 A striking example is the personal language history of François Grosjean, whose direction of dominance switched back and forth between his L1 and L2 over the course of fifty plus years (Grosjean, 2010). We note as well that the preceding characterizations of the dynamic nature of dominance relationships encompass the possibility of periods of symmetry, or balance, where neither language is dominant.

References

Al-Hoorie, A., Oga-Baldwin, W. L. Q., Hiver, P., & Vitta, J. P. (2022). Self-determination mini-theories in second language learning: A systematic review of three decades of research. *Language Teaching Research*, *0*(0). https://doi.org/10.1177/13621688221102686

Anderson, J. A. E., Yurtsever, A., Fisher-Skau, O., Cherep, L. A., MacPhee, I., Luk, G., & Grundy, J. G. (2023). Consistency in bilingualism and white matter research: A meta-analysis. Bilingualism – Mind – Brain Lab Talk, University of California-Irvine, June 19, 2023.

Ando, J., Ono, Y., & Wright, M. J. (2001). Genetic structure of spatial and working memory. *Behavior Genetics*, *31*(6), 615–624. https://doi.org/10.1023/A:1013353613591

Andrews, E., Eierud, C., Banks, D., Harshbarger, T., Michael, A., & Rammell, C. (2021). Effects of lifelong musicianship on white matter integrity and cognitive brain reserve. *Brain Science*, *11*(67). https://doi.org/10.3390/brainsci11010067

Asano, R., & Boecks, C. (2015). Syntax in language and music: What is the right level of comparision. *Frontiers in Psychology*, ***6***. https://doi.org/10.3389/fpsyg.2015.00942

Baddeley, A. D. (2012). Working memory: Theories, models and controversies. *Annual Review of Psychology*, *63*, 1–30. https://doi.org/10.1146/annurev-psych-120710-100422

Baddeley, A. D. (2015). Working memory in second language learning. In Z. Wen, M. Mota, & A. McNeill (Eds.), *Working Memory in Second Language Acquisition and Processing* (pp. 17–28). Multilingual Matters.

Baddeley, A. D., & Hitch, M. (1974). Working memory. *Psychology of Learning and Motivation*, *8*, 47–89. https://doi.org/10.1016/S0079-7421(08)60452-1

Baddeley, A., Hitch, G., & Allen, R. (2021). A multicomponent model of working memory. In R. Logie, V. Camos, & N. Cowan (Eds.), *Working Memory: The State of the Science* (pp. 10–43). Oxford Academic.

Balcom, P. (2003). Cross-linguistic influence of L2 English on middle constructions in L1 French. In V. Cook (Ed.), *Effects of the Second Language on the First* (pp. 168–192). Multilingual Matters.

Ballinger, E., Ananth, M., Talmage, D. A., & Role, L. (2016). Basal forebrain cholinergic circuits and signaling in cognition and cognitive decline. *Neuron*, *91*(6), 1199–1218. https://doi.org/10.1016/j.neuron.2016.09.006

Barbeau, E. B., Kousaie, S., Brass, K., Descoteaux, M., Petrides, D., & Klein, D. (2023). The importance of the dorsal branch of the arcuate fasciculus in phonological working memory, *Cerebral Cortex*, *33(*16), 9554–9565. https://doi.org/10.1093/cercor/bhad226

Bauer, J.-R., Martinez, J. E., Roe, M. A., & Church, J. A. (2017). Consistent performance differences between children and adults despite manipulation of cue-target variables. *Frontiers in Psychology*, *8*. https://doi.org/10.3389/fpsyg.2017.01304

Berry, A. S., Blakely, R. D., Sarter, M., & Lustig, C. (2015). Cholinergic capacity mediates prefrontal engagement during challenges to attention: Evidence from imaging genetics. *NeuroImage, 108,* 386–395. https://doi.org/10.1016/j.neuroimage.2014.12.036

Besson, D., & Schön, D. (2001). Comparison between language and music. *Annals of the New York Academy of Sciences, 930*(1), 232–258. https://doi.org/10.1111/j.1749-6632.2001.tb05736.x

Bialystok, E., & Hakuta, K. (1999). Confounded age: Linguistic and cognitive factors in age differences for second language acquisition. In D. Birdsong (Ed.), *Second Language Acquisition and the Critical Period Hypothesis* (pp. 161–181). Erlbaum.

Bidelman, G. M., Gandour, J. T., & Krishnan, A. (2010). Cross-domain effects of music and language experience on the representation of pitch in the human auditory brainstem. *Journal of Cognitive Neuroscience, 23,* 425–434. https://doi.org/10.1162/jocn.2009.21362

Bidelman, G. M., Hutka, S., & Moreno, S. (2013). Tone language speakers and musicians share enhanced perceptual and cognitive abilities for musical pitch: Evidence for bidirectionality between the domains of language and music. *PLoS ONE, 8,* e60676. https://doi.org/10.1371/journal.pone.0060676

Biedroń, A., & Birdsong, D. (2019). Highly proficient and gifted bilinguals. In L. Ortega & A. De Houwer (Eds.), *Cambridge Handbook of Bilingualism* (pp. 307–323). Cambridge University Press.

Birdsong, D. (1992). Ultimate attainment in second language acquisition. *Language, 68,* 706–755. https://doi.org/10.2307/416851

Birdsong, D. (2005). Interpreting age effects in second language acquisition. In J. Kroll & A. DeGroot (Eds.), *Handbook of Bilingualism: Psycholinguistic Perspectives* (pp. 109–127). Oxford University Press.

Birdsong, D. (2006). Age and second language acquisition and processing: A selective overview. *Language Learning, 56,* 9–49. https://doi.org/10.1111/j.1467-9922.2006.00353.x

Birdsong, D. (2009). Age and the end state of second language acquisition. In W. Ritchie & T. Bhatia (Eds.), *The New Handbook of Second Language Acquisition* (pp. 401–424). Emerald.

Birdsong, D. (2012). Three perspectives on non-uniform linguistic attainment. *Linguistic Approaches to Bilingualism, 2,* 255–259. https://doi.org/10.1075/lab.2.3.02bir

Birdsong, D. (2014a). Dominance and age in bilingualism. *Applied Linguistics, 35,* 374–392. https://doi.org/10.1093/applin/amu031

Birdsong, D. (2014b). The Critical Period Hypothesis for second language acquisition: Tailoring the coat of many colors. In M. Pawlak & L. Aronin (Eds.), *Essential Topics in Applied Linguistics and Multilingualism. Studies in Honor of David Singleton* (pp. 43–50). Springer.

Birdsong, D. (2016). Dominance in bilingualism: Foundations of measurement, with insights from the study of handedness. In C. Silva-Corvalán & J. Treffers-

Daller (Eds.), *Language Dominance in Bilinguals: Issues of Operationalization and Measurement* (pp. 85–105). Cambridge University Press.

Birdsong, D. (2018). Plasticity, variability and age in second language acquisition and bilingualism. *Frontiers in Psychology, 9*: 81. https://doi.org/10.3389/fpsyg.2018.00081

Birdsong, D. (2021). Analyzing variability in L2 ultimate attainment. *Language, Interaction and Acquisition, 12,* 133–156. https://doi.org/10.1075/lia.21001.bir

Birdsong, D. (2022). *Critical Periods.* In M. Aronoff (Ed.), *Oxford Bibliographies in Linguistics.* Oxford University Press. www.oxfordbibliographies.com/browse?module_0=obo-9780199772810

Birdsong, D. (2023). Whither bilinguals, natives, and variability? A commentary on Janet van Hell "The neurocognitive underpinnings of second language processing: Knowledge gains from the past and future outlook." *Language Learning, 73*(S2), 147–150.

Birdsong, D. (Ed). (1999). *Second Language Acquisition and the Critical Period Hypothesis.* Lawrence Erlbaum.

Birdsong, D., & Gertken, L. M. (2013). In faint praise of folly: A critical review of native/non-native comparisons, with examples from native and bilingual processing of French complex syntax. *Language, Interaction and Acquisition, 4,* 107–133. https://doi.org/10.1075/lia.4.2.01bir

Birdsong, D., & Molis, M. (2001). On the evidence for maturational effects in second language acquisition. *Journal of Memory and Language, 44,* 235–249. https://doi.org/10.1006/jmla.2000.2750

Birdsong, D., & Vanhove, J. (2016). Age of second language acquisition: Critical periods and social concerns. In E. Nicoladis & S. Montanari (Eds.), *Bilingualism across the Lifespan* (pp. 163–181). APA/De Gruyter.

Bolibaugh, C., & Foster, P. (2021). Implicit statistical learning in naturalistic and instructed morphosyntactic attainment: An aptitude-treatment interaction design. *Language Learning, 71*(4), 959–1003. https://doi.org/10.1111/lang.12465

Bouchard, T. J., & McGue, M. (1981). Familial studies of intelligence: A review. *Science, 212*(4498), 1055–1059. https://doi.org/10.1126/science.7195071

Bowden, H. W., Sanz, C., & Stafford, C. A. (2005). Individual differences: Age, sex, working memory, and prior knowledge. In C. Sanz (Ed.), *Mind and Context in Adult Second Language Acquisition: Methods, Theory, and Practice* (pp. 105–140). Georgetown University Press.

Bueno, D. (2019) Genetics and learning: How the genes influence educational attainment. *Frontiers in Psychology, 10,* 1622. https://doi.org/10.3389/fpsyg.2019.01622

Buttelmann, F., Könen, T., Hadley, L. V., Meaney, J. A., Auyeung, B., Morey, C. C., Chevalier, N., & Karbach, J. (2020). Age-related differentiation in verbal and visuospatial working memory processing in childhood. *Psychological Research, 84*(8), 2354–2360. https://doi.org/10.1007/s00426-019-01219

Caplan, D., & Waters, G. (1999). Verbal working memory and sentence comprehension. *Behavioral and Brain Sciences, 22*(1), 77–94. https://doi.org/10.1017/S0140525X99001788

Carstensen, L. L., & Hartel, C. R. (Eds.) (2006). *When I'm 64. National Research Council (US) Committee on Aging Frontiers in Social Psychology, Personality, and Adult Developmental Psychology*. National Academies Press (US).

Cesarini, D., & Visscher, P. M. (2017). Genetics and educational attainment. *npj Science of Learning, 2*(4). https://doi.org/10.1038/s41539-017-0005-6

Chan, A. S., Ho, Y. C., & Cheung, M. C. (1998). Music training improves verbal memory. *Nature, 396(*6707), 128. https://doi.org/10.1038/24075

Chang, C. B. (2012). Rapid and multifaceted effects of second-language learning on first-language speech production. *Journal of Phonetics, 40*(2), 249–268.

Chang, D., Hedberg, N., & Wang, Y. (2016). Effects of musical and linguistic experience on categorization of lexical and melodic tones. *Journal of the Acoustical Society of America, 139*, 2432–2447. https://doi.org/10.1121/1.4947497

Chen, S., Yang, Y., & Wayland, R. (2021). Categorical perception of Mandarin pitch directions by Cantonese-speaking musicians and non-musicians. *Frontiers in Psychology, 12*. https://doi.org/10.3389/fpsyg.2021.713949

Chobert, J., & Besson, M. (2013). Musical expertise and second language learning. *Brain Sciences, 3*(2), 923–940. https://doi.org/10.3390/brainsci3020923

Choi, W. (2021). Musicianship influences language effect on musical pitch perception. *Frontiers in Pyschology, 12*. https://doi.org/10.3389/fpsyg.2021.712753

Conway, A. R. A., Kane, M. J., Bunting, M. F., Hambrick, D. Z., Wilhelm, O., & Engle, R. W. (2005). Working memory span tasks: A methodological review and user's guide. *Psychonomic Bulletin & Review, 12*, 769–786. https://doi.org/10.3758/BF03196772

Cowan, N. (2016). Working memory maturation: Can we get at the essence of cognitive growth? *Perspectives on Psychological Science: A Journal of the Association for Psychological Science, 11*(2), 239–264. https://doi.org/10.1177/1745691615621279

Craik, F. I. M., & Salthouse, T. A. (Eds.) (2000). *The Handbook of Aging and Cognition II*. Erlbaum.

Csizér, K., & Dörnyei, Z. (2005). The internal structure of language learning motivation and its relationship with language choice and learning effort. *Modern Language Journal, 89*, 19–36. https://doi.org/10.1111/j.0026-7902.2005.00263.x

Cunnings, I. (2017). Parsing and working memory in bilingual sentence processing.*Bilingualism: Language and Cognition*, 20(4), 659–678. https://doi.org/10.1017/S1366728916000675

Dąbrowska, E. (2018). Experience, aptitude and individual differences in native language ultimate attainment. *Cognition, 178*, 222–235. https://doi.org/10.1016/j.cognition.2018.05.018

Dąbrowska, E. (2019). Experience, aptitude, and individual differences in linguistic attainment: A comparison of native and nonnative speakers. *Language Learning, 69*, 72–100. https://doi.org/10.1111/lang.12323

Dani, J. A., & Bertrand, D. (2007). Nicotinic acetylcholine receptors and nicontinic cholinergic mechanisms of the central nervous system. *Annual Review of*

Pharmacology and Toxicology, 47, 699-729. https://doi.org/10.1146/annurev.pharmtox.47.120505.105214

Daselaar, S., & Cabeza, R. (2013). Age-related-decline in working memory and episodic memory: Contributions of the prefrontal cortex and medial temporal lobes. In K. N. Ochsner & S. Kosslyn (Eds.), *The Oxford Handbook of Cognitive Neuroscience, Volume 1: Core Topics* (pp. 456–472). Oxford University Press.

De Bot, K., Lowie, W., & Verspoor, M. (2007). A dynamic systems theory approach to second language acquisition. *Bilingualism: Language and Cognition, 10*(1), 7–21. https://doi.org/10.1017/S1366728906002732

DeKeyser, R. (2012). Age effects in second language learning. In S. M. Gass & A. Mackey (Eds.), *The Routledge Handbook of Second Language Acquisition* (pp. 442–460). Routledge.

DeKeyser, R., Alfi-Shabtay, I., & Ravid, D. (2010). Cross-linguistic evidence for the nature of age effects in second language acquisition. *Applied Psycholinguistics, 31*(3), 413–438. https://doi.org/10.1017/S0142716410000056

DeLuca, V., Miller, D., Pliatsikas, C., & Rothman, J. (2019). Brain adaptations and neurological indices of processing in adult second language acquisition: Challenges to the critical period hypothesis. In J. W. Schwieter & M. Paradis (Eds.), *The Handbook of the Neuroscience of Multilingualism* (1st ed., pp. 170–196). Wiley.

Dijkstra, T., & van Heuven, W. J. B. (2002). The architecture of the bilingual word recognition system: From identification to decision. *Bilingualism: Language and Cognition*, 5(3), 175–197. https://doi.org/10.1017/S1366728902003012

Dörnyei, Z. (2005). *Psychology of the Language Learner: Individual Differences in Second Language Acquisition*. Erlbaum.

Dörnyei, Z. (2009). The L2 motivational self system. In Z. Dörnyei & E. Ushioda (Eds.), *Motivation, Language Identity and the L2 Self* (pp. 9–42). Multilingual Matters.

Dörnyei, Z. (2019). From integrative motivation to directed motivational currents: The evolution of the understanding of L2 motivation over three decades. In M. Lamb et al. (Eds.), *The Palgrave Handbook of Motivation for Language Learning* (pp. 39–69). Palgrave Macmillan.

Dörnyei, Z., & Mentzelopoulos, K. (2022). *Lessons from Exceptional Language Learners Who Have Achieved Nativelike Proficiency.* Multilingual Matters.

Dörnyei, Z., & Ushioda, E. (2021). *Teaching and Researching Motivation* (3rd ed.) Routledge.

Dörnyei, Z., MacIntyre, P. D., & Henry, A. (2015). Introduction: Applying complex dynamic systems principles to empirical research on L2 motivation. In Z. Dörnyei, P. D. MacIntyre, & A. Henry (Eds.), *Motivational Dynamics in Language Learning* (pp. 1–7). Multilingual Matters.

English, B. A., Hahn, M. K., Gizer, I. R., Mazei-Robison, M., Steele, A., Kurnik, D. M., Stein, M. A., Waldman, I. D., & Blakely, R. D. (2009). Choline transporter gene variation is associated with attention-deficit hyperactivity disorder. *Journal of Neurodevelopmental Disorders, 1*, 252–263. https://doi.org/10.1007/s11689-009-9033-8

Ettlinger M., Bradlow A., & Wong, P. C. M. (2014). Variability in the learning of complex morphophonology. *Applied Psycholinguistics, 35*, 807–831. https://doi.org/10.1017/S0142716412000586

Farmer, T. A., Misyak, J. B., & Christiansen, M. H. (2012). Individual differences in sentence processing. In M. J. Spivey, K. McRae, & M. F. Joanisse (Eds.), *The Cambridge Handbook of Psycholinguistics* (pp. 353–364). Cambridge University Press.

Felser, C., & Roberts, L. (2007). Processing *wh*-dependencies in a second language: A cross-modal priming study. *Second Language Research, 23*(1), 9–36. https://doi.org/10.1177/0267658307071600

Feng, L., & Papi, M. (2020). Persistence in language learning: The role of grit and future self-guides. *Learning and Individual Differences, 81*, 101904. https://doi.org/10.1016/j.lindif.2020.101904

Fenk-Oczlon, G. (2022). Iconic associations between vowel acoustics and musical patterns, and the Musical Protolanguage Hypothesis. *Frontiers in Communication, 7*. https://doi.org/10.3389/fcomm.2022.887739

Fenk-Oczlon, G., & Fenk, A. (2009). Some parallels between language and music from a cognitive and evolutionary perspective. *Musicae Scientiae, 13*, 201–226. https://doi.org/10.1177/1029864909013002101

Flege, J. E., & Bohn, O.-S. (2021). The revised speech learning model (SLM-r). In R. Wayland (Ed.), *Second Language Speech Learning: Theoretical and Empirical Progress* (pp. 3–83). Cambridge University Press.

Flege, J. E., & Eefting, W. (1987a). Cross-language switching in stop consonant perception and production by Dutch speakers of English. *Speech Communication, 6*, 185–202. https://doi.org/10.1016/0167-6393(87)90025-2

Flege, J. E., & Eefting, W. (1987b). Production and perception of English stops by native Spanish speakers. *Journal of Phonetics, 15*, 67– 83.

Flege, J. E., Schirru, C., & MacKay, I. R. A. (2003). Interaction between the native and second language phonetic subsystems. *Speech Communication, 40*(4), 467–491. https://doi.org/10.1016/S0167-6393(02)00128-0

Flege, J. E., Yeni-Komshian, G. H., & Liu, S. (1999). Age constraints on second-language acquisition. *Journal of Memory and Language, 41*, 78–104. https://doi.org/10.1006/jmla.1999.2638

François, C., Chobert, J., Besson, M., & Schön, D. (2012) Music training for the development of speech segmentation. *Cerebral Cortex, 23*(9), 2038–2043. https://doi.org/10.1093/cercor/bhs180

Franklin, M. S., Sledge Moore, K., Yip, C.-Y., Jonides, J., Rattray, K., & Moher, J. (2008). The effects of musical training on verbal memory. *Psychology of Music, 36*(3), 353–365. https://doi.org/10.1177/0305735607086044

Freeman, G. B., & Gibson, G. E. (1988). Dopamine, acetylcholine, and glutamate interactions in aging. Behavioral and neurochemical correlates. *Annals of the New York Academy of Sciences, 515*, 191–202. https://doi.org/10.1111/j.1749-6632.1988.tb32984.x.

Fritz, J., Poeppel, D., Trainor, L. et al. (2013). The neurobiology of language, speech, and music. In M. A. Arbib (Ed.), *Language, Music, and the Brain:*

A Mysterious Relationship. MIT Press. https://doi.org/10.7551/mitpress/9780262018104.003.0017

Gallo, F., Bermudez-Margaretto, B., Shtyrov, Y., Abutalebi, J., Kreiner, H., Chitaya, T., Petrova, A., & Myachykov, A. (2021). First language attrition: What it is, what it isn't and what it can be. *Frontiers in Human Neurosicence, 07*. https://doi.org/10.3389/fnhum.2021.686388

Gardner, R. C., & Lambert, W. E. (1972). *Attitudes and Motivation in Second-Language Learning*. Newbury House.

Gathercole, S. E., Pickering, S. J., Ambridge, B., & Wearing, H. (2004). The structure of working memory from 4 to 15 years of age. *Developmental Psychology, 40*, 177–190. https://doi.org/10.1037/0012-1649.40.2.177

Gottfried, T. L., Staby, A. M., & Ziemer, C. J. (2004). Musical experience and Mandarin tone discrimination and imitation. *Journal of the Acoustical Society of America, 115*, 2545. https://doi.org/10.1121/1.4783674

Gottfried, T. L., & Ouyang, G. Y.-H. (2005). Production of Mandarin tone contrasts by musicians and non-musicians. *Journal of the Acoustical Society of America, 118*, 2025. https://doi.org/10.1121/1.4785767

Graham, S. A., & Fisher, S. E. (2013). Decoding the genetics of speech and language. *Current Opinion in Neurobiology, 23*(1), 43–51. https://doi.org/10.1016/j.conb.2012.11.006

Grosjean, F. (1989). Neurolinguists, beware! The bilingual is not two monolinguals in one person. *Brain and Language, 36*, 3–15. https://doi.org/10.1016/0093-934X(89)90048-5

Grosjean, F. (2010). *Bilingual: Life and Reality*. Harvard University Press.

Hakuta, K., Bialystok, E., & Wiley, E. (2003). Critical evidence: A test of the critical-period hypothesis for second-language acquisition. *Psychological Science, 14*, 31–38. https://doi.org/10.1111/1467-9280.01415

Halliday, D. W. R., Gawryluk, J. R., Garcia-Barrera, M. A., & MacDonald, S. W. S. (2019). White matter integrity is associated with intraindividual variability in neuropsychological test performance in healthy older adults. *Frontiers in Human Neuroscience, 13*, 352. https://doi.org/10.3389/fnhum.2019.00352

Hamrick, P., Graff, C., & Finch, B. (2019). Contributions of episodic memory to novel word learning. *The Mental Lexicon, 14*(3), 381–398. https://doi.org/10.1075/ml.19019.ham

Han, J.-I., Kim, J.-Y., & Tsukada, K. (2023). Foreign accent in L1 (first language): Case of Korean immigrants in North America. *Linguistic Approaches to Bilingualism, 14*(5), 740–758. https://doi.org/10.1075/lab.22028.han

Harrington, M. (2006). The lexical decision task as a measure of L2 proficiency. *EUROSLA Yearbook, 6*, 147–168. https://doi.org/10.1075/eurosla.6.10har

Hartshorne, J. K. (2022). When do children lose the language instinct? A critical review of the critical periods literature. *Annual Review of Linguistics, 8*, 143–151. https://doi.org/10.1146/annurev-linguistics-032521-053234

Hartshorne, J. K., Tenenbaum, J. B., & Pinker, S. (2018). A critical period for second language acquisition: Evidence from 2/3 million English speakers. *Cognition, 177*, 263–277. https://doi.org/10.1016/j.cognition.2018.04.007

Harvey, L. (2017). Language learning motivation as ideological becoming. *System, 65*, 69–77. https://doi.org/10.1016/j.system.2016.12.009

Hasselmo, M. E. (2006) The role of acetylcholine in learning and memory. *Current Opinion in Neurobiology, 16*(6), 710–715. https://doi.org/10.1016/j.conb.2006.09.002

Havik, E., Roberts, L., Van Hout, R., Schreuder, R., & Haverkort, M. (2009). Processing subject-object ambiguities in the L2: A self-paced reading study with German L2 learners of Dutch. *Language Learning, 59*, 73–112. https://doi.org/10.1111/j.1467-9922.2009.00501.x

Haworth, C. M., Wright, M. J., Luciano, M., et al. (2010). The heritability of general cognitive ability increases linearly from childhood to young adulthood. *Molecular Psychiatry, 15*(11),1112–1120. https://doi.org/10.1038/mp.2009.55

Henry, A. (2017). L2 motivation and multilingual identities. *The Modern Language Journal, 101*(3), 548–565. https://doi.org/10.1111/modl.12412

Hernandez, A. E., Bodet III, J. P., Gehm, K., & Shen, S. (2021). What does a critical period for second language acquisition mean? Reflections on Hartshorne et al. (2018). *Cognition, 206*, 104478. https://doi.org/10.1016/j.cognition.2020.104478

Hernandez, A. E., Claussenius-Kalman, H. L., Ronderos, J., Castilla-Earls, A. P., Sun, L., Weiss, S. D., et al. (2019). Neuroemergentism: Response to commentaries. *Journal of Neurolinguistics, 49*, 258–262. https://doi.org/10.1016/j.jneuroling.2018.06.001

Ho, Y.-C., Cheung, M.-C., & Chan, A. S. (2003). Music training improves verbal but not visual memory: Cross-sectional and longitudinal explorations in children. *Neuropsychology, 17*(3), 439–450. https://doi.org/10.1037/0894-4105.17.3.439

Hopp, H. (2014). Working memory effects in the L2 processing of ambiguous relative clauses. *Language Acquisition, 21*(3), 250–278. https://doi.org/10.1080/10489223.2014.892943

Huang, Q., Liao, C., Ge, F., Aoi, J., & Liu, T. (2022). Acetylcholine bidirectionally regulates learning and memory. *Journal of Neurorestoratology, 10*(2), 100002. https://doi.org/10.1016/j.jnrt.2022.100002

Hulstijn, J. H., van Gelderen, A., & Schoonen, R. (2009). Automatization in second language acquisition: What does the coefficient of variation tell us? *Applied Psycholinguistics, 30*(4), 555–582. https://doi.org/10.1017/S0142716409990014

Hyltenstam, K. (Ed.) (2016). *Advanced Proficiency and Exceptional Ability in Second Language*. Mouton de Gruyter.

Hyltenstam, K., & Abrahamsson, N. (2000). Who can become native-like in a second language? All, some, or none? On the maturational constraints controversy in second language acquisition. *Studia Linguistica, 54*, 150–166. https://doi.org/10.1111/1467-9582.00056

Indefrey, P. (2006). It is time to work toward explicit processing models for native and second language speakers. *Applied Psycholinguistics, 27*, 66–69. https://doi.org/10.1017/S0142716406280032

Jänke, L. (2012). The relationship between music and language. *Frontiers in Psychology, 3*. https://doi.org/10.3389/fpsyg.2012.00123

Jantzen, M. G., Howe, B. M., & Jantzen, K. J. (2014) Neurophysiological evidence that musical training influences the recruitment of right hemispheric homologues for speech perception. *Frontiers in Psychology, 3*(5), 171. https://doi.org/10.3389/fpsyg.2014.00171

Jiang, N., & Forster, K. (2001). Cross-language priming asymmetries in lexical decision and episodic recognition. *Journal of Memory and Language, 44*, 32–51. https://doi.org/10.1006/jmla.2000.2737

Johnson, J. S., & Newport, E. L. (1989). Critical period effects in second language learning: The influence of maturational state on the acquisition of English as a second language. *Cognitive Psychology, 21*, 60–99. https://doi.org/10.1016/0010-0285(89)90003-0

Juffs, A., & Harrington, M. (2011). Aspects of working memory in L2 learning. *Language Teaching, 44*, 137–166. https://doi.org/10.1017/S0261444810000509

Just, M. A., & Carpenter, P. A. (1992). A capacity theory of comprehension: Individual differences in working memory. *Psychological Review, 99*(1), 122–149. https://doi.org/10.1037/0033-295X.99.1.122

Kane, M. J., Conway, A. R. A., Hambrick, D. Z., & Engle, R. W. (2007). Variation in working memory capacity as variation in executive attention and control. In A. R. A. Conway, C. Jarrold, M. J. Kane, A. Miyake, & J. N. Towse (Eds.), *Variation in Working Memory* (pp. 21–49). Oxford University Press.

Karlsgodt, K. H., Bachman, P., Winkler, A. M., Bearden, C. E., & Glahn, D. C. (2011). Genetic influence on the working memory circuitry: Behavior, structure, function and extensions to illness. *Behavioral and Brain Research, 225*(2), 610–622. https://doi.org/10.1016/j.bbr.2011.08.016

Karlsgodt, K. H., Kochunov, P., Winkler, A. M, ... & Glahn, D. C. (2010). A multimodal assessment of the genetic control over working memory. *Journal of Neuroscience, 30*(24), 8197–8202. https://doi.org/10.1523/JNEUROSCI.0359-10.2010

Kidd, E., Donnelly, S., & Christiansen, M. H. (2018). Individual differences in language acquisition and processing. *Trends in Cognitive Sciences, 22*(2), 154–169. https://doi.org/10.1016/j.tics.2017.11.006

Kim, A. E., Oines, L., & Miyake, A. (2018). Individual differences in verbal working memory underlie a tradeoff between semantic and structural processing difficulty during language comprehension: An ERP investigation. *Journal of Experimental Psychology: Learning, Memory, and Cognition, 44*(3), 406–420. https://doi.org/10.1037/xlm0000457

Kim, J. H., & Christianson, K. (2013). Sentence complexity and working memory effects in ambiguity resolution. *Journal of Psycholinguistic Research, 42*(5), 393–411. https://doi.org/10.1007/s10936-012-9224-4

Klencklen, G., Lavenex, P. B., Grandner, C., & Lavenex, P. (2017). Working memory decline in normal aging: Is it really worse in space than in color? *Learning and Motivation, 57*, 48–60. https://doi.org/10.1016/j.lmot.2017.01.007

Knowles, E. E., Mathias, S. R., McKay, D. R., Sprooten, E., Blangero, J., Almasy, L., & Glahn, D. C. (2014). Genome-wide analyses of working-memory ability:

A review. *Current Behavioral Neuroscience Reports, 1*(4), 224–233. https://doi.org/10.1007/s40473-014-0028-8

Köpke, B., & Genevska-Hanke, D. (2018). First language attrition and dominance: Same same or different? *Frontiers in Psychology, 9,* 1963. https://doi.org/10.3389/fpsyg.2018.01963

Köpke, B., & Schmid, M. S. (2004). Language attrition: The next phase. In M. S. Schmid, B. Köpke, M. Keijzer, & L. Weilemar (Eds.), *First Language Attrition: Interdisciplinary Perspectives on Methodological Issues* (pp. 1–45). John Benjamins. https://doi.org/10.1075/sibil.28.02kop

Kormos, J., & Csizér, K. (2014). The interaction of motivation, self-regulatory strategies, and autonomous learning behavior in different learner groups. *TESOL Quarterly, 48*(2), 275–299. https://doi.org/10.1002/tesq.129

Kornder, L., & Mennen, I. (2021). Longitudinal developments in bilingual second language acquisition and first language attrition of speech: The case of Arnold Schwarzenegger. *Languages, 6*(2), 61, 1–25. https://doi.org/10.3390/languages6020061

Kousaie S., Chen, J. K., Baum, S. R., Phillips, N. A., Titone, D., & Klein, D. (2021). Bilingual language experience and the neural underpinnings of working memory. *Neuropsychologia, 163,* 1–9. https://doi.org/10.1016/j.neuropsychologia.2021.108081

Kovas, Y., Voronin, I., Kaydalov, A., Malykh, S. B., Dale, P. S., & Plomin, R. (2013). Literacy and numeracy are more heritable than intelligence in primary school. *Psychological Science, 24*(10), 2048–2056. https://doi.org/10.1177/0956797613486982

Kraus, N., & Chandrasekaran, B. (2010). Music training for the development of auditory skills. *Nature Reviews Neuroscience, 11,* 599–605.

Kremen, W. S., Jacobsen, K. C., Xian, H., Eisen, S. A., Eaves, L. J., Tsuang, M. T., & Lyons, M. J. (2007). Genetics of verbal working memory processes: A twin study of middle-aged men. *Neuropsychology, 21*(5), 569–580. https://doi.org/10.1037/0894-4105.21.5.569

Kroll, J. F., Bogulski, C. A., & McClain, R. (2012). Psycholinguistic perspectives on second language learning and bilingualism: The course and consequence of cross-language competition. *Linguistic Approaches to Bilingualism, 2*(1), 1–24. https://doi.org/10.1075/lab.2.1.01kro

Lai, C. S. L., Fisher, S. E., Hurst, J. A., Vargha-Khadem, F., & Monaco, A. P. (2001). A forkhead-domain gene is mutated in a severe speech and language disorder. *Nature, 413,* 519–523. https://doi.org/10.1038/35097076

Lamb, M., Csizér, K., Henry, A., & Ryan, S. (2019). *The Palgrave Handbook of Motivation for Language Learning.* Palgrave Macmillan.

Laufer, B. (2003). The influence of L2 on L1 collocational knowledge and on L1 lexical diversity in free written expression. In V. Cook (Ed.), *Effects of the Second Language on the First* (pp. 19–31). Multilingual Matters.

Leaderbrand, K., Chen, H. J., Corcoran, K. A., Guedea, A. L., Jovasevic, V., Wess, J., & Radulovic, J. (2016). Muscarinic acetylcholine receptors act in synergy to facilitate learning and memory. *Learning & Memory, 23*(11), 631–638. https://doi.org/10.1101/lm.043133.116

Lee, C.Y., & Hung, T. H. (2008). Identification of Mandarin tones by English-speaking musicians and non-musicians. *Journal of the Acoustical Society of America, 124,* 3235–3248. https://doi.org/10.1121/1.2990713

Li, S., Hiver, P., & Papi, M. (2022). *The Routledge Handbook of Second Language Acquisition and Individual Differences.* Routledge. https://doi.org/10.4324/9781003270546

Li, W., & Hartshorne, J. K. (2022). Even simultaneous bilinguals do not reach monolingual levels of proficiency in syntax. *Languages, 7,* 293. https://doi.org/10.3390/languages7040293

Linck, J. A., Osthus, P., Koeth, J. T., & Bunting, M. F. (2014). Working memory and second language comprehension and production: A meta-analysis. *Psychonomic Bulletin & Review, 21*(4), 861–883. https://doi.org/10.3758/s13423-013-0565-2

Logie, R. H., Belletier, C., & Doherty, J. M. (2021). Integrating theories of working memory. In R. Logie, V. Camos, & N. Cowan (Eds.), *Working Memory: The State of the Science* (pp. 389–430). Oxford Academic. https://doi.org/10.1093/oso/9780198842286.003.0014

Long, M. H. (1990). Maturational constraints on language development. *Studies in Second Language Acquisition, 12,* 251–285. https://doi.org/10.1017/S0272263100009165

Lybeck, K. (2002). Cultural identification and second language pronunciation of Americans in Norway. *The Modern Language Journal, 86*(2), 174–191. https://doi.org/10.1111/1540-4781.00143

MacIntyre, P. D., & Serroul, A. (2014). Motivation on a per-second timescale: Examining approach-avoidance motivation during L2 task performance. In Z. Dörnyei, P. D. Macintyre, & A. Henry (Eds.), *Motivational Dynamics in Language Learning* (pp. 109–138). Multilingual Matters.

MacWhinney, B. (2001). The competition model: The input, the context, and the brain. In P. Robinson (Ed.), *Cognition and Second Language Instruction* (pp. 69–90). Cambridge University Press. https://doi.org/10.1017/CBO9781139524780.005

Madden, D. J., Bennett, I. J., Burzynska, A., Potter, G. G., Chen, N. K., & Song, A. W. (2012) Diffusion tensor imaging of cerebral white matter integrity in cognitive aging. *Biochimica et Biophysica Acta (BBA): Molecular Basis of Disease, 1822*(3), 386–400. https://doi.org/10.1016/j.bbadis.2011.08.003

Magnusson, J. E., & Stroud, C. (2012). High proficiency in markets of performance: A sociocultural approach to nativelikeness. *Studies in Second Language Acquisition, 34*(2), 321–345. https://doi.org/10.1017/S0272263112000071

Mamiya, P. C., Richards, T. L., Coe, B. P., & Kuhl, P. K. (2016). Brain white matter structure and *COMT* gene are linked to second-language learning in adults. *Proceedings of the National Academy of Sciences, 113*(26), 7249–7254. https://doi.org/10.1073/pnas.1606602113

Mankel, K., & Bidelman, G. M. (2018). Inherent auditory skills rather than formal music training shape the neural encoding of speech. *Proceedings of the National Academy of Sciences, 115*(51), 13129–13134. https://doi.org/10.1073/pnas.1811793115

Marie, C., Delogu, F., Lampis, G., Olivetti Belardinelli, M., & Besson, M. (2011). Influence of musical expertise on segmental and tonal processing in Mandarin Chinese. *Journal of Cognitive Neuroscience, 23*, 2701–2715. https://doi.org/10.1162/jocn.2010.21585

Marques, C., Moreno, S., Luís Castro, S., & Besson, M. (2007). Musicians detect pitch violation in a foreign language better than nonmusicians: Behavioral and electrophysiological evidence. *Journal of Cognitive Neuroscience, 19*,1453–1463. https://doi.org/10.1162/jocn.2007.19.9.1453

Masgoret, A. M., & Gardner, R. C. (2003). Attitudes, motivation and second language learning: A meta-analysis of studies conducted by Gardner and his associates. *Language Learning, 53*(1), 123–163. https://doi.org/10.1111/1467-9922.00212

Mashburn, C.A., Tsukahara, J. S., & Engle, R. W. (2021). Individual differences in attention control: Implications for the relationship between working memory capacity and fluid intelligence (pp. 175–211). Oxford Academic. https://doi.org/10.1093/oso/9780198842286.003.0007

Mayberry, R. I., & Kluender, R. (2018). Rethinking the critical period for language: New insights into an old question from American Sign Language. *Bilingualism: Language and Cognition, 21*, 886–905. https://doi.org/10.1017/S1366728917000724

McMullen, E., & Saffran, J. R. (2004). Music and language: A developmental comparison. *Music Perception, 21*(3), 289–311. https://doi.org/0.1525/mp.2004.21.3.289

McNab, F., Zeidman, P., Rutledge, R. B. ..., & Dolan, R. J. (2015). Age-related changes in working memory and the ability to ignore distraction. *Publications of the National Academy of Sciences, 112*(20), 6515–6518. https://doi.org/10.1073/pnas.1504162112

Meulman, N., Wieling, M., Sprenger, S. A., Stowe, L. A., & Schmid, M. S. (2015). Age effects in L2 grammar processing as revealed by ERPs and how (not) to study them. *PLoS ONE, 10*(12). https://doi.org/10.1371.journal.pone.0143328

Michalczyk, K., Malstädt, N., Worgt, M., Könen, T., & Hasselhorn, M. (2013). Age differences and measurement invariance of working memory in 5- to 12-year-old children. *European Journal of Psychological Assessment, 29*(3), 220–229. https://doi.org/10.1027/1015-5759/a000149

Mountford, H. S., Braden, R., Newbury, D. F., & Morgan, A. T. (2022). The genetic and molecular basis of developmental language disorder: A review. *Children, 9*(5), 586. https://doi.org/10.3390/children9050586

Muñoz-Pradas, R., Díaz-Palacios, M., Rodriguez-Martinez, E., & Gómez, C. M. (2021). Order of maturation of the components of working memory from childhood to emerging adulthood. *Current Research in Behavioral Sciences, 2*, 100062. https://doi.org/10.1016/j.crbeha.2021.100062

Naiman, N., Fröhlich, M., Stern, H. H., & Todesco, A. (1978). *The Good Language Learner.* Ontario Institute for Studies in Education (OISE).

Newbury, D. F., & Monaco, A. P. (2010). Genetic advances in the study of speech and language disorders. *Neuron, 68*(2), 309–320. https://doi.org/10.1016/j.neuron.2010.10.001

Newman, A. J., Tremblay, A., Nichols, E. S., Neville, H. J., & Ullman, M. T. (2012). The influence of language proficiency on lexical semantic processing in native and late learners of English. *Journal of Cognitive Neuroscience, 24*(5), 1205–1223. https://doi.org/10.1162/jocn_a_00143

Newman, E. L., Gupta, K., Climer, J. R., Monaghan, C. K., & Hasselmo, M. E. (2012). Cholinergic modulation of cognitive processing: Insights drawn from computational models. *Frontiers in Behavioral Neuroscience, 6*(24). https://doi.org/10.3389/fnbeh.2012.00024

Newport, E. L., Bavelier, D., & Neville, H. J. (2001). Critical thinking about critical periods: Perspectives on a critical period for language acquisition. In E. Dupoux (Ed.), *Language, Brain and Cognitive Development: Essays in Honor of Jacques Mehler* (pp. 481–502). MIT Press.

Nie, P., Wang, C., Rong, G., Du, B., Lu, J., Li, S., Putkinen, V., Tao, S., & Tervaniemi, M. (2022). Effects of music training on the auditory working memory of Chinese-speaking school-aged children: A longitudinal study. *Frontiers in Psychology, 12.* https://doi.org/10.3389/fpsyg.2021.770425

Noels, K. A., Lou, N. M., Lascano, D. I. V., Chaffee, K. E., Dincer, A., Zhang, Y. S. D., & Zhang, X. (2019). Self-determination and motivated engagement in language learning. In M. Lamb, K. Csizér, A. Henry, & S. Ryan (Eds.), *The Palgrave Handbook of Motivation for Language Learning* (pp. 95–115). Palgrave Macmillan.

Norton, B. (2000). *Identity and Language Learning: Gender, Ethnicity and Educational Change.* Longman.

Novén, M., Olsson, H., Helms, G., Horne, M., Nilsson, M., & Roll, M. (2021). Cortical and white matter correlates of language-learning aptitudes. *Human Brain Mapping, 42*(15), 5037–5050.

Nyberg L., Lövdén, M., Riklund K., Lindenberger U., and Bäckman L. (2012). Memory, aging and brain maintenance. *Trends in Cognitive Sciences, 16,* 292–305. https://doi.org/10.1016/j.tics.2012.04.005

Ogg, M., & Slevc, R. L. (2019). Neural mechanisms of music and language. In G. I. de Zubicary & N. O. Schiller (Eds.), *The Oxford Handbook of Neurolinguistics* (pp. 907–952). Oxford University Press. https://doi.org/10.1093/oxfordhb/9780190672027.013.35

Okbay, A., Beauchamp, J., Fontana, M., et al. (2016). Genome-wide association study identifies 74 loci associated with educational attainment. *Nature, 533,* 539–542. https://doi.org/10.1038/nature17671

Papi, M., & Hiver, P. (2020). Language learning motivation as a complex dynamic system: A global perspective of truth, control and value. *The Modern Language Journal, 104*(1), 209–232. https://doi.org/10.1111/modl.12624

Papi, M., & Teimouri, Y. (2012). Dynamics of selves and motivation: A cross-sectional study in the EFL context of Iran. *IRAL: International Journal of Applied Linguistics, 22*(3), 287–309. https://doi.org/10.1111/j.1473-4192.2012.00312.x

Parbery-Clark, A., Skoe, E., Lam, C., & Kraus, N. (2009b). Musician enhancement for speech in noise. *Ear and Hearing, 30*, 653–661. https://doi.org/10.1097/AUD.0b013e3181b412e9

Parbery-Clark, A., Strait, D. L., Anderson, S., Hittner, E., and Kraus, N. (2011b). Musical experience and the aging auditory system: Implications for cognitive abilities and hearing speech in noise. *PLoS ONE, 6*(5), e10182.https://doi.org/10.1371/journal.pone.0018082

Parbery-Clark, A., Strait, D. L., & Kraus, N. (2011a). Context-dependent encoding in the auditory brainstem subserves enhanced speech-in-noise perception in musicians. *Neuropsychologia, 49*, 3338–3345. https://doi.org/10.1016/j.neuropsychologia.2011.08.007

Parbery-Clark, A., Tierney, A., Strait, D. L., and Kraus, N. (2012). Musicians have fine-tuned neural distinction of speech syllables. *Neuroscience, 219*, 111–119. https://doi.org/10.1016/j.neuroscience.2012.05.042

Park, D. C. (1999). The basic mechanisms accounting for age-related decline in cognitive function. In D. Park & N. Schwarz (Eds.), *Cognitive Aging: A Primer* (pp. 3–21). Psychology Press. https://doi.org/10.4324/9780203727027

Patel, A. D. (2011). Why would musical training benefit the neural encoding of speech? The OPERA hypothesis. *Frontiers in Psychology, 2*, 1–14. https://doi.org/10.3389/fpsyg.2011.00142

Patel, A. D. (2014). Can nonlinguistic musical training change the way the brain processes speech? The expanded OPERA hypothesis. *Hearing Research, 308*, 98–108. https://doi.org/10.1016/j.heares.2013.08.011

Peretz, I., Nguyen, S., & Cummings, S. (2011). Tone language fluency impairs pitch discrimination. *Frontiers in Psychology, 2*.https://doi.org/10.3389/fpsyg.2011.00145

Pesnot Lerousseau, J., Hidalgo, C., & Schön, D. (2020). Musical training for auditory rehabilitation in hearing loss. *Journal of Clinical Medicine, 9*(4), 1058. https://doi.org/10.3390/jcm9041058

Pfenninger, S. E., & Singleton, D. (2019). Starting age overshadowed: The primacy of differential environmental and family support effects on second language attainment in an instructional context. *Language Learning, 69*(S1), 207–234. https://doi.org/10.1111/lang.12318

Picciotto, M. R., Higley, M. J., & Mineur, Y. S. (2012). Acetylcholine as a neuromodulator: Cholinergic signaling shapes nervous system function and behavior. *Neuron, 76*(1), 116–129. https://doi.org/10.1016/j.neuron.2012.08.036

Piller, I. (2002). Passing for a native speaker: Identity and success in second language learning. *Journal of Sociolinguistics, 6*, 179–206. https://doi.org/10.1111/1467-9481.00184

Pliatsikas, C., Veríssimo, J., Babcock, L., Pullman, M. Y., Glei, D. A., Weinstein, M., Goldman, N., & Ullman, M. T. (2019). Working memory in older adults declines with age, but is modulated by sex and education. *Quarterly Journal of Experimental Psychology, 72*(6), 1308–1327. https://doi.org/10.1177/1747021818791994

Prins, N., & Scheltens, P. (2015). White matter hyperintensities, cognitive impairment and dementia: An update. *Nature Reviews Neurology, 11,* 157–165. https://doi.org/10.1038/nrneurol.2015.10

Qi, Z., Beach, S. D., Finn, A. S., Minas, J., Goetz, C., Chan, B., & Gabrieli, J. D. E. (2017). Native-language N400 and P600 predict dissociable language-learning abilities in adults. *Neuropsychologia, 98,* 177–191. https://doi.org/10.1016/j.neuropsychologia.2016.10.005

Rebuschat, P., Rohrmeier, M., Hawkins, J. A., & Cross, I. (2012). *Language and Music as Cognitive Systems.* Oxford University Press.

Reiterer, S. M. (2019). Neuro-psycho-cognitive markers for pronunciation/speech imitation as language aptitude. In Z. Wen, P. Skehan, A. Biedroń, S. Li, & S. L. Sparks (Eds.), *Language Aptitude: Advancing Theory, Testing, Research and Practice* (pp. 277–299). Taylor & Francis.

Rimfeld, K., Malanchini, M., Krapohl, E., et al. (2018). The stability of educational achievement across school years is largely explained by genetic factors. *npj Science of Learning, 3*(16). https://doi.org/10.1038/s41539-018-0030-0

Robinson, P. (2002). Learning conditions, aptitude complexes and SLA: A framework for research and pedagogy. In P. Robinson (Ed.), *Individual Differences and Instructed Language Learning* (pp. 113–133). John Benjamins.

Rubenfeld, S., Clément, R., Lussier, D., Lebrun, M., & Auger, R. (2006). Second language learning and cultural representations: Beyond competence and identity. *Language Learning, 56(*4), 609–632. https://doi.org/10.1111/j.1467-9922.2006.00390.x

Rubin, J. (1975). What the "good language learner" can teach us. *TESOL Quarterly, 9,* 41–51. https://doi.org/10.2307/3586011

Saito, K. (2024). Age effects in spoken second language vocabulary attainment beyond the critical period. *Studies in Second Language Acquisition, 46*(1), 3–27. https://doi.org/10.1017/S0272263122000432

Salthouse, T. A. (2011) Neuroanatomical substrates of age-related cognitive decline. *Psychological Bulletin, 137*(5), 753–784. https://doi.org/10.1037/a0023262

Schmidt, R. (1983). Interaction, acculturation and the acquisition of communication competence: A case study of an adult. In N. Wolfson & E. Judd (Eds.), *Sociolinguistics and Language Acquisition* (pp. 137–174). Newbury House.

Schneiderman, E. I., & Desmarais, C. (1988). A neuropsychological substrate for talent in second-language acquisition. In L. K. Obler & D. Fein (Eds.), *The Exceptional Brain. Neuropsychology of Talent and Special Abilities* (pp. 103–126). The Guilford Press.

Schwartz, A. I., & Kroll, J. F. (2006). Bilingual lexical activation in sentence context. *Journal of Memory and Language, 55*(2), 197–212. https://doi.org/10.1016/j.jml.2006.03.004

Schwering, S. C., & MacDonald, M. C. (2020) Verbal working memory as emergent from language comprehension and production. *Frontiers in Human Neuroscience, 14*, 68. https://doi.org/10.3389/fnhum.2020.00068

Schwieter, J. W., & Wen, Z. (2022). *The Cambridge Handbook of Working Memory in Language and Linguistics.* Cambridge University Press.

Setter, J., & Jenkins, J. (2005). State-of-the-Art Review Article. *Language Teaching, 38*(1), 1–17. https://doi.org/10.1017/S026144480500251X

Shook, A., & Marian, V. (2013). The bilingual language interaction network for comprehension of speech. *Bilingualism: Language and Cognition, 16*(2), 304–324. https://doi.org/10.1017/S1366728912000466

Shook, A., Marian, V., Bartolotti, J., & Schroeder, S. R. (2013). Musical experience influences statistical learning of a novel language. *The American Journal of Psychology, 126*(1), 95–104. https://doi.org/10.5406/amerjpsyc.126.1.0095

Singleton, D., & Leśniewska, J. (2021). The critical period hypothesis for L2 acquisition: An unfalsifiable embarrassment? *Languages, 6*(3), 149. https://doi.org/10.3390/languages6030149

Skehan, P. (1998). *A Cognitive Approach to Language Learning.* Oxford University Press.

Skehan, P. (2016). Foreign language aptitude. In G. Granena, D. O. Jackson, & Y. Yilmaz (Eds.), *Cognitive Individual Differences in L2 Processing and Acquisition* (pp. 381–395). John Benjamins.

Slevc, L. R., Davey, N. S., Buschkuehl, M., & Jaeggi, S. M. (2016). Tuning the mind: Exploring the connections between musical ability and executive functions. *Cognition, 152,* 199–211. https://doi.org/10.1016/j.cognition.2016.03.017

Slevc, L. R., & Miyake, A. (2006). Individual differences in second-language proficiency: Does musical ability matter? *Psychological Science, 17*(8), 675–681. https://doi.org/s10.1111/j.1467-9280.2006.01765.x

Slevc, L. R., & Okada, B. M. (2015). Processing structure in language and music: A case for shared reliance on cognitive control. *Psychonomic Bulletin & Review, 22*(3), 637–652. https://doi.org/10.3758/s13423-014-0712-4

Smith, A. W., Holden, K. R., Dwivedi, A., Dupont, B. R., & Lyons, M. J. (2015). Deletion of 16q24.1 supports a role for the ATP2C2 gene in specific language impairment. *Journal of Child Neurology, 30,* 517–521. https://doi.org/10.1177/0883073814545113

Sparks, R. L. (2012). Individual differences in L2 learning and long-term L1-L2 relationships. *Language Learning, 64*(s2), 5–27. https://doi.org/10.1111/j.1467-9922.2012.00704.x

Sparks, R. L., Patton, J., & Ganschow, L. (2012). Profiles of more and less successful L2 learners: A cluster analysis study. *Learning and Individual Differences, 22*(4), 463–472. https://doi.org/10.1016/j.lindif.2012.03.009

Sparks, R. L., Patton, J., Ganschow, L., Humbach, N., & Javorsky, J. (2006). Native language predictors of foreign language proficiency and foreign language aptitude. *Annals of Dyslexia, 56*(1), 129–160. https://doi.org/10.1007/s11881-006-0006-2

Steinhauer, K., & Kasparian, K. (2020). Brain plasticity in adulthood: ERP evidence for L1-attrition in lexicon and morphosyntax after predominant L2 use. *Language Learning, 70*(S2), 171–193.

Stevick, E. (1989). Success with foreign languages: Seven who achieved it and what worked for them. Prentice Hall.

Stromswold, K. (1998). Genetics of spoken language disorders. *Human Biology, 70,* 297–324. www.ncbi.nlm.nih.gov/pubmed/9549241

Swanson, H. L. (2017). Verbal and visual-spatial working memory: What develops over a life span? *Developmental Psychology, 53*(5), 971–995. https://doi.org/10.1037/dev0000291

Swets, B., Desmet, T., Hambrick, D. Z., & Ferreira, F. (2007). The role of working memory in syntactic ambiguity resolution: A psychometric approach. *Journal of Experimental Psychology: General, 136*(1), 64–81. https://doi.org/10.1037/0096-3445.136.1.64

Takahashi, N., Nishimura, T., Harada, T., Okumura, A., Iwabuchi, T., Rahman, M. S., Kuwabara, H., Takagai, S., Nomura, Y., Takei, N., & Tsuchiya, K. J. (2021). Association between genetic risks for obesity and working memory in children. *Frontiers in Neuroscience, 22*(15), 749230. https://doi.org/10.3389/fnins.2021.749230

Thevenon, J., Souchay, C., Seabold, G., ..., & Faivre, L. (2016). Heterozygous deletion of the *LRFN2* gene is associated with working memory deficits. *European Journal of Human Genetics, 24,* 911–918. https://doi.org/10.1038/ejhg.2015.221

Troutman, S. B. W., & Diaz, M. T. (2020). White matter disconnection is related to age-related phonological deficits. *Brain Imaging and Behavior,* ***14***(5), 1555–1565. doi. 10.1007/s11682-019-00086-8

Tucker-Drob, E. M., Briley, D. A., & Harden, K. P. (2013). Genetic and environmental influences on cognition across development and context. *Current Directions in Psychological Science, 22*(5), 349–355. https://doi.org/10.1177/0963721413485087

Turker, S., & Reiterer, S. (2021). Brain, musicality, and language aptitude: A complex interplay. *Annual Review of Applied Linguistics, 41,* 95–107. https://doi.org/10.1017/S0267190520000148

Ullman, M. T., Miranda, R. A., & Travers, M. L. (2007). Sex differences in the neurocognition of language. In J. B. Becker et al. (Eds.), *Sex Differences in the Brain: From Genes to Behavior* (pp. 291–310). Oxford Academic. https://doi.org/10.1093/acprof:oso/9780195311587.003.0015

Unsworth, N., Heitz, R., Schrock, J. C., & Engle, R. (2005). An automated version of the operation span task. *Behavioral Research Methods, 37,* 498–505. https://doi.org/10.3758/BF03192720

Ushioda, E. (2009). A person-in-context relational view of emergent motivation, self and identity. In Z. Dörnyei & E. Ushioda (Eds.), *Motivation, Language Identity and the L2 Self* (pp. 215–228). Multilingual Matters.

Ushioda, E. (2016). Language learning motivation through a small lens: A research agenda. *Language Teaching, 49*(4), 564–577. https://doi.org/10.1017/S0261444816000173

Ushioda, E. (2019a). Motivation in second language acquisition. In C. Chapelle (Ed.), *The Encyclopedia of Applied Linguistics* (pp. 1–6). Wiley. https://doi.org/10.1002/9781405198431

Ushioda, E. (2019b). Researching L2 motivation: Past, present and future. In M. Lamb et al. (Eds.), *The Palgrave Handbook of Motivation for Language Learning* (pp. 661–682). Palgrave Macmillan.

van der Lely, H. K., & Stollwerck, L. (1996). A grammatical specific language impairment in children: An autosomal dominant inheritance? *Brain and Language, 52*(3), 484–504. https://doi.org/10.1006/brln.1996.0026

Van Hedger, S. C., Johnsrude, I. S., & Batterink, L. J. (2022) Musical instrument familiarity affects statistical learning of tone sequences. *Cognition, 218*, 104949. https://doi.org/10.1016/j.cognition.2021.104949

van Hell, J. G. (2023). The neurocognitive underpinnings of second language processing: Knowledge gains from the past and future outlook. *Language Learning, 73*(S2), 95–138.

van Hell, J., & Abdollahi, F. (2017). Individual variation in syntactic processing in the second language: behavioral and electrophysiological approaches. In E. Segers & P. van de Broek (Eds.), *Developmental Perspectives in Written Language and Literacy* (pp. 257–273). John Benjamins. https://doi.org/10.1075/z.206.16van

van Hell, J. G., & Dijkstra, T. (2002). Foreign language knowledge can influence native language performance in exclusively native contexts. *Psychonomic Bulletin & Review, 9*(4), 780–789. https://doi.org/10.3758/BF03196335

van Hell, J. G., & Tanner, D. (2012). Second language proficiency and cross-language lexical activation. *Language Learning, 62*(S2), 148–171. https://doi.org/10.1111/j.1467-9922.2012.00710.x

van Leeuwen, M., van den Berg, S. M., Hoekstra, R. A., & Boomsma, D. I. (2009). The genetic and environmental structure of verbal and visuospatial memory in young adults and children. *Neuropsychology, 23*(6), 792–802. https://doi.org/10.1037/a0016526

Vanhove, J. (2013). The critical period hypothesis in second language acquisition: A statistical critique and a reanalysis. *PLoS ONE, 8*(7), e69172. https://doi.org/10.1371/journal/.pone.0069172

Walters, K. (2011). Gendering French in Tunisia: Language ideologies and nationalism. *International Journal of the Sociology of Language, 211*, 83–111. https://doi.org/10.1515/ijsl.2011.039

Waninge, F., Dörnyei, Z., & de Bot, K. (2014). Motivational dynamics in language learning: Change, stability and context. *The Modern Language Journal, 9*(1), 704–723. https://doi.org/10.1111/modl.12118

Weiss, M. W., & Bidelman, G. M. (2015). Listening to the brainstem: Musicianship enhances intelligibility of subcortical representations for speech. *The Journal of Neuroscience: The Official Journal of the Society for Neuroscience, 35*(4), 1687–1691. https://doi.org/10.1523/JNEUROSCI.3680-1

Wen, Z. (2016). *Working Memory and Second Language Learning: An Integrated Approach.* Multilingual Matters.

Wen, Z. (2018). Working memory in first and second language: A comprehensive bibliography. Expanded and updated (November 24, 2018) from Wen (2016). www.academia.edu/12198656

Wen, Z. (2021). Working memory. In H. Mohebbi & C. Coombe (Eds.), *Research Questions in Language Education and Applied Linguistics* (pp. 279–284). Springer.

Werker, J. F., & Hensch, T. K. (2015). Critical periods in speech perception: New directions. *Annual Review of Psychology, 66,* 173–196. https://doi.org/10.1146/annurev-psych-010814-015104

White, E. J., Hutka, S. A., Williams, L. J., & Moreno, S. (2013). Learning, neural plasticity and sensitive periods: Implications for language acquisition, music training and transfer across the lifespan. *Frontiers in Psychology, 7.* https://doi.org/10.3389/fnsys.2013.00090

Wild-Wall, N., Falkenstein, M., & Gajewski, P. D. (2011). Age-related differences in working memory performance in a 2-back task. *Frontiers in Psychology, 2,* 186. https://doi.org/10.3389/fpsyg.2011.00186

Wing, H. D. (1968). *Tests of Musical Ability and Appreciation: An Investigation into the Measurement, Distribution, and Development of Musical Capacity* (2nd ed.). Cambridge University Press.

Wong, P. C. M., Kang, X., So, H.-C., & Choy, K. W. (2022). Contributions of common genetic variants to specific languages and to when a language is learned. *Nature Scientific Reports, 12,* 580. https://doi.org/10.1038/s41598-021-04163-1

Wong, P. C., Morgan-Short, K., Ettlinger, M., & Zheng, J. (2012). Linking neurogenetics and individual differences in language learning: The dopamine hypothesis. *Cortex, 48*(9), 1091–1102. https://doi.org/10.1016/j.cortex.2012.03.017

Wright, M., De Geus, E., Ando, J., Luciano, M., Posthuma, D., Ono, Y., ..., & Boomsma, D. (2001). Genetics of cognition: Outline of a collaborative twin study. *Twin Research, 4*(1), 48–56. https://doi.org/10.1375/twin.4.1.48

You, C., & Dörnyei, Z. (2016). Language learning motivation in China: Results of a large-scale stratified survey. *Applied Linguistics, 37*(4), 495–519. https://doi.org/10.1093/applin/amu046

Young, R., & Perkins, K. (1995). Cognition and conation in second language acquisition theory. *IRAL: International Review of Applied Linguistics in Language Teaching, 33*(2), 142–164.

Zhang, F., Roland, C., Rasul, D., Cahn, S., Liang, C., & Valencia, G. (2019.) Comparing musicians and non-musicians in signal-in-noise perception. *International Journal of Audiology, 58*(11), 717–723. https://doi.org/10.1080/14992027.2019.1623424

Zhang, L., Xie, S., Li, Y., Shu, H., & Zhang, Y. (2020). Perception of musical melody and rhythm as influenced by native language experience. *Journal of the Acoustical Society of America, 147,* EL385–EL390. https://doi.org/10.1121/10.0001179

Zhang, R., & Zou, D. Z. (2022) Self-regulated second language learning: A review of types and benefits of strategies, modes of teacher support, and pedagogical implications. *Computer Assisted Language Learning, 37*(4), 720–765. https://doi.org/10.1080/09588221.2022.2055081

Zhang, Y., & Papi, M. (2021). Motivation and second language pragmatics: A regulatory focus perspective. *Frontiers in Psychology, 12*, 753605. https://doi.org/10.3389/fpsyg.2021.753605

Zhang, Y., Ridchenko, M., Hayashi, A., & Hamrick, P. (2021). Episodic memory contributions to second language lexical development persist at higher proficiencies. *Applied Cognitive Psychology, 35*(2), 1356–1361. https://doi.org/10.1002/acp.3865

Zhao, X., Xiao, W., & Zhang, J. (2022). L2 motivational self system, international posture and the sustainable development of L2 proficiency in the COVID-19 era: A case of English majors in China. *Sustainability, 14*, 8087. https://doi.org/10.3390/su14138087

Zheng, C., Liang, J. C., Li, M., & Tsai, C. C. (2018). The relationship between English language learners' motivation and online self-regulation: A structural equation modelling approach. *System, 76*, 144–157. https://doi.org/10.1016/j.system.2018.05.003

Zhou, H., Rossi, S., & Chen, B. (2017). Effects of working memory capacity and tasks in processing L2 complex sentences: Evidence from Chinese-English bilinguals. *Frontiers in Psychology, 8*, 595. https://doi.org/10.3389/fpsyg.2017.00595

12

Longitudinal fMRI Analyses of Second Language Acquisition

Expanding the Repertoire of Imaging Techniques and Proficiency Measures in Future Studies

Edna Andrews and Alexandria Swaine

12.1 Introduction: Findings from Longitudinal Studies of L2/L3 Acquisition

It comes as no surprise that there are very few longitudinal fMRI studies of second/third language acquisition in the field of neuroimaging. Reasons for the dearth of longitudinal research include significant expense, difficulty in recruitment and retention of subjects, and continuity in lab settings to sustain the research to name a few. However, the data generated from longitudinal studies, where a new language is being acquired or intensive deepening of proficiency in a known language, are unmatched in their ability to examine individual variation and group-level system commonalities across time, and track dynamic change in neurological processing of languages as proficiency increases, while providing a unique opportunity for each subject be their own control.

The initial focus of our discussion is the six longitudinal blood oxygen level dependent (BOLD) fMRI studies of second language acquisition, immersion, and lexical development published in the past thirteen-year period (Andrews et al., 2013; Brice et al., 2021; Ghazi-Saidi & Ansaldo, 2017; Ghazi-Saidi et al., 2013; Grant et al., 2015, Raboyeau et al., 2010). The studies discussed here focus on different levels of linguistic processing in second languages – from single word tokens to ecologically valid discourse-level listening comprehension and reading. The analysis by García-Pentón and colleagues (2016) of cross-sectional imaging studies of bilingualism brings into focus key issues in comparing and contrasting these two types of approaches.

The term "longitudinal" may refer to any study that collects data on a cohort of subjects over a specific period of time ranging from weeks to years. The studies considered in the introductory section include time periods of nineteen days to thirty-six months. Section 12.2 will explore the research outcomes of these studies and their relevance in developing future studies. Special attention is given to the convergence of evidence from these studies, as well as unique insights that can be gained from longitudinal fMRI studies that are not found in cross-sectional studies.

A working definition of proficiency is also needed to contextualize the discussion and analysis of these six studies. If we look to the existing proficiency standards, frameworks, and testing procedures (including the interagency language roundtable scale (ILR), the Common European Framework of Reference for languages (CEFR), American Council on the Teaching of Foreign Languages (ACTFL)), the focus is not on a specific definition of proficiency itself, but rather on measurable empirical outcomes of linguistic ability across different categories (reading, writing, speaking, listening comprehension, grammar/lexicon) at different levels of demonstrable mastery that excludes self-reporting (Breiner-Sanders et al., 2002; Child et al. 1993; Davis et al., 2018; North, 2000). Thus, we would define proficiency as *the functional ability to produce and comprehend language in different modalities that can be assessed.* In addition, there is an important field of "proficiency-based learning" that focuses on supporting deep learning through equitable and rigorous learning opportunities (Fitzsimmons et al., 2017).[1]

12.1.1 Methods across Longitudinal Studies

We divide the studies into two sets – the first three focus on vocabulary acquisition using pictures and self-reporting by subjects, and the second three study broader categories of language processing with listening and/or reading components and include proficiency and other testing methods. There are important reasons for separating into two sets based on (1) the level of linguistic focus (vocabulary vs. discourse/speech acts) and (2) the length of time of the studies (vocabulary-based lasting from nineteen to thirty days vs. discourse-level lasting between four months to thirty-six months) and (3) studies that exclusively focus on vocabulary use self-reporting, while discourse level uses different types of behavioral testing (including CEFR, Gray oral reading tests (GORT-4), test of world reading efficiency (TOWRE-2), multilingual naming test (MINT), and Test de Vocabulario de Imagenes, Peabody (TVIP)). The CEFR framework is the only internationally recognized battery of proficiency tests represented in this group and includes over forty languages of the world. GORT-4 and TOWRE-2 focus on reading, MINT is a multilingual naming test used for healthy subjects and in dementia, TVIP is a vocabulary test for Spanish bilinguals based on the Peabody Picture Vocabulary Test.

The limitations of hemodynamic neuroimaging are important to keep in mind when attempting to interpret activations. Researchers always keep in mind that fMRI activations do not distinguish between inhibitory and excitatory cellular activity (Raichle, 2001: 12), and are actually correlates of correlates, that is, there is no direct neuronal measurement, but rather using deoxygenated hemoglobin as a correlate of blood flow, which in turn is a correlate of neuronal activity. As Huettel and colleagues (2004: 127–128) clearly state: "How does fMRI create images of neuronal activity? The short answer is that is does not! Instead, fMRI creates images of physiological activity that is correlated with neuronal activity."

12.1.2 L2 Lexical/Vocabulary Acquisition Studies

We begin our discussion with the longitudinal studies that focus on vocabulary acquisition – Raboyeau and colleagues (2010), and Ghazi-Saidi and colleagues (2013 and 2017). Each of these studies lasted between nineteen to thirty days and included twelve participants for Ghazi-Saidi and colleagues (2013), twenty-four participants for Ghazi-Saidi and colleagues (2017), and ten participants for Raboyeau and colleagues (2010). Ghazi-Saidi and colleagues (2013 and 2017) both use Persian or Spanish as the L1 and French as L2 for both, while Raboyeau and colleagues (2010) have L1/French, L2/English, and L3/Spanish. All three of these studies used picture stimuli in the scanner for evocation of vocabulary in the languages under study and conducted a total of two scans per study. Each scan had two runs per two languages. Other stimuli used in the scanner were nonlinguistic. Ghazi-Saidi and colleagues (2013) and Ghazi-Saidi and colleagues (2017) have the first scan one week after the beginning of the study, and the second a month later. Raboyeau and colleagues (2010) have the first scan five days after the beginning of the study, and the second between twelve to fifteen days later. Classroom learning or time studying was fifteen to twenty minutes per day for all three studies. All behavioral measures were based on participants' self-reporting.

Activation results in Raboyeau and colleagues (2010) were divided into cognate vs. noncognate in two time phases – the early-learning phase and the consolidation phase (2010: 2854). Regions of interest (ROIs) include both left and right hemisphere significant activations (including lIFC, lMFC, l/rACC, rVLPFC) (2010: 2854–2855). Mean response times for the L1 (French) showed no differences across the two phases, but L2 cognates and noncognates demonstrate faster retrieval–response times in the consolidation phase (2010: 2856).

Regions of interest in Ghazi-Saidi and colleagues (2013) follow Abutalebi and Green (2007) and include lFG, l/rPostCG, rSUP Parietal lobule, lACC, rIFG, rlimbic lobe, rparahippocampal gyrus, lFL, rSFG (Ghazi-Saidi et al., 2013: 60). Ghazi-Saidi and colleagues (2017) focus on functional networks and how they change between the learning phase (week 1) and the consolidation phase (week 4). Results show improvement in accuracy rates and response times, as well as increased network integration between pairs of lexemes that were close or distant in terms of languages. The two networks are the "language" network and the "cognitive control" network (2017: 5–6). Results show a decrease in functional integration/flow of information in the network and less reliance on the cognitive control network with repetition (2017: 10–11).

12.1.3 L2 Language Acquisition

There are three longitudinal studies of L2 acquisition that incorporate immersion (Andrews et al., 2013), transitional components (Grant et al., 2015), or both (Brice et al., 2021). In these studies, immersion is defined as intensive formal L2 learning that occurs in the country of the L1 and L2, while transitional includes formal L2 learning prior and/or during the period of the longitudinal study. The Andrews and colleagues (2013) longitudinal fMRI study of L2 acquisition lasted eleven months and was the first to combine fMRI scan data with internationally recognized proficiency testing (specifically Common European Framework Reference (CEFR)) to measure acquisition in a novel second language. The CEFR-based proficiency tests include five areas – listening comprehension, speaking, reading, writing, and grammar/lexicon. All subjects were L1 English speakers (a group of five subjects (three female/two male, nineteen to twenty years of age)) acquiring a new L2 – Russian. Acquisition of the L2 was accompanied by class-based language instruction and in-country immersion. All students had studied at least one other language in their educational background. All completed an initial fMRI scan before beginning acquisition of the new language.[2]

While the first three studies discussed used pictures to evoke individual lexemes in multiple languages, Andrews and colleagues (2013) use both auditory and visual linguistic stimuli in the scanner. Digitized auditory segments used thirty-second recordings of digitized auditory segments of discourse level speech between two speakers in four different languages (English, Georgian, Russian, Spanish) appearing in random order during the session. The stimuli are presented in the audio task in the following way: Language A (30 s), Rest (10 s), Language B (30 s). This sequence is repeated four times in each run.

Reading segments were presented on one screen at a time in black letters on a pale blue background with a minimum of fifteen seconds and a maximum of thirty seconds with ten seconds in between each reading segment. Texts were multiple paragraphs from a variety of sources on a set of themes. This self-paced procedure combined with eye-tracking provided more information on the subject's active reading for each screen of text.

Regions of interest (ROIs) were selected based on the neuroimaging literature for fMRI studies of language(s) (cf. Price, 2010), important cognitive neuroscience and linguistic research on language(s) and brain (cf. Poeppel, Hickok, Bolinger, de Bot, Paradis, and others) and also took into account findings from the extensive research using cortical stimulation mapping (CSM) (cf. Calvin & Ojemann, 1994; Corina et al., 2010; Ojemann et al.,

2010). The ROIs for the auditory condition were bilateral for six regions: MTG (medial temporal gyrus), STG (superior temporal gyrus), MFG (medial frontal gyrus), IFG (inferior frontal gyrus), PoG (postcentral gyrus), and PrG (precentral gyrus). The ROIs for the reading condition were bilateral for five regions: IFG, STG, MTG, lingual gyrus, and precuneus.

Using a multivariate analysis of covariance (MANCOVA) based on proficiency testing measurements and fMRI ROI measurements (six bilateral ROIs for the auditory task and five bilateral ROIs for the reading task) under the conditions of English-rest and Russian-rest, it was possible to correlate the change in proficiency with the change in activations found at each time period across specific regions of interest (Andrews, 2014: 164–191; Andrews et al., 2013).[3]

Results for auditory tasks show: (1) a significant score effect for the repeated measure of Russian-rest across scans. Pillai's trace has a p-value: $p = 0.01$; (2) a significant time effect where activation levels change across the three sets of longitudinal measurements in the language of acquisition condition only – Russian-rest (not English-rest); and (3) an internal statistical test of a region that should not be relevant in language acquisition confirmed the lack of effect. Similarly with the auditory task, Pillai's trace for the reading task for the Russian-rest condition has a significant p value $= 0.017$ but not for the English-rest condition.

The Brice and colleagues (2021) study focuses on reading and auditory lexeme (tokens) based on isolated single nouns and pronounceable pseudo-words in two contexts of language immersion using BOLD fMRI – L1 English speakers learning Hebrew in Israel and L1 Hebrew speakers learning English in the United States (2021: 107796/1–14). Behavioral tasks for Hebrew included printed word naming, pseudo-word reading, and cross-modal visual lexical decision with auditory priming, Hebrew oral passage reading, and comprehension.

English proficiency was measured using a test of word reading efficiency (2nd edition) and the Gray Oral reading test (4th edition) (2021: 4). All subjects did the Multilingual Naming Test (MINT) and some testing of general cognitive function (ibid.). Actual proficiency testing was not used in the study (e.g. ILR, CEFR, American Council on the Teaching of Foreign Languages (ACTFL)). However, the researchers selected subjects who were seen to have "moderate proficiency" in the L2 (2021: 9).

The selection of stimuli was focused on eliciting and monitoring lexical processing, not overall L2 proficiency, and the lexemes chosen were accessible to beginning L2 learners in both English and Hebrew.

Results from fMRI showed a significant effect in response times, where L2 responses were slower than L1 responses (2021:5). The English L1 group had slower L2 responses than the Hebrew L1, which was not unexpected because the HL1 group had more exposure to English than the EL1 group had to Hebrew (ibid.). At a more general level, Brice and colleagues note a reduction in activation in EL1 group across three regions (IF, bilateral insula, and left STG), while the HL1 group showed an increase in activations. In terms of brain activation patterns, Brice and colleagues (2021: 6–8) note the following and divide based on "speech" vs. "print":

(1) listening to speech: (a) In comparison to what was observed in L1, slightly greater activation was observed in the bilateral IFG and left anterior STG in the earlier stages of L2 proficiency. (b) Longitudinally, ROI effects include reduction of bilateral IFG, aMCC, rIPL, rCerebellum for L1.

(2) reading print text: (a) For L2, greater activation was observed in various regions of the ventral system of the brain including the lIFG, precentral gyrus, left posterior inferior temporal gyrus, right inferior gyrus, and the middle occipital gyrus. (b) Lower activation was observed in dorsal regions for L2 to include the left superior parietal cortex, bilateral superior occipital gyrus, right angular gyrus, and right supramarginal gyrus.

Three different physical sites were utilized in fMRI scanning (see Appendix B). Scanner differences are an additional challenge in understanding neuroimaging data. This problem has been discussed in the literature on fMRI (cf. Demirci et al., 2008, and Van Horn & Toga, 2009). While multiple sites provide the opportunity for a larger subject pool, it also brings with it additional challenges.

Brice and colleagues (2021) include the following points in their conclusions: (1) the L1 reading system is utilized by proficient L2 readers for reading in the L2, (2) the impact on L1 due to reduction in exposure and the introduction of new characteristics of the L2, (3) the lack of strong changes in activations in the L2 of participants and how that may be related to their beginning proficiency when they began the immersion program (which was considered to be "moderate" proficiency since the L2 learners were outside of the early stages of L2 acquisition), and (4) the nonstatic nature of processing L1 throughout the life cycle (not just the L2) (2021: 12).[4]

The Grant and colleagues (2015: 35–47) study focuses on lexical development and learning in L2. The primary goals included (1) tracking neurocognitive changes in L2 learners during lexical processing, and (2) investigating the potential impact on lexical processing patterns based on cognitive control. The study included twenty-four subjects (twenty females, four males), all L1 speakers of English with L2 Spanish. Subject differences included experience in study abroad (six subjects) and exposure to a different language other than the languages of the study (four subjects). All had learned Spanish as an L2 in the classroom. Grant and colleagues determined subject proficiency as "intermediate" using self-reporting and the TVIP Spanish vocabulary test.

Using analysis of variance (ANOVA), results suggest no main effect of session, but a significant interaction between session and contrast type ($p < .05$ FWE corrected) for left IFC (inferior frontal cortex), insula, cingulate gyrus, cuneus, lingual gyrus, cerebellum, bilateral MFG (middle frontal gyrus) and SFG (superior frontal gyrus), and right PrG (precentral gyrus), PoG (postcentral gyrus), precuneus and MTG (medial temporal gyrus) (2015: 41). The behavioral measures (TVIP, letter-number sequencing test (LNS), Flanker) were treated as covariates in the analysis, and no relationship was found between the vocabulary testing and letter–number sequencing. However, there was a significant correlation with the flanker effect for homograph and unambiguous Spanish words at both sessions (2015: 42). The analysis includes commentary on differences in the learning process for ambiguous vs. unambiguous lexical items in the L2, where the selected ROIs were found in the unambiguous items, but ambiguous items recruited more broadly. Note that Grant and colleagues include homographs under "ambiguous" (2015: 42).

Six ROIs were selected based on earlier studies of cognitive control: MFG, PFC (prefrontal cortex), IFG, ACC (anterior cingulate cortex), CN (caudate nucleus), MTG. Results showed significant differences between the two sessions of acquisition, and in keeping with the convergence hypothesis, the underlying neural mechanism for processing L2 lexical items after increased exposure to the L2 results in greater similarities between L1 and L2 processing. This includes a reliance on frontal and temporal networks in more advanced L2 lexical processing, similar to L1. Additionally, Grant and colleagues note that the role of cognitive control in interacting with the L2 system is more relevant in lower proficiency and earlier stages than in more advanced learning (2015: 45).

For specific details of each study, see Tables A–G in Appendix A, which give a summary of the defining parameters of each of these six studies: (a) imaging techniques, (b) study classification, (c) participants and scans, (d) behavioral testing measures, (e) languages included in each study, (f) longitudinal time frames, L2 training, timing of scans, and (g) stimuli in scanner. All studies except for Brice and colleagues (2021) used one scanner for the entire study. For details, see Appendix B.

12.2 Synthesis of Outcomes Found in L2 Acquisition Longitudinal Studies

The six studies referenced above show some of the unique contributions of longitudinal studies that include important data for understanding proficiency levels and modalities, how individual proficiency may change in different types of learning, and the importance of individual variation and experiential differences in the learning process.

12.2.1 Comparing Longitudinal and Cross-Sectional Studies

In considering research design and analysis for studies of L2 acquisition, it is important to acknowledge the unique perspectives and valuable insights that longitudinal studies (data collected of a cohort of subjects over a specific period of time ranging from weeks to years) contribute, that cross-sectional studies (data collected at one point in time) do not provide on their own. Key insights resulting from longitudinal studies, including increased proficiency outcomes and a spectrum of both converging and nonconverging evidence, are enhanced by the inclusion of data from cross-sectional studies.

García-Pentón and colleagues (2016) make an important contribution by analyzing a large number of cross-sectional fMRI studies of bilingualism that show variable findings across imaging studies that focus on GM and/or WM analysis (2016: 307). A variety of analytical approaches, scanners, and software packages are included in the studies under review: (a) analytical approaches: VBM (voxel-based morphometry), ROI (regions of interest), CT (cortical thickness); (b) scanners: 1.5T and 3T; (c) software packages: SPM, FSL/FMRIB. In the group of studies that compare mono- and bilingual neural processing of language(s), García-Pentón and colleagues note the following results using whole brain FWE (family-wise error) correction (2016: 310): (1) Regions of difference include left IPL, cerebellum, left aITG, ACC (using voxel-based or cluster-based FWE, where they differ in levels of sensitivity); (2) Three studies show no difference between these two cohorts (ibid.). They also note that different levels of analysis were used, namely voxel-based or cluster-based. However, when the analyses were limited to a specific ROI, there was more convergence.

García-Pentón and colleagues (2016) also address the problems that may arise in using an ROI analysis, as well as challenges when using voxel-based morphometry (2016: 311). The authors point out that that there are additional problems connected to the lack of uniform participant groups. Four different groups are noted – mono- vs. bilinguals, bi- vs. multilinguals, mono- vs. multilinguals, and unimodal monolinguals vs. bimodal bilinguals (2016: 311). One must take into account that the differences in these four groups could be a source of the variation found in the results of these studies. Their important observation speaks to a more general issue concerning individual differences and variation, namely that individual subject differences, including those that develop over time, greatly affect proficiency and acquisition and are not observable in cross-sectional studies. Thus, it is difficult to determine which differences between subjects are due to individual variations or the conditions or design of the study itself.

Longitudinal studies, on the other hand, can provide a more accurate representation of developmental learning outcomes since each subject can act as his or her own control. As language acquisition follows a nonlinear trajectory, longitudinal studies demonstrate and account for fluctuations over time due to variable factors like motivation, the consistency of language use, the need to use the L2, questions of identity and increased interactions with speakers of the L2 (de Bot, 2008; Pavlenko, 2005). Furthermore, with the addition of proficiency and behavioral testing in tandem with neuroimaging of subjects in longitudinal studies, a multivariate analysis of covariance (MANCOVA) can be conducted to further augment the meanings of activation patterns acquired during the scans.

Furthermore, longitudinal studies allow for the observation of long-term effects that result from continued L2 learning versus remaining at a lower level of proficiency. For instance, Grant and colleagues (2015: 45) demonstrate that the networks upon which processing relies vary as L2 proficiency increases. In this case, a longitudinal design provides the potential to examine whether L2 acquisition effects that are detected earlier in the language acquisition process continue to be significant as L2 proficiency increases. This allows researchers to gain more insights into the mechanisms of L2 acquisition. The Brice and colleagues study (2021) also highlights the influence L2 can have on L1, in addition to the impact of environmental exposure, something cross-sectional studies are unable to show.

In using a longitudinal design, a more thorough examination of growth and differences at both individual and group levels can be performed, which points to the dynamicity in the L1/L2 interaction. As early as 2013, Andrews and colleagues observed nonlinear differentiation in neural processing in subjects across L1 and L2. We will return to this topic in the discussion about dynamic processing hypotheses and theories.

12.2.2 Outcomes of Convergence, Differentiation and Variability

The six longitudinal studies examined also provide an interesting convergence of evidence and follow the division into two groups. It is important to note that statistical significance for whole brain scanning requires that fewer ROIs are hypothesized for the analysis. The greater the number of ROIs included, the weaker the statistical significance.

Andrews and colleagues (2013), Brice and colleagues (2021), and Grant and colleagues (2015) all focus on cortical gray matter neuronal activation patterns and replicate findings of other cross-sectional studies for these areas (including Mårtensson et al., 2012; Stein et al., 2012 for IFG, MFG; Price, 2010 for STG). The number of ROIs in these studies is restricted, and bilaterality is noted (Andrews – six bilateral ROIs in listening/five bilateral ROIs in reading, Brice – seven bilateral ROIs for listening/nine bilateral for reading, Grant – six bilateral ROIs for reading). For important research on bilaterality of language processing, see Schirmer and colleagues (2012), Poeppel and Hickok (2004), Hickok and Poeppel (2004, 2007).

The vocabulary studies include a larger number of ROIs and some bilaterality. The two Ghazi-Saidi studies use thirty-two ROIs: twenty-one ROIs from Price (2010) and eleven ROIs from Abutalebi and Green (2016). Raboyeau and colleagues (2010: 2853–2855) use twenty-nine ROIs and specify two ROIs for cognate vocabulary vs. six ROIs for noncognate vocabulary. As noted above, the overall number of ROIs posited has an effect on the reliability of statistical significance.

In the three language processing studies, a convergence in bilateral ROIs was observed to include the IFG (all three), the STG (Brice and Andrews), and the MFG and MTG (Andrews and Grant). These three studies show the dynamic interaction of L1 and L2 in reading outcomes that point to similarities in processing in L1 and L2 reading, shifting of cognitive control as proficiency increases, and improved speed of reading at higher L2 proficiency levels. Lastly, the nonstatic processing of both L1 and L2 across the lifecycle was noted in two of these studies (Brice, 2015: 5, 12; Andrews, 2014: 177–186).

The dynamic interaction between L1 and L2 lexical processing and nonstatic processing of L1 and L2 are extremely important when these data are contextualized both in the longitudinal context, and in the theoretical knowledge acknowledged by the field (including dynamic developmental network models of bilingualism [Abutalebi & Green, 2016; Claussenius-Kalman et al., 2021; Dash et al., 2022; de Bot, 2009; Pliatsikas, 2019; Pliatsikas et al., 2020]). Bilingualism-induced neuroplasticity is the explanation offered by Pliatsikas (2019), Pliatsikas et al. (2020) and Claussenius-Kalman and colleagues (2021) for their models. In the longitudinal studies discussed above, we see the emergent nonconvergence of results outside of ROI analysis, and while there is no single answer, we propose that it is due to a multiplicity of factors that might include the different linguistic levels being examined (vocabulary vs. language processing studies), the different protocols and designs, and the subject pools.

An additional advantage of longitudinal data is expressed in scan outcomes that demonstrate nonlinear individual variation in the percentage of nonzero voxels across ROIs and hemispheres (cf. Andrews et al., 2013 and Andrews, 2014: 183–185). Interpretation of these types of data would have benefited from contextualization in the dynamic models mentioned above. Even though earlier research findings show interesting patterns of variation between longitudinal learner/subjects, the dynamic developmental and restructuring and adaptive system-level models, which are in many ways complementary, have become much more recognized in the past several years and have amplified the research analyses and interpretations. And these models highlight the relevance of fine-tuning our empirical baseline data on participant groups.

Finally, given the spectral and nonlinear nature of bilingualism and the uniqueness of individual acquisition and maintenance, there remain compelling challenges to determining which outcomes of fMRI imaging studies of bilingualism are due to individual variation or the conditions or design of the study itself. This point is central to the conclusions in García-Pentón and colleagues (2016: 321) and we will return to it in the section on proficiency testing.

12.3 The Importance in Neuroimaging of Brain and Language(s) and Linguistic Approaches

Given the ongoing issues with interpretation of activations in fMRI experiments, as well as the need to provide more empirical measurements of subject linguistic abilities in bi/multilingual studies, it is important to identify salient parameters that yield similarities between the L1 and L2 of users as well as important environmental variables that play a significant role in bi- and multilingualism. It is to this point that several analyses have shown that proficiency, not age of acquisition, is one of the most important factors in predicting robust outcomes in an L2 (Abutalebi et al., 2013; Birdsong, 2006; Claussenius-Kalman et al., 2021; Pliatsikas, 2020; Pliatsikas et al., 2020). Birdsong notes the importance of proficiency as the "strongest predictor" in terms of the similarities in processing of a user's first and second languages. This is substantiated even in studies where L2 acquisition began at 12 or older (2006: 24–26). Birdsong notes hemodynamic (fMRI, PET) and electrophysio logical studies (ERP) that also show these results (ibid.). These findings dovetail with a call for more rigorous proficiency measurements in fMRI studies (Abutalebi et al., 2013; Abutalebi & Green, 2016; Andrews, 2014; Claussenius-Kalman et al., 2021; Dash et al., 2022; de Bot, 2008, 2009, 2019). The inclusion of proficiency does not mean that AoA is not relevant in modeling acquisition of languages. Rather, it points to the importance of recognizing multiple factors, which include both proficiency and AoA as contributing to embodied cognition and lifelong bi- and multilingualism.

While they exhibit important differences in focus and design, the three longitudinal studies of second language acquisition presented here using BOLD fMRI share some important findings. Both show (1) a trend where higher proficiency in L2 results in neural processing and networks for L1 and L2 are more similar, and (2) the need for additional measurements, including empirical behavioral data, to facilitate more rigorous interpretations of activations acquired in fMRI task-based experiments (MANCOVA is a good example). We will suggest below how additional structural and fMRI techniques can be added to enhance data generation in both cross-sectional and longitudinal studies.

It is important to note key influential research that has fundamentally changed how cognitive neuroscientists and neurolinguists study the processing of language(s) in the human brain using a variety of neuroimaging techniques. While we see the roots of these changes clearly in the earlier period of fMRI and PET studies of language processing in the 1990s and across the past two decades, the most notable moments are the ones that bring theory, design, and analysis together in (1) ecologically valid protocol designs and (2) contextualization of the sociocultural context of *languages in contact* (including a significant array of important empirical markers like intensity, social networks, where and how languages are used, levels of complexity, thematic exposures, etc.).[5]

Central theoretical principles that have emerged in the fields of neurolinguistics and cognitive neuroscience that focus on brain and languages include: (1) a shift toward network models of language processing that are more consistent with multimodalities than modular approaches, (2) a view of neural regions as more heterogeneous and dynamic than monolithic and static, (3) building a shared vocabulary of analysis across linguistics and the cognitive neurosciences, (4) recognition of the importance of ecological validity and its implementation at all stages of experimental design and interpretation, (5) shifting research outcomes from description (and in many cases the overuse of localization) to an explanatory basis, (6) taking into account the interactive nature of speech perception/comprehension and production, (7) the importance of bilaterality of language processing in the brain at both the cortical and subcortical levels, (8) defining the system-level processing of languages to include resting state networks and functional connectivity, and (9) incorporating research focusing on neurological changes across the lifecycle that can occur in bilingualism/multilingualism. These principles have been generally acknowledged by the field of neuroimaging of language(s) and continue to be incorporated into the fabric of current and ongoing research (cf. Abutalebi et al., 2009, 2012, 2013; Bialystok, 2011; Claussenius-Kalman et al., 2021; Dash et al., 2022; Gallo et al., 2021, 2022; García-Pentón et al., 2016; Hickok & Poeppel, 2004, 2007; Luk et al., 2010, 2011; Pliatsikas, 2020; Pliatsikas et al., 2015, 2020, 2021; Poeppel, 2008; Poeppel & Hickok, 2004; Schirmer et al., 2012).

12.4 Critical Approaches to Understanding Neuroimaging Studies of Language(s)

> It appears the localization of psychological functions has become the epistemological driving force, and that brain images are assumed to constitute a privileged type of evidence that forms the basis for the explanation of mental life. In contrast, it is argued here that cartography is immensely useful but intensely limited *vis-à-vis* the explanation of function. Poeppel (2008: 1)

The challenges of rigorous interpretation of activations found in neuroimaging studies of language processing continue to be one of the most significant problems in the field, especially when viewed without appreciating Poeppel's characterization above. Marek and colleagues (2022) focus on what they see as common replication failures in the neuroimaging literature from multisite studies. Eklund and colleagues (2016) focus on a different set of problems, mainly the different error rates and "false positives" emerging from comparisons of the same data with the use of different types of fMRI software packages, namely SPM, FSL, and AFNI. In this case, Cox (2019) provides an important response with changes to AFNI that produce more robust results for future analysis.

On another level, there are very specific challenges that arise between different communities of practice that are invested in and conduct neuroimaging research on the acquisition, maintenance, and loss of language(s). One of the issues that is discussed in neurolinguistics in the criticism of neuroimaging experiments focuses on so-called *single word studies*. Paradis (2019: xxxiv) clearly repeats his critique of many decades concerning the inability to generalize results from single word studies to language systems:

> Single words are the least likely candidates for investigating *language* representation, given that what makes language most specific as a cognitive function, namely the language system (phonology, morphology, syntax), is supported by procedural memory, whereas isolated words, being explicitly

> known form-meaning associations, are supported by declarative memory ... Not only can results obtained with single-word stimuli not be generalized to the representation and processing of language ..., but experiments that use such stimuli address a component that differs radically from the language system.[6]

When one considers the important linguistic research on lexical meaning, the relationship of words and thought, polysemy, ecologically valid speech acts, and the important role of context in stabilizing lexical outcomes (Vygotsky, Bolinger, Jakobson for a few examples), it becomes clear that special attention must be paid to the role of context in stabilizing outcomes of meanings. For example, if a protocol presents on lists of words, then it would be difficult to know what the listener/reader is imaging (e.g., "book" as noun and verb and adjective [book it, book him, John Book in the film "Witness," book a ticket, a book, etc.). This calls for protocol designs that can facilitate less ambiguity, and that is what we find in studies that use pictures to go with vocabulary. However, ecological language usage in the real world is not a list of terms, but dynamic speech acts that require interpretation and negotiation (cf. Jakobson, 1987). We suggest that protocols for task-based fMRI should include language(s) as they occur in actual usage. Listening and reading tasks (not speaking) are better for fMRI due to motion artifacts that result from movement of any kind. If using vocabulary lists, they should be augmented with ecologically valid discourse.

The cognitive neuroscience community interested in language processing in the brain has clearly articulated for the past twenty years the importance of distinguishing between *functional* and *structural* changes in acquisition of language(s), as well as where to seek more substantial evidence that explains language acquisition and processing (cf. Andrews, 2011, 2014, 2019; Andrews et al., 2019; Gallo et al., 2021, 2022; Green et al., 2006; Hagoort, 2006; Pliatsikas, 2020, Pliatsikas et al., 2015, 2020, 2021).

12.5 Understanding Activations: Best Practices for Future Neuroimaging Studies of Bi/Multilingualism

Robust interpretations of activations in task-based fMRI and resting state fMRI (rsfMRI) functional connectivity are more grounded and reliable when coupled with appropriate evidence-based empirical behavioral data, including proficiency testing in the case of bi/multilingualism. Given the nonlinear relationship between activations and any potential correlation with "knowledge" or "proficiency" (i.e., more signal does not mean more knowledge), it is imperative to provide an ecologically valid, nonessentialist context for protocol design and implementation. Understanding variation across brains and the dynamic nature of language processing as an *embodied process* are foundational constructs that facilitate how we conduct our research and provide a more integrated model of variation that embraces the important findings of sociolinguistic research and its significance in the design, contextualization, and implementation of neuroimaging studies of language(s).

The addition of DTI structural imaging and rsfMRI to studies of bi/multilingualism is critical to improving the interpretability of results, as well as explain how bilingualism can enhance cognitive reserve across the lifecycle and yield stronger resting state functional connectivity (Dash et al., 2022; Jafari et al., 2021; Liu et al., 2021; Wang et al., 2020). As we have seen in this analysis, ROI analysis, for example, has never been fully satisfactory in bilingualism research. Examination of the networks using rsfMRI that are essential to processing languages, as well as seeing the enhanced functional connectivity between specific networks in a non-task-based scan, provides a more powerful set of data that can be used to explain, not just describe, neural processing.

The longitudinal studies discussed here were sensitive to individual variation as a key aspect of longitudinal design across subjects that included both the individual and group levels. The ability to look at the changes in language processing at the individual level in intensive acquisition of a second language, the unique individual structural anatomy where each subject is their own control, and the multivariate analysis of covariance of changes in proficiency and changes in activations is a rewarding exercise that yields vibrant empirical outcomes.

12.6 Language Proficiency Measurements and Their Role in Neuroimaging Studies

There have been significant advancements in understanding the importance of proficiency and interactions between proficiency in age over the past two decades of neuroimaging research. As noted earlier, it is proficiency, not age of acquisition, that is the "strongest predictor" of degrees of similarity between a user's first and second languages (Birdsong, 2006: 24–26). The time has come for more rigorous proficiency measurements to be a central part in all neuroimaging studies of bi/multilingualism and second language acquisition.

The Abutalebi research group began testing for proficiency in subjects as early as 2007 using translation tasks (Abutalebi and Green, 2007). Andrews and colleagues (2013) was the first longitudinal fMRI study to include internationally recognized CEFR proficiency testing (Andrews, 2014: 162–167). The Andrews lab continues to use CEFR proficiency testing in all of their language studies (Andrews et al., 2019). While the other longitudinal studies discussed here used a variety of testing batteries, proficiency testing was not included. The inclusion of CEFR or other frameworks of proficiency testing requires additional time and resources to collect and evaluate proficiency in a way that is consistent across subjects and studies.

One of the first publications recognizing the significance of linguistic proficiency in understanding the language control network in fMRI studies is found in Abutalebi and colleagues (2013):

> The importance of our data is that they show for the first time that differences in language proficiency, defined within participants, differentially modulate activity in core regions of the language control network during language switching. Switching language increases demand on a region associated with monitoring the language in use (pre-SMA/ACC) and switching to the least proficient language increases demand in the region implicated in selecting the language in use (left caudate). Such demands change with proficiency and are likely to underlie the adaptive response to gray matter in these regions.
>
> *(Abutalebi et al., 2012, 2013: 910; Zou et al., 2012)*[7]

All these findings set the stage for the continuation of more rigorous proficiency testing measurements in ongoing and future analyses of neuroimaging studies of language(s) processing.

The ability to relate protocol design, stimuli, and tasks to behavioral measurements (like proficiency testing and other language-based testing) in a study produces meaningful empirical results that may include statistical significance as well. The advantage of using a proficiency-based testing framework (CEFR, ILR, ACTFL) brings us closer to the inclusion of ecological validity in our design and analysis. As we understand more about the multimodalities and networks of the processing of languages in the brain, we are able to utilize proficiency-based testing not only as indicators of specific levels of proficiency, but also to clarify specific modalities of language usage, including reading, listening comprehension, writing, speaking, grammar/lexicon. Note that vocabulary is always incorporated into a larger grammatical context in proficiency frameworks, and as such can have a place in future neuroimaging studies of brain and languages.

While the inclusion of proficiency testing requires more work in screening subjects and collecting data in neuroimaging studies of languages (including time, effort, and costs), its benefits far outweigh the negatives. Benefits include a more inclusive subject pool, where subjects are selected based on what they know and can do (increased diversity), and not based on origin. Longitudinal studies that incorporate proficiency testing can select for not only specific levels of proficiency (from zero knowledge to a particular level) but can also target specific intervals of proficiency (e.g., B1 to C1). Proficiency testing can also be very useful in cross-section studies, where subjects would be tested prior to their single scan.

12.7 Neuroimaging and the Executive Function

Executive function and the general executive control network are particularly important in neuroimaging studies that examine the effects of bilingualism. A large body of research has demonstrated that the complex sensory-motor activities required for lifelong bilingualism are also integrated with the building of cognitive reserve and cognitive flexibility. This body of research also demonstrates that bi/multilingualism creates higher demands on cognitive control and executive function in the constant interaction of the languages, working memory, and increased levels of inhibition (including inhibition of response and suppression of language interference; Abutalebi et al., 2012; Abutalebi & Green, 2016; Bialystok, 2011; Bialystok et al., 2012; Grundy et al., 2017; Luk et al., 2010; Reverberi et al., 2015).

In both the Grant and colleagues (2015) and Brice and colleagues (2021) studies, subjects underwent tasks that measured executive control, specifically in the areas of response inhibition and working memory. Brice and colleagues (2021) also examined task-switching in testing subjects for executive control. Bilinguals are known to show stronger functional connectivity to areas recruited in attention and selection, inhibition, working memory, and task-switching during task-evoked brain activity and greater frontal-to-posterior connections at rest, while monolinguals show greater connectivity within the frontal lobes at rest (Grundy et al., 2017). Thus, it is vital to test and account for the possibility of enhanced executive function in monolinguals who are involved in neuroimaging studies in order to control for them.

Language switching, a parallel to task-switching, has been shown to activate portions of the brain that belong to the executive control network (Bialystok et al., 2012). This coordinated use of multiple executive control functions has been observed to be more efficient in bilinguals than in monolingual controls (Bialystok, 2011). In addition, as a result of language proficiency and exposure, bilinguals experience organizational and structural changes in the brain – notably increased white matter integrity (including the corpus callosum and specific subcortical white matter fiber tracks), allowing information to travel faster between the two hemispheres in bilingual brains. Grundy and colleagues (2017) suggest that highly proficient bilinguals have higher neural efficiency and cognitive flexibility in recruiting the frontal and executive regions primarily in cognitively demanding tasks, bilinguals use the posterior and subcritical regions of the brain in nonverbal executive tasks, while monolinguals use frontal and executive regions in those same tasks. In short, the effects of an enhanced executive control network are evident in both structural and functional tasks in neuroimaging studies of bi/multilinguals.

Another important contribution to the study of executive function comes from Raichle's work on resting state neural networks in the human brain (Raichle, 2010, 2011, 2015). The first network discovered by Raichle was the default mode network (DMN). Raichle noticed that there were brain regions that showed less activation during task-based stimuli than in resting state. This observation led him to the discovery of seven different resting-state neural networks in the human brain: default mode, auditory, visual, sensorimotor, executive control, salience, dorsal attention (Raichle, 2010, 2015). The review of fMRI studies of executive control in bilinguals by Pliatsikas and Luk (2016) shows important synergies between executive control and language control regions, regardless of age and languages. They also identify the importance of including not only task-based fMRI, but DTI structural imaging (Pliatsikas et al., 2015) and resting state fMRI analysis (Dash et al., 2022) in order to reveal the full range of networks involved in executive function and control.

12.8 The Unique Contribution of Longitudinal Studies of L2 Acquisition: Using Structural and Functional MRI and Resting State Functional Connectivity

The six studies discussed in this analysis show how longitudinal studies can provide important data that are not derivable from cross-sectional studies, including a more precise view of developmental learning outcomes, variation and individual differences, the important role of proficiency in understanding changes over time, the ability to examine both individual and group-level phenomena, the dynamic interactions between L1 and L2 processing, and a more powerful empirical inclusion of behavioral data using a multivariate analysis of covariance.

The addition of a broader spectrum of imaging technologies, including structural and functional imaging, task-based and resting state, to future longitudinal studies of L2 acquisition can result in important findings that include not only cortical areas and GM, but also subcortical fiber tracts and WM integrity (fractional anisotropy (FA) and mean diffusivity (MD) measurements) and functional connectivity of the networks that support language processing (Pliatsikas, 2020, 2023; Pliatsikas et al., 2015). In so doing we can now measure important changes that occur in short periods of time at multiple neurological levels. These outcomes are expressed even more profoundly in studies that involve intensive learning and also dovetail with dynamic developmental network modeling of bilingualism.[8]

12.9 Future Directions

Neuroimaging studies, and specifically studies of second language acquisition and bi/multilingualism, make a significant contribution to the research on neural processing of languages in the brain and their role in neurological and behavioral cognitive brain reserve. It is clear that a broader repertoire of imaging techniques is needed to amplify fMRI task-based studies, especially structural DTI and resting state fMRI and functional connectivity. In order to move the field forward, there needs to be a strengthening of the empirical basis that takes into account the neural networks that support higher cognitive processing integral to bi- and multilingualism. The expansion of neuroimaging techniques also requires more empirically reliable behavioral data, and language proficiency data will be a critical piece of the puzzle as new advancements are made in clarifying the processing of languages by the human brain. The inclusion of proficiency testing in the screening process of study subjects is a reasonable and important component for attempting to improve on the reliability and transparency of study results.

Future neuroimaging of language(s) should be designed to strengthen and clarify the connections between the fields of theoretical linguistics and the cognitive neurosciences. The paradigm shifts that give preference to full brain analysis (including GM, WM, cortical/subcortical, networks, etc.) allow the research to engage with new, more nuanced questions, including a more interactive perspective of multimodalities and collaboration of neural networks and systems (Gallese & Lakoff, 2005; Weiss & Bidelman, 2015), and appreciating and grappling with neurodiversity and variation across and within brains (Abutalebi, 2008; Abutalebi et al., 2009, 2013, 2016; de Bot, 2009, 2019, Pliatsikas, 2019, 2020; Pliatsikas & Luk, 2016; Pliatsikas et al., 2021).[9]

We recommend consideration of the CEFR of proficiency testing because it is not only an internationally recognized tool that includes a large range of world languages, but is also a user-friendly and affordable way to include proficiency measurements into future neuroimaging studies of bilingualism. The rigorous addition of proficiency-based behavioral empirical data are critical for clarifying important differences in participant pools and improving the ability to synthesize neuroimaging results across both cross-sectional and longitudinal designs will enhance the researcher's ability to design and conduct more powerful neuroimaging experiments of second language acquisition and processing, as well as reveal interactions between first and second languages in order to achieve new understandings in the modeling and implications of bi/multilingualism in the human brain.

APPENDIX A: TABLE 12.1(A–G) OF SIX LONGITUDINAL STUDIES (2010–2021)

Table 12.1A *Imaging technique – all used BOLD fMRI*

	Block design	Event-related design	Time in scanner	Protocol design
Andrews et al., 2013	X		Approx. 50 min.	8 runs, 4 languages in random order; audio/reading in separate runs
Brice et al., 2021	X		Total duration not provided	10 runs, 2 languages, audio/ reading together in runs
Ghazi-Saidi et al., 2013		X	47 min.	2 runs; 2 languages
Ghazi-Saidi & Ansaldo, 2017		X	47 min.	2 runs; 2 languages
Grant et al., 2015		X	25 min.	3 runs of 90 trials; 2 languages and interlingual homographs
Raboyeau et al., 2010		X	Time duration in scanner was not provided.	2 runs; 2 languages

Table 12.1B *Study classification*

	L2-acquisition	Immersion	Transitional	Vocabulary acquisition
Andrews et al., 2013	X	X		
Brice et al., 2021	X	X	X	
Ghazi-Saidi et al., 2013				X
Ghazi-Saidi & Ansaldo, 2017				X
Grant et al., 2015	X		X	
Raboyeau et al., 2010				X

Table 12.1C *Participants/scans*

	# of participants	# of scans/participant	Total # of scans
Andrews et al., 2013	5	3	15
Brice et al., 2021	80	2	160
Ghazi-Saidi et al., 2013	12	2	24
Ghazi-Saidi & Ansaldo, 2017	24	2	48
Grant et al., 2015	24	2	48
Raboyeau et al., 2010	10	2	20

Table 12.1D *Behavioral testing measures*

	Proficiency testing	Self-reporting	Vocabulary test/ picture naming	Post-scan interviews	Course-based learning and testing
Andrews et al., 2013	CEFR/TRKI			X	X
Brice et al., 2021	GORT-4, TOWRE-2, MINT, Oral and pseudoword reading, priming				
Ghazi-Saidi et al., 2013		X			
Ghazi-Saidi & Ansaldo, 2017		X			
Grant et al., 2015			TVIP		X
Raboyeau et al., 2010		X			

Table 12.1E *Languages included in each study*

	English	French	Spanish	Russian	Persian	Hebrew	Prior L2/L3 experience?
Andrews et al., 2013	L1			L2			No
Brice et al., 2021	L1/L2					L1/L2	Yes, but no significant L2 immersion
Ghazi-Saidi et al., 2013		L2			L1		Yes; matched for elementary level
Ghazi-Saidi & Ansaldo, 2017		L2	L1		L1		Yes; recruited from same first-level L2 immersion course
Grant et al., 2015	L1		L2				Yes; recruited from intermediate/advanced L2 courses
Raboyeau et al., 2010	L2	L1	L3				No

Table 12.1F *Longitudinal time frames, L2 training, timing of scans*

	Length of study	Time in L2 learning	Timing of Scans Scan 1	Scan 2	Scan 3
Andrews et al., 2013	11 months	6 contact hrs/wk for 13 weeks, 20 contact hrs/wk for 3 weeks, 120 contact hrs over 7 weeks; 8 intensive months total	Apr. 2011 (zero level)	Dec. 2011	Mar. 2012
Brice et al., 2021	14 months min., 36 months max.	Information was not provided	At inception (intermediate L2 before immersion)	14 months min.; 36 months max.	N/A
Ghazi-Saidi et al., 2013	30 days	15 minutes per day	1 week after inception	30 days after inception	N/A
Ghazi-Saidi & Ansaldo, 2017	28 days	15 minutes per day	1 week after inception	4 weeks after inception	N/A
Grant et al., 2015	4 months	Classroom learners of Spanish; contact hours were not provided	November	Following March	N/A
Raboyeau et al., 2010	19 days	20 minutes per day	5 days after inception	12 to 15 days	N/A

Table 12.1G *Stimuli in scanner*

	Listening	Reading	Discourse/ dialogue	Single words	Images	Semantic categorization	Stimuli languages
Andrews et al., 2013	X	X	X				Spanish, English, Russian, Georgian (listening only)
Brice et al., 2021	X	X		X		X	English, Hebrew
Ghazi-Saidi et al., 2013					X		N/A*
Ghazi-Saidi & Ansaldo, 2017					X		N/A*
Grant et al., 2015		X		X		X	Spanish, English
Raboyeau et al., 2010					X		N/A*

*Stimuli provided was non-linguistic.

APPENDIX B: PARAMETERS OF FMRI ACQUISITION

Andrews et al. (2013: 856); Andrews (2014)

Imaging was performed using a General Electric LX 3T Signa scanner using head restraints to reduce motion. All scans were conducted on the same BIAC3 scanner at Duke University Hospital. The subject's head is positioned along the canthomeatal line. For functional scans, a total of thirty slices with a

5 mm thickness are obtained using a TR of 1500 ms, a TE of 35 ms, a 64 × 64 matrix, and an echoplanar pulse sequence. Slices are axial, taken parallel to the plane of the anterior and posterior commissures with the most inferior image level with the top of the pons. High resolution dual echo proton density and T2-weighted anatomical images are acquired in the same slice planes as the functional scans. A high-resolution 3D fast spoiled GRASS T1-weighted scan covering the whole brain with isotropic 1 mm cubic voxels is acquired to allow structural visualization in any orientation (256 × 256 × 166 voxels).

Brice et al. (2021: 4)

Imaging was performed using three different scanners and a 12-channel head coil at different sites. At the New York University Centre for Brain Imaging, a 3T Siemens Allegra MRI scanner was used. At the Ein Karem University Hospital campus, a 3T Siemens Trio MRI scanner was used. At the Givat Ram campus of the Hebrew University, a 3T Siemens Skyra MRI scanner was used. Participants' heads were supported by foam wedges to decrease movement. For functional scans, thirty-two T2*-weighted image slices with a 4 mm thickness were obtained using a TR of 2000 ms, a TE of 30 ms, and a 64 × 64 matrix using single-shot echo planar imaging. For anatomical scans, images were acquired in an axial-oblique orientation parallel to the inter-commissural line with a TR of 2530 ms, a TE of 3.66 ms, a 256 × 256 mm matrix, and a voxel size of 1 × 1 × 1 mm.

Ghazi-Saidi & Ansaldo (2017: 4–5)

Imaging was performed using a 3T Siemens MRI scanner using cushions and belts to reduce head motion. All scans were completed in the Unité Neuroimagerie Fonctionnelle at the Institut Universitaire de Gériatrie de Montréal. For functional scans, a total of 28 slices with a 5 mm thickness were obtained using a TR of 3 s, a TE of 40 ms, and a 64 × 64 matrix. A high-resolution anatomical scan of 76 slices was acquired in the axial plane with a TR of 13 ms, a TE of 4.92 ms, a 256 × 256 mm matrix, a voxel size of 1 × 1 × 1 mm, and a 3D T1-weighted pulse sequence.

Ghazi-Saidi et al. (2013: 59)

Imaging was performed using a 3T MRI Siemens scanner. For functional scans, a total of twenty-eight slices with a 5 mm thickness were obtained using a TR of 3 s, a TE of 40 ms, and a 64 × 64 matrix. A high-resolution anatomical scan of seventy-six slices was acquired in the axial plane with a TR of 13 ms, a TE of 4.92 ms, a 256 × 256 mm matrix, a voxel size of 1 × 1 × 1 mm, and a 3D T1-weighted pulse sequence.

Grant et al. (2015: 39)

All fMRI scans were done on a Siemens Magnetom Trio 3T scanner at the SLEIC center using T_{2*} weighted gradient-echo EPI sequence with TR (repetition time) = 2000 ms, TE (echo time) = 25 ms, 64 × 64 matrix, twenty-two slices/5mm thickness. T_1 –weighted MPRAGE 3D gradient-echo sequence was used to acquire anatomical images (1 × 1 × 1 mm^3).

Raboyeau et al. (2010: 2852)

Imaging was performed using a 3T MRI Siemens scanner with an 8-channel head coil, using pillows to stabilize the head. For functional scans, a total of twenty-eight slices with a 4.55 mm thickness were obtained using a TR of 2 s, a TE of 30 ms, a 64 × 64 matrix, and a T2*-weighted pulse sequence. A high-resolution anatomical scan of seventy-six slices was acquired in the axial plane with a TR of 13 ms, a TE of 4.92 ms, a 256 × 256 mm matrix, a voxel size of 1 × 1 × 1 mm, and a 3D T1-weighted pulse sequence.

Notes

1 For the ILR language modalities and level descriptions, see www.govtilr.org/Skills/ILRscale2.htm. For ACTFL guidelines and languages, see www.actfl.org/educator-resources/actfl-proficiency-guidelines. For CEFR level descriptions and other information, see www.coe.int/en/web/common-european-framework-reference-languages/level-descriptions.

2 For the full study, see Andrews and colleagues (2013). Additional information and analysis can be found in Andrews (2014: 164–191). The original publication in 2013 included the time period April 2011 through June 30, 2012. Scanning began before exposure to the new L2 (April through August, 2011), and was repeated twice more. CEFR/TRKI proficiency testing for Russian included four sessions in this period – November, 2011; December, 2011; January, 2012; June, 2012.

3 The MANCOVA in this analysis used the vector of responses (the percept of nonzero voxels by time, region, hemisphere, and subject) modeled as the additive effects of two categorical variables (hemisphere and region) and one continuous covariate (proficiency testing scores): Y = mean + region effect + hemisphere effect + score effect. Using Pillai's trace, the primary result yields $p = 0.01$ for the Russian-rest condition but shows no such effect for the English-rest condition.

4 Hickok and Poeppel (2007) propose the dual stream model of functional anatomy of language processing based on ventral and dorsal streams and bilaterality of the ventral stream. See also Schirmer and colleagues (2012) for more information about the ventral and dorsal streams in the processing of sound in the temporal lobe.
5 For more on the importance and complexities of understanding *contact,* see Weinreich (1953/1968) and de Bot (2019: 6–7).
6 For additional sources about the problems with single word studies, see Paradis (2000, 2004), Stowe and colleagues (2005), Andrews (2014). For important theoretical linguistic analyses of speech acts, the metalingual function, the interactions between languages and cultures, and approaches to the relationship between linguistic form and meaning, see Jakobson (1956/1985, 1965/1985, 1987) and Bolinger (1965, 1986, 1989).
7 For more information on language networks, see Hodgson and colleagues (2021). For in-depth studies on resting state networks, see Raichle (2001, 2006, 2010, 2011, 2015).
8 The research using DTI (diffusion tensor imaging – a structural imaging technique) in analyzing bilingualism and musicianship is very widely cited and is an important contribution to understanding the role of subcortical white matter in bi- and multilingualism (cf. Luk et al., 2011; Pliatsikas, 2020, 2023; Pliatsikas et al., 2015) and in musicianship (cf. Andrews et al., 2021, 2023; Eierud et al., 2023; Elmer & Jäncke, 2018). The most common DTI scalar used is FA (franctional anisotropy), which is a measurement of white matter integrity sensitive to microstructural changes. Other DTI scalars include MD (mean diffusivity), AD (axial diffusivity), and RD (radial diffusivity). Resting state fMRI (abbreviated as rsfMRI or rfMRI or rs-fMRI) has become very popular and is a powerful way to measure functional connectivity in networks. DTI acquisition takes approximately 5.5 minutes, and resting state acquisition recommends at least 8 minutes. The addition of DTI and rsfMRI to scan protocols is easily done and they can still fit into one hour in the scanner.
9 There has been important work on cognitive brain reserve and the protective effects that may manifest in forms of dementia (see Bialystok et al., 2007; Craik et al., 2010; Guzmán-Vélez et al., 2015; Luk et al. 2010, 2011; Valenzuela et al., 2006). Gallese and Lakoff (2005) is of particular relevance to multimodalities and sensory-motor mappings in the brain, including languages, as well as critiques of *embodied cognition* (cf. Mahon & Caramazza, 2008).

References

Abutalebi, J. (2008). Neural processing of second language representation and control. *Acta Psychologica, 128,* 466–478. http://dx.doi.org/10.1016/j.actpsy.2008.03.014

Abutalebi, J., Della Rosa, P. A., Ding, G., Weekes, B., Costa, A., & Green, D. W. (2013). Language proficiency modulates the engagement of cognitive control areas in multilinguals. *Cortex, 49*, 905–911. https://doi.org/10.1016/j.cortex.2012.08.018

Abutalebi, J., Della Rosa, P. A., Green, D. W., Hernandez, M., Scifo, P., Keim, R., et al. (2012). Bilingualism tunes the anterior cingulate cortex for conflict monitoring. *Cerebral Cortex, 22*(9), 2076e2086.

Abutalebi, J., & Green, D. W. (2007). Bilingual language production: the neurocognition of language representation and control. *Journal of Neurolinguistics, 20*, 242–275. https://doi.org/10.1016/j.jneuroling.2006.10.003

Abutalebi, J., & Green, D. W. (2016). Neuroimaging of language control in bilinguals: Neural adaptation and reserve. *Bilingualism: Language and Cognition, 19*(4), 689–698. https://doi.org/10.1017/S1366728916000225

Abutalebi, J., Tettamanti, M., & Perani, D. (2009). The bilingual brain: Linguistic and non-linguistic skills. *Brain and Language, 109*, 51–54. http://dx.doi.org/10.1016/j.bandl.2009.04.001

Andrews, E. (2011). Language and brain: Recasting meaning in the definition of human language. *Semiotica, 184*(1/4), 11–32. https://doi.org/10.1515/semi.2011.020

Andrews, E. (2014). *Neuroscience and Multilingualism.* Cambridge University Press.

Andrews, E. (2019). Cognitive neuroscience and multilingualism. In J. W. Schwieter (Ed.), *The Handbook of The Neuroscience of Multilingualism.* Wiley Blackwell.

Andrews, E., Eierud, C., Banks, D., Harshbarger, T., Michael, A., & Rammell, C. (2021). Effects of lifelong musicianship on white matter integrity and cognitive brain reserve. *Brain Sci, 11*, 67.

Andrews, E., Eierud, C., Banks, D., Harshbarger, T., Michael, A., & Rammell, C. (2023). DTI analysis of white matter integrity and cognitive brain reserve in lifelong musicians and controls. *J Psychiatry Psychiatric Disord, 7*(2), 67–79.

Andrews, E., Frigau, L., Voyvodic-Casabo, C., Voyvodic, J., & Wright, J. (2013). Multilingualism and fMRI: Longitudinal study of second language acquisition. *Brain Sciences, 3*(2), 849–876. https://doi.org/10.3390%2Fbrainsci3020849

Andrews, E., Harshbarger, T., & Rammell, C. S. (2019). Multilingual listening and reading: An fMRI study of Russian/ English and Spanish/English bilinguals. *Glossos, 14.*

Bialystok, E. (2011). Coordination of executive functions in monolingual and bilingual children. *Journal of Experimental Child Psychology, 110*, 461–468. https://doi.org/10.1016%2Fj.jecp.2011.05.005

Bialystok, E., Craik, F. I. M., & Freedman, M. (2007). Bilingualism as a protection against the onset of symptoms of dementia. *Neuropsychologia, 45*, 459–464. https://doi.org/10.1016/j.neuropsychologia.2006.10.009

Bialystok, E., Craik, F. I. M., & Luk, G. (2012). Bilingualism: Consequences for mind and brain. *Trends in Cognitive Sciences, 16*(4), 240–250. https://doi.org/10.1016%2Fj.tics.2012.03.001

Birdsong, D. (2006). Age and second language acquisition and processing: A selective overview. *Language Learning, 56*, 9–49. https://doi.org/10.1111/j.1467-9922.2006.00353.x

Bolinger, D. (1949/1965). The sign is not arbitrary. Reprinted in I. Abe & T. Kanekiyo (Eds.), *Forms of English: Accent, Morpheme, Order.* Harvard University Press.

Bolinger, D. (1986). *Intonation and Its Parts: Melody in Spoken English.* Stanford University Press.

Bolinger, D. (1989). *Intonation and Its Uses: Melody in Grammar and Discourse.* Stanford University Press.

Breiner-Sanders, K. E., Swender, E., & Terry, R. (2002). ACTFL proficiency guidelines (revised). *Foreign Language Annuals, 35*(1), 9–15.

Brice, H., Frost, S., Bick, A. S., Molfese, P. J., Rueckl, J. G., Pugh, K. R., & Frost, R. (2021). Tracking second language immersion across time: Evidence from a bi-directional longitudinal cross-linguistic fMRI study. *Neuropsychologia, 154*(3), 107796. https://doi.org/10.1016/j.neuropsychologia.2021.107796

Calvin, W. H., & Ojemann, G. A. (1994). *Conversations with Neil's Brain: The Neural Nature of Thought and Language.* Addison-Wesley/Harper & Row.

Child, J., Clifford, R.T., & Lowe, P. Jr. (1993). Proficiency and performance in language testing. *Applied Language Learning,* 4(1&2), 19–54.

Claussenius-Kalman, H., Hernandez, A. E., & Li, P. (2021). Expertise, ecosystem, and emergentism: Dynamic developmental bilingualism. *Brain and Language, 222,* 105013. https://doi.org/10.1016/j.bandl.2021.105013

Corina, D. P, Loudermilk, B. C., Detwiler, L., Martin, R. F., Brinkley, J. F., & Ojemann, G. A. (2010). Analysis of naming errors during cortical stimulation mapping: Implications for models of language representation. *Brain and Language, 115,* 101–112. https://doi.org/10.1016/j.bandl.2010.04.001

Cox, R. (2019). Equitable thresholding and clustering: A novel method for fMRI clustering in AFNI. *Brain Connectivity, 9*(7). https://doi.org/10.1089/brain.2019.0666

Craik, F. I. M, Bialystok, E., & Freedman, M. (2010). Delaying the onset of Alzheimer disease: Bilingualism as a form of cognitive reserve. *Neurology, 75*(19), 1726–1729. https://doi.org/10.1212/wnl.0b013e3181fc2a1c

Dash, T., Joanette, Y., Ansaldo, A. I. (2022). Exploring attention in the bilingualism continuum: A resting-state functional connectivity study. *Brain and Language, 224,* 105048. https://doi.org/10.1016/j.bandl.2021.105048

Davis, J. McE., Norris, J. M., Malone, M. E., McKay, T. H., & Son, Y.-A. (Eds.) (2018). *Useful Assessment and Evaluation in Language Education.* Georgetown University Press. https://doi.org/10.2307/j.ctvvngrq

de Bot, K. (2008). The imaging of what in the bilingual mind? *Second Language Research, 24*(1), 111–133. http://dx.doi.org/10.1177/0267658307083034

de Bot, K. (2009). Multilingualism and aging. In W. C. Ritchie & T. K. Bhatia (Eds.), *The New Handbook of Second Language Acquisition* (pp. 425–442). Emerald Group Publishing Ltd.

de Bot, K. (2019). Defining and assessing multilingualism. In J. W. Schwieter (Ed.), *The Handbook of The Neuroscience of Multilingualism.* Wiley Blackwell.

Demirci, O., Clark, V. P., Magnotta, V. A., et al. (2008). A review of challenges in the use of fMRI for disease classification / characterization and a projection pursuit application from a multi-site fMRI schizophrenia study. *Brain Imaging and Behavior, 2,* 207–226. https://psycnet.apa.org/doi/10.1007/s11682-008-9028-1

Eierud, C., Michael, A., Banks, D., & Andrews, E. (2023). Resting-state functional connectivity in lifelong musicians. *Psychoradiology, 3,* 1–8.

Eklund, A., et al. (2016). Cluster failure: Why fMRI inferences for spatial extent have inflated false-positive rates. *PNAS, 113*(28), 7900–7905. https://doi.org/10.1073/pnas.1602413113

Elmer, S. & Jäncke, L. (2018). Relationships between music training, speech processing, and word learning: A network perspective. *Ann. N. Y. Acad. Sci., 2018*(1423), 10–18.

Fitzsimmons, P., et al. (2017). What is proficiency-based learning? https://education.vermont.gov/sites/aoe/files/documents/edu-proficiency-based-learning-what-is-proficiency-based-learning.pdf

Gallese, V., & Lakoff, G. (2005). The brain's concepts: The role of the sensory-motor system in conceptual knowledge. *Cognitive Neuropsychology, 22*(3/4), 455–479. https://doi.org/10.1080/02643290442000310

Gallo, F., DeLuca, V., Prystauka, Y., Voits, T., Rothman, J., & Abutalebi, J. (2022). Bilingualism and aging: Implications for (delaying) neurocognitive decline. *Frontiers in Human Neuroscience, 16.*

Gallo, F., Novitskiy, N., Myachykov, A., & Shtyrov, Y. (2021). Individual differences in bilingual experience modulate executive control network and performance: Behavioral and structural neuroimaging evidence. *Bilingualism: Language and Cognition, 24*(2), 293–304. https://doi.org/DOI:10.1017/S1366728920000486

García-Pentón, L., Fernández García, Y., Costello, B., Andoni Duñabeitia, J., & Carreiras, M. (2016). The neuroanatomy of bilingualism: How to turn a hazy view into the full picture. *Language, Cognition and Neuroscience, 31*(3), 303–327. https://doi.org/10.1080/23273798.2015.1068944

Ghazi-Saidi, L., & Ansaldo, A. I. (2017). Second language word learning through repetition and imitation: Functional networks as a function of learning phase and language distance. *Frontiers in Human Neuroscience, 11*(463). https://doi.org/10.3389/fnhum.2017.00463

Ghazi-Saidi, L., Perlbarg, V., Marrelec, G., Pelegrini-Issac, M., Benali, H., & Ansaldo, A. I. (2013). Functional connectivity changes in second language vocabulary learning. *Brain and Language, 124,* 56–65. https://doi.org/10.1016/j.bandl.2012.11.008

Grant, A., et al. (2015). Second language lexical development and cognitive control: A longitudinal fMRI study. *Brain and Language, 144,* 35–47. https://doi.org/10.1016/j.bandl.2015.03.010

Green, D. W., Crinion, J., & Price, C. J. (2006). Convergence, degeneracy and control. *Language Learning, 56*(1), 99–125. https://doi.org/10.1111%2Fj.1467-9922.2006.00357.x

Grundy, J. G., Anderson, J. A. E., & Bialystok, E. (2017). Neural correlates of cognitive processing in monolinguals and bilinguals. *Annals of the New York Academy of Sciences, 1396*(1), 183–201. https://doi.org/10.1111/nyas.13333

Guzmán-Vélez, E., & Tranel, D. (2015). Does bilingualism contribute to cognitive reserve? Cognitive and neural perspectives. *Neuropsychology*, *29*, 139–150. https://doi.org/10.1037%2Fneu0000105

Hagoort, P. (2006). What we cannot learn from neuroanatomy about language learning and language processing: Commentary on Uylings. In M. Gullberg & P. Indefrey (Eds.), *The Cognitive Neuroscience of Second Language Acquisition*. Blackwell Publishers. https://doi.org/10.1111/j.1467-9922.2006.00356.x

Hickok, G., & Poeppel, D. (2004). Dorsal and ventral streams: A framework for understanding aspects of the functional anatomy of language. *Cognition*, *92*, 67–99. https://doi.org/10.1016/j.cognition.2003.10.011

Hickok, G., & Poeppel, D. (2007). The cortical organization of speech processing. *Nature Reviews*, *8*, 393–402. https://doi.org/10.1038/nrn2113

Hodgson, V. J., Ralph, M. A. L., & Jackson, R. L. (2021). Multiple dimensions underlying the functional organization of the language network. *NeuroImage*, *241*, 118444. https://doi.org/10.1016/j.neuroimage.2021.118444

Huettel, S., Song, A. W., & McCarthy, G. (2004). *Functional Magnetic Resonance Imaging*. Sinauer Associates.

Jafari, Z., Perani, D., Kolb, B. E, & Mohajerani, M. H. (2021). Bilingual experience and intrinsic functional connectivity in adults, aging, and Alzheimer's disease. *Annals of the New York Academy of Sciences*, *1*, nyas.14666. https://doi.org/10.1111/nyas.14666

Jakobson, R. (1956/1985). Metalanguage as a linguistic problem. In S. Rudy (Ed.), *Selected Writings VII* (pp. 113–121). Mouton.

Jakobson, R. (1957/1987). Linguistics and poetics. In K. Pomorska & S. Rudy (Eds.), *Language in Literature* (pp. 62–94). Belknap Press of Harvard University Press.

Jakobson, R. (1967/1985). Language and culture. In S. Rudy (Ed.), *Selected Writings VII* (pp. 101– 112). Mouton.

Liu, X., Tu, L., Chen, X., Wang, J., Li, M., Lu, Z., & Huang, R. (2021). Effects of AoA-L2 on L1 and L2 networks in early and late bilinguals. *International Journal of Bilingualism*, *25*(6). https://doi.org/10.1177/13670069211033026

Luk, G., Anderson J. A., Craik F. I., Grady, C., & Bialystok, E. (2010). Distinct neural correlates for two types of inhibition in bilinguals: Response inhibition versus interference suppression. *Brain and Cognition*, *74*(3), 347–357. https://doi.org/10.1016/j.bandc.2010.09.004

Luk, G., Bialystok, E., Craik, F. I. M., & Grady, C. L. (2011). Lifelong bilingualism maintains white matter integrity in older adults. *J. Neurosci.*, *31*, 16808–16813. https://doi.org/10.1523/JNEUROSCI.4563-11.2011

Mahon, B. Z., & Caramazza, A. (2008). A critical look at the embodied cognition hypothesis and a new proposal for grounding conceptual content. *Journal of Physiology*, *102*(1–3), 59–70. https://doi.org/10.1016/j.jphysparis.2008.03.004

Marek, S., et al. (2022). Reproducible brain-wide association studies require thousands of individuals. *Nature*, *603*, 654–660. https://doi.org/10.1038/s41586-022-04492-9

Mårtensson J,, Eriksson J,, Bodammer, N. C., Lindgren, M., Johansson, M., Nyberg, L., & Lövdén, M. (2012). Growth of language-related brain areas after foreign language learning. *Neuroimage, 63*(1), 240–244. https://doi.org/10.1016/j.neuroimage.2012.06.043

North, B. (2000). *The Development of a Common Framework Scale of Language Proficiency.* P. Lang. https://doi.org/10.3726/978-1-4539-1059-7

Ojemann, G. A, Corina, D. P., Corrigan, N., Schoenfield-McNeill, J., Poliakov, A., Zamora, L., & Zanos, S. (2010). Neuronal correlates of functional magnetic resonance imaging in human temporal cortex. *Brain, 33*(1), 46–59. https://doi.org/10.1093%2Fbrain%2Fawp227

Paradis, M. (2000). The neurolinguistics of bilingualism in the next decades. *Brain and Language, 71*, 178–180. https://doi.org/10.1006/brln.1999.2245

Paradis, M. (2004). *The Neurolinguistics of Bilingualism.* John Benjamins.

Paradis, M. (2019). Special forward. In J. W. Schwieter (Ed.), *The Handbook of the Neuroscience of Multilingualism.* Wiley Blackwell.

Pavlenko, A. (2005). *Emotions and Multilingualism.* Cambridge University Press.

Pliatsikas, C. (2019). Multilingualism and brain plasticity. In J. W. Schwieter & M. Paradis (Eds.), *The Handbook of the Neuroscience of Multilingualism* (pp. 230–251). Wiley Publishers.

Pliatsikas, C. (2020). Understanding structural plasticity in the bilingual brain: The dynamic restructuring model. *Bilingualism: Language and Cognition, 23*(2), 459–471. https://doi.org/10.1017/S1366728919000130

Pliatsikas, C. 2023. Bilingualism and brain structure: Insights from healthy ageing and progressive neurodenegerative disease. In G. Luk et al. (Eds.), *Understanding Language and Cognition through Bilingualism: In Honor of Ellen Bialystok* (pp. 301–317). John Benjamins.

Pliatsikas, C., Deluca, V., & Volts, T. (2020). The many shades of bilingualism: Language experiences modulate adaptations in brain structure. *Language Learning, 70*(52), 133–149. https://doi.org/10.1111/lang.12386

Pliatsikas, C., & Luk, G. (2016). Executive control in bilinguals: A concise review on fMRI studies. *Bilingualism: Language and Cognition, 19*, 699–705. https://doi.org/10.1017/S1366728916000249

Pliatsikas, C., Moschopoulou, E., & Saddy, J. D. (2015). The effects of bilingualism on the white matter structure of the brain. *PNAS, 112*(5). https://doi.org/10.1073/pnas.1414183112

Pliatsikas, C., Pereira Soares, S. M., Volts, T., Deluca, V., & Rothman, J. (2021). Bilingualism is a long-term cognitively challenging experience that modulates metabolite concentrations in the healthy brain. *Scientific Reports, 11*(1), 1–12. https://doi.org/10.1038/s41598-021-86443-4

Poeppel, D. (2008). The cartographic imperative: Confusing localization and explanation in human brain mapping. *Bildwelten des Wissens (Ikonographie des Gehirns)*, 6(1), 1–21. https://doi.org/10.1515/9783110548778-003

Poeppel, D., & Hickok, G. (2004). Towards a new functional anatomy of language. *Cognition, 92*, 1–12. https://doi.org/10.1016/j.cognition.2003.11.001

Price, C. (2010). The anatomy of language: A review of 100 fMRI studies published in 2009. *Annals of the New York Academy of Sciences, 1191*, 62–88. https://doi.org/10.1111/j.1749-6632.2010.05444.x

Raboyeau, G., Marcotte, K., Adrover-Roig, D., & Ansaldo, A. I. (2010). Brain activation and lexical learning: The impact of learning phase and word type. *NeuroImage, 49*, 2850–2861. https://doi.org/10.1016/j.neuroimage.2009.10.007

Raichle, M. (2001). Functional neuroimaging: A historical and physiological perspective. In R. Cabeza & A. Kingstone (Eds.), *Handbook of Functional Neuroimaging of Cognition* (pp. 3–26). MIT Press.

Raichle, M. (2006). The brain's dark energy. *Science, 314*, 1249–1250. https://doi.org/10.1126/science.1134405

Raichle, M. (2010). Two views of brain function. *Trends in Cognitive Sciences, 14*(4), 180–190. https://doi.org/10.1016/j.tics.2010.01.008

Raichle, M. (2011). The restless brain. *Brain Connectivity, 1*, 3–12. https://doi.org/10.1089%2Fbrain.2011.0019

Raichle, M. (2015). The brain's default mode network. *Annual Review of Neuroscience, 38*, 433–447. https://doi.org/10.1146/annurev-neuro-071013-014030

Reverberi, C., Kuhlen, A., Abutalebi, J., Greulich, R. S., Costa, A., Seyed-Allaei, S., Haynes, J. D. (2015). Language control in bilinguals: Intention to speak vs. execution. *Brain & Language, 144*, 1–9. https://doi.org/10.1016/j.bandl.2015.03.004

Schrimer, A., Fox, P. M., & Grandjean, D. (2012). On the spatial organization of sound processing in the human temporal lobe: A meta-analysis. *Neuroimage, 63*(1), 137–147. https://doi.org/10.1016/j.neuroimage.2012.06.025

Stein, M., Federspiel, A., Koenig, T., Wirth, M., Strik, W., Wiest, R., Brandeis, D., & Dierks, T. (2012). Structural plasticity in the language system related to increased second language proficiency. *Cortex, 48*(4), 458–465. https://doi.org/10.1016/j.cortex.2010.10.007

Stowe, L. A., Haverkort, M., & Zwarts, F. (2005). Rethinking the neurological basis of language. *Lingua, 115*, 997–1042. https://doi.org/10.1016/j.lingua.2004.01.013

Valenzuela, M. J., & Sachdev, P. (2006). Brain reserve and dementia: A systematic review. *Psychological Medicine, 36*, 441–454. https://doi.org/10.1017/s0033291705006264

Van Horn, J. D., & Toga, A. W. (2009). Multisite neuroimaging trials. *Current Opinion in Neurology, 22*(4), 370–378. https://doi.org/10.1097/wco.0b013e32832d92de

Vygotsky, L. S. (1934/1987). Thinking and speech. In R. W. Rieber & A. S. Carton (Eds.), *The Collected Works of L. S. Vygotsky*. Plenum Press.

Wang, R., Ike, S., Zhang, Q., Zhou, K., Li, P., & Yang, J. (2020). Functional and structural neuroplasticity associated with second language proficiency: An MRI study of Chinese-English bilinguals. *Journal of Neurolinguistics, 56*, 100940. https://doi.org/10.1016/j.jneuroling.2020.100940

Weinreich, U. (1953/1968). *Languages in Contact: Findings and Problems*. Mouton.

Weiss, M. W., & Bidelman, G. M. (2015). Listening to the brainstem: Musicianship enhances intelligibility of subcortical representations for speech. *The Journal of Neuroscience, 35*(4), 1687–1691.

Zou, L., Guosheng, D., Abutalebi, J., Shu, H., & Peng, D. (2012). Structural plasticity of left caudate in bimodal bilinguals. *Cortex, 48*(9), 1197e1206.

13

Cognitive and Neural Aspects of the Multilingual Mental Lexicon

John W. Schwieter and Gabrielle Klassen

13.1 Introduction

Individuals who have competence in more than one language can often name an object using more than one word. For instance, an English-Czech speaker may call a cat either a *cat* or *kočka* and an Italian-French speaker may call it a *gatto* or *chat*. This multiple-word-to-one-concept mapping not only has important implications for the brain as we will discuss in this chapter, but it also reveals critical neural activities that underpin human language and cognition.

In the upcoming sections, we will address several key questions reflecting the state of the literature on the mental lexicon and brain activity among multilinguals. In Section 13.2, we discuss research examining whether languages have separate, integrated, or partially integrated mental representations, and how words are processed in different languages. In Section 13.3, we ask whether words are best viewed as mental representations or patterns of brain activity and how we can examine words and lexical processing in the brain. In Section 13.4, we report on how research on the multilingual lexicon has advanced our understanding of brain function and cognition. On this backdrop, in Section 13.5, we transition to discussions on how perspectives of the mental lexicon can be conceptualized and some of the implications for models of the multilingual lexicon. Finally, we conclude the chapter in Section 13.6.

We would like to note that we refer to *multilinguals* as individuals with competence in two or more languages, including individuals who also have been referred to as *bilinguals*, *trilinguals*, and *polyglots*. Multilinguals can range from those who regularly and equally use their languages in various contexts to exclusively using one language in certain settings and other languages in different settings. Similarly, multilinguals may be more, less, or equally proficient in each of their languages. In this chapter, we will not delve further into the variable nature of multilingual profiles as this goes beyond our scope (but see de Bot, 2019 for a review). However, we will argue that multilinguals, regardless of how, where, and how much of their languages they use and know, offer opportunities for studying the mental lexicon and shaping our understanding of human cognition. This said, however, we use the terms *bilinguals* and the *bilingual brain* or *lexicon*, as most of the previous work presented here has focused on how two languages exist and interact in a single mind or brain. We acknowledge that the participants included in past studies are not typically identified as speaking only two languages. It is possible that they may speak more than two languages to some degree, and that these details may not have been considered important in previous work. Therefore, we specify for multilingual speakers whenever it is possible to do so, but the term bilingual is used throughout when referencing specific studies who identify their participants as such.

13.2 How Does the Brain Represent and Process Words in Different Languages?

Early research on cognitive and neural aspects of the multilingual lexicon was largely focused on examining whether words in the languages were represented and processed separately or in parallel (see Bobb & Kroll, 2018; Kroll & Tokowicz, 2005, for reviews). However, by the early 2000s, numerous studies had provided evidence that multilinguals' languages are constantly active (i.e., readily available) to some degree, regardless of whether they are speaking, listening, reading, or writing, and irrespective of their intention to do so in only one of their languages (Colomé, 2001; Costa et al., 1999; Dijkstra, 2005; Hermans, 2004; Spivey & Marian, 1999). To illustrate, in a study by Marian and Spivey (2003), participants were asked to view a computer screen which presented four objects and to choose the object whose name was played through earphones. The participants' eye movements were tracked while making their decision. In the critical condition of the experiment, the auditory second language (L2) target label (e.g., English "marker") had a phonological similarity with one of the irrelevant object's names in the first language (L1) (e.g., *marka*, Russian "stamp"). The findings demonstrated that the participants significantly looked more often to the cross-language distracter than unrelated distracter objects, indicating that words in both languages were coactivated. Further tests revealed support for coactivation in the opposite direction in which the L2 was activated during L1 processing. Evidence for coactivation of languages has also been found among multilingual children (Poarch & van Hell, 2012; Von Holzen & Mani, 2012), suggesting that coactivation in the multilingual lexicon exists from the earliest stages of additional language learning and development. This accumulating evidence across the lifespan suggests that the dynamics of languages in the multilingual mental lexicon characterizes a highly penetrable and interactive system.

The coactivation of words in multiple languages does not appear to be sensitive to their typological similarity, but rather to the mere presence of the languages themselves (Bobb & Kroll, 2018). For instance, studies have found evidence for cross-language interactions among languages with different written scripts (Deng et al., 2024; Hoshino & Kroll, 2008; Thierry & Wu, 2007; Wu & Thierry, 2010) and modalities (i.e., signed vs. spoken or written language; Emmorey et al., 2008; Morford et al., 2011). These findings illustrate the openness and permeability of multiple languages in one brain.

Given that a multilingual's languages are constantly activated in parallel, it is appropriate to ask how and to what extent the languages influence each other. It should not be surprising that the more dominant language (e.g., the L1) often impacts the less-dominant languages. A case in point is a multilingual who has a "foreign accent" in one of their languages. In fact, these sound clues may permit the listener to "detect" what the speaker's L1 may be. What might be more surprising, however, is that less-dominant languages can also leave their mark on more-dominant languages. For instance, a study by Ameel and colleagues (2005) demonstrated the bidirectional nature of cross-language influence at the lexical and conceptual levels. In the study, Dutch-French multilinguals were asked to name common household objects in the two languages. Compared to Dutch- and French-speaking monolinguals, the naming patterns of the multilinguals converged on a common naming pattern. This finding suggests that multilinguals become less sensitive to the linguistic properties of each language when labeling common objects, and that the category boundaries in each language merge and thus diverge from the boundaries observed in monolinguals of those same languages.

The revelation that for multilinguals, words in all their languages are active and competing for selection, led to a shift in research foci to examine issues such as how words in the intended language are ultimately selected and the nature of the language control processes that are recruited to enable accurate usage. There have been at least three important theoretical models that have elaborated explanations of these issues. The Bilingual Interactive Activation+ (BIA+) model (Dijkstra & van Heuven, 2002) is an account of visual word recognition in which bottom-up information (i.e., perceptual information in the form of written language) serves to cue a given language based on language-specific properties (i.e., phonological, orthographic). Top-down attentional processes simultaneously support the selection of the appropriate linguistic features.

Another proposed model critical to the bilingual lexicon is Green's (1998) inhibitory control model (ICM), which argues that competing linguistic features and words are inhibited in accordance with the goals of the speaker. In this model, top-down mechanisms are responsible for judging lexical candidates for selection and inhibiting those that are irrelevant to the speaker's intent. Inhibitory control is reactive in that it suppresses nontarget words as they become activated.

Finally, the Bilingual Language Interaction Network for Comprehension of Speech (BLINCS) model (Shook & Marian, 2013) posits that a multilingual's languages have shared phonological representations, integration at the lexical level, and a single semantic system in which languages share conceptual representations. The model captures developmental changes to the language system as a result of learning, with the interconnected levels of processing being created by using dynamic, self-organizing maps. These three models have developed alongside innovative ways of identifying and measuring words in the multilingual brain, which we will discuss in the next two sections.

13.3 How Can We Examine Words and Lexical Processing in the Brain?

Several methodologies have been employed to investigate lexical activation in the bilingual brain, and these methodologies differ in terms of their advantages and disadvantages. These methodologies can be divided into two broad categories: those that measure behavior and those that measure brain activity, which include neuroimaging and electrophysiological measures.

13.3.1 Measuring Behavior

In this section, we describe the methodologies that measure the behavior of bilinguals and how these methodologies can provide insight into the bilingual brain. While they do not measure brain activity directly, the behavior of bilinguals can indexically serve as an indirect measure of processing difficulty, which can then be extrapolated to theories about how language is represented in the multilingual mind.

Behavioral studies on the bilingual lexicon have centered around measuring two processes: lexical activation and inhibition. Linguistic priming is a common method to measure activation. Priming is the influence of a recently processed stimulus on a currently processed one, usually indicating that the prime has activated the target item and, therefore, there is a meaningful link associating the two. In other words, the prime and the target are related in some meaningful way (semantically, phonologically, syntactically, and/or morphologically), and so the activation of the prime prompts the activation of the target word or structure. This may be either employed as masked or unmasked priming. In masked priming studies, the prime is presented too quickly for the participant to consciously perceive it, but slowly enough that the information is still processed. In unmasked priming, the preceding prime is presented long enough for it to be consciously perceived. Note that some priming studies have presented the prime at the same time as the target stimulus (e.g., Meyer et al., 1975) but this methodology is not common as one is not able to control for the time the prime is presented. As a result, it is difficult to draw conclusions about the order of processing of the prime and the target, given that they are presented simultaneously, and the argument of "priming" at all becomes difficult to prove, even if robustly supported by results. Additionally, error rates may also be reported in these studies as a measure of difficulty (i.e., the more errors produced, the more difficult the stimulus is to process in that condition).

Inhibition is typically measured by providing participants with competing words or structures which they must supress or inhibit in one language while responding in another. Again, this is a means of demonstrating how words are related in the bilingual brain, particularly when switching between languages. Switching from a particular language to another may be more costly than the reverse (e.g., Costa & Santesteban, 2004; Schwieter & Sunderman, 2008), showing a higher cost of inhibiting the competing items in one language more than the other.

Activation and inhibition are two processes that can be observed in behavioral studies by measuring indirect responses to linguistic stimuli, for example, responding by pressing a button or by providing an oral response. The time between the onset of the stimulus and the response is measured, providing an indirect measure of processing difficulty. If it takes longer to respond to a stimulus in some condition compared to a neutral condition, this is attributed to some processing difficulty experienced by the participant, such as lexical competition and subsequent inhibition of a competing word in the other language. Likewise, if a participant responds faster to a stimulus in some primed condition compared to a neutral condition, there is strong evidence of primed activation.

As mentioned above, it is generally accepted that when bilinguals speak one language, lexical representations from both languages are activated (e.g., Abutalebi & Green, 2007; Branzi et al., 2014; Costa & Caramazza, 1999; Costa & Santesteban, 2004; de Bot, 1992; Finkbeiner et al., 2006; Green, 1998; Kroll et al., 2006, 2008). A large body of work on how words from two languages are represented and interrelated in bilingual minds reports on priming in multilingual populations. Priming studies involve presenting the participant with a stimulus that is similar to the target stimulus and is therefore intended to prime the activation of the upcoming stimulus. The participant is then asked to respond to the stimulus. A common example is a lexical decision task, where the participant is asked to press a specific button if the word presented is a real word in a given language and another button if it is not. Participants respond to target words related to the prime (in form, meaning, etc.) faster than if it were unrelated (e.g., Martin et al., 2009), indicated by their faster button presses, oral responses, or eye gaze shifts.

In priming experiments, the nature of how words are stored in the bilingual brain and how words from both languages might interact can be examined by varying the type of prime presented. This could include variation in how similar the two words are phonetically, phonologically, syntactically, or semantically. Additionally, the exact parameters of the task (e.g., the amount of time each stimulus is presented) can provide insight into the steps in lexical activation in bilinguals. For example, Keatley and de Gelder (1992) examined the role of semantic priming in bilingual speakers across a series of experiments, modulating the difficulty of the task (neutral or speeded) and the type of prime (direct translations or semantically related words). In the experiments, participants were instructed to press a specific button to indicate whether the word they viewed was a real word in a given language or not. The purpose of this study was to determine whether priming across languages occurred as a prelexical strategy, a processing strategy employed to help the participant respond faster. This is a predictive strategy for timely lexical selection or activation (e.g., Grainger & Beauvillain, 1988). Alternatively, it may be a post-lexical integration process, whereby the meaning is integrated after accessing the target and, as a result, primed stimuli are more biased toward a positive response in a decision task, resulting in faster reaction times (e.g., de Groot & Nas, 1991). The authors found that the speeded tasks erased any cross-linguistic priming effects, except with direct translations. They concluded that this provides evidence for priming as a post-lexical process.

Above and beyond when priming takes place in a processing sequence, there are the central questions of how the lexical items in two languages are related; specifically, at what levels, and the directionality of the association (whether there is a stronger activation from one language to another). A common result, specifically for sequential bilinguals or L2 learners, is that priming is stronger from the first or more dominant language to the second or less dominant language than vice versa (e.g., L1 → L2 greater than L2 → L1; Altarriba, 1992; Chen & Ng, 1989; Frenck & Pynte, 1987; Jin, 1990; Keatley et al., 1994). This begs the question of whether the activation is stronger in one direction than another, or if one language has overall faster processing than the other. The answer to questions like these provides insight into the representation of lexical items and how they are associated between languages. A case in point, for example, comes from Jiang (1999) in which the author investigated the priming of translated words in Chinese-English bilinguals across five experiments that manipulated the directionality of the prime (L1 → L2 vs. L2 → L1) and the presentation conditions of the primes and targets, as a measure of processing difficulty. Namely, the timing of presentation of the prime, the timing between the prime and the target, and the distribution of the items across the experiment were all varied. These manipulations should act to enhance processing of the L2 prime in particular (e.g., giving more time between the prime and the target allows more time to process the L2 prime). Jiang found consistent results across all manipulations of the stimuli: No matter how the processing task was adjusted, there were strong priming effects in the expected direction (L1 → L2) and little to no effects were observed in the opposite direction (L2 → L1). The author argued that the results support representational accounts of priming (e.g., Kroll & Stewart's (1994) revised hierarchical model), where the strength of association between the form and the meaning is stronger in the L1 than the L2, leading to priming effects from the L1 to the L2 (but not from the L2 to L1), through the shared semantic meaning.

Several studies have found that priming effects may depend on various factors beyond those specifically described above, including proficiency in the two languages (e.g., Francis et al., 2014), age of acquisition (e.g., Veríssimo et al., 2018), directionality of the languages (e.g., Altarriba, 1990; Jin, 1990; Keatley et al., 1994), greater priming within languages than across languages (e.g., Kirsner et al., 1984), or with direct cognates rather than semantically related words (e.g., de Groot & Nas, 1991). Many of these factors are interrelated, as well. In particular, proficiency may be impacted by other factors, such as difficulty of the task, age of acquisition, task modality, and experience. For example, Lee and colleagues (2018) observed that lower proficiency may impact the degree to which L1 priming affects L2 activation, but that it is modulated by the details of the task itself. Low proficiency Korean-English bilinguals showed the typical L2 → L1 priming effect when the prime was presented for 150 ms, but not when presented for 60 ms. This relatively longer prime presentation allowed the unbalanced bilinguals sufficient processing time in their L2 to activate the L2 word, permitting it to prime the L1 equivalent. When there was not sufficient time to process the L2 form, it was not available to prime the L1 form. In other words, the low proficiency in this group might have initially been interpreted to mean that low proficiency learners or unbalanced bilinguals do not activate their L2 while using their L1. Instead, the results show that they do activate their L2, but that it is constrained by increased processing cost. Therefore, any method seeking to show priming effects must account for this increased need for processing time. In sum, additional factors like proficiency and age of acquisition must be considered carefully both in the experimental design and in the results.

Highly related to proficiency is the factor of age of acquisition. Simultaneous bilinguals may show different behaviors than sequential bilinguals, which may interact with the relative proficiency or dominance that these two groups tend to have in their two languages. Unlike in studies with sequential bilinguals, simultaneous bilinguals show priming effects in both directions (i.e., L1 → L2 and L2 → L1; Duñabeitia et al., 2010; Sabourin et al., 2014; for a computational simulation, see Zhao & Li, 2013). For example, Sabourin and colleagues investigated masked semantic priming in English-French bilinguals who began learning their two languages simultaneously, early or late, with similar end-state proficiency. They observed priming effects only in the early and simultaneous bilinguals, but not in the late bilinguals. These results highlight the importance of considering the many extraneous factors that may impact how a bilingual's lexicon is structured.

Although it is not the focus of this chapter, it is interesting to note that this methodology has informed not only our understanding of the bilingual lexicon in terms of how lexical items are stored and accessed, but also of how the syntactic rules/patterns are stored and accessed. There is much evidence from priming studies demonstrating that not only lexical items but also syntactic rules may be activated in both a bilingual's languages, given the influence of one language's syntax on another (e.g., Bernolet et al., 2013; Hartsuiker et al., 2016; Kantola & van Gompel, 2011; Loebell & Bock, 2003; for reviews, see

Hartsuiker & Pickering, 2008; van Gompel & Arai, 2018). For a discussion on various theories of whether syntactic details are shared or separate in the bilingual brain, see Muylle and colleagues (2021).

13.3.2 Measuring Inhibitory Control

While priming is measured as a general increase in processing (faster responses and/or fewer errors), inhibition is a measure of the other side of the same coin. Priming shows that words in both languages are activated because the activation of a word prompts the activation of another before it is presented. Measuring inhibitory control, on the other hand, estimates the inhibition of a competing word while another is being processed, resulting in processing difficulty (slower responses and/or more errors). This result provides evidence for the same types of questions: How are words from two languages represented and related in the bilingual brain?

There is some debate about the specific role of inhibition of competing lexical items, including whether it is a language-specific phenomena (language nonspecific models: Abutalebi & Green, 2007; de Bot, 1992; Green, 1998; Kroll et al., 2006; language-specific models: e.g., Costa & Caramazza, 1999; Finkbeiner et al., 2006) and the degree to which relative proficiency in each language modulates inhibitory control (Branzi et al., 2014; Costa & Santesteban, 2004; Jia, 2022; Kroll et al., 2008). For example, Filippi and colleagues (2014) investigated the inhibition of cognate words in late Italian-English bilinguals. The participants read words from a computer screen in a specific language, indicated by a color cue. The reaction measure is the time between the presentation of the stimulus and the onset of the speech production. Overall, the results showed a slower production for cognates than for noncognates, indicating that there was competition for selection between languages. In another study, Broersma and colleagues (2016) investigated effects of inhibition and dominance in English-Welsh bilinguals by using a picture-naming task. The authors found that English-Welsh bilinguals who were more dominant in English experience slower naming times with cognates in Welsh compared to noncognates. This result suggests that dominance may affect the degree to which related words are activated and/or inhibited in bilinguals.

Another study that measured competition and inhibition was conducted by Martín and colleagues (2010), who presented Spanish-English bilinguals with a semantic relatedness task. Participants viewed pairs of words in English and had to determine whether they were semantically related or not by pressing a button. Participants were presented with pairs that included a homograph in English and Spanish, whose translation would be semantically related (e.g., *pie*-toe, where *pie* means *foot* in Spanish). This was followed by the translation in English (e.g., foot-toe). Martín and colleagues found that both the homograph and the subsequent translation trial evoked longer response times, indicating inhibition of the Spanish word and its translation in English in the subsequent trial. This offered evidence not only that there is inhibition of competitor lexical items in bilinguals, but that this inhibition may linger, as the already activated English form (e.g., foot) was not only activated but also inhibited in the task. They observe that the inhibition of the English translation is temporary and fades within a 175 ms interval.

A third methodology that can elucidate our understanding of bilingual processing is eye-tracking, a measure which records eye movements and fixations in real time. This methodology has been used to determine the nature of competitors in bilingual lexical access. Evidence has shown that there is an effect of competition for words that are semantically related (e.g., Martin et al., 2009) or phonologically related (e.g., Marian et al., 2008) in the two languages. In these studies, bilingual participants look more to the competing items and fixate longer on the target items when there is a competitor item on the screen, including both within- and cross-language competition (e.g., Canseco-Gonzalez et al., 2010; Cutler et al., 2006; Ju & Luce, 2004; Marian & Spivey, 2003; Spivey & Marian, 1999; Weber & Cutler, 2004). These results show that both the target and the competitor are activated, even before the word is completed. For example, Spivey and Marian (1999) investigated the inhibition of phonologically similar words across languages in Russian-English bilinguals. The participants saw a target image (e.g., a marker), a picture that would be a phonological competitor in Russian (e.g., a stamp, which is *marka* in Russian), as well as some unrelated distracter items. The authors found that the bilinguals looked significantly more at the cross-language competitor than at the unrelated items, showing evidence of activation in both languages.

The methodologies described in this section have been instrumental in the development of various representational models of the bilingual lexicon. A clear advantage of behavioral methodologies is that they are accessible and noninvasive. It is possible for researchers to employ these tasks with little formal training and many programs exist that can accurately and affordably collect and store behavioral data. The tasks are noninvasive and can be done quickly and largely without physical contact. As such, participants may feel more comfortable in these tasks. A clear disadvantage of these methodologies is that they cannot provide insight into where in the brain bilingual processing occurs or characterize the brain activity that takes place. These latter measures are essential to understanding the bilingual brain. They can serve to provide evidence for the relationship between the proposed components of the bilingual lexicon, but this evidence is still indirect. Additionally, there is more temporal distance between the brain's reaction and the measurement taken. For example, in a lexical decision task, a participant's brain may respond to a difficult stimulus in a specific way, causing a delay in the response of pressing the button. Between this brain activity and the pressing of the button is the execution of a motor response (e.g., the finger responding). These intermediate steps are potential areas in which variability may occur.

13.3.3 Measuring Brain Activity

Methods of directly measuring brain activity can be largely divided into neuroimaging and electrophysiological research. These methods have the benefit of measuring brain activity more precisely through the measure of a subconscious physiological reaction.

13.3.3.1 Neuroimaging

Neuroimaging, including positron emission tomography (PET) and functional magnetic resonance imaging (fMRI), measure the blood flow to the different regions of the brain. Increases in regional cerebral blood flow have been shown to be closely related to changes in neural activity. Typically, this is measured in two steps: measuring blood flow in a control condition and again in an experimental condition. Increased blood flow to a specific region in the experimental condition compared to the control condition provides evidence for which areas of the brain are responsible for language processing. For this reason, they are excellent indications of the localization of brain activity and can be used to answer questions about which areas of the brain are recruited during language processing in bilinguals. The disadvantage of this methodology is that it is not as temporally sensitive as other methods, given that there is a delay between the onset of a stimulus and the blood flow to the relevant region, constrained by the natural rate of blood flow. Another disadvantage of fMRI imaging is that the orbitofrontal and inferior temporal regions and the temporal pole may not be well visualized due to interference with the magnetic field. This is caused by the air in the middle ear and the mastoid bone creating interference with the magnetic field. For an overview, see Abutalebi and colleagues (2005).

A key question in bilingual neuroimaging studies has centered on whether the same area of the brain is employed to process a bilingual's two languages. Early neuroimaging work on bilinguals showed that the same areas of the brain supported the function of the bilingual's two languages, given a relatively high level of proficiency in the two languages (e.g., Abutalebi et al., 2005; Briellmann et al., 2004; De Bleser et al., 2003; Hasegawa et al., 2002; Hernandez & Meschyan, 2006; Klein et al., 2006; Luke et al., 2002; Meschyan & Hernandez, 2006; Rüschemeyer et al., 2006; Vingerhoets et al., 2003; Xue et al., 2004; Yetkin et al., 1996; see also Correia et al., 2014; Li et al., 2021; Van de Putte et al., 2017; Yang et al., 2017). Differences in brain activity can be attributed to a relatively low proficiency or dominance in one of the languages, resulting in the need to employ more cognitive control mechanisms (Abutalebi et al., 2001; Abutalebi & Green, 2007; Hesling et al., 2012; Linck et al., 2012; Van Hell & Dijkstra, 2002), late age of acquisition of the L2 (Kim et al., 1997; Perani et al., 1998; Wartenburger et al., 2003; for discussion see Hernandez & Li, 2007), the level of analysis (e.g., early vs. later stages of processing: Marian et al., 2003), and the linguistic similarity between the L1 and L2 (Jeong, Sugiura, Sassa, Haji, et al., 2007; Jeong, Sugiura, Sassa, Yokoyama, et al., 2007; Kovelman et al., 2008).

Several studies have probed deeper into the related variables of proficiency and age of acquisition, given that early bilinguals tend to have more balanced proficiency in their two languages compared to late bilinguals. In a meta-analysis of bilingual neuroimaging studies, Liu and Cao (2016) observed that some areas of the brain (e.g., insula and frontal cortex areas) were activated more in an L2 than an L1, especially in late bilinguals. For instance, Indefrey (2006) found increased activation in the L2 compared to the L1, especially when the individual was a late learner or had low proficiency in or exposure to their L2. Likewise, Sebastian and colleagues (2011) conducted a meta-analysis on the role of proficiency in bilingual neuroimaging studies. The authors found that highly proficient bilinguals demonstrated more similar networks in their L1 and L2 than less-proficient bilinguals, who showed greater differences between their L1 and L2.

Similarly, some authors have argued that differences in neuroimaging results for the two languages is due to the increased requirement of control for one language (the more dominant) relative to the other (the weaker language; Abutalebi & Green, 2007). Kroll and colleagues (2013) provide an overview of the evidence showing that the area of the brain associated with cognitive control is not only the same area that is engaged for bilingual processing but that bilingual speakers employ cognitive control networks more efficiently than monolingual speakers. For example, Abutalebi and colleagues (2012) observed reduced activity in the anterior cingulate cortex, which has been identified as an area involved in cognitive control, during a nonverbal conflict monitoring task. This result suggests that bilingual speakers require less effort in employing cognitive control, resulting in less blood flow to this crucial area, even for nonlinguistic tasks.

Another question that has been investigated using fMRI is whether bilingual lexical activation is language selective or nonselective. In other words, when processing in one language, does the other language compete for selection? This methodology is useful to determine whether words with competitors across languages, such as homographs, elicit evidence of processing difficulty (i.e., increased blood flow to the relevant region). For example, van Heuven and colleagues (2008) used fMRI to investigate the processing of English-Dutch homographs by bilinguals in two lexical decision tasks. For example, *boom* is a real word in both languages but refers to an onomatopoeic word representing a loud noise in English and *tree* in Dutch. In a lexical decision task, the authors found evidence of increased blood flow, specifically in the left prefrontal cortex, and behavioral evidence of processing difficulty (longer reaction times) in another task. Together, the evidence suggests that the lexical items in both languages were activated, even when instructed to complete the task in only one language. Another example comes from De Blesser and colleagues (2003) who investigated the activation of cognates in Flemish-Dutch-French multilinguals using PET. The authors observed a different activation pattern (left inferior frontal and temporoparietal areas) when processing L2 noncognates, compared to cognates named in the L1 or

L2. The authors conclude that producing words in a less proficient language may require recruitment of additional neural areas.

While PET and fMRI provide interesting and important results about the localization of bilingual processing, there are some key disadvantages to using these methods. It is possible that the overlapping areas observed for both languages could differ in terms of degree of activation or by recruiting actions at different time courses (see van Heuven & Dijkstra, 2010, for a summary). Even when languages appear to be processed in overlapping regions, it is also possible that they may not overlap at higher levels of brain organization (Hernandez et al., 2005; Xu et al., 2017). Neuroimaging, therefore, is most useful for broad localization but not for time-course specific questions or higher levels of brain organization.

13.3.3.2 Electrophysiological Research

Another method that directly measures brain activity utilizes electroencephal ography. This is often conducted using electroencephalogram (EEG), which typically measures intervals related to a given stimulus. An EEG device measures the electrical activity (event-related potentials (ERPs)) of the brain through electrodes placed along the scalp. Unlike blood flow, electrical activity is an instantaneous measure of the brain which offers high temporal resolution, but low spatial resolution. Moreover, ERPs are typically used to answer questions that concern time-locked processing, such as when cognitive processes are employed in bilingual processing.

For the bilingual lexicon, ERPs are particularly useful to answer questions such as whether lexical activation is language selective or nonselective. For example, orthographic "neighbours" have been used to test the coactivation of words with similar spelling, even in monolingual populations, to examine whether orthographic details of a word are connected, and thus stored in a shared or integrated lexicon (Holcomb et al., 2002; Massol et al., 2010; Midgley et al., 2008). These neighbours have been found to influence the N400 ERP component (i.e., a heightened negativity between 300 and 500 ms after the onset of the word, typically peaking around 400 ms). For example, Thierry and Wu (2004, 2007) investigated the processing of translated neighbors in Chinese-English bilinguals. Participants were presented with semantically related pairs in English only. Half of these pairs had translations that shared a character (were neighbours in Chinese) and half had translations that did not. A larger ERP response was observed for the word pairs with translation neighbors than for those without. As the participants were reading exclusively in English, the only explanation for this difference is that the L1 equivalent words were also activated, and this prompted the expected neighbor processing effect. Similar results were observed in Martin and colleagues (2009), who found an effect of semantic priming across languages, even though these results were not reflected in the offline behavioral data collected. These findings support the notion that lexical access is nonselective: both languages are active, even when processing in a single language.

Kerkhofs and colleagues (2006) also investigated whether bilingual lexical access is language-nonselective by analyzing ERPs. In this study, the authors presented Dutch-English bilinguals with semantically related or nonrelated pairs of words, where the preceding English prime was semantically related or nonrelated to the homograph in Dutch. In this way, effects of priming can only be explained by the activation of the Dutch homograph by the English prime. The authors observed different degrees of negativity in the N400 component depending on the language and the degree of frequency. These findings point to strong evidence for cross-language activation, which was also impacted by word frequency. The study, and many others, provide evidence that bilingual processing is nonselective; both languages are activated in parallel (for recent reviews, see Kroll, Bogulski, & McClain, 2012; Kroll, Dussias, et al., 2012).

More recent studies have emerged demonstrating that the degree of language selectivity may be modulated by external factors (Casaponsa et al., 2014; Casaponsa & Duñabeitia, 2016; de León Rodríguez et al., 2022; Hoversten et al., 2017). De León Rodríguez and colleagues investigated the role of L2 exposure in highly proficient bilinguals using ERPs in a semantic categorization task. The authors found that the participants with more exposure to their L2 were more accurate and displayed neurophysiological differences compared to the participants with less exposure to their L2, specifically in the N1 component. These results suggest a difference in prelexical and lexical processing in bilinguals that depends on exposure to the L2, but not the level of proficiency, as all learners were equally highly proficient. In sum, EEG results suggest that language selectivity may not be a categorical phenomenon but, rather, a gradient one, which may depend on other external factors, such as the degree of exposure to the L2.

The use of ERPs as a time-locked methodology is particularly useful for investigating the processing delay typically assigned to bilingual or L2 learner populations. The N400 component has been argued to represent lexical and semantic processing (see review in Kutas & Federmeier, 2000). Typically, the N400 effect is delayed for bilinguals with less proficiency in the L2 (Ardal et al., 1990; Elston-Güttler & Friederici, 2005; Hahne, 2001; Kutas & Kluender, 1994; Moreno & Kutas, 2005; Phillips et al., 2006; Weber-Fox & Neville, 1996; but see Kotz, 2001). For a review of N400 results in bilinguals, see Moreno and colleagues (2008).

While there is a wealth of knowledge to be gleaned from previous works in the previous two sections, it is worth noting that earlier works on neuroimaging and electrophysiological studies and the methodologies employed have endured heavy criticism. Paradis (2004) criticized these studies along several parameters, arguing that conflicting results, in particular with the localization of language in the brain, may be due to inconsistent methodologies. One criticism is about the presentation of one single word as a representation of language processing. Paradis argues that these are not linguistic tasks given that they do not involve the implicit procedural operations concerned in processing a verbal or written message. In isolation, without any context, a single word may be processed using declarative memory, rather than accessing procedural memory, where a language's implicit system is supported. In this sense, the conclusions drawn about language may not be reliable or informative, given that only a single word was accessed. Another criticism is that the varying methodologies employed make comparing across studies difficult, if not impossible. Unless a more systematic analysis is employed, it is not prudent to compare across different tools (e.g., PET vs. fMRI, or EEG vs. MEG), different populations, and different linguistic tasks or modalities. Paradis' call for caution and systematicity, as well as more careful and rigorous investigation, suggests that these methodologies should not be abandoned but rather prove to be a potentially enlightening means of understanding the bilingual brain, given sufficient cooperation and rigour from the scientific community.

13.4 How Has Research on the Multilingual Lexicon Advanced Our Understanding of Brain Function and Cognition?

Significant strides in our understanding and tracing of cognitive activities in the brain have been fueled by research examining the mental lexicon among multilinguals. In the previous section, we summarized the work in behavioral and neurological methodologies and what they have shown us about the bilingual lexicon. In this section, we review research employing various research methods among both healthy individuals and those diagnosed with the acquired language disorder known as aphasia, and what bilingualism in these populations has shown us about the human brain and mind.

13.4.1 Evidence from Aphasia and Dementia

Participants who have suffered physiological damage to the areas of the brain involved in language often experience difficulty with language use. The clinical term for this type of language impairment is aphasia. In the case of bilinguals, the affected language can vary. Bilingual aphasics have provided much information about how the bilingual brain is organized. Studies on this topic may also provide insight into the amount of overlap between the two languages, given the different patterns of impairment and recovery. The various lesion sites and the corresponding symptoms offer insight into which processes are localized in certain areas (and in which language(s)). Lastly, the differences between monolingual and bilingual aphasiacs allow researchers to understand how the bilingual brain may be structurally different from the monolingual brain.

Behavioral and neurological data from healthy individuals has provided ample evidence regarding the neural overlap of a bilingual's two languages (e.g., Abutalebi et al., 2001; Fabbro, 1995; Liu et al., 2010; Meykadeh et al.,

2021; Perani & Abutalebi, 2005; Tan et al., 2003; Wartenburger et al., 2003). Evidence from bilinguals suffering from aphasia calls into question the degree of overlap that occurs. If there were complete overlap, one would expect that injury or insult to one area of the brain would result in the same level of impact for both languages. A large body of evidence shows that this is not the case (e.g., Fabbro & Paradis, 1995; Hernandez, 2013; Paradis, 1977; Paradis et al., 1982). Indeed, a common observation is that there are heterogeneous language recovery patterns in polyglot aphasiacs. Fabbro (2001) followed the recovery of twenty Friulian-Italian bilinguals, all right-handed and with lesions in the left hemisphere. The author observed that 65 percent of the patients experienced parallel recovery of their two languages, 20 percent greater impairment of the L2, and 15 percent greater impairment of the L1. A similar distribution of recovery patterns was observed in a review by Paradis (2001). Aside from the order of acquisition, relative proficiency in the two languages may impact recovery patterns. Several authors have observed that the more proficient language may be less impaired than the weaker language (e.g., Druks & Weekes, 2013; Edmonds & Kiran, 2006; Gray & Kiran, 2013).

Many other factors have been found to influence the trajectory of recovery for each language, including the quantity of input from and amount of use of each language following the injury, and an individual's emotional experience (Aglioti et al., 1996; Goral et al., 2006; Green, 2005; Minkowski, 1927/1983; see also Połczyńska & Bookheimer, 2021, for a review of neural overlap and external contributing factors). An amalgamated view of language separation, then, best supports these findings, suggesting that while a bilingual's language system may overlap, there may remain some neurological pathways that develop in distinct ways.

Our understanding of the difference between implicit/procedural memory and explicit/declarative memory has also been impacted by work on bilingual aphasiacs. Implicit memory relates to automatic processes, whereas explicit memory relates to conscious, controlled processes (Ullman, 2001). These two memory systems employ different areas of the brain: Generally, implicit memory involves the left frontal region and basal ganglia, and explicit memory involves the bilateral temporal lobes (for discussion, see Voss & Paller, 2008). The difference between these processes is argued to be related to language learning modality. Bilinguals may differ in the way their L2s are learned as an interaction with age (e.g., learning in a naturalistic environment as a child vs. learning in a traditional classroom setting with explicit instruction) and proficiency (attainment of native-like proficiency or not). As a result of these factors, the neural mapping of the languages may be impacted by means of accessing different memory stores (Paradis, 2004; Ullman, 2001). Results from bilinguals with aphasia generally support this distinction between more or less employment of different memory systems given earlier/later age of acquisition and higher/lower relative proficiency. In particular, when there is damage to the subcortical region (which includes the basal ganglia, associated with implicit memory), better recovery of the weaker L2 is observed (Aglioti et al., 1996; Paradis, 2004). This is argued to be caused by a reliance on explicit memory and the assumption that the L2 relies more on explicit memory. In other words, if implicit memory is compromised, the reliance on explicit memory naturally leads to higher levels of L2 use, for which explicit memory is typically employed more often.

Another area where research on bilingual aphasia has contributed to our understanding of the bilingual brain comes from work on brain areas responsible for nonlinguistic or general cognitive function. Bilingual patients with damage to the prefrontal cortex (associated with executive control), or frontal lobe more generally, show an inability to control which language they use, resulting in "pathological switching" (e.g., Abutalebi et al., 2000; Fabbro et al., 2000; Kong et al., 2014). This is evidence that the prefrontal cortex is responsible for inhibiting one language while using the other, as damage in this area results in an inability to control language selection.

Lastly, evidence from dementia and Alzheimer's disease in bilingual patients has also enlightened our understanding of the bilingual brain. Bilinguals generally experience symptoms of dementia later than monolinguals (see the review by Bialystok et al., 2009). Additionally, conditions that are caused by deterioration of the brain, like Alzheimer's disease and dementia, have been found to manifest later (Alladi et al., 2013), causing bilinguals to function at high levels longer than monolinguals (e.g., Schweizer et al., 2012). These findings show that bilinguals may be coping better with brain damage and have later onset of symptoms for brain-related diseases, indicating that the linguistic and nonlinguistic components employ the same areas of the brain, and practice or use of one area enhances the endurance/perseverance/strength of that area (i.e., it is malleable).

13.4.2 Evidence from Electrophysiological Research

Evidence from electrophysiological research on bilingual speakers has advanced our understanding of the human brain as well. In particular, it has provided insight into how knowing or learning a second language impacts brain function in general. Results have largely followed two trends: evidence showing an impact on one's native language, causing behavioral and/or neurological differences between monolingual and bilingual speakers, and evidence showing general cognitive changes in nonlinguistic areas, compared to monolingual speakers.

Several behavioral and cognitive/neurological studies have found that learning an L2 or being bilingual may result in behavioral changes when using one's L1 (e.g., Ameel et al., 2009; Kroll & Gollan, 2014). Although this has been largely considered an effect observed in highly proficient L2 learners, Bice and Kroll (2015) observe differences in L1 processing at early stages of L2 learning. Using ERPs, the authors tested whether cognates in the L2 would impact processing in the L1 at early and intermediate levels of proficiency. The authors observed an effect of L2 priming of these cognates (increased N400 negativity compared to the monolingual group), particularly for intermediate speakers. These results suggest that at early to intermediate levels of proficiency, there is a bidirectional influence of the two languages, affecting processing in the L1 (see also McLaughlin et al., 2004; Zou et al., 2012).

Aside from consequences for the mechanisms involved in linguistic processes, bilinguals experience some cognitive benefits (termed "bilingual advantage"), wherein they fare better on nonlinguistic cognitive tasks involving executive control (Bialystok et al., 2005; Gold et al., 2013; Kroll & Bialystok, 2013; Luk et al., 2010). Given that bilinguals are tasked with controlling which language they use, which involves selection of the target language and inhibition of the competing language, they have more experience engaging this type of control, and, as a result, employ this process with greater efficiency. Studies using fMRI have demonstrated that the frontal areas involved in executive control in language processing (switching between languages) are the same as those employed for nonverbal executive function tasks (e.g., testing selective attention; see Luk et al., 2012, for a meta-analysis). For example, Abutalebi and colleagues (2012) compared the brain activity of bilinguals and monolinguals performing a nonverbal conflict-monitoring task. Bilinguals showed less activation than monolinguals of the anterior cingulate cortex, an area of the brain associated with cognitive control. The authors concluded that this was evidence that the bilingual group engaged in the same level of conflict monitoring more efficiently.

Compared to monolinguals, the frequent recruitment of executive functions among bilinguals increases later in adulthood, even though cognitive resources are more difficult to recruit later in life (Gold et al., 2013). Namely, increased employment of executive function throughout the lifespan of an adult leads to an accommodation of this need in the structure of the brain. It is clear that being bilingual may result in the reorganization of the processes associated with language as well as those not associated with language (see Abutalebi & Green, 2007, for a review).

Beyond the employment of cognitive mechanisms, and the subsequent neural accommodation of these mechanisms, evidence for the impact of bilingualism on the brain comes from the neurological changes observed in bilinguals. This is not specific to language, with other areas of expertise showing neurological changes, such as taxi driving (Maguire et al., 2000), playing video games (Gong et al., 2015), and juggling, even after only seven days of training (Driemeyer et al., 2008). Moreover, EEG studies have provided evidence of greater neural efficiency in bilinguals (Calvo et al., 2023; Kałamała et al., 2018; Kousaie & Phillips, 2012; Morales et al., 2015). One example comes from research on the N2 ERP component. The N2 is associated with conflict monitoring (Nieuwenhuis et al., 2003) and inhibition (Falkenstein et al., 2002). Some studies have found larger N2 responses for bilinguals performing conflict monitoring tasks (such as go/no go tasks (Fernandez et al., 2013; Moreno et al., 2014) and AX-CPT tasks (Morales et al., 2015). These suggest that bilinguals engage in more conflict monitoring and/or inhibition than monolinguals. Other studies (e.g., Kousaie & Phillips, 2012, 2017) found smaller and earlier N2 responses during a Stroop task for bilinguals compared to monolinguals. These results are consistent with other findings, suggesting that bilinguals are more efficient at tasks employing conflict monitoring and/or inhibition, despite the same level or superior performance (Kousaie et al., 2017). It appears that from these conflicting results, the type of task may elicit processing mechanisms that are more specific than cognitive control and inhibition and/or that the groups in these studies may have differed in some other respect.

Lastly, evidence from the resting state of the brain shows that bilingualism has an impact on brain activity (e.g., Berken et al., 2016; Grady et al., 2015; Gullifer et al., 2018; Kousaie et al., 2017). Resting-state studies examine the human brain when no stimulus is presented to the participant and no task is given (i.e., participants are simply awake and at rest). This is a measure of brain activity that is not specific for language but, rather, can help determine whether bilingualism has a broad impact on the human brain. For example, Bice and colleagues (2020) used ERPs to characterize the resting state of bilinguals compared to monolinguals. They observed overall greater alpha power and greater coherence in alpha and beta frequencies in bilinguals compared to monolinguals. They also observed a correlation between higher alpha frequencies and language control, such as L2 use, higher L1 proficiency, and earlier L2 age of acquisition. This is not surprising, given that alpha activity has been found to correspond with language processing (Bornkessel et al., 2004) and inhibitory control (Klimesch et al., 2007; Strauß et al., 2014). While it is clear that bilingualism leaves a mark on the human brain, further research is merited to better understand the cognitive and neural effects of the various experiences of bilinguals on the human brain.

13.5 How Can We Change How the Mental Lexicon Is Conceptualized?

The many models that have been proposed for the bilingual lexicon are supported by both behavioral and neurological data. The continuous tension between connecting the human brain and the mind is not lost in these discussions. Neurological studies provide evidence as to what occurs in the brain, where, and at which point in time. Behavioral findings provide additional support as to how words are related to each other, within and across languages, when tasks require lexical activation and inhibition. Together, they paint a clear picture of what happens and when. However, these results have not yet led to an all-encompassing, unified theory. Some models of bilingual processing (BIA+; ICM) propose a clear distinction between language processing and more general task/decision processing. If we assume that different processes are likely represented in different areas of the brain, there is clear evidence for this distinction.

Neurological models of language processing also assume that language processing is hierarchical and temporally structured. Neurological results show that the processing of a single linguistic component (e.g., visual word recognition) is temporally (Bentin et al., 1999; Pulvermüller et al., 2009) and structurally complex (Dehaene et al., 2005). In other words, neurological models were posited with the same assumption that the processing components involved in language processing are recruited in a hierarchical and temporally constrained manner. There is still some disconnect between the neurological or cognitive processes involved in language processing and the models built primarily on behavioral data, and then tested with neurological evidence.

More recently, some models of language processing have emerged from neurological evidence. Friederici (2002) proposed a neurocognitive model of auditory sentence processing, where the task and decision components are not specified, but rather, there is a distinction between the temporality of working memory (observed in the prefrontal cortex) and language representations (observed in the occipital-temporal-parietal regions). Likewise, Hagoort (2005) proposed a memory-unification-control framework, where the mental lexical is associated with the left temporal cortex (long-term working memory), and language unification occurs in the left inferior frontal gyrus (syntactic, phonological, or semantic integration), and general control engages the dorsolateral prefrontal cortex. These distinctions are based on previous observations that associate specific areas of the brain with specific components of language processing.

Although not specific to language processing, it is relevant to briefly introduce a more recent approach to brain mapping summarized by Cabeza and colleagues (2018), which is a conceptualization of brain function where many areas of the brain may work together to complete tasks in a sort of "team," which they refer to as process specific alliances (PSAs). These PSAs are described as flexible and opportunistic, in the sense that some brain areas may work together with many other brain areas to complete different tasks. In this approach there is less of a one-to-one mapping between brain area and function. Rather, one brain area may be involved in many functions, depending on which other areas it teams up with. To illustrate this concept, the authors compare the mediation of attentional modulation in visual processing and motor performance. In visual processing, the prefrontal cortex may form a PSA with the visual cortex. In monitoring motor performance, the prefrontal cortex may form a PSA with the striatum. In this sense, the prefrontal cortex plays the same role in both tasks, that of attentional control, but the control processes may differ because the content that is being controlled is different. The relationship between these brain areas to form PSAs may be hierarchical in nature, indicating that the activation of one will lead to the activation of the other (Cabeza et al., 2012). A PSA, then, is defined by three criteria: the area of the brain that is activated is linked to a broad cognitive process that is involved in the PSA; the various regions activated in the PSA should perform suboperations of the PSA; there should be evidence of these areas interacting during the employment of the PSA (e.g., white matter). While this approach is not specific for language processing, nor has it been systematically applied to problems of bilingual processing, it is a promising approach for solving the problem of how to best characterize brain functions that are complex and interactive. As a result, it may be useful in characterizing the bilingual brain, where a more interactive approach may better capture the additional processing demands of bilingual processing.

Some ongoing questions remain before a proper unification of the various accounts can be obtained. To begin with, it is important to recognize that many theories are not mutually exclusive, but rather focus on different areas of importance. Relatedly, if a study finds a null result, this does not necessarily mean that the hypothesis is proven false, given the low level of power typically found in these types of studies (for a discussion, see Kroll & Bialystok, 2013). Lastly, neither bilingualism nor cognition is categorical in nature; therefore, when comparing across studies, there may be extraneous factors contributing to results about which we are not aware. A multifaceted account of the bilingual lexicon that considers linguistic and nonlinguistic factors, as well as external and internal characteristics, and the differences in processing tasks and modalities, may be the most efficient way to begin unifying the various linguistic and neurological approaches to the bilingual lexicon.

13.6 Conclusion

The bilingual brain is a relatively new area of research, which has benefited greatly from a range of methodological approaches. In this chapter we have reviewed previous work on the bilingual lexicon and have outlined the different methodological approaches to various questions about the bilingual lexicon, including inquires probing the nature of the relationship between the two languages. We have observed the advantages and disadvantages to each and have also summarized how research on the bilingual brain has informed our understanding of the human brain, including how acquiring more than one language alters the brain in the short and long term. Lastly, we have considered some models that attempt to combine what we know from behavioral studies and what we know from neurological studies into more unified approaches, ending with some outstanding issues.

References

Abutalebi, J., Cappa, S., & Perani, D. (2001). The bilingual brain as revealed by functional neuroimaging. *Bilingualism: Language and Congnition*, *4*, 179–190.

Abutalebi, J., Cappa, S., & Perani, D. (2005). What can functional neuroimaging tell us about the bilingual brain? In J. Kroll & A. de Groot (Eds.), *The Handbook of Bilingualism: Psycholinguistic Approaches* (pp. 497–515). Oxford University Press.

Abutalebi, J., Della Rosa, P., Green, D., Hernandez, M., Scifo, P., Keim, R., Cappa, S., & Costa, A. (2012). Bilingualism tunes the anterior cingulate cortex for conflict monitoring. *Cerebral Cortex, 22*, 20762086.

Abutalebi, J., & Green, D. (2007). Bilingual language production: The neurocognition of language representation and control. *Journal of Neurolinguistics, 20*, 242–275.

Abutalebi, J., Miozzo, A., & Cappa, S. (2000). Do subcortical structures control "language selection" in polyglots? Evidence from pathological language mixing. *Neurocase, 6*, 5156.

Aglioti, S., Beltramello, A., Girardi, F., & Fabbro, F. (1996). Neurolinguistic and follow-up study of an unusual pattern of recovery from bilingual subcortical aphasia. *Brain, 119*, 1551–1564.

Alladi, S., Bak, T., Duggirala, V., Surampudi, B., Shailaja, M., Shukla, A., Chaudhuri, J., & Kaul, S. (2013). Bilingualism delays age at onset of dementia, independent of education and immigration status. *Neurology*, 81, 1938–1944.

Altarriba, J. (1990). Constraints on interlingual facilitation effects in priming in Spanish-English bilinguals. Unpublished doctoral dissertation, Vanderbilt University.

Altarriba, J. (1992). The representation of translation equivalents in bilingual memory. In R. Harris (Ed.), *Cognitive Processing in Bilinguals* (pp. 157–174). Elsevier.

Ameel, E., Malt, B., Storms, G., & Van Assche, F. (2009). Semantic convergence in the bilingual lexicon. *Journal of Memory and Language, 60*, 270–290.

Ameel, E., Storms, G., Malt, B., & Sloman, S. (2005). How bilinguals solve the naming problem. *Journal of Memory and Language, 53*, 60–80.

Ardal, S., Donald, M., Meuter, R., Muldrew, S., & Luce, M. (1990). Brain responses to semantic incongruity in bilinguals. *Brain and Language, 39*, 187–205.

Bentin, S., Mouchetant-Rostaing, Y., Giard, M., Echallier, J., & Pernier, J. (1999). ERP manifestations of processing printed words at different psycholinguistic levels: Time course and scalp distribution. *Journal of Cognive Neuroscience, 11*, 235–260.

Berken, J., Chai, X., Chen, J.-K., Gracco, V., & Klein, D. (2016). Effects of early and late bilingualism on resting-state functional connectivity. *Journal of Neuroscience, 36*(4), 1165–1172.

Bernolet, S., Hartsuiker, R., & Pickering, M. (2013). From language-specific to shared syntactic representations: The influence of second language proficiency on syntactic sharing in bilinguals. *Cognition, 127*(3), 287–306.

Bialystok, E., Craik, F., Green, D., & Gollan, T. (2009). Bilingual minds. *Psychological Science in the Public Interest, 10*, 89–129.

Bialystok, E., Martin, M., & Viswanathan, M. (2005). Bilingualism across the lifespan: The rise and fall of inhibitory control. *International Journal of Bilingualism, 9*, 109–119.

Bice, K., & Kroll, J. (2015). Native language change during early stages of second language learning. *Neuroreport, 26*(16), 966–971.

Bice, K., Yamasaki, B., & Prat, C. (2020). Bilingual language experience shapes resting-state brain rhythms. *Neurobiology of Language, 1*(3), 288–318.

Bobb, S., & Kroll, J. (2018). Words on the brain: The bilingual mental lexicon. In D. Miller, F. Bayram, J. Rothman, & L. Serratrice (Eds.), *Bilingual Cognition and Language: The State of the Science across Its Subfields* (pp. 307–324). Benjamins.

Bornkessel, I., Fiebach, C., Friederici, A., & Schlesewsky, M. (2004). "Capacity" reconsidered: Interindividual differences in language comprehension and individual alpha frequency. *Experimental Psychology, 51*(4), 279–289.

Branzi, F., Martin, C., Abutalebi, J., & Costa, A. (2014). The after-effects of bilingual language production. *Neuropsychologia, 52,* 102–116.

Briellmann, R., Saling, M., Connell, A., Waites, A., Abbott, D., & Jackson, G. (2004). A high-field functional MRI study of quadri-lingual subjects. *Brain and Language, 89,* 531–542.

Broersma, M., Carter, D., & Acheson, D. (2016). Cognate costs in bilingual speech production: Evidence from language switching. *Frontiers in Psychology, 7,* 1461.

Cabeza, R., Ciaramelli, E., & Moscovitch, M. (2012). Cognitive contributions of the ventral parietal cortex: An integrative theoretical account. *Trends in Cognitive Sciences, 16*(6), 338–352. https://doi.org/10.1016/j.tics.2012.04.008

Cabeza, R., Stanley, M. L., & Moscovitch, M. (2018). Process-specific alliances (PSAs) in cognitive neuroscience. *Trends in Cognitive Sciences, 22*(11), 996–1010. https://doi.org/10.1016/j.tics.2018.08.005

Calvo, N., Grundy, J., & Bialystok, E. (2023). Bilingualism modulates neural efficiency at rest through alpha reactivity. *Neuropsychologia, 180,* 108486.

Canseco-Gonzalez, E., Brehm, L., Brick, C., Brown-Schmidt, S., Fischer, K., & Wagner, K. (2010). Carpet or cárcel: The effect of age of acquisition and language mode on bilingual lexical access. *Language and Cognitive Processes, 25,* 669–705.

Casaponsa, A., Carreiras, M., & Duñabeitia, J. (2014). Discriminating languages in bilingual contexts: The impact of orthographic markedness. *Frontiers in Psychology, 5,* 424–424.

Casaponsa, A., & Duñabeitia, J. (2016). Lexical organization of language-ambiguous and language-specific words in bilinguals. *Quarterly Journal of Experimental Psychology, 69*(3), 589–604.

Chen, H-C., & Ng, M-L. (1989). Semantic facilitation and translation priming effects in Chinese-English bilinguals. *Memory & Cognition, 17,* 454–462.

Colomé, À. (2001). Lexical activation in bilinguals' speech production: Language-specific or language-independent? *Journal of Memory and Language, 45,* 721–736.

Correia, J., Formisano, E., Valente, G., Hausfeld, L., Jansma, B., & Bonte, M. (2014). Brain-based translation: fMRI decoding of spoken words in bilinguals reveals language-independent semantic representations in anterior temporal lobe. *The Journal of Neuroscience, 34*(1), 332–338.

Costa, A., & Caramazza, A. (1999). Is lexical selection in bilingual speech production language-specific? Further evidence from Spanish-English and English-Spanish bilinguals. *Bilingualism: Language and Cognition, 2*(3), 231–244.

Costa, A., Miozzo, M., & Caramazza, A. (1999). Lexical selection in bilinguals: Do words in the bilingual's two lexicons compete for selection? *Journal of Memory and Language, 41*, 365–397.

Costa, A., & Santesteban, M. (2004). Lexical access in bilingual speech production: Evidence from language switching in highly proficient bilinguals and L2 learners. *Journal of Memory and Language, 50*, 491–511.

Cutler A., Weber A., & Otake, T. (2006). Asymmetric mapping from phonetic to lexical representations in second-language listening. *Journal of Phonetics, 34*, 269–284.

De Bleser, R., Dupont, P., Postler, J., Bormans, G., Speelman, D., Mortelmans, L., & Debrock, M. (2003). The organisation of the bilingual lexicon: A PET study. *Journal of Neurolinguistics, 16*(4), 439–456.

de Bot, K. (1992). A bilingual production model: Levelt's "Speaking" model adapted. *Applied Linguistics, 13*, 1–24.

de Bot, K. (2019). Defining and assessing multilingualism. In J. W. Schwieter (Ed.), *The Handbook of the Neuroscience of Multilingualism* (pp. 3–18). Wiley-Blackwell.

de Groot, A., & Nas, G. (1991). Lexical representation of cognates and noncognates in compound bilinguals. *Journal of Memory and Language, 30*, 90–123.

de León Rodríguez, D., Mouthon, M., Annoni, J.-M., & Khateb, A. (2022). Current exposure to a second language modulates bilingual visual word recognition: An EEG Study. *Neuropsychologia, 164*, 108109–108109.

Dehaene, S., Cohen, L., Sigman, M., & Vinckier, F. (2005). The neural code for written words: A proposal. *Trends in Cognitive Science, 9*, 335–341.

Deng, T., Schwieter, J. W., Huan, L., Zhang, Y., Yuan, J., & Wang, R. (2024). Cross-language activation of translation ambiguous words in semantic judgment task of Chinese-English bilinguals. International Journal of Bilingualism, 0(0). https://doi.org/10.1177/13670069241233388

Dijkstra, T. (2005). Bilingual word recognition and lexical access. In J. Kroll & A. de Groot (Eds.), *The Handbook of Bilingualism: Psycholinguistic Approaches* (pp. 179–201). Oxford University Press.

Dijkstra, T., & van Heuven, W. (2002). The architecture of the bilingual word recognition system: From identification to decision. *Bilingualism: Language and Cognition, 5*, 175–197.

Driemeyer, J., Boyke, J., Gaser, C., Büchel, C., & May, A. (2008). Changes in gray matter induced by learning revisited. *PLoS ONE, 3*, e2669.

Druks, J., & Weekes, B. (2013). Parallel deterioration to language processing in a bilingual speaker. *Cognitive Neuropsychology, 30*(7–8), 578–596.

Duñabeitia, J., Perea, M., & Carreiras, M. (2010). Masked translation priming effects with highly proficient simultaneous bilinguals. *Experimental Psychology, 57*(2), 98–107.

Edmonds, L., & Kiran, S. (2006). Effect of semantic naming treatment on cross-linguistic generalization in bilingual aphasia. *Journal of Speech, Language, and Hearing Research, 49*(4), 729–748.

Elston-Güttler, K., & Friederici, A. (2005). Native and L2 processing of homonyms in sentential context. *Journal of Memory and Language, 52*, 256–283.

Emmorey, K., Borinstein, H., Thompson, R., & Gollan, T. (2008). Bimodal bilingualism. *Bilingualism: Language and Cognition, 11*, 43–61.

Fabbro, F. (1995). *The Neurolinguistics of Bilingualism: An Introduction.* Psychology Press.

Fabbro, F. (2001). The bilingual brain: bilingual aphasia. *Brain and Language, 79*, 201–210.

Fabbro, F., & Paradis, M. (1995). Differential impairments in four multilingual patients with subcortical lesions. In M. Paradis (Ed.), *Aspects of Bilingual Aphasia* (pp. 139–176). Pergamon.

Fabbro, F., Skrap, M., & Aglioti, S. (2000). Pathological switching between languages after frontal lesions in a bilingual client. *Journal of Neurology, Neurosurgery, and Psychiatry, 68*, 650–652.

Falkenstein, M., Hoormann, J., & Hohnsbein, J. (2002). Inhibition-related ERP components: Variation with modality, age, and time-on task. *Journal of Psychophysiology, 16*(3), 167–175.

Fernandez, M., Tartar, J., Padron, D., & Acosta, J. (2013). Neurophysiological marker of inhibition distinguishes language groups on a non-linguistic executive function test. *Brain and Cognition, 83*, 330–336.

Filippi, R., Karaminis, T., & Thomas, M. (2014). Language switching in bilingual production: Empirical data and computational modelling. *Bilingualism: Language and Cognition, 17*, 294–315.

Finkbeiner, M., Gollan, T., & Caramazza, A. (2006). Lexical access in bilingual speakers: What's the (hard) problem? *Bilingualism: Lang. Cognit., 9*, 153–166.

Francis, W., Tokowicz, N., & Kroll, J. (2014). The consequences of language proficiency and difficulty of lexical access for translation performance and priming. *Memory & Cognition, 42*, 27–40.

Frenck, C., & Pynte, J. (1987). Semantic representation and surface forms: A look at across-language priming in bilinguals. *Journal of Psycholinguistic Research, 16*, 383–396.

Friederici, A. (2002). Towards a neural basis of auditory sentence processing. *Trends in Cognitive Science, 6*, 78–84.

Gold, B., Kim, C., Johnson, N., Kriscio, R., & Smith, C. (2013). Lifelong bilingualism maintains neural efficiency for cognitive control in aging. *Journal of Neuroscience, 33*, 387–396.

Gong, D., He, H., Liu, D., Ma, W., Dong, L., Luo, C., & Yao, D. (2015). Enhanced functional connectivity and increased gray matter volume of insula related to action video game playing. *Scientific Reports, 5*, 9763.

Goral, M., Levy, E., Obler, L., & Cohen, E. (2006). Cross-language lexical connections in the mental lexicon: Evidence from a case of trilingual aphasia. *Brain and Language, 98*, 235–247.

Grady, C., Luk, G., Craik, F., & Bialystok, E. (2015). Brain network activity in monolingual and bilingual older adults. *Neuropsychologia, 66*, 170–181.

Grainger, J., & Beauvillain, C. (1988). Associative priming in bilinguals: Some limits of interlingual facilitation effects. *Canadian Journal of Psychology, 42*, 261–273.

Gray, T., & Kiran, S. (2013). A theoretical account of lexical and semantic naming deficits in bilingual aphasia. *Journal of Speech, Language, and Hearing Research, 56*(4), 1314–1327.

Green, D. (1998). Mental control of the bilingual lexico-semantic system. *Bilingualism: Language and Cognition, 1*, 67–81.

Green, D. (2005). The neurocognition of recovery patterns in bilingual speakers with aphasias. In J. Kroll & A. de Groot (Eds.), *Handbook of Bilingualism: Psycholinguistic Perspectives* (pp. 516–530). Oxford University Press.

Gullifer, J., Chai, X., Whitford, V., Pivneva, I., Baum, S., Klein, D., & Titone, D. (2018). Bilingual experience and resting-state brain connectivity: Impacts of L2 age of acquisition and social diversity of language use on control networks. *Neuropsychologia, 117*(April), 123–134.

Hagoort, P. (2005). On Broca, brain, and binding: A new framework. *Trends in Cognitive Science, 9*, 416–423.

Hahne, A. (2001). What's different in second-language processing? Evidence from event-related brain potentials. *Journal of Psycholinguistic Research, 30*, 251–266.

Hartsuiker, R., & Pickering, M. (2008). Language integration in bilingual sentence production. *Acta Psychologica, 128*(3), 479–489.

Hartsuiker, R., Beerts, S., Loncke, M., Desmet, T., & Bernolet, S. (2016). Crosslinguistic structural priming in multilinguals: Further evidence for shared syntax. *Journal of Memory and Language, 90*, 14–30.

Hasegawa, M., Carpenter, P., & Just, M. (2002). An fMRI study of bilingual sentence comprehension and workload. *NeuroImage, 15*, 647–660.

Hermans, D. (2004). Between-language identity effects in picture-word interference tasks: A challenge for language-nonspecific or language-specific models of lexical access? *International Journal of Bilingualism, 8*, 115–125.

Hernandez, A. (2013). What factors influence how two languages are coded in one brain: Comment on "The bilingual brain: Flexibility and control in the human cortex" by Buchweitz and Prat. *Physics of Life Reviews, 10*(4), 450–451.

Hernandez, A., & Li, P. (2007). Age of acquisition: Its neural and computational mechanisms. *Psychological Bulletin, 133*, 638–650.

Hernandez, A., Li, P., & MacWhinney, B. (2005). The emergence of competing modules in bilingualism. *Trends in Cognitive Science, 9*, 220–225.

Hernandez, A., & Meschyan, G. (2006). Executive function is necessary to enhance lexical processing in a less proficient L2: Evidence from fMRI during picture naming. *Bilingualism: Language and Cognition, 9*, 177–188.

Hesling, I., Dilharreguy, B., Bordessoules, M., & Allard, M. (2012). The neural processing of second language comprehension modulated by the degree of proficiency: A listening connected speech FMRI study. *The Open Neuroimaging Journal, 6*, 44–54.

Holcomb, P., Grainger, J., & O'Rourke, T. (2002). An electrophysiological study of the effects of orthographic neighborhood size on printed word perception. *Journal of Cognitive Neuroscience, 14*, 938–950.

Hoshino, N., & Kroll, J. (2008). Cognate effects in picture naming: Does cross-language activation survive a change of script? *Cognition, 106,* 501–511.

Hoversten, L., Brothers, T., Swaab, T., & Traxler, M. (2017). Early processing of orthographic language membership information in bilingual visual word recognition: Evidence from ERPs. *Neuropsychologia, 103,* 183–190.

Indefrey, P. (2006). A meta-analysis of hemodynamic studies on first and second language processing: Which suggested differences can we trust and what do they mean? *Language Learning, 56,* 279–304.

Jeong, H., Sugiura, M., Sassa, Y., Haji, T., Usui, N., Taira, M., Horie, K., Sato, S., & Kawashima, R. (2007). Effects of syntactic similarity on cortical activation during second language processing: A comparison of English and Japanese among native Korean trilinguals. *Hum. Brain Mapp., 28,* 194–204.

Jeong, H., Sugiura, M., Sassa, Y., Yokoyama, S., Horie, K., Sato, S., Taira, M., Kawashima, R. (2007). Cross-linguistic influence on brain activation during second language processing: An fMRI study. *Biling-Lang. Cogn., 10,* 175–187.

Jia, F. (2022). Effect of second language proficiency on inhibitory control in the Simon task: An fMRI study. *Frontiers in Psychology, 21*(13), 812322.

Jiang, N. (1999). Testing processing explanations for the asymmetry in masked cross-language priming. *Bilingualism: Language and Cognition, 2*(1), 59–75.

Jin, Y-S (1990). Effects of concreteness on cross-language priming in lexical decisions. *Perceptual and Motor Skills, 70,* 1139–1154.

Ju, M., & Luce, P. (2004). Falling on sensitive ears: Constraints on bilingual lexical activation. *Psychological Science, 15,* 314–318.

Kałamała, P., Drożdżowicz, A., Szewczyk, J., Marzecová, A., & Wodniecka, Z. (2018). Task strategy may contribute to performance differences between monolinguals and bilinguals in cognitive control tasks: ERP evidence. *Journal of Neurolinguistics, 46,* 78–92.

Kantola, L., & van Gompel, R. (2011). Between- and within-language priming is the same: Evidence for shared bilingual syntactic representations. *Memory & Cognition, 39*(2), 276–290.

Keatley, C., & de Gelder, B. (1992). The bilingual primed lexical decision task: Cross-language priming disappears with speeded responses. *European Journal of Cognitive Psychology, 4*(4), 273–292.

Keatley, C., Spinks, J., & de Gelder, B. (1994). Asymmetrical cross-language priming effects. *Memory and Cognition, 22,* 70–84.

Kerkhofs, R., Dijkstra, T., Chwilla, D., & De Bruijn, E. (2006). Testing a model for bilingual semantic priming with interlingual homographs: RT and N400 effects. *Brain Research, 1068,* 170–183.

Kim, K., Relkin, N., Lee, K-M, & Hirsch, J. (1997). Distinct cortical area associated with native and second languages. *Nature, 388,* 171–174.

Kirsner, K., Smith, M., Lockhart, R., King, M., & Jain, M. (1984). The bilingual lexicon: Language specific units in an integrated network. *Journal of Verbal Learning and Verbal Behavior, 23,* 519–539.

Klein, D., Watkins, K., Zatorre, R., & Milner, B. (2006). Word and nonword repetition in bilingual subjects: A PET study. *Human Brain Mapping, 27,* 153–161.

Klimesch, W., Sauseng, P., & Hanslmayr, S. (2007). EEG alpha oscillations: The inhibition-timing hypothesis. *Brain Research Reviews, 53*(1), 63–88.

Kong, A., Abutalebi, J., Lam, K., & Weekes, B. (2014). Executive and language control in the multilingual brain. *Behavioural Neurology*, 527951.

Kotz, S. (2001). Neurolinguistic evidence for bilingual language representation: A comparison of reaction times and event-related brain potentials. *Bilingualism: Language and Cognition, 4*, 143–154.

Kousaie, S., Chai, X., Sander, K., & Klein, D. (2017). Simultaneous learning of two languages from birth positively impacts intrinsic functional connectivity and cognitive control. *Brain and Cognition, 117*, 49–56.

Kousaie, S., & Phillips, N. (2012). Conflict monitoring and resolution: Are two languages better than one? Evidence from reaction time and event-related brain potentials. *Brain Research, 1446*, 71–90.

Kousaie, S., & Phillips, N. (2017). A behavioural and electrophysiological investigation of the effect of bilingualism on aging and cognitive control. *Neuropsychologia, 94*, 23–35.

Kovelman, I., Baker, S., & Petitto, L. (2008). Bilingual and monolingual brains compared: A functional magnetic resonance imaging investigation of syntactic processing and a possible "neural signature" of bilingualism. *Journal of Cognitive Neuroscience, 20*(1), 153–169.

Kroll, J., & Bialystok, E. (2013). Understanding the consequences of bilingualism for language processing and cognition. *Journal of Cognitive Psychology, 25*(5), 497–514.

Kroll, J., Bobb, S., Misra, M., & Guo, T. (2008). Language selection in bilingual speech: Evidence for inhibitory processes. *Acta Psychologica, 128*, 416–430.

Kroll, J., Bobb, S., & Wodniecka, Z. (2006). Language selectivity is the exception, not the rule: Arguments against a fixed locus of language selection in bilingual speech. *Bilingualism: Language and Cognition, 9*, 119–135.

Kroll, J., Bogulski, C., & McClain, R. (2012). Psycholinguistic perspectives on second language learning and bilingualism: The course and consequence of cross-language competition. *Linguistic Approaches to Bilingualism, 2*, 124.

Kroll, J., Dussias, P., Bogulski, C., & ValdesKroff, J. (2012). Juggling two languages in one mind: What bilinguals tell us about language processing and its consequences for cognition. In B. Ross (Ed.), *The Psychology of Learning and Motivation* (pp. 229–262). Academic Press.

Kroll, J., & Gollan T. (2014). Speech planning in two languages: what bilinguals tell us about language production. In V. Ferreira, M. Goldrick, & M. Miozzo (Eds.), *The Oxford Handbook of Language Production* (pp. 165–181). Oxford University Press.

Kroll, J., Gullifer, J., & Rossi, E. (2013). The multilingual lexicon: The cognitive and neural basis of lexical comprehension and production in two or more languages. *Annual Review of Applied Linguistics, 33*, 102–127.

Kroll, J., & Stewart, E. (1994). Category interference in translation and picture naming: Evidence for asymmetric connections between bilingual memory representations. *Journal of Memory and Language, 33*, 149–174.

Kroll, J., & Tokowicz, N. (2005). Models of bilingual representation and processing. In J. Kroll & A. de Groot (Eds.), *The Handbook of Bilingualism: Psycholinguistic Approaches* (pp. 531–553). Oxford University Press.

Kutas, M., & Federmeier, K. (2000). Electrophysiology reveals semantic memory use in language comprehension. *Trends in Cognitive Sciences*, *4*(12), 463–470.

Kutas, M., & Kluender, R. (1994). What is who violating? A reconsideration of linguistic violations in light of event-related brain potentials. In H. Heinze, T. Münte, & G. Mangun (Eds.), *Cognitive Electrophysiology* (pp. 183–210). Springer Science.

Lee, Y., Jang, E., & Choi, W. (2018). L2-L1 translation priming effects in a lexical decision task: Evidence from low proficient Korean-English bilinguals. *Frontiers in Psychology*, *2*(9), 267.

Li, Q., Pasquini, L., Del Ferraro, G., Gene, M., Peck, K., Makse, H., & Holodny, A. (2021). Monolingual and bilingual language networks in healthy subjects using functional MRI and graph theory. *Scientific Reports*, *11*(1), 10568.

Linck, J., Schwieter, J. W., & Sunderman, G. (2012). Inhibitory control predicts language switching performance in trilingual speech production. *Bilingualism: Language and Cognition*, *15*(3), 651–662.

Liu, H., & Cao, F. (2016). L1 and L2 processing in the bilingual brain: A meta-analysis of neuroimaging studies. *Brain and Language*, *159*, 60–73.

Liu, H., Hu, Z., Guo, T., & Peng, D. (2010). Speaking words in two languages with one brain: Neural overlap and dissociation. *Brain Research*, *1316*, 75–82.

Loebell, H., & Bock, K. (2003). Structural priming across languages. *Linguistics*, *41*(5), 791–824.

Luk, G., Anderson, J., Craik, F., Grady, C., & Bialystok, E. (2010). Distinct neural correlates for two types of inhibition in bilinguals: Response inhibition versus interference suppression. *Brain and Cognition*, *74*, 347357.

Luk, G., Green, D., Abutalebi, J., & Grady, C. (2012). Cognitive control for language switching in bilinguals: A quantitative meta-analysis of functional neuroimaging studies. *Language and Cognitive Processes*, *27*, 14791488.

Luke, K., Liu, H., Wai, Y., Wan, Y., & Tan, L. (2002). Functional anatomy of syntactic and semantic processing in language comprehension. *Human Brain Mapping*, *16*(3), 133–45.

Maguire, E., Gadian, D., Johnsrude, I., Good, C., Ashburner, J., Frackowiak, R., & Frith, C. (2000). Navigation-related structural change in the hippocampi of taxi drivers. *PNAS*, *97*, 4398–4403.

Marian, V., Blumenfeld, H., & Boukrina, O. (2008). Sensitivity to phonological similarity within and across languages. *Journal of Psycholinguistic Research*, *37*(3), 141–170.

Marian, V., & Spivey, M. (2003). Competing activation in bilingual language processing: Within-and between-language competition. *Bilingualism: Language and Cognition*, *6*, 97–115.

Marian, V., Spivey, M., & Hirsch, J. (2003). Shared and separate systems in bilingual language processing: Converging evidence from eyetracking and brain imaging. *Brain and Language*, *86*, 70–82.

Martin, C., Dering, B., Thomas, E., & Thierry, G. (2009). Brain potentials reveal semantic priming in both the "active" and the "non-attended" language in early bilinguals. *NeuroImage, 47,* 326–333.

Martín, M., Macizo, P., & Bajo, T. (2010). Time course of inhibitory processes in bilingual language processing. *British Journal of Psychology,* 101(4), 679–693.

Massol, S., Grainger, J., Dufau, S., & Holcomb, P. (2010). Masked priming from orthographic neighbors: An ERP investigation. *Journal of Experimental Psychology. Human Perception and Performance, 36*(1), 162–174.

McLaughlin, J., Osterhout, L., & Kim, A. (2004). Neural correlates of second-language word learning: Minimal instruction produces rapid change. *Nature Neuroscience, 7,* 703–704.

Meschyan, G., & Hernandez, A. (2006). Impact of language proficiency and orthographic transparency on bilingual word reading: An fMRI investigation. *NeuroImage, 29,* 1135–1140.

Meyer, D., Schvaneveldt, R., & Ruddy, M. (1975). Loci of contextual effects in visual word recognition. In P. Rabbitt & S. Dornic (Eds.), *Attention and Performance V.* Academic Press.

Meykadeh, A., Golfam, A., Batouli, S., & Sommer, W. (2021). Overlapping but language-specific mechanisms in morphosyntactic processing in highly competent L2 acquired at school entry: fMRI evidence from an alternating language switching task. *Frontiers in Human Neuroscience, 15,* 728549–728549.

Midgley, K., Holcomb, P., van Heuven, W., & Grainger, J. (2008). An electrophysiological investigation of cross-language effects of orthographic neighborhood. *Brain Research, 1248,* 123–135.

Minkowski, M. (1983). A clinical contribution to the study of polyglot aphasia especially with respect to Swiss-German. In M. Paradis (Ed.), *Readings on Aphasia in Bilinguals and Polyglots* (pp. 205–232). Didier. (Original work published 1927)

Morales, J., Yudes, C., Gómez-Ariza, C., & Bajo, M. (2015). Bilingualism modulates dual mechanisms of cognitive control: Evidence from ERPs. *Neuropsychologia, 66,* 157–169.

Moreno, E., & Kutas, M. (2005). Processing semantic anomalies in two languages: An electrophysiological exploration in both languages of Spanish-English bilinguals. *Cognitive Brain Research, 22,* 205–220.

Moreno, E., Rodríguez-Fornells, A., & Laine, M. (2008). Event-related potentials (ERPs) in the study of bilingual language processing. *Journal of Neurolinguistics, 21,* 477–508.

Moreno, S., Wodniecka, Z., Tays, W., Alain, C., & Bialystok, E. (2014). Inhibitory control in bilinguals and musicians: Event-related potential (ERP) evidence for experience-specific effects. *PLoS ONE, 9,* e94169.

Morford, J., Wilkinson, E., Villwock, A., Piñar, P., & Kroll, J. (2011). When deaf signers read English: Do written words activate their sign translations? *Cognition, 118,* 286–292.

Muylle, M., Bernolet, S., & Hartsuiker, R. (2021). The development of shared syntactic representations in late L2-learners: Evidence from structural priming in an artificial language. *Journal of Memory and Language, 119,* 104233.

Nieuwenhuis, S., Yeung, N., van den Wildenberg, W., & Ridderinkhof, K. (2003). Electrophysiological correlates of anterior cingulate function in a go/no-go task: Effects of response conflict and trial type frequency. *Cognitive, Affective, & Behavioral Neuroscience, 3*, 17–26.

Paradis, M. (1977). Bilingualism and aphasia. In H. Whitaker & H. Whitaker (Eds.), *Studies in Neurolinguistics, Vol. 3*. Academic Press.

Paradis, M. (2001). Bilingual and polyglot aphasia. In R. Berndt (Ed.), *Handbook of Neuropsychology* (2nd ed., pp. 69–91). Elsevier.

Paradis, M. (2004). *A Neurolinguistic Theory of Bilingualism*. Benjamins.

Paradis, M., Goldblum, M., & Abidi, R. (1982). Alternate antagonism with paradoxical translation behavior in two bilingual aphasic patients. *Brain and Language, 15*, 55–69.

Perani, D., & Abutalebi, J. (2005). The neural basis of first and second language processing. *Current Opinion in Neurobiology, 15*(2), 202–206.

Perani, D., Paulesu, E., Galles, N., Dupoux, E., Dehaene, S., Bettindardi, V., Cappa, S., Fazio, F., & Mehler, J. (1998). The bilingual brain: Proficiency and age of acquisition of the second language. *Brain, 121*, 1841–1852.

Phillips, N., Klein, D., Mercier, J., & de Boysson, C. (2006). ERP measures of auditory word repetition and translation priming in bilinguals. *Brain Research, 1125*, 116–131.

Poarch, G., & van Hell, J. (2012). Cross-language activation in children's speech production: Evidence from second language learners, bilinguals, and trilinguals. *Journal of Experimental Child Psychology, 111*, 419–438.

Połczyńska, M., & Bookheimer, S. (2021). General principles governing the amount of neuroanatomical overlap between languages in bilinguals. *Neuroscience and Biobehavioral Reviews, 130*, 1–14.

Pulvermüller, F., Shtyrov, Y., & Hauk, O. (2009). Understanding in an instant: Neurophysiological evidence for mechanistic language circuits in the brain. *Brain and Language, 110*, 81–94.

Rüschemeyer, S-A., Zysset, S., & Friederici, A. (2006). Native and non-native reading of sentences: An fMRI experiment. *NeuroImage, 31*, 354–365.

Sabourin, L., Brien, C., & Burkholder, M. (2014). The effect of age of L2 acquisition on the organization of the bilingual lexicon: Evidence from masked priming. *Bilingualism, 17*(3), 542–555.

Schweizer, T., Ware, J., Fischer, C., Craik, F., & Bialystok, E. (2012). Bilingualism as a contributor to cognitive reserve: Evidence from brain atrophy in Alzheimer's disease. *Cortex, 48*, 991–996.

Schwieter, J. W., & Sunderman, G. (2008). Language switching in bilingual speech production: In search of the language-specific selection mechanism. *The Mental Lexicon, 3*(2), 214–238.

Sebastian, R., Laird, A., & Kiran, S. (2011). Meta-analysis of the neural representation of first language and second language. *Applied Psycholinguistics, 32*, 799–819.

Shook, A., & Marian, V. (2013). The bilingual language interaction network for comprehension of speech. *Bilingualism: Language and Cognition, 16*, 304–324.

Spivey, M., & Marian, V. (1999). Cross talk between native and second languages: Partial activation of an irrelevant lexicon. *Psychological Science, 10,* 281–284.

Strau?, A., Wöstmann, M., & Obleser, J. (2014). Cortical alpha oscillations as a tool for auditory selective inhibition. *Frontiers in Human Neuroscience, 8,* 350.

Tan, L., Spinks, J., Feng, C.-M., Siok, W., Perfetti, C., Xiong, J., Fox, P., & Gao, J.-H. (2003). Neural systems of second language reading are shaped by native language. *Human Brain Mapping, 18*(3), 158–166.

Thierry, G., & Wu, Y. (2004). Electrophysiological evidence for language interference in late bilinguals. *Neuroreport, 15,* 1555–1558.

Thierry, G., & Wu, Y. (2007). Brain potentials reveal unconscious translation during foreign-language comprehension. *Proceedings of the National Academy of Sciences, 104,* 12530–12535.

Ullman, M. (2001). A neurocognitive perspective on language: The declarative/procedural model. *Nature Reviews, 2,* 717–726.

Van de Putte, E., De Baene, W., Brass, M., & Duyck, W. (2017). Neural overlap of L1 and L2 semantic representations in speech: A decoding approach. *NeuroImage, 162,* 106–116.

van Gompel, R., & Arai, M. (2018). Structural priming in bilinguals. *Bilingualism: Language and Cognition, 21*(3), 448–455.

Van Hell, J., & Dijkstra, T. (2002). Foreign language knowledge can influence native language performance in exclusively native contexts. *Psychonomic Bulletin & Review, 9,* 780–789.

van Heuven, W., & Dijkstra, T. (2010). Language comprehension in the bilingual brain: fMRI and ERP support for psycholinguistic models. *Brain Research Reviews, 64*(1), 104–122.

van Heuven, W., Schriefers, H., Dijkstra, T., & Hagoort, P. (2008). Language conflict in the bilingual brain. *Cerebral Cortex, 18,* 2706–2716.

Veríssimo, J., Heyer, V., Jacob, G., & Clahsen, H. (2018). Selective effects of age of acquisition on morphological priming: Evidence for a sensitive period. *Language Acquisition, 25*(3), 315–326.

Vingerhoets, G., Van Borsel, J., Tesink, C., van den Noort, M., Deblaere, K., Seurinck, R., Vandemaele, P., & Achten, E. (2003). Multilingualism: An fMRI study. *NeuroImage, 20,* 2181–2196.

von Holzen, K., & Mani, N. (2012). Language nonselective lexical access in bilingual toddlers. *Journal of Experimental Child Psychology, 113,* 569–586.

Voss, J., & Paller, K. (2008). Brain substrates of implicit and explicit memory: The importance of concurrently acquired neural signals of both memory types. *Neuropsychologia, 46*(13), 3021–3029.

Wartenburger, I., Heekeren, H., Abutalebi, J., Cappa, S., Vilringer, A., & Perani, D. (2003). Early setting of grammatical processing in the bilingual brain. *Neuron, 37,* 159–170.

Weber, A., & Cutler, A. (2004). Lexical competition in non-native spoken-word recognition. *Journal of Memory and Language, 50,* 1–25.

Weber-Fox, C., & Neville, H. (1996). Maturational constraints on functional specializations for language processing: ERP and behavioral evidence in bilingual speakers. *Journal of Cognitive Neuroscience, 8*, 231–256.

Wu, Y., & Thierry, G. (2010). Chinese-English bilinguals reading English hear Chinese. *Journal of Neuroscience, 30*(22), 7646–7651.

Xu, M., Baldauf, D., Chang, C., Desimone, R., & Tan, L. (2017). Distinct distributed patterns of neural activity are associated with two languages in the bilingual brain. *Science Advances, 3*(7), e1603309–e1603309.

Xue, G., Dong, Q., Jin, Z., Zhang, L., & Wang, Y. (2004). An fMRI study with semantic access in low proficiency second language learners. *NeuroReport, 15*, 791–796.

Yang, Y., Wang, J., Bailer, C., Cherkassky, V., & Just, M. (2017). Commonalities and differences in the neural representations of English, Portuguese, and Mandarin sentences: When knowledge of the brain-language mappings for two languages is better than one. *Brain and language, 175*, 77–85.

Yetkin, O., Yetkin, Z., Haughton, V., & Cox, R. (1996). Use of functional MR to map language in multilingual volunteers. *American Journal of Neuroradiology, 17*, 473–477.

Zhao, X., & Li, P. (2013). Simulating cross-language priming with a dynamic computational model of the lexicon. *Bilingualism: Language and Cognition, 16*, 288–303.

Zou, L., Abutalebi, J., Zinszer, B., Yan, X., Shu, H., Peng, D., & Ding, G. (2012). Second language experience modulates functional brain network for the native language production in bimodal bilinguals. *NeuroImage, 62*, 1367–1375.

Part IVB

Aphasia and Autism Spectrum Disorders

14

Language Control in Bilingual Aphasia

Teresa Gray and Arpita Bose

14.1 Introduction

Multilingualism refers to individuals who speak more than one language on a daily basis (Grosjean, 1992). With already more than half of the world's population being bilingual (Ardila & Ramos, 2007; hereafter also used to refer to multilinguals) and with an ever-growing bilingual population due to globalization and migration, an increasing number of people who develop aphasia after neurological damage are bilingual (Ansaldo & Saidi, 2014). Bilingual aphasia can be defined as an impairment of one or both languages in bilingual persons following a neurological event. Studying bilingual aphasia provides a unique opportunity to untangle the implications of bilingualism for manifestation of the aphasic symptoms, for diagnosis, and for rehabilitation. Importantly, bilingual aphasia provides a window to explore mechanisms of language control and how these processes present in aphasia and influence the recovery process. In this chapter we aim to: a) provide an overview of bilingual language processing and language control in the non-injured brain; b) discuss language impairment and recovery patterns and the role of language control, as well as various bilingualism factors, for example, proficiency and linguistic similarity across language combinations; c) review how bilingual language control interacts with cognitive control; and d) discuss the role of language control and cross-language generalization in treatment.

14.2 Bilingual Language Processing and Language Control

14.2.1 Bilingual Language Processing

Models of bilingual language processing account for the association between lexical and semantic representation across two or more languages. For the purpose of this chapter, a brief overview is beneficial because it can help us better understand the connection between two languages, which informs how these processes present in bilingual aphasia. Word association and concept mediation models of bilingual language processing propose that first language (L1) and second language (L2) access one conceptual memory (Potter et al., 1984; de Groot, 1992; de Groot et al., 1994; Kroll & Stewart, 1994). For instance, the revised hierarchical model (Kroll & Stewart, 1994) posits that L1 and L2 have bidirectional connections to the semantic system and to each other. These connections are modulated by proficiency and affect lexical access and translation. Based on this model's cross-language connections, it is assumed that access to the target language affects the nontarget language. In another model that accounts for visual word recognition, the bilingual interactive activation model (Dijkstra & Van Heuven, 2002) posits that word recognition is nonselective and L1 and L2 words exist within an integrated lexicon, so word activation involves cross-language competition. Thus, in regard to both the revised hierarchical model and bilingual interactive activation model, we can consider that some element of control is required to overcome the competition and gain retrieval to the target language. In sum, models of bilingual language processing support our examination of how two or more languages are represented and interact in the brain which, in turn, can help elucidate how two or more languages are affected in bilingual aphasia and why they may or may not respond to treatment.

14.2.2 Bilingual Language Control

General cognitive control is the integration of internal thought and intention to achieve a purposeful outcome, requiring the ability to stay on task and make decisions while suppressing distracting, task-irrelevant information (Braver et al., 2003; Miller, 2000; Miller & Cohen 2001). Cognitive control has been linked to how bilingual persons manage (or control) languages in everyday life. In the bilingual brain, two or more languages are active simultaneously (Colomé, 2001; Costa & Santesteban, 2004; Hermans et al., 1998; Kaushanskaya & Marian, 2007; Kroll & Stewart, 1994; Zied et al., 2004), yet bilingual persons can freely translate, speak in one language, inhibit distracting information, and switch between languages based on desire or environmental factors. Bilingual language control refers to the ability to select the appropriate language for a specific task or environment. Because two (or more) languages access one semantic system, two languages are linked, and managing two languages is hypothesized to be a competitive process.

There are a variety of theoretical models that attempt to account for how two or more languages are controlled in the brain (e.g., Blanco-Elorrieta & Carramazza, 2021; Bloem & La Heij, 2003; Costa et al., 1999; Costa & Santesteban, 2004; Green & Abutalebi, 2013). One of the dominant bilingual language control frameworks is the inhibitory control model proposed by Green (1986, 1998). The inhibitory control model posits that in order to select the target language, it must be activated while the nontarget language is suppressed. Specifically, lemmas are associated with language tags, and when we intend to speak, schemas are formed, and their activity is modulated by the supervisory attention system. For instance, based on environmental cues, L1 may be the target language which initiates schemas that activate L1 language tags and suppress L2 tags. In other words, bilingual language control is considered a competitive process (e.g., Michael & Gollan, 2005; Van Heuven et al., 1998; Wheeldon & Monsell, 1994; although, see Blanco-Elorrieta & Carramazza, 2021 for an alternative framework).

Brain imaging studies have examined bilingual language control in healthy adults, identifying regions that overlap with domain general cognitive control (also referred to as domain general executive control). Abutalebi and Green (2007) identified the cognitive control network, such that during bilingual language processing, specific brain areas are associated with particular tasks. The left prefrontal cortex is responsible for response selection, response inhibition, decision making, and executive functions, the anterior cingulate cortex is involved with attention, conflict monitoring, and error detection, and the basal ganglia modulates language selection, set switching, language planning, and lexical selection (Abutalebi & Green, 2007, 2008; for review, see Calabria et al., 2018). Bilingual language control requires conflict monitoring (e.g., observing the language used in the environment) and conflict resolution which includes response inhibition (e.g., inhibiting the nontarget language), and response selection (e.g., selecting the target language) (Abutalebi & Green, 2007; Calabria et al., 2018). In addition, there is evidence showing that the recruitment of control mechanisms to support bilingual language control is modulated by proficiency (Abutalebi et al., 2013), offering more support for the role that cognitive control has on language use. In sum, there is ample evidence showing an overlap between neural networks that support bilingual language control and domain general cognitive control, and it is reasonable to explore how cognitive control interacts with language control.

Up to this point, we have discussed how bilingual language processing and control mechanisms present in the healthy bilingual brain. This is important because it helps us to understand how language processing and language control might break down in bilingual aphasia. In what follows, we characterize the critical ways in which language processing and control present in bilingual aphasia regarding: how they relate to each other after brain injury; how they recover after brain injury; how they may or may not respond to treatment within hypothesized outcomes. In addition, terminology has been consolidated to decrease confusion presented in the literature. Specifically, inhibitory control is a type of cognitive control, and the term nonlinguistic cognitive control is synonymous with nonverbal cognitive control, executive control, and cognitive control, and verbal control is synonymous with language control and linguistic cognitive control.

14.3 Language Impairment and Recovery Patterns in Bilingual Aphasia: The Role of Language Control and Language-Related Factors

14.3.1 Types of Language Impairment and Recovery Patterns in Bilingual Aphasia

An important aspect of bilingual aphasia includes how languages present after stroke or other brain injury. Specific types of impairment and recovery patterns have been identified in bilingual persons with aphasia (BPWA; for a review, see Paradis, 2004). Language impairment identifies language lost from pre-injury to post-injury. Two types of impairment patterns have been identified: (1) *parallel impairment* describes a pattern where both languages have lost similar amounts of skill (e.g., the stronger pre-injury language remains the stronger post-injury language), and (2) *differential impairment* describes a pattern where there is more language loss in one language compared to the other language. To identify language impairment patterns, it is obligatory that the examiner obtain pre-injury language proficiency information.

Language recovery patterns identify how BPWA language improves after brain injury. In addition to *parallel* (making equal gains across both languages) and *differential* (one language makes greater gains compared to the other language) recovery patterns, there are a few more. Specifically, *alternating antagonistic recovery* describes a pattern where one language resolves and the other does not, but as the resolved language begins to lose skill, the unresolved language begins to improve. *Selective recovery* describes a pattern where one language resolves and the other does not, *successive recovery* describes a pattern where one language resolves followed by the other language resolving, and *blended recovery* describes a pattern when BPWA involuntarily mix languages. Previous work has shown that parallel and differential patterns of language recovery and impairment are the most common types of clinical profiles observed in BPWA (Fabbro, 2001; Gray & Kiran, 2013; Paradis, 2004; Peñaloza et al., 2020; for a review, see Peñaloza & Kiran, 2019).

14.3.2 Language Control Interacts with Language Recovery Patterns

Paradis and colleagues (1982) speculated that alternating antagonistic recovery could be due to the unrecovered language being "inhibited" rather than lost. More recently, the literature has begun to revisit the nature of recovery and impairment patterns in relation to language control. It has been suggested that loss of access to language (e.g., selective or differential recovery) could be a manifestation of language control impairment in bilingual aphasia (Abutalebi & Green, 2007; Green & Abutalebi, 2008; Peñaloza & Kiran, 2019; Verreyt et al., 2013). For instance, a selective loss of one language may be due to its increased inhibition or the inability to activate it to the appropriate level for selection to take place (Abutalebi & Green, 2007; Green & Abutalebi, 2008;

Green et al., 2010; Paradis et al., 1982). According to this view, linguistic presentations are not lost by brain damage but the control processes behind language activation and inhibition are affected.

Similarly, linguistic behaviors such as pathological language mixing, pathological language switching, and crosslinguistic intrusions have all been argued to reflect difficulties in language control (e.g., Abutalebi et al., 2000; Aglioti et al., 1996; Ansaldo et al, 2010; Keane & Kiran, 2015; Kong et al., 2014). Pathological language mixing refers to the uncontrolled mixing of elements of two languages within a single utterance (e.g., Ansaldo et al., 2010; Fabbro et al., 2000; Goral et al., 2019), whilst pathological language switching entails involuntarily alternating languages across different utterances (Ansaldo & Marcotte, 2007). From the limited neural data from clinical cases, it has been speculated that damage to language control networks, especially in the subcortical areas (basal ganglia and subthalamic regions) and/or the frontal areas and their connections with striatal structures, are responsible for such behaviors (Fabbro et al., 2000; Keane & Kiran, 2015). Researchers have tried to disentangle how specific measures of language control relate to language mixing or switching behaviors (e.g., Kong et al., 2014). Kong and colleagues (2014) examined these processes in Cantonese in a Cantonese-English-Mandarin trilingual person with aphasia presenting with a lesion in the left frontal lobe. The patient showed impairment in executive control tasks testing inhibition and shifting abilities. The authors concluded that linguistic and nonlinguistic control processes are closely linked and could be responsible for the switching behaviors. Although there exist case studies exploring the relationship between specific linguistic behaviors (e.g., language mixing, crosslinguistic intrusions) and impairment patterns (e.g., selective impairment) with language control in bilingual aphasia, further research in this domain would inform our understanding of the interaction of language impairments, recovery patterns and language control in greater depth.

Although this chapter is focused on language control in bilingual aphasia, it is important to highlight that language recovery patterns and specific impairments are also influenced by bilingualism variables and language-specific characteristics of the two languages. We will briefly provide the summary for these two domains in the following section.

14.3.3 Bilingualism Variables Influence Post-Injury Language Outcomes

Bilingualism variables, such as age-of-acquisition, language exposure, usage, and proficiency play important roles in post-injury language recovery. Age of acquisition refers to the age at which people learn a language. It is well established that earlier acquired words are better preserved in aging and neurological disorders (e.g., Brysbaert & Ellis, 2016). Age of acquisition has also been identified as a contributor to pre-injury language proficiency (Peñaloza et al., 2020). Language proficiency is a multidimensional construct (Van Hell & Tanner, 2012) that modulates language processing in bilinguals with and without aphasia (Blumenfeld et al., 2016; Gray & Kiran 2013; Kiran et al., 2014; Kroll & Stewart, 1994). Because BPWA have experienced a stroke or other brain injury, clinicians must consider pre-injury proficiency to determine post-injury skill level. Pre-injury language proficiency is influenced by a variety of factors (e.g., daily use, educational history, confidence; Peñaloza & Kiran, 2019; Peñaloza et al., 2020), and measuring it remains a challenging task which is mostly completed by subjective rating scales (Kastenbaum et al., 2019; Muñoz & Marquardt, 2003). Consistency of rating proficiency with objective measures is still contended (Tomoschuk et al., 2019), and self-ratings have the potential to overestimate ability (MacIntyre et al., 1997; Veríssimo, 2021). However, measures of self-report have also been shown to provide valid indices (Delgado et al., 1999; Kohnert et al., 1998; Li et al., 2006). Informative for clinical implications, pre-injury language proficiency has been shown to predict post-injury language impairment, and the strongest pre-injury language is typically the strongest post-injury language, regardless of L1 or L2 status (Gray & Kiran, 2013; Peñaloza et al., 2020). It is possible that variations from this post-injury language impairment and recovery pattern could be indicative of a pre- to post-injury change in activation or inhibition of language representations. Finally, depending on communication needs, pre- vs post-injury language use and exposure can change (Peñaloza & Kiran, 2019), and in terms of language control in bilingual aphasia, living in an environment where one language is predominantly used can lead to inhibition of less used languages (Linck et al., 2009).

In addition to bilingualism variables, linguistic similarities and differences between language combinations and how these factors interact with language control are important because they affect language access. The following section unpacks these issues, exploring the effects in bilingual aphasia, especially how processing and shared structures between languages interact in the context of language control.

14.3.4 Linguistic Similarities and Differences between the Languages and Language Control in Bilingual Aphasia

In the 1980s, Paradis proposed the "structural distance hypothesis" to explain manifestation of symptoms in bilingual aphasia. The main tenet of this hypothesis is that shared linguistic structures (or lack thereof) between languages impacts the specific features that are impaired in bilingual aphasia. For example, Fabbro (2001) examined Italian-Friulian speakers with aphasia who exhibited parallel recovery but demonstrated different agrammatic manifestation in Italian and Friulian due to different levels of pronoun omissions in Italian and Friulian. Italian is a partial pro-drop language allowing speakers to "drop" the subject pronoun if the subject can be inferred from the context. In contrast, Friulian, like English, is a non-pro-drop language, meaning speakers must use the subject pronoun regardless of the referent's transparency. The bilingual speakers with agrammatism in Fabbro's study made more frequent pronoun deletion errors in Friulian than in Italian, suggesting that the structural distance between languages can influence linguistic impairment in bilingual aphasia.

A body of lexical level research has exploited cognate vs non-cognate words to study the impact of shared structures at lexical level in bilinguals. Cognates are words that have the same meaning and a similar phonological form in both languages (e.g., English-Spanish cognate pair: *cat* in English, *gato* in Spanish); non-cognate words have the same meaning in both languages but different forms (e.g., English-Spanish non-cognate pair: *apple* in English, *manzana* in Spanish). The majority of the studies involving bilinguals without aphasia have shown cognate facilitation: cognates are processed and produced faster and more accurately than non-cognates (Costa, 2005; Lalor & Kirsner, 2001; Marte et al., 2023; Roberts & Deslauriers, 1999). However, opposing findings also have been reported, ranging from no effect of cognate status (Grasso et al., 2019) to poorer performance on cognate words (Broersma et al., 2016; Kurland & Falcon, 2011; Verreyt et al., 2013).

Although the non-aphasic bilingual literature has suggested that executive control can influence cognate effects, there are only two studies that experimentally investigate the relationship between executive control and cognate effects in bilingual aphasia (Van der Linden et al., 2018; Patra et al., in-review). Van der Linden and colleagues (2018) tested seven patients with differential aphasia, eight patients with parallel aphasia, and nineteen bilingual matched controls on a lexical decision task to examine the cognate effect in these groups. Additionally, executive control was assessed using a nonlinguistic flanker task. The authors observed cognate facilitation for all three groups with a larger cognate facilitation effect for the differential group compared to the parallel group and controls. On the flanker task, the differential group made more errors on the incongruent trials compared to congruent trials. Greater cognate facilitation effect and poorer performance in the executive control task was attributed to a control deficit in bilinguals with differential aphasia.

Similarly, Patra and colleagues (in-review) investigated the interaction and influence of cognate status, language dominance, and executive control using a picture-naming task for cognate vs. non-cognate words from seven Bengali-English BPWA and eight age-matched adults without aphasia. Preliminary findings revealed cognate facilitation for BPWA only in their nondominant language. Importantly, from the perspective of language control, the magnitude of cognate facilitation was inversely related to executive control, that is, better switching abilities (indicated by smaller trail making test ratio) were associated with a smaller cognate facilitation effect. This relationship might appear counter intuitive, but these findings are consistent with the inhibitory control model of language production (Green, 1998). According to the inhibitory control model, at the lexical level, greater activation of the cognate word in the nontarget language may lead to increased cross-language competition while naming the word in the target language. It can be argued that participants with better executive control abilities can supress that cross-language interference and may not be able (or may not need) to use any advantage of the cognate word. This is in line with Linck and colleagues' (2008) study involving healthy adults, which reported greater cognate facilitation for participants with poorer inhibitory control abilities. The findings from both Van der Linden and colleagues' (2018) and Patra and colleagues' (in-review) studies show that processing of shared linguistic structure is modulated by executive control. In addition, linguistic distance between languages has been shown to impact cross-linguistic treatment generalization and intrusion patterns (e.g., Conner et al., 2018; Goral et al., 2010). These lines of research investigating the interaction between language control and shared structure processing have the potential to inform our understanding of the interaction between language control and type of language combinations used by bilinguals.

14.4 Bilingual Language Control and Cognitive Control

14.4.1 Domain General vs Domain Specific Cognitive Control

There is a small group of studies that examine the relationship between linguistic control and nonlinguistic control mechanisms. They attempt to identify a possible association (aka overlap) between the two domains, which is indicative of domain general cognitive control, or identify a possible dissociation (aka no overlap) between the two domains, which is indicative of domain specific cognitive control. This type of work requires systematic research that accounts for type of cognitive control within and across linguistic and nonlinguistic domains. Inhibition is a type of cognitive control, and Friedman and Miyake (2004) define three types of inhibition: (1) *prepotent response inhibition* is the ability to suppress a dominant or automatic response (e.g., Stroop task); (2) *resistance to distracter interference* is the ability to suppress or resolve interference from competing, nontarget stimuli (e.g., flanker task); and (3) *resistance to proactive interference* is the ability to resist information that is no longer the target (e.g., cued recall task). This framework is useful because it identifies types of inhibitory control which is an integral component to systematically examining bilingual language control by way of experimental paradigms (flanker, Stroop, etc.).

When examining whether control mechanisms are domain general or domain specific, it is critical that the experimental paradigms employ the same type of inhibition across linguistic and nonlinguistic paradigms. In addition, it is important that the paradigms isolate linguistic control from nonlinguistic control as best they can. This limits confounding results, allowing for a methodical comparison of control mechanisms across linguistic and nonlinguistic domains. A small group of studies have explored this line of work (Dash & Kar, 2014; Faroqi-Shah et al., 2018; Gray & Kiran, 2016, 2019; Green et al., 2010; Verreyt et al., 2013; for a review see Nair et al., 2021). Some studies use traditional flanker tasks whereas others use linguistic and nonlinguistic triad tasks, lexical decision tasks, or Stroop tasks paired with other linguistic tasks. The results have been wide-ranging such that some show an overlap, partial overlap or no overlap between linguistic control and nonlinguistic control.

14.4.2 The Relationship between Language Control and Cognitive Control in Bilingual Aphasia: Case Studies and Experiments

There are behavioral studies that employ a variety of methodologies to explore how linguistic and nonlinguistic control processes present in bilingual aphasia, for example: fluency tasks (Carpenter et al., 2020; Patra et al., 2020), language switching (Grunden et al., 2020; Radman et al., 2016), the examination of task complexity (Gray & Kiran, 2019), and the investigation of semantic and phonological control (Calabria et al., 2019; Calabria et al., 2020; Gray, 2020). By having a variety of approaches, we can better understand the complexity of language control and the relationship between language control and nonlinguistic control. In what follows, we discuss studies that examine how language control processes interact with nonlinguistic cognitive control.

Two recent studies have examined language control via verbal fluency tasks, exploring the interaction between executive control and its impact on how BPWA manage fluency in their two languages. Traditional fluency tasks include letter (or phonemic) fluency (i.e., generating words with a particular letter) and semantic (or category) fluency (i.e., generating words that belong to a particular category) tasks. Both tasks require cognitive control (e.g., inhibition, conflict monitoring); however, because semantic fluency resembles the everyday life experience that relies on semantic associations (e.g., buying items for a party or what to bring to the beach) and letter fluency is not a typical strategy used for word recall, semantic fluency is more automatic and considered an easier task that produces more exemplars than the letter fluency task (Gollan et al., 2002; Luo et al., 2010; Mathuranath et al., 2003; Shao et al., 2014). Patra and colleagues (2020) asked eight Bengali-dominant Bengali-English BPWA and eight age-matched adults without aphasia to complete semantic and letter fluency tasks in Bengali and in English, in addition to the Stroop task, trail making test, and backward digit span task. Results showed that BPWA who performed poorly on the letter fluency task also performed poorly on the executive control tasks. Another finding revealed that participants without aphasia performed better on the semantic fluency task in their dominant language, whereas BPWA scored similarly on Bengali fluency tasks and English fluency tasks. These results suggest that in addition to executive control, language control interacts with type of fluency task. In a second study, Carpenter and colleagues (2020) manipulated cognitive control demands of a fluency task to investigate how BPWA manage language control when cognitive control demands increase. Thirteen Spanish-English BPWA and twenty-two Spanish-English age-matched adults without aphasia completed a semantic fluency task and a letter fluency task. L1 was the dominant language for most participants. The semantic fluency task included four conditions that manipulated cognitive control demands: No-Switch-L1 and No-Switch-L2 were administered in single language contexts and Self-Switch and Forced-Switch were administered in dual language contexts. The No-Switch L1 and Self-Switch conditions were considered low cognitive control demand trials, and No-Switch L2 and Forced-Switch conditions were considered high cognitive control demand trials. Participants also completed a single language context letter fluency task. Results revealed that on the semantic fluency task, BPWA performed worse than the participants without aphasia on the more cognitive control demanding Forced-Switch condition. On the letter fluency task, participants without aphasia named the same number of items in L1 and L2, whereas BPWA named more items in L1 compared to L2, revealing a naming dissociation between the more dominant L1 and less dominant L2. These findings highlight the link between language control and cognitive control demands, showing that bilingual language control in BPWA is negatively affected when cognitive control demands increase.

Another study has also shown that as cognitive control demands of the task increase, BPWA performance diverges from age-matched bilingual adults. Gray and Kiran (2019) investigated the relationship between linguistic and nonlinguistic cognitive control using tasks that increased in complexity. Thirteen Spanish-English BPWA and twenty Spanish-English age-matched adults completed two low complexity tasks (a linguistic and nonlinguistic flanker) and two high complexity tasks (a linguistic and nonlinguistic triad), all designed to tap resistance to distracter interference (Friedman & Miyake, 2004). Results revealed that both groups performed similarly on the low complexity tasks. However, on the high complexity tasks, findings from the age-matched adults were indicative of domain general cognitive control, whereas BPWA data was indicative of domain specific cognitive control. Additional analyses showed that for BPWA, greater bilingual language control impairment was associated with more impaired language ability. Thus, even though bilingual language control and language impairment (i.e., aphasia) are distinct processes, they are associated, and this underscores the challenges involved in better understanding the dynamic and complex nature of their relationship.

Other concerns related to language control in bilingual aphasia include the heterogeneous nature of bilingualism (e.g., simultaneous vs sequential language learning, various proficiency levels, language use experience) and how these factors are accounted for in study methodologies. Because BPWA present with various proficiencies across two languages, as well as a range of impairment levels and variety of aphasia types, these factors must also be addressed, sometimes with individual analyses from the larger sample size. Grunden and colleagues (2020) examined voluntary language switching in bilingual aphasia. Seven Catalan-Spanish BPWA and ten age-matched bilingual adults were asked to complete a voluntary picture-naming switching task that included two single-language blocks (single trials) and one dual-language block (repeat and switch trials). Reactive control was measured by switch costs (the difference between switch trials and repeat trials), and proactive control was measured by mixing costs (the difference between repeat and single trials). Results were examined at the group and individual level. BPWA and age-matched bilingual adult data showed effects of reactive control, but there was no difference in switch cost magnitude between the two groups. For proactive control, there were no significant findings aside from age-matched bilingual adults showing the mixing cost for cognates in their nondominant language. Individual analyses revealed that two BPWA exhibited increased cross-language errors in the single-language condition. Cross-language intrusion errors can be indicative of language control impairment, and both BPWA with increased cross-language errors presented with moderate aphasia, whereas the other BPWA presented with mild aphasia. Additionally, of the two BPWA with language control impairment, one BPWA presented with fluent aphasia and the other BPWA presented with nonfluent aphasia. Taken together, these findings suggest that rather than type of aphasia, severity of aphasia may be more indicative of language control deficits. These results are in line with previous studies that also found that aphasia severity is correlated with cognitive abilities (Fucetola et al., 2009; Gray, 2020; Gray & Kiran, 2019; Lee & Pyun, 2014; Patra et al., 2020).

As previously discussed, bilingual lexical selection requires inhibition of the nontarget language. Another type of control associated with lexical selection is semantic control. Retrieving a target word (e.g., banana), also activates semantically related competitors (e.g., cherry; Collins & Loftus, 1975), and these lexical entries compete for selection (Schnur et al., 2006; Schriefers et al., 1990). Thus, it is logical that a new body of research that examines bilingual language control and semantic control in aphasia is emerging (Calabria et al., 2019; Gray, 2020). Calabria and colleagues (2019) asked eleven Catalan-Spanish BPWA and thirteen Catalan-Spanish age-matched adults without aphasia to complete three tasks. The first was a semantically blocked cyclical naming task that included within category and mixed category conditions. It is hypothesized that within category conditions generate more interference than mixed category conditions. The second task was a word-picture match task where the distracter was semantically related to the target, the trials were within language, and the blocks were mixed language. The third task was a nonlinguistic flanker task. Results showed an overlap between semantic control and nonlinguistic control, and that for BPWA, semantic interference presented differently across the dominant and nondominant languages, revealing more interference on the nondominant language condition. In another study designed to examine the possible overlap in bilingual language control, semantic control and nonlinguistic control mechanisms (Gray, 2020), twelve Spanish-English BPWA and twenty Spanish-English age-matched adults without aphasia completed three tasks designed to tap resistance to distracter interference: a nonlinguistic color shape triad task, a language control triad task, and a semantic control triad task. In these triad tasks, three images are presented (a given item, target, and distracter), and with a button press, the participant selects the target or distracter that "matches" the given item. Results showed that for age-matched adults without aphasia, semantic control and nonlinguistic control tasks were significantly correlated, suggesting that these two types of control are associated, similar to Calabria and colleagues' (2019) findings. For BPWA, language control and nonlinguistic control tasks were significantly correlated. This was unexpected considering previous findings that showed no correlation between language control and nonlinguistic control (e.g., Gray & Kiran, 2019). The author theorized that the result might be capturing increased cognitive load needed to simply complete the task, rather than capturing language control processes, underscoring the importance of considering aphasia severity and how it may interact with cognitive control. In sum, the Calabria and colleagues (2019) and Gray (2020) studies extend the examination of bilingual language control and nonlinguistic control to include semantic control, another facet of how control mechanisms operate in the brain, and findings show that semantic control may partially overlap with nonlinguistic control.

This section has discussed a body of literature that examines the relationship between linguistic, nonlinguistic, and semantic control processes and how they present in BPWA. These processes are complex, and the studies have employed a variety of methodologies which enables a wide range of interpretations to better understand how these control processes function in the bilingual brain. In addition, examining the role of aphasia severity has the potential to offer meaningful interpretations of how these control processes function across linguistic and nonlinguistic domains. Finally, this work will further our knowledge on how these control processes impact post-injury language presentation (e.g., impairment and recovery patterns) which has the potential to inform best practice for rehabilitation treatment programs for this BPWA.

14.5 Language Control in Context of Therapy

14.5.1 Cross-Language Generalization in Bilingual Aphasia

In efforts to advance bilingual aphasia treatment interventions, there has been a focus to identify optimal conditions to observe cross-language generalization. This is a complicated endeavor because many factors (e.g., age of acquisition, language combinations, proficiency) that influence bilingual language processing, may also impact language rehabilitation in bilingual aphasia. Currently, the literature findings are varied regarding how and whether cross-language generalization can be achieved (e.g., Amberber, 2012; Edmonds & Kiran, 2006; Goral et al., 2012; Goral et al., 2013; Gray et al., 2023; Kiran & Iakupova, 2011; Kiran & Roberts, 2010; Kiran et al., 2013; Knoph, 2013; Kurland & Falcon, 2011; Lerman et al., 2022; Miertsch et al., 2009;

Sandberg et al., 2021; for a review see Goral, 2023). A complete review of factors that influence cross-language generalization is outside the scope of this chapter; however, we will discuss how control mechanisms may affect this process. As discussed in Section 14.2, neural correlates of language control overlap with cognitive control, and pertinent to BPWA language rehabilitation, damage to the bilingual language control network has the potential to affect treatment outcomes (Abutalebi & Green, 2007; Abutalebi et al., 2009; Keane & Kiran, 2015; Radman et al., 2016). For instance, Keane and Kiran (2015) delivered a naming treatment to a trilingual BPWA with subcortical damage that included the basal ganglia. A nonlinguistic flanker task that examines cognitive control was also administered. Results showed that the patient presented with impaired cognitive control on the flanker task. Regarding treatment outcomes, the patient did not show effects of cross-language generalization, and the therapy also appeared to cause language control problems characterized by language intrusion errors. These results suggest that to achieve cross-language generalization, the bilingual language control network must be intact.

A few models that account for within- and cross-language generalization have been proposed. The lingering suppression theory hypothesizes that by training the dominant language, minimal suppression is exerted on the untrained less-dominant language, so after treatment, both languages can be accessed, and cross-language generalization can be achieved (Goral et al., 2013; Lerman et al., 2022). Another framework posits that based on the revised hierarchical model (Kroll & Stewart, 1994), training the nondominant language supports cross-language generalization because the nondominant language accesses semantic representations via the dominant language (Edmonds & Kiran, 2006; Kiran & Iakupova, 2011; Kurland & Falcon, 2011). Extending this concept, the "integrative framework" posits that two mechanisms, spreading activation and nonlinguistic cognitive control (i.e., inhibitory control), are essential to achieve within- and cross-language generalization on semantic-based naming interventions (Kiran et al., 2013). More recently, the treatment effects in aphasia in multilingual (TEAM) model has been proposed (Goral & Lerman, 2020), extending the integrative framework to include multilingualism variables (e.g., language use and exposure, pre-stroke language abilities) and stroke-related variables (e.g., time post-onset, lesion site which may affect the language network and/or language control), all of which interact with treatment outcomes (e.g., response to directly trained items, within-language generalization, and cross-language generalization). All these models require some level of intact facilitation and suppression of lexical representation, and ultimately, these models are valuable tools when investigating cross-language generalization in bilingual aphasia because they can guide hypotheses and help us understand BPWA treatment outcomes.

There is a growing body of treatment studies that specifically investigate the role of cognitive control and how it modulates cross-language generalization, showing that to achieve cross-language generalization by treating the nondominant language, nonlinguistic cognitive control must be intact (Gray et al., 2023; Keane & Kiran, 2015; Kiran et al., 2013; Sandberg et al., 2021). Kiran and colleagues (2013) examined within- and cross-language generalization patterns in seventeen Spanish-English BPWA who participated in an intensive naming study. Results showed that the majority of BPWA improved on directly trained items, whereas some showed within-language generalization and/or cross-language generalization. The authors found that outcomes are variable and rely on a delicate interaction between facilitation and inhibition. In another study, Sandberg and colleagues (2021) administered bilingual abstract semantic associative network training (BAbSANT) to a Polish dominant Polish-English BPWA. The intervention included a Polish phase and an English phase. The participant also completed a nonlinguistic flanker task. Results revealed cross-language generalization to the untrained language only when the dominant language was trained. This finding was unexpected; however, the participant also presented with impaired cognitive control as evidenced by flanker results. The authors discuss these findings within the "integrative framework," suggesting that without intact control mechanisms, the lexical connections between L1 and L2 are disrupted so the hypothesized cross-language generalization to the dominant language cannot be achieved. Most recently, Gray and colleagues (2023) delivered BAbSANT to Dmitri, a Russian-English BPWA. Dmitri also completed two nonlinguistic cognitive control tasks that examined inhibitory control. Results revealed that Dmitri exhibited cross-language generalization only when the less dominant language was trained, and he presented with intact nonlinguistic cognitive control abilities as shown by his performance on the cognitive control tasks. These findings offer support for the integrative framework, such that both spreading activation and intact cognitive control mechanisms are required for cross-language generalization to occur.

This is an exciting time for bilingual aphasia treatment because we are on the forefront of better understanding the factors that influence bilingual language processing, how to best interpret post-injury bilingual aphasia treatment outcomes, and how to integrate these variables to optimize language rehabilitation programs. It is understood that language control plays a role in bilingual aphasia language recovery. We are at a point where we are learning how it functions and how it relates to other critical variables that also influence recovery. Both the "integrative framework" and TEAM are innovative models that account for cross-language generalization and have the potential to push the field forward so we can better understand treatment recovery patterns.

14.6 Conclusion

For over 150 years, there have been reports of bilingual aphasia in the literature, and the field has progressed with regards to identifying and characterizing recovery patterns, determining critical factors that contribute to recovery, and the development of clinical tools for better assessments. As seen in this review, despite progress in the discipline, experimental research exploring the mechanisms and interactions of language control in bilingual aphasia is still in its infancy and filled with opportunities. This chapter focused on the emerging themes in the field of language control and bilingual aphasia. First, an overview of language processing and mechanisms of language control in the healthy bilingual brain was provided. Understanding this foundation is critical to exploring how these processes manifest in bilingual aphasia. Next, language impairment and recovery patterns were discussed, and evidence suggests that these profiles can be interpreted in the context of language control. Then, bilingualism variables (age of acquisition, proficiency across languages, language combinations, etc.) and how they influence post-injury language use were reviewed. Given the heterogeneity of bilingual profiles, identifying meaningful patterns can be challenging, but nevertheless, evidence suggests these patterns have the potential to inform treatment agendas. Next, a body of work that examines whether language control and nonlinguistic (or nonverbal) cognitive control are associated, which is indicative of domain general cognitive control, or if these processes are dissociated, which is indicative of domain specific cognitive control, was discussed. Methodologically, this line of work is challenging due to the wide-range use of experimental tasks and paradigms across linguistic and nonlinguistic domains, however, the literature is slowly adapting so that experimental paradigms systematically examine control mechanisms across linguistic and nonlinguistic domains. Finally, treatment and the role of language control was discussed, with data suggesting that language control mechanisms must be intact to observe cross-language generalization. In sum, the investigation of language control in bilingual aphasia within the context of behavioral and imaging studies and treatment has the potential to inform best practice to improve clinical models and rehabilitation programs for BPWA.

References

Abutalebi, J., Della Rosa, P. A., Ding, G., Weekes, B., Costa, A., & Green, D. W. (2013). Language proficiency modulates the engagement of cognitive control areas in multilinguals. *Cortex, 49*(3), 905–911.

Abutalebi, J., Della Rosa, P. A., Tettamanti, M., Green, D. W., & Cappa, S. F. (2009). Bilingual aphasia and language control: A follow-up fMRI and intrinsic connectivity study. *Brain and Language, 109*(2), 141–156.

Abutalebi, J., & Green, D. (2007). Bilingual language production: The neurocognition of language representation and control. *Journal of Neurolinguistics, 20*(3), 242–275.

Abutalebi, J., & Green, D. W. (2008). Control mechanisms in bilingual language production: Neural evidence from language switching studies. *Language and Cognitive Processes, 23*(4), 557–582.

Abutalebi, J., Miozzo, M., & Cappa, S. F. (2000). Do subcortical structures control "language selection" in polyglots? Evidence from pathological language mixing. *Neurocase, 6,* 51–56.

Aglioti, S., Beltramello, A., Girardi, F., & Fabbro, F. (1996). Neurolinguistic and follow-up study of an unusual pattern of recovery from bilingual subcortical aphasia. *Brain: A Journal of Neurology, 119,* 1551–1564.

Amberber, A. M. (2012). Language intervention in French–English bilingual aphasia: Evidence of limited therapy transfer. *Journal of Neurolinguistics, 25*(6), 588–614.

Ansaldo, A. I., & Marcotte, K. (2007). Language switching and mixing in the context of bilingual aphasia. In J. G. Centeno, L. K. Obler, & R. T. Anderson (Eds.), *Studying Communication Disorders in Spanish Speakers: Theoretical, Research, and Clinical Aspects.* Multilingual Matters.

Ansaldo, A. I., & Saidi, L. G. (2014). Aphasia therapy in the age of globalization: cross-linguistic therapy effects in bilingual aphasia. *Behavioural Neurology,* 603085. https://doi.org/10.1155/2014/603085

Ansaldo, A. I., Saidi, L. G., & Ruiz, A. (2010). Model-driven intervention in bilingual aphasia: Evidence from a case of pathological language mixing. *Aphasiology, 24*(2), 309–324.

Ardila, A., & Ramos, E. (2007). *Speech and Language Disorders in Bilinguals.* Nova Science Publishers.

Blanco-Elorrieta, E., & Caramazza, A. (2021). A common selection mechanism at each linguistic level in bilingual and monolingual language production. *Cognition, 213,* 104625.

Bloem, I., & La Heij, W. (2003). Semantic facilitation and semantic interference in word translation: Implications for models of lexical access in language production. *Journal of Memory and Language, 48*(3), 468–488.

Blumenfeld, H. K., Bobb, S. C., & Marian, V. (2016). The role of language proficiency, cognate status and word frequency in the assessment of Spanish–English bilinguals' verbal fluency. *International Journal of Speech-Language Pathology, 18*(2), 190–201. https://doi.org/10.3109/17549507.2015.1081288

Braver, T. S., Reynolds, J. R., & Donaldson, D. I. (2003). Neural mechanisms of transient and sustained cognitive control during task switching. *Neuron, 39*(4), 713–726.

Broersma, M., Carter, D., & Acheson, D. J. (2016). Cognate costs in bilingual speech production: Evidence from language switching. *Frontiers in Psychology, 7,* 1461.

Brysbaert, M., & Ellis, A. W. (2016). Aphasia and age of acquisition: are early-learned words more resilient? *Aphasiology, 30,* 1240–1263.

Calabria, M., Costa, A., Green, D. W., & Abutalebi, J. (2018). Neural basis of bilingual language control. *Annals of the New York Academy of Sciences, 1426*(1), 221–235.

Calabria, M., Grunden, N., Iaia, F., & García-Sánchez, C. (2020). Interference and facilitation in phonological encoding: Two sides of the same coin? Evidence from bilingual aphasia. *Journal of Neurolinguistics, 56*, 100935. https://doi.org/10.1016/j.jneuroling.2020.100935

Calabria, M., Grunden, N., Serra, M., García-Sánchez, C., & Costa, A. (2019). Semantic processing in bilingual aphasia: Evidence of language dependency. *Frontiers in Human Neuroscience, 13*, 205. https://doi.org/10.3389/fnhum.2019.00205

Carpenter, E., Rao, L., Peñaloza, C., & Kiran, S. (2020). Verbal fluency as a measure of lexical access and cognitive control in bilingual persons with aphasia. *Aphasiology, 34*(11), 1341–1362.

Collins, A. M., & Loftus, E. F. (1975). A spreading-activation theory of semantic processing. *Psychological review, 82*(6), 407.

Colomé, À. (2001). Lexical activation in bilinguals' speech production: Language-specific or language-independent? *Journal of Memory and Language, 45*(4), 721–736.

Conner, P. S., Goral, M., Anema, I., Borodkin, K., Haendler, Y., Knoph, M., Mustelier, C., Paluska, E., Melnikova, Y., & Moeyaert, M. (2018). The role of language proficiency and linguistic distance in cross-linguistic treatment effects in aphasia. *Clinical Linguistics & Phonetics, 32*(8), 739–757.

Costa, A. (2005). Lexical access in bilingual production. In J. F. Kroll, & A. M. B. De Groot (Eds.), *Handbook of Bilingualism: Psycholinguistic Approaches* (pp. 308–325). Oxford University Press.

Costa, A., Miozzo, M., & Caramazza, A. (1999). Lexical selection in bilinguals: Do words in the bilingual's two lexicons compete for selection? *Journal of Memory and Language, 41*(3), 365–397. https://doi.org/10.1006/jmla.1999.2651

Costa, A., & Santesteban, M. (2004). Lexical access in bilingual speech production: Evidence from language switching in highly proficient bilinguals and L2 learners. *Journal of Memory and Language, 50*(4), 491–511.

Dash, T., & Kar, B. R. (2014). Bilingual language control and general purpose cognitive control among individuals with bilingual aphasia: Evidence based on negative priming and flanker tasks. *Behavioural Neurology, 2014*(1), 679–706.

de Groot, A. M. (1992). Determinants of word translation. *Journal of Experimental Psychology: Learning, Memory, and Cognition, 18*(5), 1001–1018. https://doi.org/10.1037/0278-7393.18.5.1001

de Groot, A. M., Dannenburg, L., & Vanhell, J. G. (1994). Forward and backward word translation by bilinguals. *Journal of Memory and Language, 33*(5), 600–629.

Delgado, P., Guerrero, G., Goggin, J. P., Ellis, B. B., Goggin, J. P., & Ellis, B. B. (1999). Behavioral sciences self-assessment of linguistic skills. *Hispanic Journal of Behavioral Sciences, 21*, 31–46.

Dijkstra, T., & Van Heuven, W. J. (2002). The architecture of the bilingual word recognition system: From identification to decision. *Bilingualism: Language and Cognition, 5*(3), 175–197.

Edmonds, L. A., & Kiran, S. (2006). Effect of Semantic Naming Treatment on Crosslinguistic Generalization in Bilingual Aphasia. *Journal of Speech, Language, and Hearing Research, 49*, 729–749.

Fabbro, F. (2001). The bilingual brain: Bilingual aphasia. *Brain and Language, 79*(2), 201–210. https://doi.org/10.1006/brln.2001.2480

Fabbro, F., Skrap, M., & Aglioti, S. (2000). Pathological switching between languages after frontal lesions in a bilingual patient. *Journal of Neurology, Neurosurgery, and Psychiatry, 68(5*), 650–652.

Faroqi-Shah, Y., Sampson, M., Pranger, M., & Baughman, S. (2018). Cognitive control, word retrieval and bilingual aphasia: Is there a relationship? *Journal of Neurolinguistics, 45*, 95–109.

Friedman, N. P., & Miyake, A. (2004). The relations among inhibition and interference control functions: A latent-variable analysis. *Journal of Experimental Psychology: General, 133*(1), 101–135.

Fucetola, R., Connor, L. T., Strube, M. J., & Corbetta, M. (2009). Unravelling nonverbal cognitive performance in acquired aphasia. *Aphasiology, 23*(12), 1418–1426.

Gollan, T. H., Montoya, R. I., & Werner, G. A. (2002). Semantic and letter fluency in Spanish-English bilinguals. *Neuropsychology, 16*(4), 562–576. https://doi.org/10.1037/0894-4105.16.4.562

Goral, M., & Lerman, A. (2020). Variables and mechanisms affecting response to language treatment in multilingual people with aphasia. *Behavioral Sciences, 10*(9), 144.

Goral, M., Levy, E. S., & Kastl, R. (2010). Cross-language treatment generalisation: A case of trilingual aphasia. *Aphasiology, 24*(2), 170–187.

Goral, M., Naghibolhosseini, M., & Conner, P. S. (2013). Asymmetric inhibitory treatment effects in multilingual aphasia. *Cognitive Neuropsychology, 30*(7–8), 564–577.

Goral, M., Norvik, M. I., Antfolk, J., Agrotou, I., & Lehtonen, M. (2023). Cross-language generalization of language treatment in multilingual people with post-stroke aphasia: A meta-analysis. *Brain and Language, 246*, 105326.

Goral, M., Norvik, M., & Jensen, B. U. (2019). Variation in language mixing in multilingual aphasia. *Clinical Linguistics & Phonetics, 33*(10–11), 915–929.

Goral, M., Rosas, J., Conner, P. S., Maul, K. K., & Obler, L. K. (2012). Effects of language proficiency and language of the environment on aphasia therapy in a multilingual. *Journal of neurolinguistics, 25*(6), 538–551.

Grasso, S. M., Cruz, D. F., Benavidez, R., Peña, E. D., & Henry, M. L. (2019). Video-implemented script training in a bilingual Spanish-English speaker with aphasia. *Journal of Speech, Language, and Hearing Research, 62*(7), 2295–2316.

Gray, T. (2020). The relationship between language control, semantic control and nonverbal control. *Behavioral Sciences, 10*(11), 169. https://doi.org/10.3390/bs10110169

Gray, T., & Kiran, S. (2013). A theoretical account of lexical and semantic naming deficits in bilingual aphasia. *Journal of Speech, Language, and Hearing Research, 56*(4), 1314–1327.

Gray, T., & Kiran, S. (2016). The relationship between language control and cognitive control in bilingual aphasia. *Bilingualism: Language and Cognition, 19*(3), 433–452.

Gray, T., & Kiran, S. (2019). The effect of task complexity on linguistic and non-linguistic control mechanisms in bilingual aphasia. *Bilingualism: Language and Cognition*, *22*(2), 266–284.

Gray, T., Palevich, J., & Sandberg, C. (2023). Bilingual abstract semantic associative network training (BAbSANT): A Russian-English case study. *Bilingualism: Language and Cognition*, *27*(2), 1–17. https://doi.org/10.1017/S1366728923000500

Green, D. W. (1986). Control, activation, and resource: A framework and a model for the control of speech in bilinguals. *Brain and Language*, *27*(2), 210–223.

Green, D. W. (1998). Mental control of the bilingual lexico-semantic system. *Bilingualism: Language and Cognition*, *1*(2), 67–81.

Green, D. W., & Abutalebi, J. (2008). Understanding the link between bilingual aphasia and language control. *Journal of Neurolinguistics*, *21*(6), 558–576.

Green, D. W., & Abutalebi, J. (2013). Language control in bilinguals: The adaptive control hypothesis. *Journal of Cognitive Psychology*, *25*(5), 515–530.

Green, D. W., Grogan, A., Crinion, J., Ali, N., Sutton, C., & Price, C. J. (2010). Language control and parallel recovery of language in individuals with aphasia. *Aphasiology*, *24*(2), 188–209.

Grosjean, F. (1992). Another view of bilingualism. In R. Harris (Ed.), *Cognitive Processing in Bilinguals*, vol. 13 (pp. 51–62). North-Holland.

Grunden, N., Piazza, G., García-Sánchez, C., & Calabria, M. (2020). Voluntary language switching in the context of bilingual aphasia. *Behavioral Sciences*, *10*(9), 141.

Hermans, D., Bongaerts, T., De Bot, K., & Schreuder, R. (1998). Producing words in a foreign language: Can speakers prevent interference from their first language? *Bilingualism: Language and Cognition*, *1*(3), 213–229.

Kastenbaum, J. G., Bedore, L. M., Peña, E. D., Sheng, L., Mavis, I., Sebastian-Vaytadden, R., ... & Kiran, S. (2019). The influence of proficiency and language combination on bilingual lexical access. *Bilingualism: Language and Cognition*, *22*(2), 300–330.

Kaushanskaya, M., & Marian, V. (2007). Bilingual language processing and interference in bilinguals: Evidence from eye tracking and picture naming. *Language Learning*, *57*(1), 119–163.

Keane, C., & Kiran, S. (2015). The nature of facilitation and interference in the multilingual language system: Insights from treatment in a case of trilingual aphasia. *Cognitive Neuropsychology*, *32*(3–4), 169–194.

Kiran, S., Balachandran, I., & Lucas, J. (2014). The nature of lexical-semantic access in bilingual aphasia. *Behavioural Neurology*. https://doi.org/10.1155/2014/389565

Kiran, S., & Iakupova, R. (2011). Understanding the relationship between language proficiency, language impairment and rehabilitation: Evidence from a case study. *Clinical Linguistics & Phonetics*, *25*(6–7), 565–583.

Kiran, S., & Roberts, P. M. (2010). Semantic feature analysis treatment in Spanish-English and French-English bilingual aphasia. *Aphasiology*, *24*(2), 231–261.

Kiran, S., Sandberg, C., Gray, T., Ascenso, E., & Kester, E. (2013). Rehabilitation in bilingual aphasia: Evidence for within- and between-language generalization. *American Journal of Speech Language Pathology*, *22*(2), 298–310.

Knoph, M. I. K. (2013). Language intervention in Arabic-English bilingual aphasia: A case study. *Aphasiology*, *27*(12), 1440–1458.

Kohnert, K. J., Hernandez, A. E., & Bates, E. (1998). Bilingual performance on the Boston Naming Test: Preliminary norms in Spanish and English. *Brain and Language*, *65*(3), 422–440.

Kong, A. P. H., Abutalebi, J., Lam, K. S. Y., & Weekes, B. (2014). Executive and language control in the multilingual brain. *Behavioural Neurology*. https://doi.org/10.1155/2014/527951

Kroll, J. F., & Stewart, E. (1994). Category interference in translation and picture naming: Evidence for asymmetric connections between bilingual memory representations. *Journal of Memory and Language*, *33*(2), 149–174.

Kurland, J., & Falcon, M. (2011). Effects of cognate status and language of therapy during intensive semantic naming treatment in a case of severe nonfluent bilingual aphasia. *Clinical Linguistics & Phonetics*, *25*(6–7), 584–600.

Lalor, E., & Kirsner, K. (2001). The role of cognates in bilingual aphasia: Implications for assessment and treatment. *Aphasiology*, *15*(10–11), 1047–1056.

Lee, B., & Pyun, S. B. (2014). Characteristics of cognitive impairment in patients with post-stroke aphasia. *Annals of Rehabilitation Medicine*, *38*(6), 759–765.

Lerman, A., Goral, M., Edmonds, L. A., & Obler, L. K. (2022). Strengthening the semantic verb network in multilingual people with aphasia: Within- and cross-language treatment effects. *Bilingualism: Language and Cognition*, *25*(4), 645–659.

Li, P., Sepanski, S., & Zhao, X. (2006). Language history questionnaires: A web-based interface for bilingual research. *Behavior Research Methods*, *38*(2), 202–210.

Linck, J. A., Hoshino, N., & Kroll, J. F. (2008). Cross-language lexical processes and inhibitory control. *The Mental Lexicon*, *3*(3), 349–374.

Linck, J. A., Kroll, J. F., & Sunderman, G. (2009). Losing access to the native language while immersed in a second language: Evidence for the role of inhibition in second-language learning. *Psychological Science*, *20*(12), 1507–1515.

Luo, L., Luk, G., & Bialystok, E. (2010). Effect of language proficiency and executive control on verbal fluency performance in bilinguals. *Cognition*, *114*(1), 29–41.

MacIntyre, P. D., Noels, K. A., & Clément, R. (1997). Biases in self-ratings of second language proficiency: The role of language anxiety. *Language Learning*, *47*(2), 265–287.

Marte, M. J., Peñaloza, C., & Kiran, S. (2023). The cognate facilitation effect on lexical access in bilingual aphasia: Evidence from the Boston Naming Test. *Bilingualism: Language and Cognition*, *26*(5), 1009–1025. https://doi.org/10.1017/S1366728923000251

Mathuranath, P. S., George, A., Cherian, P. J., Alexander, A. L., Sarma, S. G., & Sarma, P. S. (2003). Effects of age, education and gender on verbal fluency. *Journal of Clinical and Experimental Neuropsychology*, *25*(8), 1057–1064.

Michael, E. B., & Gollan, T. H. (2005). Being and becoming bilingual: Individual differences and consequences for language production. In J. F. Kroll & A. M. B. de Groot (Eds.), *Handbook of Bilingualism: Psycholinguistic Approach* (pp. 389–407). Oxford University Press.

Miertsch, B., Meisel, J. M., & Isel, F. (2009). Non-treated languages in aphasia therapy of polyglots benefit from improvement in the treated language. *Journal of Neurolinguistics, 22*(2), 135–150.

Miller, E. K. (2000). The prefontral cortex and cognitive control. *Nature Reviews Neuroscience, 1*(1), 59–65.

Miller, E. K., & Cohen, J. D. (2001). An integrative theory of prefrontal cortex function. *Annual Review of Neuroscience, 24*(1), 167–202.

Muñoz, M., & Marquardt, T. (2003). Picture naming and identification in bilingual speakers of Spanish and English with and without aphasia. *Aphasiology, 17*(12), 1115–1132.

Nair, V. K., Rayner, T., Siyambalapitiya, S., & Biedermann, B. (2021). Domain-general cognitive control and domain-specific language control in bilingual aphasia: A systematic quantitative literature review. *Journal of Neurolinguistics, 60*, 101021.

Paradis, M. (2004). *A Neurolinguistic Theory of Bilingualism*. John Benjamins Publishing.

Paradis, M., Goldblum, M., & Abidi, R. (1982). Alternate anatagonism with paradoxical translation behaviour in two bilingual aphasic patients. *Brain and Language, 15*, 55–69.

Patra, A., Bose, A., & Marinis, T. (2020). Lexical and cognitive underpinnings of verbal fluency: Evidence from Bengali-English bilingual aphasia. *Behavioral Sciences, 10*(10), 155.

Patra, A., Marinis, T., & Bose, A. (in-review). Cognate picture naming in bilingual aphasia. *International Journal of Language and Communication Disorders.*

Peñaloza, C., & Kiran, S. (2019). Recovery and rehabilitation patterns in bilingual and multilingual aphasia. *The Handbook of the Neuroscience of Multilingualism*, 553–571.

Peñaloza, C., Barrett, K., & Kiran, S. (2020). The influence of prestroke proficiency on poststroke lexical-semantic performance in bilingual aphasia. *Aphasiology, 34*(10), 1223–1240.

Potter, M. C., So, K. F., Von Eckardt, B., & Feldman, L. B. (1984). Lexical and conceptual representation in beginning and proficient bilinguals. *Journal of Verbal Learning and Verbal Behavior, 23*(1), 23–38.

Radman, N., Mouthon, M., Di Pietro, M., Gaytanidis, C., Leemann, B., Abutalebi, J., & Annoni, J. M. (2016). The role of the cognitive control system in recovery from bilingual aphasia: A multiple single-case fMRI study. *Neural Plasticity*. http://dx.doi.org/10.1155/2016/8797086

Roberts, P. M., & Deslauriers, L. (1999). Picture naming of cognate and non-cognate nouns in bilingual aphasia. *Journal of Communication Disorders, 32*(1), 1–23.

Sandberg, C. W., Zacharewicz, M., & Gray, T. (2021). Bilingual abstract semantic associative network training (BAbSANT): A Polish-English case study. *Journal of Communication Disorders, 93*, 106143. https://doi.org/10.1016/j.jcomdis.2021.106143

Schnur, T. T., Schwartz, M. F., Brecher, A., & Hodgson, C. (2006). Semantic interference during blocked-cyclic naming: Evidence from aphasia. *Journal of Memory and Language, 54*(2), 199–227.

Schriefers, H., Meyer, A. S., & Levelt, W. J. (1990). Exploring the time course of lexical access in language production: Picture-word interference studies. *Journal of Memory and Language, 29*(1), 86–102.

Shao, Z., Janse, E., Visser, K., & Meyer, A. S. (2014). What do verbal fluency tasks measure? Predictors of verbal fluency performance in older adults. *Frontiers in Psychology, 5*, 772. https://doi.org/10.3389/fpsyg.2014.00772

Tomoschuk, B., Ferreira, V. S., & Gollan, T. H. (2019). When a seven is not a seven: Self-ratings of bilingual language proficiency differ between and within language populations. *Bilingualism: Language and Cognition, 22*(3), 516–536.

Van der Linden, L., Verreyt, N., De Letter, M., Hemelsoet, D., Mariën, P., Santens, P., Stevens, M., Szmalec, A., & Duyck, W. (2018). Cognate effects and cognitive control in patients with parallel and differential bilingual aphasia. *International Journal of Language & Communication Disorders, 53*(3), 515–525.

Van Hell, J. G., & Tanner, D. (2012). Second language proficiency and cross-language lexical activation. *Language Learning, 62(2)*, 148–171. https://doi.org/10.1111/j.1467-9922.2012.00710.x

Van Heuven, W. J. B., Dijkstra, T., & Grainger, J. (1998). Orthographic neighborhood effects in bilingual word recognition. *Journal of Memory & Language, 39*, 458–483.

Veríssimo, J. (2021). Analysis of rating scales: A pervasive problem in bilingualism research and a solution with Bayesian ordinal models. *Bilingualism: Language and Cognition, 24*(5), 842–848.

Verreyt, N., De Letter, M., Hemelsoet, D., Santens, P., & Duyck, W. (2013). Cognate effects and executive control in a patient with differential bilingual aphasia. *Applied Neuropsychology: Adult, 20*(3), 221–230.

Wheeldon, L. R., & Monsell, S. (1994). Inhibition of spoken word production by priming a semantic competitor. *Journal of Memory & Language, 33*, 332–356.

Zied, K. M., Phillipe, A., Karine, P., Valerie, H. T., Ghislaine, A., & Arnaud, R. (2004). Bilingualism and adult differences in inhibitory mechanisms: Evidence from a bilingual Stroop task. *Brain and Cognition, 54*(3), 254–256.

15

Neural Foundations of Language in Autism Spectrum Disorder

Meredith Pecukonis, Lindsay Butler, Charlotte Gaynor, and Helen Tager-Flusberg

15.1 Language in Autism Spectrum Disorder

Autism spectrum disorder (ASD) is diagnosed on the basis of challenges with social communication and the presence of repetitive behaviors, sensory sensitivities, and narrowly focused interests that emerge during the second or third year of life (APA, 2013). Social communication challenges encompass difficulties with social reciprocity, developing and maintaining social relationships, nonverbal communication, and conversational skills. Although language disorder is no longer included in the primary symptoms for ASD, most individuals with ASD have impairments in language that go beyond pragmatic difficulties. Furthermore, most infants and children who are later diagnosed with ASD exhibit delays in the onset and early milestones of language development. Indeed, for parents of children with ASD, delays in the emergence of language are usually the first signs they worry about (Talbott et al., 2015). Thereafter, trajectories in language development vary considerably across individuals (Pickles et al., 2014), which has consequences for long-term academic, vocational, and quality of life outcomes.

Different language subtypes can be identified by the time children with ASD reach school age (Kjelgaard & Tager-Flusberg, 2001). Some children, perhaps one-quarter of the population, have typical language skills as measured on standardized assessments, though these numbers are changing as more children are diagnosed earlier and have access to early, high-quality behavioral interventions. In contrast, about half of all children with ASD who acquire spoken language lag behind their typically developing peers in the development of phonological processing skills, vocabulary, and grammar in ways that parallel children with developmental language disorder (DLD), but these delays in language are not easily explained by more primary difficulties with social communication (Norbury, 2013). Some researchers have argued that autistic children with language impairment have comorbid DLD as evidenced by their shared difficulties with language, particularly in phonological working memory and morphosyntax (e.g., Tager-Flusberg & Joseph, 2003; Tomblin, 2011), though this hypothesis remains controversial (e.g., Taylor et al., 2014). The remaining children with ASD, up to one third of the population, enter school having acquired little to no spoken language skills, remaining minimally verbal into adulthood, using just a limited number of words and fixed phrases. Relatively little is known about this group, as until recently, they had not been the focus of much research (Tager-Flusberg & Kasari, 2013). Most minimally verbal children and adults with ASD also have intellectual disability, as do some verbal individuals with milder language impairments. Across the spectrum, both gross and fine motor difficulties are common, and are associated with rate of language development in the early years (Hwang & Lee, 2022). Recent studies have found that more than half of all minimally verbal children with ASD have cooccurring childhood apraxia of speech or other speech-motor impairments that are related to their expressive vocabulary limitations (Chenausky et al., 2019), and some have significant auditory processing challenges that are related to their receptive vocabulary (Schwartz et al., 2020). Finally, we note that subtle language difficulties, such as challenges with pragmatics, are sometimes found in first degree relatives of autistic individuals, reflecting the high heritability of ASD and language-related disorders (Lindgren et al., 2009; Ruser et al., 2007).

In this chapter, we review what is known about the neural bases of language in ASD, focusing on structural and functional investigations in studies of infants, children, and adults. We begin by summarizing the range of methods used to investigate brain structure and function in ASD and the challenges in conducting neuroimaging research with a population that experiences sensory sensitivities and other behavioral challenges. We then survey the research on children and adults to highlight some of the major findings that characterize the neural underpinnings of language in ASD. Given that ASD is classified as a neurodevelopmental disorder, there is growing interest in understanding the developmental origins of heterogeneous language profiles. Thus, we also provide a detailed review of this literature, which highlights the very early emergence of atypical neural structure and function in ASD. We end by drawing some tentative conclusions and identifying gaps in the literature that point to future directions for research on language in ASD.

15.2 Studying the Brain in Autism Spectrum Disorder

A variety of safe, noninvasive neuroimaging methodologies have been used to study brain structure and brain function in infants, children, and adults with ASD. In this section, we briefly review six commonly used neuroimaging methodologies in autism research – MRI, DTI, fMRI, fNIRS, EEG, and MEG.

15.2.1 Structural Neuroimaging Methodologies

MRI: Magnetic resonance imaging (MRI) is one of the most used neuroimaging methodologies that provides information about brain structure. To collect MRI data, the research participant is placed inside of a large magnetic tube that creates a magnetic field. Computational approaches, such as voxel-based morphometry, are used to segment gray matter and white matter to create structural brain images in fine detail. This allows researchers to gather information about the volume, cortical thickness, and cortical surface area of different brain regions.

DTI: Diffuse tensor imaging (DTI) is another MRI-based methodology that is used to characterize the structural connectivity between different regions of the brain. DTI measures the movement or "diffusion" of water molecules throughout the body. Fractional anisotropy is one outcome measure obtained from DTI that assesses the directionality of the diffusion of water molecules. An increase in fractional anisotropy typically reflects an increase in myelination of white matter tracts, demonstrating greater structural connectivity between brain regions. Increased fractional anisotropy is associated with decreased mean diffusivity and axial diffusivity, two other measures obtained from DTI, which provide information about intracellular and extracellular space, as well as variations in amount of cerebral spinal fluid.

15.2.2 Functional Neuroimaging Methodologies

fMRI: Like structural MRI, functional magnetic resonance imaging (fMRI) utilizes magnetism and radio waves to gather information about the brain, although it focuses on characterizing brain function rather than brain structure. Moreover, fMRI allows researchers to assess changes in brain function by measuring changes in the concentration of deoxygenated hemoglobin (deoxyHb) and oxygenated hemoglobin (oxyHb) in different brain regions. These changes in blood flow are reflected in the fMRI blood oxygen level dependent (BOLD) signal. Information is provided by fMRI about the strength or magnitude of brain activation in specific regions, which is particularly useful when comparing activation patterns across groups or stimulus conditions.

fNIRS: Functional near-infrared spectroscopy (fNIRS) is a newer neuroimaging methodology that also measures brain function by detecting changes in concentration of oxyHb and deoxyHb. However, fNIRS utilizes the varying *optical* properties of these molecules with near-infrared light sources of different wavelengths, and detectors that are strategically placed on a cap or headband to cover certain brain regions of interest. One limitation of fNIRS is that neural activity can only be measured in regions of the brain's cortical surface and not in subcortical structures.

EEG: Electroencephalography (EEG) measures the extracellular electrical activity from millions of neurons using electrodes placed in a cap on the scalp. Spontaneous or "resting state" EEG measures the naturally occurring electrical activity of the brain. It can detect the spectral power of various frequency bands of electrical activity (delta, theta, alpha, beta, gamma), which are associated with different cognitive processes. When electrical activity is measured in response to a given stimulus, this is called an event-related potential (ERP). Studies commonly report the latency and/or amplitude of ERPs, which provide valuable information about brain function with high resolution.

MEG: Magnetoencephalography (MEG) is another functional neuroimaging methodology with excellent temporal resolution. However, unlike EEG, MEG also has high spatial resolution. Moreover, MEG measures the magnetic field that is created by neurons when they are activated. Like EEG, MEG can be used to measure spontaneous neural oscillations in different frequency bands or event-related fields (ERFs), which are the MEG-equivalent of ERPs. MEG utilizes a large helmet-like structure with detectors that are placed around, but not directly on, the head.

Functional Connectivity: Functional neuroimaging methodologies (fMRI, fNIRS, EEG, MEG) can be used to measure brain function at rest or in response to a given stimulus. They can also be used to measure functional connectivity (FC), which assesses the extent to which two brain regions are activated simultaneously using "statistical dependencies" or correlations between the time series of neural activity in different regions of the brain. Moreover, FC studies provide important information about the function of neural networks rather than independent brain regions. FC can be used to assess short range (local), as well as more long range (global), connections between brain regions or networks.

15.2.3 Challenges in Neuroimaging Research on ASD

There is now a considerable body of research on the neural bases of language in ASD that has utilized these neuroimaging methodologies (see Cermak et al., 2022 for review). Nevertheless, the scanning environment for MRI, DTI, and fMRI poses considerable challenges during data collection, especially for autistic individuals. For example, the MRI scanner is loud and physically restrictive, which can lead to sensory overstimulation, claustrophobia, and anxiety. Additionally, because moving even a few millimeters can compromise data quality, participants must lay motionless during data collection. This sensitivity to motion makes data collection with infants and young children particularly difficult considering the attentional and motor demands of the scanning environment. Thus, nearly all MRI, DTI, and fMRI studies on infants and young children with ASD have collected data while participants were asleep or sedated.

Research has demonstrated that those with ASD are more likely to move during scanning when compared to typically developing (TD) individuals without ASD. One study reported a 56 percent scanning success rate for children and adolescents with ASD, far lower than the 91 percent scanning success rate for TD children and adolescents (Yerys et al., 2009). For these reasons, findings from MRI, DTI, and fMRI studies should be interpreted carefully, as research participants who are able to tolerate the restrictive scanning environment do not accurately represent the entire heterogeneous ASD population. Furthermore, excessive motion artifacts that are not removed or corrected appropriately may impact the validity of findings (Van Dijk et al., 2012). Other neuroimaging techniques, like fNIRS and EEG, also pose challenges when collecting data with autistic individuals, as the caps may not be tolerated by those with sensory sensitivities.

Findings across neuroimaging studies are often inconsistent, making it difficult to draw firm conclusions about how language skills are linked to brain data in ASD. One key concern that may be contributing to these inconsistencies is the wide heterogeneity across the autism spectrum, especially in varying language phenotypes. Other participant factors that are known to influence findings from both structural and functional studies include age, gender, handedness, IQ, attentional factors, hair color and texture, co-occurring disorders or conditions, and the use of medications. Another important factor in the design and interpretation of functional studies that may contribute to inconsistencies across studies is the choice of a paradigm and stimuli, as well as the precise procedures for implementing the paradigm in the imaging environment. Even small differences in task design can influence results. Finally, the most significant problem is that most studies include relatively small numbers of participants which often lack the statistical power to detect significant differences in the neural bases of language in ASD.

15.3 Structural Neuroimaging Studies of Language in ASD

Structural imaging studies of primary language regions in individuals with ASD give us a deeper understanding of the neural network differences between the neurotypical and autistic brain. Using MRI and diffusion tensor imaging (DTI), researchers have been able to map the volume, cortical thickness, and structural connectivity of different brain regions in children and adults with ASD.

MRI: Several studies have demonstrated that the total brain volume in autistic individuals is increased relative to neurotypical individuals without ASD (Stanfield et al., 2008). Many different regions within the brain have been identified as structurally "atypical," with some regions showing increased volume and others decreased volume. Moreover, MRI has revealed that enlargement of the frontotemporal cortices is associated with language skills (Li et al., 2017), and that this enlargement is due to increased gray and white matter volume. Carper and colleagues (2002) used MRI to study young children with ASD and found increased white matter volume in regions of the frontal lobe and increased gray matter volume in regions of the frontal and temporal lobes. Longitudinal MRI studies, including those by Schumann and colleagues (2010) and Hazlett and colleagues (2006), similarly found increased gray matter volume and white matter volume in the cerebrum, notably in the frontal and temporal lobes.

One of the earliest MRI studies on brain regions associated with language processing in adults with ASD found decreased gray matter volume relative to neurotypical adults without ASD, specifically in left inferior frontal gyrus (Abell et al., 1999). Atypical cortical asymmetry in language regions has consistently been found across studies of autistic children and adults, with most studies reporting either reduced left hemisphere volume in Broca's area and Wernicke's area, or reversed asymmetry characterized by larger volumes of analogous regions in the right hemisphere (Lindell & Hudry, 2013). A recent large-scale study found that extreme rightward asymmetry patterns were associated with lower language skills, suggesting that atypical structural lateralization with varied patterns across individuals may be a hallmark feature of language impairments in ASD (Floris et al., 2021).

Fewer studies have investigated links between language skills and subcortical brain structures in ASD. Arutiunian and colleagues (2023) measured gray matter volume in the amygdala and hippocampus in school-aged children with ASD and found reduced gray matter volume in these structures. Interestingly, reduced bilateral amygdala volume and left hippocampal volume were both associated with lower language skills.

In addition to brain volume, cortical thickness has also been measured in those with ASD. One small-scale study found that total cerebral sulcal and gyral thickness were larger in children with ASD compared to typically developing children without ASD (Hardan et al., 2006). When examining all lobes of the brain, this increased sulcal and gyral thickness was found in the temporal and parietal lobes but not in the frontal and occipital lobes. Hardan and his colleagues suggest that increased cortical thickness may contribute to increased total brain volume in autistic individuals, particularly in the temporal lobe. Similar findings were reported in a more recent and large-scale study of children with ASD, suggesting that atypical patterns of cortical thickness emerge early in development (Duan et al., 2023). Interestingly, greater cortical thickness has been found to be associated with better language skills in ASD (Crutcher et al., 2018; Zhang, Qin, et al. 2022).

DTI: Several studies using DTI have found evidence of atypical structural connectivity between key language regions of the brain in autistic individuals. In a study of adolescents with ASD and co-occurring language impairment, Nagae and colleagues (2012) reported increased mean diffusivity in the left superior longitudinal fasciculus, which includes the arcuate fasciculus, connecting Broca's and Wernicke's areas. In a similar study of school-aged children with ASD, increased mean diffusivity was found throughout the left hemisphere (Peterson et al., 2015). Following this trend, Barnea-Goraly and colleagues (2010) observed decreased fractional anisotropy in the frontal and temporal lobes of those with ASD. These findings of increased mean diffusivity and decreased fractional anisotropy indicate reduced structural connectivity between cortical tracts in language-affiliated regions of the brain. This reduced connectivity is one possible explanation for why many autistic individuals have difficulties understanding and/or using spoken language.

This hypothesis is supported by a few studies that explicitly explored the associations between measures of structural connectivity and language skills in ASD. Naigles and colleagues (2017) looked at structural connectivity in preschool-aged children with ASD. White matter tracts with the lowest fractional anisotropy values were found in the subgroup of children with the lowest language skills. Across all participants, fractional anisotropy in the left inferior longitudinal fasciculus (ILF), a tract involved in lexical and semantic processing, was positively associated with both expressive and receptive language skills. Lu and colleagues (2016) also used DTI to measure white matter structural connectivity in children with ASD. Relative to typically developing children without ASD, children with ASD showed altered white matter microstructure in the temporoparietal portion of the left arcuate fasciculus and in the occipitotemporal portion of the right ILF, as indicated by reduced fractional anisotropy and increased radial diffusivity. In addition, the structural integrity of the right ILF was positively correlated with phonological working memory abilities in the group of autistic children. Minnigulova and colleagues (2023) investigated the structure of the corpus callosum, a white matter tract that integrates networks across the left and right hemispheres, in children with and without ASD. Interestingly, children with ASD demonstrated reduced axial diffusivity across all segments of the corpus callosum, and this reduced axial diffusivity was related to lower language skills.

Taken together, studies on the structural foundations of language in ASD have identified several differences between autistic and neurotypical individuals. Studies suggest that there is variability in brain structure across individuals on the autism spectrum, particularly in global and local brain volume, cortical thickness, degree of atypical asymmetry, and structural connectivity between core language regions. Additionally, many of these measures of brain structure appear to be associated with language skills in children and adults with ASD. Despite this heterogeneity, one consistent pattern that emerges within the literature is that ASD may be characterized by reduced or reversed volumetric asymmetry in frontal and posterior language regions, and reduced connectivity across the language network. Although there have been a few studies that included minimally verbal individuals with ASD in their samples (Olivé et al., 2022; Wan et al., 2012), very little is known about the structural brain differences that underlie the limited spoken language abilities in this subgroup of the autism spectrum.

15.4 Functional Neuroimaging Studies of Language in ASD

Functional neuroimaging studies focusing on speech and language in individuals with ASD have spanned the domains of auditory processing, speech perception and production, semantics, grammar, and prosody. Findings in these areas, however, diverge due to the heterogeneity of language abilities across the autism spectrum. Another factor contributing to inconsistent findings across studies is adequate characterization of ASD with co-occurring language impairment compared to language impairment only. Regardless of the cause, these conflicting findings highlight the importance of methodological rigor in terms of experimental design, sample size, and accounting for the heterogeneity of autism characteristics.

15.4.1 Auditory Processing

Language use and understanding rely on the ability to perceive incoming speech. Differences in auditory processing are common in individuals with ASD. Estimates suggest that 50–70 percent of autistic individuals experience decreased sound tolerance (Williams et al., 2021b). Other auditory perceptual differences in ASD include increased loudness perception (Dunlop et al., 2016; Khalfa et al., 2004), decreased speech-in-noise perception (Alcántara et al., 2004), temporal processing challenges (Alcántara et al., 2012; Foss-Feig et al., 2017), and impaired auditory-visual integration (Feldman et al., 2018). These differences are hypothesized to contribute to the core symptoms of ASD by influencing the ways in which autistic individuals learn from their environment (Cascio et al., 2016; Thye et al., 2018).

The neural underpinnings of auditory processing are often studied using auditory event-related potentials (ERPs) measured by EEG and event-related magnetic fields (ERFs) measured by MEG. Studies on auditory ERP/ERF responses comparing autistic and neurotypical individuals have yielded conflicting results. Some studies have reported delayed P1/M50 and N1/M100 latencies, which are responses produced by neural activity from the primary and association auditory cortices, in children and adults with ASD (Matsuzaki et al., 2020; Roberts et al., 2010; Roberts et al., 2020), while other studies have reported no significant group differences (Donkers et al., 2020; Gage et al., 2003).

Schwartz and colleagues (2018) carried out a systematic review and meta-analysis exploring the amplitude and latency in EPR/ERF responses to auditory mismatch negativity (MMN), an auditory event-related potential that occurs when a sequence of repetitive sounds is interrupted by an occasional "oddball" sound differing in frequency or duration, in individuals with ASD. Most studies that Schwartz and colleagues reviewed did not counterbalance stimuli, so their analysis was confined to a small number of studies that used acoustically identical stimuli in both standard and deviant contexts. They found no major group differences for MMN amplitude or latency. However, group differences in amplitude changed as a factor of age and stimulus type. When compared to those without ASD, younger children with ASD tended to show reduced amplitude responses while adults with ASD tended to show equal or larger amplitude responses. In addition, autistic individuals showed a significantly reduced MMN amplitude response to non-speech, but not speech, deviants.

Williams and colleagues (2021a) conducted a meta-analysis of auditory cortical ERP/ERF responses to tone stimuli and sensory gating performance in paired-click paradigms in children and adolescents with ASD. This meta-analysis revealed that autistic individuals showed small but significant latency delays for P1/M50 and M100, components associated with neural processing within the primary and secondary auditory cortices, reduced N2 amplitude, a component associated with attention and cognitive control, and reduced P1/N1b, a visual evoked potentials associated with sensory gating. These reduced responses were most often found in studies of individuals with ASD and co-occurring intellectual disability. Group differences in ERP/ERF components remained even when accounting for variations in sample demographics, characteristics of the stimuli, and methodological differences in data processing (e.g., filter settings), and therefore likely reflect true differences in brain function (Williams et al., 2021a).

15.4.2 Speech Perception and Production

Language use and understanding also rely on the ability to parse incoming speech streams to form representations of speech sounds in memory and further manipulate those sounds to produce words. Several studies have found atypical brain response to speech compared to nonspeech sounds in autistic individuals. For example, Lai and colleagues (2012) used fMRI and DTI to evaluate frontoparietal activation and functional connectivity in response to song stimuli compared to speech stimuli in children, adolescents, and adults with and without ASD. The song and speech stimuli were individualized to present each participant's favorite song and natural conversational speech from each participant's caregiver. Focusing on the language network, researchers found that individuals with ASD showed increased activation in the left inferior frontal gyrus when listening to song stimuli compared to speech stimuli. Several studies have also reported increased neural activation in the language network for nonspeech sounds compared to speech sounds in those with ASD (e.g., Funabiki et al., 2012; Lloyd-Fox et al., 2018). These findings suggest that autistic individuals exhibit decreased neural activation in response to speech compared to song and other nonspeech sounds.

Fewer studies have examined the neural correlates of speech production in ASD. The movements of the jaw and mouth associated with speech production present challenges for common neuroimaging techniques like fMRI (Butler et al., 2020). Pang and colleagues (2016) used MEG to investigate phoneme sequencing (e.g., producing single syllables such as /pa/ versus syllable sequences such as /pa/-/da/-/ka/) in children with ASD and typically developing (TD) children without ASD. In the single syllable task, autistic children showed delayed latencies in frontal (Broadman Area 47) and temporal (Broadman Area 40) language regions compared to TD children. In the phoneme sequencing task, autistic children showed increased magnitude and delayed latency in the left insula, a region responsible for sensory integration, and the right superior temporal gyrus, suggesting that some individuals with ASD may experience difficulties in sensorimotor and oral motor sequencing during speech production.

Producing speech sounds also relies heavily on phonological working memory, often measured behaviorally using the nonword repetition task, which requires a participant to repeat back pseudowords; words composed of one to five syllables. O'Brien and colleagues (2023) used fMRI to examine neural activation during nonword repetition in functional brain networks comparing speech perception, working memory, and speech production in children with ASD and TD children without ASD. They found that children with ASD showed reduced activation in the supplementary motor area in response to stimuli with increased phonological working memory load (five syllables vs. two syllables), and the degree of activation was positively associated with performance on the nonword repetition task.

15.4.3 Phonological Processing

In the domain of phonology, several studies have investigated the neural correlates of vowel and syllable processing in ASD (see Key & D'Ambrose Slaboch, 2021, for an overview). Using MEG, two studies used a passive oddball paradigm to examine the processing of English vowels /a/ and /u/ in school-aged autistic children with or without language impairment (Berman et al, 2016; Matsuzaki et al., 2019). Both studies reported that all children with ASD (including those who were minimally verbal, verbal with language impairment, and verbal without language impairment) showed atypical right lateralization in response to vowel stimuli; TD children without ASD showed left lateralized activation. Minimally verbal children with ASD showed delayed mismatch field latencies bilaterally, and these delayed mismatch field latencies were associated with lower language and communication scores.

15.4.4 Semantic Processing

In the domain of semantics, the findings from neuroimaging studies are somewhat conflicting. Some children with ASD experience difficulties with semantic processing (Kamio and Toichi, 2000; Toichi and Kamio, 2003). Studies that have used EEG to investigate neural responses to semantically incongruent words or sentences have reported that children with ASD have a lower N400, a component associated with semantic processing (Dunn and Bates, 2005; McCleery et al., 2010; Pijnacker et al., 2010; Ribeiro et al., 2013). Studies that have used other functional neuroimaging methodologies, like fMRI, to study the neural bass of semantic processing in ASD have produced inconsistent findings, with some studies reporting decreased activation within the left inferior frontal gyrus (e.g., Just et al., 2004) and others reporting increased activation within this region (Knaus et al., 2008).

A recent meta-analysis of over 400 research participants examined semantic relatedness in ASD. Semantic relatedness was determined based on whether two words were synonyms, whether two words belonged to the same category, and whether a word matched a previously presented passive versus active sentence. This meta-analysis identified three clusters of active brain regions during semantic processing in individuals with ASD: the left inferior frontal gyrus and insula, the left medial and superior temporal gyri, and the medial and superior frontal gyri. While these regions were also found to be active in neurotypical individuals without ASD, a few differences emerged. First, those without ASD had greater activation in the right-lateralized core language regions and in prefrontal regions associated with executive functions compared to those with ASD. And second, individuals with ASD had more activation in the left hemisphere language network regions compared to individuals without ASD (Phan et al., 2021). Another meta-analysis, however, reported that individuals with ASD exhibited more activation in the right hemisphere language network regions (i.e., superior temporal gyrus and inferior frontal gyrus) when considering tasks involving semantic processing, sentence comprehension, processing figurative language, and speech production (Herringshaw et al., 2016).

Looking outside of the canonical language network, several studies have investigated the role of the occipital lobe during semantic processing in ASD (e.g., Dunn & Bates, 2005; Kamio & Toichi, 2000; Toichi & Kamio, 2002, 2003; Wang et al., 2019). These findings suggest that visual brain regions are uniquely recruited during semantic processing in those with ASD (see also Lai et al., 2012; Gao et al., 2019), a functional pattern that is rarely observed in those without ASD.

15.4.5 Grammar

In the domain of grammar, a handful of studies have explored differences in brain function across groups of individuals with ASD and neurotypical individuals without ASD. For example, Jouravlev and colleagues (2020) used fMRI to measure neural responses while adults with and without ASD listened to sentences or lists of nonwords. Researchers found that those with ASD showed reduced activation in regions of the left hemisphere. Using a sentence comprehension task, Eigsti and colleagues (2016) examined differences in neural activation in adolescents with ASD, adolescents without ASD, and adults who were diagnosed with ASD in childhood but who no longer met diagnostic criteria for ASD. Interestingly, adolescents with ASD and adults who no longer met diagnostic criteria for ASD showed similar brain activation in the left middle frontal gyrus, the left supramarginal gyrus, the right superior temporal gyri, and the posterior cingulate, suggesting that the neural bases of grammatical processing in ASD are similar regardless of the behavioral manifestations of ASD symptoms.

15.4.6 Prosody

A fundamental aspect of language processing which spans the levels of words, sentences, and discourse is prosody. Prosody refers to the pitch (fundamental frequency), intensity (amplitude), and durational aspects of speech. Grammatical prosodic cues signal syntactic information, such as whether an utterance has a declarative or an interrogative function. Affective prosodic cues indicate the speaker's emotional state. Behavioral studies on prosody in ASD have yielded conflicting results as to whether individuals with ASD perceive prosody differently than neurotypical individuals without ASD (e.g., Briend et al., 2023; Franich et al., 2021) or not (Icht et al., 2021).

In the processing of prosodic cues, studies have found that the posterior superior temporal regions of the brain are active in neurotypical individuals, but not in autistic individuals (Just et al., 2004). Eigsti and colleagues (2012) found that children with and without ASD did not show group differences in speed or accuracy during prosody tasks. However, researchers did find significant differences in neural activation when using fMRI, even when controlling for scores on a standardized language assessment. Relative to typically developing children without ASD, children with ASD showed reduced activation in the left inferior frontal gyrus and increased activation in the right superior temporal gyrus (STG), right middle frontal gyrus, and precentral gyrus when processing affective prosody. When processing grammatical prosodic cues, children with ASD showed reduced activation in the right STG and increased activation in the right anterior cingulate cortex and the right superior frontal gyrus. In sum, autistic children appear to use more distributed brain regions associated with language and executive control when processing prosodic stimuli. This finding parallels studies on other domains of language that have demonstrated that neural activation during language processing is more distributed across brain regions.

15.4.7 Summary

Research into the functional neural bases of language in ASD spans all domains of language. While studies vary widely in methodology and scope, some consistent findings have emerged. One key difference in neural activation patterns between those with and without ASD is atypical lateralization for language characterized by decreased activation in left hemisphere language network regions and increased activation in analogous right hemisphere regions. Another key difference is reduced language network connectivity, which corroborates the findings from structural studies discussed in the previous section. Two major challenges complicate this area of research. One is adequate inclusion and characterization of the heterogeneous cognitive, social, and language profiles of autistic individuals participating in functional neuroimaging studies. The second is differentiating ASD with language impairment from non-ASD language impairment. Going forward, tackling these challenges will be integral to improving our understanding of the neural bases of language in ASD.

15.5 Early Development of the Neural Foundations of Language in ASD

15.5.1 Studying Infant Siblings

Various neuroimaging methodologies have been used to study early brain development in infants who have at least one older sibling with ASD. Here, we refer to these infants as both "infant siblings" and infants at high familial risk for ASD (HR infants). When compared to infants at low familial risk for ASD (LR infants), who have no family history of ASD, HR infants are more likely to receive an ASD diagnosis themselves: an estimated 20 percent of HR infants are diagnosed with ASD compared to 1–2 percent in the general population (Messinger et al., 2015). Studying infant siblings is important because it allows researchers to assess early brain development, even before the behavioral signs of autism become apparent with the possibility of identifying predictive biomarkers for later autism, and importantly, language impairment in infants with or without a later autism diagnosis. In this section, we review studies that have measured the brains of infant siblings using a variety of neuroimaging methodologies.

15.5.2 Structural Neuroimaging Studies of Infant Siblings

Several studies have used MRI to assess structural brain development in infant siblings. Similar to what has been reported in the literature on children and adults with ASD, HR infants with ASD exhibit greater total brain volume when compared to LR infants and HR infants without ASD beginning around twelve months of age (Shen et al., 2013) and continuing until twenty-four months of age (Hazlett et al., 2017). These group differences in total brain volume may not emerge until twelve months of age, as Hazlett and colleagues (2012) found that HR infants and LR infants did not show significant differences in total brain volume at six months of age (although also see Pote et al., 2019 who reported group differences in total brain volume in a smaller sample of six-month-old infants). This overgrowth in total brain volume appears to be a unique feature of autism, as infants with other developmental delays exhibit "typical" total brain volume relative to LR infants (Shen et al., 2013). Unlike what has been found in the adult literature, HR infants with ASD do not show increased cortical thickness. However, they do show greater cortical surface area at twenty-four months of age relative to LR infants and HR infants without ASD, especially within more posterior regions of the brain (e.g., occipital lobe; Hazlett et al., 2017). Hazlett and colleagues suggest that an overgrowth in cortical surface area between six and twelve months of age may contribute to an overgrowth in total brain volume between twelve and twenty-four months of age for HR infants later diagnosed with ASD.

Other researchers have explored volumetric differences in specific brain structures. Swanson and colleagues (2017) studied the volumes of subcortical brain structures, including the thalamus and amygdala, in groups of HR infants with ASD and HR infants with language delay. HR infants with ASD showed positive associations between the volumes of these subcortical structures at six months and language skills at twenty-four months, while HR infants with language delay showed negative associations, indicating that the relation between brain structure and language outcomes may vary across infant siblings based on diagnostic outcome.

In summary, by twelve months of age, infant siblings who are diagnosed with ASD have larger brains than those without ASD. This neural endophenotype of autism seems to be distinct from the neural endophenotype of HR infants without ASD as well as infants with other developmental delays. The volume of subcortical brain structures like the amygdala and thalamus may be a potential biomarker for language outcomes in HR infants with ASD.

DTI has been used to study structural connectivity in the brains of infant siblings during the first two years of life. Liu and colleagues (2019) demonstrated that the brains of infant siblings show differences in structural connectivity of language-specific tracts as early as six weeks of age. This study examined two dorsal white matter tracts involved in language processing: the superior longitudinal fasciculus (SLF) and the arcuate fasciculus (AF). Results from this study showed that LR infants had higher fractional anisotropy in the left SLF while HR infants had higher fractional anisotropy in the right SLF. Thus, LR infants had greater connectivity between left-lateralized language regions and HR infants had greater connectivity between right-lateralized language regions. This study also found that across both groups, higher fractional anisotropy in the left SLF and left AF at six weeks of age was related to better language skills at eighteen months of age, indicating that stronger left-lateralized structural connectivity predicts language outcomes later in development (Liu et al., 2019; see also Naigles et al., 2017).

In summary, by six months of age, infant siblings who are later diagnosed with ASD begin exhibiting atypical structural connectivity characterized by reduced efficiency of global connections between brain regions, reduced efficiency of local connections within brain regions, and a slower rate of myelination over time. Infant siblings also show reduced connectivity in left-lateralized tracts and increased connectivity in right-lateralized tracts as early as six weeks of age. Fractional anisotropy in tracts associated with language processing, such as the left SLF and AF, may serve as a potential biomarker for language outcomes. Taken together, findings from MRI and DTI studies support the hypothesis that the brains of those with autism exhibit hyper expansion and overgrowth that originates during the first year of life

(Courchesne, 2004). These differences in brain structure likely reflect differences in processes of myelination that occur at the cellular level.

15.5.3 Functional Neuroimaging Studies of Infant Siblings

A variety of functional neuroimaging methodologies have been used to study the neural bases of basic speech and phonological processing in infant siblings. Blasi and colleagues (2015) used fMRI to assess how the brains of four- to seven-month-old infants responded to speech versus nonspeech and found that LR infants showed greater activation to speech compared to nonspeech in the bilateral superior and middle temporal and frontal regions of the brain, while HR infants showed greater activation to speech compared to nonspeech in the right inferior parietal lobule. This finding suggests that HR infants do not appear to show functional specialization for speech processing in traditional language regions of the brain (i.e., frontal and temporal lobes), but they do show functional specialization for speech processing in the right inferior parietal lobe. This study also reported that when compared to LR infants, HR infants exhibited reduced activation in regions of the frontal and temporal lobes, indicating that infant siblings exhibit reduced brain response to speech in traditional language regions of the brain.

Other studies that have focused on identifying the neural bases of phonological processing in infant siblings have found similar results. A study by Pecukonis and colleagues (2021) used fNIRS to investigate how the brains of six-month-old infants process speech syllables. As expected, LR infants showed activation within bilateral anterior regions of the frontal and temporal lobes. However, HR infants did not show evidence of this functional specialization; rather, brain response to speech was similar across anterior and posterior regions of the brain. A small group of HR infants with ASD showed reduced brain response to speech within anterior regions when compared to LR infants and HR infants without ASD, indicating that these regions may be even less active during phonological processing for infant siblings who are later diagnosed with ASD. Interestingly, a fNIRS study by Edwards and colleagues (2017) reported no significant differences in brain response to speech between groups of three-month-old HR infants and LR infants, suggesting that this group difference does not emerge until six months of age. However, when accounting for sex in statistical models, three-month-old female HR infants showed greater brain response to speech syllables in bilateral anterior regions of the brain when compared to male HR infants and female and male LR infants, suggesting a potential "female protective effect" in which the brains of female HR infants respond more strongly to speech during the first three months of life.

An ERP study by Seery and colleagues (2013) also found evidence of atypical functional specialization in infant siblings. In this study, infants listened to recordings of consonant-vowel speech syllables. At nine and twelve months of age, LR infants showed a larger, more negative amplitude later negative-going slow wave (LSW) response in left hemisphere central electrodes compared to right hemisphere central electrodes, indicating functional specialization for phonological processing in the left hemisphere of the brain. Moreover, HR infants showed similar LSW amplitudes in both hemispheres, indicating a lack of left-lateralized functional specialization; this finding seemed to be driven by a subsample of HR infants with ASD (Finch et al., 2017). A follow-up ERP study assessed how brain response to repeated consonant-vowel speech stimuli at nine months of age related to language outcomes at eighteen months of age (Seery et al., 2014). When compared to LR infants, HR infants without ASD showed a larger amplitude P150 response to repeated phonemes over central electrodes. This amplitude was positively related to expressive language outcomes at eighteen months of age for HR infants only, suggesting that greater brain response to speech predicts better language outcomes later in development.

Liu and colleagues (2021) used fMRI to examine the neural bases of prosodic processing and language learning in nine-month-old infant siblings using recordings of speech syllables with varying statistical and prosodic cues. While both LR infants and HR infants showed activation in the bilateral temporal lobes when listening to speech, HR infants demonstrated reduced brain response to speech in the left amygdala, a region that is often associated with prosodic processing. Researchers also explored how brain response changed over time as infants were exposed to repeated speech. LR infants showed increased activation over time within the left superior temporal gyrus and Heschl's gyrus, which is indicative of learning. As a group, HR infants did not show this change in activation over time, although this was not true for all HR infants. When exploring individual differences, infants who demonstrated this learning-related increase in brain response had higher expressive language skills at thirty-six months. Thus, HR infants who exhibited evidence of learning at the neural level had better language outcomes later in development.

In summary, functional neuroimaging studies have demonstrated that the neural bases of basic speech and phonological processing differ for infant siblings. Beginning in the first year of life, the brains of HR infants show reduced brain response to speech in regions of the frontal and temporal lobes and the amygdala, "atypical" functional specialization (e.g., similar brain response to speech across regions and hemispheres), and a lack of learning-related increases in brain response to speech over time. Function of brain regions outside of the frontal and temporal lobes, such as the right inferior parietal lobule, may play an important role in speech processing for HR infants. Moreover, P150 amplitude and learning-related increases in brain response to speech over time may be potential biomarkers for language outcomes in infant siblings. It is important to note that most of these studies did not separate HR infants with ASD from HR infants without ASD in their analyses due to small sample sizes, and thus, we cannot draw conclusions about whether these functional differences are characteristic of autism or the broader autism phenotype.

Several EEG studies have assessed resting state electrical activity in the brains of infant siblings. Levin and colleagues (2017) and Tierney and colleagues (2012) reported that HR infants show reduced power compared to LR infants as early as three and six months of age, respectively. This reduced power seems to be present in high alpha and beta bands over frontal electrodes at three months (Levin et al., 2017) and extends to all frequency bands by six months (Tierney et al., 2012). Early alpha power is one potential biomarker for language outcomes, as infants who had greater high alpha power at three months of age had greater expressive language skills at twelve months of age (Levin et al., 2017). Other work has shown that the power of various frequency bands measured at six months of age is related to language skills at twenty-four months of age. Wilkinson and colleagues (2020) reported that six-month theta and delta power were positively associated with twenty-four-month language outcomes for both HR infants with ASD and HR infants without ASD. While six-month beta and gamma power were also related to twenty-four-month language outcomes, the direction of these associations differed based on diagnostic outcome; greater power predicted higher language skills in HR infants without ASD but lower language skills in HR infants with ASD (Wilkinson et al., 2020). This finding suggests that potential biomarkers for language outcomes may vary across groups of HR infants with ASD, HR infants without ASD, and LR infants (also see Romeo et al., 2022).

Several studies have characterized neural networks in infant siblings by using fMRI, fNIRS, and EEG to measure resting state functional connectivity (FC). These studies have shown that differences in FC between HR infants and LR infants emerge as early as the first few weeks of life. Indeed, newborn HR infants show greater short range, local FC (i.e., "regional homogeneity") than LR infants within various networks, including somatosensory/motor, auditory, temporal, frontal, parietal, visual, and default mode networks (Ciarrusta et al., 2020). Zhang and colleagues (2022), who used fNIRS to measure FC in infant siblings, found that five-month-old HR infants demonstrated greater FC within regions of the frontal and temporal lobes when compared to LR infants.

Other infant sibling work has focused on characterizing long range, global FC between cortical regions of the brain. For example, Liu and colleagues (2020) used fMRI to study long-range FC between cortical regions of the brain associated with language processing. Relative to LR infants, HR infants demonstrated greater FC between auditory and somatosensory regions, but reduced FC between auditory and temporal regions at nine months of age. Nair and colleagues (2021) also used fMRI to assess FC in infant siblings but focused instead on long-range FC between the thalamus and cortical regions of the brain. Researchers found that newborn HR infants exhibited hypoconnectivity between the thalamus and prefrontal regions, but hyperconnectivity between the thalamus and occipital and motor regions. Okada and colleagues (2022) used fMRI to explore FC outside of the cortex, examining cerebro-cerebellar networks in infant siblings. HR infants and LR infants who showed weaker FC in the RCrusI network in the cerebellum, which is involved in language processing, at nine months of age had delayed language outcomes at three years of age. Taken together, these findings suggest that within the first year of life, infant siblings exhibit atypical short-range FC within cortical regions, as well as atypical long-range FC between different cortical regions, cortical regions and subcortical regions, and cortical regions and cerebellar regions. This atypical FC is characterized by both hyperconnectivity and hypoconnectivity, depending on the brain regions and networks assessed.

Longitudinal studies on infant siblings have reported developmental changes in FC during the first year of life. A fNIRS study by Keehn and colleagues (2013) reported that at three months of age, HR infants demonstrated greater long-range FC than LR infants between anterior and posterior brain regions in the left hemisphere. However, by twelve months of age, FC in HR infants became lower than FC in LR infants. Righi and colleagues (2014) used EEG to assess FC between frontal and temporoparietal electrodes and reported similar findings. By twelve months of age, both HR infants with ASD and HR infants without ASD had reduced intrahemispheric FC relative to LR infants in both the left and right hemispheres. As suggested by Liu and colleagues (2020), these developmental changes in FC may be explained by the fact that HR infants do not show the typical developmental pattern of an increase in long-range FC between frontal and temporal regions of the brain during the first year of life, resulting in a somewhat flattened growth trajectory in functional connections that is surpassed by that of LR infants. Taking all these findings into consideration, HR infants appear to show hyperconnectivity between three and five months of age, followed by hypoconnectivity by twelve months of age. Differences in synaptic plasticity, which is the process by which connections between neurons strengthen or weaken over time, may be contributing to these FC differences that are observed in infant siblings.

15.6 Conclusions and Future Directions

For over two decades now, researchers have sought to understand the neural mechanisms that underlie the highly variable language outcomes among individuals with ASD. Considerable progress has been made in identifying key differences between individuals with ASD and neurotypical individuals without ASD in studies of infants, children, and adults. Across many studies, several differences have been found, among which the most consistent findings include atypical structural and functional lateralization of language functions and reduced structural and functional connectivity across language networks. These differences emerge very early in development, during the first year of life. While some studies have linked these neural patterns to behavioral measures of language, these relationships are not always apparent, suggesting that for some individuals, these hallmark features of neural functioning in ASD serve a compensatory mechanism. But this idea has yet to receive much empirical support, particularly because there is a dearth of longitudinal research beyond the earliest years of development. So far, neuroimaging studies of language in ASD have focused exclusively on the individual without considering the context – linguistic, demographic, cultural – in which that individual was raised (but see Romeo et al., 2022 for one exception). If we are to make progress in understanding the wide variability in brain structure, brain function, and language outcomes, future studies must incorporate measures of diverse environments and related experiences of all individuals on the autism spectrum. Finally, we still know little about why nearly one-third of children with ASD make little to no progress in acquiring language and remain minimally verbal into adulthood, most likely due to the significant barriers that researchers face when conducting neuroimaging research with this subgroup. In the coming decades, as new technologies emerge that address current challenges and limitations of existing neuroimaging methodologies, research will lead us to a better understanding of how neural development shapes the emergence and growth of language across a wide range of contexts. In turn, this research will lead to new directions in how we might provide interventions that will optimize the language outcomes for every individual with ASD.

Notes

Preparation of this chapter was supported by the following grants: P50 DC 18006 (PI: Tager-Flusberg); RO1 DC 10290 (PI: Tager-Flusberg); RO1 NS 12098 (PI: Tager-Flusberg) and F31 DC019562 (PI: Pecukonis).

References

Abell, F., Krams, M., Ashburner, J., Passingham, R., Friston, K., Frackowiak, R., ... & Frith, U. (1999). The neuroanatomy of autism: A voxel-based whole brain analysis of structural scans. *NeuroReport, 10*, 1647–1651.

Alcántara, J. I., Cope, T. E., Cope, W., & Weisblatt, E. J. (2012). Auditory temporal-envelope processing in high-functioning children with autism spectrum disorder. *Neuropsychologia, 50*, 1235–1251.

Alcántara, J. I., Weisblatt, E. J., Moore, B. C., & Bolton, P. F. (2004). Speech-in-noise perception in high-functioning individuals with autism or Asperger's syndrome. *Journal of Child Psychology and Psychiatry, 45*, 1107–1114.

American Psychiatric Association, D. S. M. T. F., & American Psychiatric Association. (2013). *Diagnostic and Statistical Manual of Mental Disorders: DSM-5*. American Psychiatric Association.

Arutiunian, V., Gomozova, M., Minnigulova, A., Davydova, E., Pereverzeva, D., Sorokin, A., ... & Dragoy, O. (2023). Structural brain abnormalities and their association with language impairment in school-aged children with autism spectrum disorder. *Scientific Reports, 13*, 1172.

Barnea-Goraly, N., Lotspeich, L. J., & Reiss, A. L. (2010). Similar white matter aberrations in children with autism and their unaffected siblings: A diffusion tensor imaging study using tract-based spatial statistics. *Archives of General Psychiatry, 67*, 1052–1060.

Berman, J. I., Edgar, J. C., Blaskey, L., Kuschner, E. S., Levy, S. E., Ku, M., Dell, J., & Roberts, T. P. L. (2016). Multimodal diffusion-MRI and MEG assessment of auditory and language system development in autism spectrum disorder. *Frontiers in Neuroanatomy*, 10.

Blasi, A., Lloyd-Fox, S., Sethna, V., Brammer, M. J., Mercure, E., Murray, L., ... & Johnson, M. H. (2015). Atypical processing of voice sounds in infants at risk for autism spectrum disorder. *Cortex, 71*, 122–133.

Briend, F., David, C., Silleresi, S., Malvy, J., Ferré, S., & Latinus, M. (2023). Voice acoustics allow classifying autism spectrum disorder with high accuracy. *Translational Psychiatry, 13*, 250.

Butler, L. K., Kiran, S., & Tager-Flusberg, H. (2020). Functional near-infrared spectroscopy (fNIRS) in the study of speech and language impairment across the lifespan: A systematic review. *American Journal of Speech-Language Pathology, 29*, 1674–1701.

Carper, R. A., Moses, P., Tigue, Z. D., & Courchesne, E. (2002). Cerebral lobes in autism: Early hyperplasia and abnormal age effects. *Neuroimage, 16*, 1038–1051.

Cascio, C. J., Woynaroski, T., Baranek, G. T., & Wallace, M. T. (2016). Toward an interdisciplinary approach to understanding sensory function in autism spectrum disorder. *Autism Research, 9*, 920–925.

Cermak, C. A., Arshinoff, S., Ribeiro de Oliveira, L., Tendera, A., Beal, D. S., Brian, J., ... & Sanjeevan, T. (2022). Brain and language associations in autism spectrum disorder: A scoping review. *Journal of Autism and Developmental Disorders, 52,* 725–737.

Chenausky, K., Brignell, A., Morgan, A., & Tager-Flusberg, H. (2019). Motor speech impairment predicts expressive language in minimally verbal, but not low verbal, individuals with autism spectrum disorder. *Autism & Developmental Language Impairments, 4*. https://doi.org/2396941519856333

Ciarrusta, J., Dimitrova, R., Batalle, D., O'Muircheartaigh, J., Cordero-Grande, L., Price, A., ... & McAlonan, G. (2020). Emerging functional connectivity differences in newborn infants vulnerable to autism spectrum disorders. *Translational Psychiatry, 10*, 131.

Courchesne, E. (2004). Brain development in autism: Early overgrowth followed by premature arrest of growth. *Mental Retardation and Developmental Disabilities Research Reviews, 10*, 106–111.

Crutcher, J., Martin, A., & Wallace, G. L. (2018). Dissociations in the neural substrates of language and social functioning in autism spectrum disorder. *Autism Research, 11*, 1175–1186.

Donkers, F. C., Carlson, M., Schipul, S. E., Belger, A., & Baranek, G. T. (2020). Auditory event-related potentials and associations with sensory patterns in children with autism spectrum disorder, developmental delay, and typical development. *Autism, 24,* 1093–1110.

Duan, H., Eyler, L., Pierce, K., Lombardo, M., Datko, M., Hagler, D. et al. (2023). Language, social and face regions are affected in toddlers with autism and predictive of language outcome. *Res Sq* [Preprint]. https://doi.org/10.21203/rs.3.rs-2451837/v1

Dunlop, W. A., Enticott, P. G., & Rajan, R. (2016). Speech discrimination difficulties in high-functioning autism spectrum disorder are likely independent of auditory hypersensitivity. *Frontiers in Human Neuroscience, 10.*

Dunn, M. A., & Bates, J. C. (2005). Developmental change in neutral processing of words by children with autism. *Journal of Autism and Developmental Disorders, 35,* 361–376.

Edwards, L. A., Wagner, J. B., Tager-Flusberg, H., & Nelson, C. A. (2017). Differences in neural correlates of speech perception in 3 month olds at high and low risk for autism spectrum disorder. *Journal of Autism and Developmental Disorders, 47,* 3125–3138.

Eigsti, I. M., Schuh, J., Mencl, E., Schultz, R. T., & Paul, R. (2012). The neural underpinnings of prosody in autism. *Child Neuropsychology, 18,* 600–617.

Eigsti, I. M., Stevens, M. C., Schultz, R. T., Barton, M., Kelley, E., Naigles, L., ... & Fein, D. A. (2016). Language comprehension and brain function in individuals with an optimal outcome from autism. *NeuroImage: Clinical, 10,* 182–191.

Feldman, J. I., Dunham, K., Cassidy, M., Wallace, M. T., Liu, Y., & Woynaroski, T. G. (2018). Audiovisual multisensory integration in individuals with autism spectrum disorder: A systematic review and meta-analysis. *Neuroscience Biobehavioral Reviews, 95,* 220–234.

Finch, K. H., Seery, A. M., Talbott, M. R., Nelson, C. A., & Tager-Flusberg, H. (2017). Lateralization of ERPs to speech and handedness in the early development of autism spectrum disorder. *Journal of Neurodevelopmental Disorders, 9,* 1–14.

Floris, D. L., Wolfers, T., Zabihi, M., Holz, N. E., Zwiers, M. P., Charman, T., ... & Wooldridge, C. (2021). Atypical brain asymmetry in autism: A candidate for clinically meaningful stratification. *Biological Psychiatry: Cognitive Neuroscience and Neuroimaging, 6,* 802–812.

Foss-Feig, J. H., Adkinson, B. D., Ji, J. L., Yang, G., Srihari, V. H., McPartland, J. C., Krystal, J. H., Murray, J. D., & Anticevic, A. (2017). Searching for cross-diagnostic convergence: Neural mechanisms governing excitation and inhibition balance in schizophrenia and autism spectrum disorders. *Biological Psychiatry, 81,* 848–861.

Franich, K., Wong, H. Y., Yu, A. C., & To, C. K. (2021). Temporal coordination and prosodic structure in autism spectrum disorder: Timing across speech and non-speech motor domains. *Journal of Autism and Developmental Disorders, 51,* 2929–2949.

Funabiki, Y., Murai, T., & Toichi, M. (2012). Cortical activation during attention to sound in autism spectrum disorders. *Research in Developmental Disabilities, 33,* 518–524.

Gage, N. M., Siegel, B., Callen, M., & Roberts, T. P. L. (2003). Cortical sound processing in children with autism disorder: An MEG investigation. *NeuroReport, 14,* 2047–2051.

Gao, Y., Linke, A., Jao Keehn, R. J., Punyamurthula, S., Jahedi, A., Gates, K., ... & Müller, R. A. (2019). The language network in autism: Atypical functional connectivity with default mode and visual regions. *Autism Research, 12,* 1344–1355.

Hardan, A. Y., Muddasani, S., Vemulapalli, M., Keshavan, M. S., & Minshew, N. J. (2006). An MRI study of increased cortical thickness in autism. *American Journal of Psychiatry, 163,* 1290–1292.

Hazlett, H. C., Gu, H., McKinstry, R. C., Shaw, D. W., Botteron, K. N., Dager, S. R., ... & IBIS network. (2012). Brain volume findings in 6-month-old infants at high familial risk for autism. *American Journal of Psychiatry, 169,* 601–608.

Hazlett, H. C., Gu, H., Munsell, B. C., Kim, S. H., Styner, M., Wolff, J. J., ... & Piven, J. (2017). Early brain development in infants at high risk for autism spectrum disorder. *Nature, 542,* 348–351.

Hazlett, H. C., Poe, M. D., Gerig, G., Smith, R. G., & Piven, J. (2006). Cortical gray and white brain tissue volume in adolescents and adults with autism. *Biological Psychiatry, 59,* 1–6.

Herringshaw, A. J., Ammons, C. J., DeRamus, T. P., & Kana, R. K. (2016). Hemispheric differences in language processing in autism spectrum disorders: A meta-analysis of neuroimaging studies. *Autism Research, 9,* 1046–1057.

Hwang, B. H., & Lee, D. (2022). Association between motor and language skills development in children with autism spectrum disorder: A scoping review. *International Journal of Disability, Development and Education, 71*(2), 1–15.

Icht, M., Zukerman, G., Ben-Itzchak, E., & Ben-David, B. M. (2021). Keep it simple: Identification of basic vs. complex emotions in spoken language in individuals with autism spectrum disorder without intellectual disability: A meta-analysis study. *Autism Research, 14,* 1948–1964.

Jouravlev, O., Kell, A. J., Mineroff, Z., Haskins, A. J., Ayyash, D., Kanwisher, N., & Fedorenko, E. (2020). Reduced language lateralization in autism and the broader autism phenotype as assessed with robust individual-subjects analyses. *Autism Research, 13,* 1746–1761.

Just, M. A., Cherkassky, V. L., Keller, T. A., & Minshew, N. J. (2004). Cortical activation and synchronization during sentence comprehension in high-functioning autism: Evidence of underconnectivity. *Brain, 127,* 1811–1821.

Kamio, Y., & Toichi, M. (2000). Dual access to semantics in autism: Is pictorial access superior to verbal access? *Journal of Child Psychology and Psychiatry, 41,* 859–867.

Keehn, B., Wagner, J. B., Tager-Flusberg, H., & Nelson, C. A. (2013). Functional connectivity in the first year of life in infants at-risk for autism: A preliminary near-infrared spectroscopy study. *Frontiers in Human Neuroscience, 7,* 444.

Key, A. P., & D'Ambrose Slaboch, K. (2021). Speech processing in autism spectrum disorder: An integrative review of auditory neurophysiology findings. *Journal of Speech, Language and Hearing Research, 64,* 4192–4212.

Khalfa, S., Bruneau, N., Rogé, B., Georgieff, N., Veuillet, E., Adrien, J.-L., Barthélémy, C., & Collet, L. (2004). Increased perception of loudness in autism. *Hearing Research, 198*, 87–92.

Kjelgaard, M. M., & Tager-Flusberg, H. (2001). An investigation of language impairment in autism: Implications for genetic subgroups. *Language and Cognitive Processes, 16*, 287–308.

Knaus, T., Silver, A., Lindgren, K., Hadjikhani, N., & Tager-Flusberg, H. (2008). fMRI activation during a language task in adolescents with autism spectrum disorder. Journal of the International Neuropsychological Society, 14, 967–979.

Lai, G., Pantazatos, S. P., Schneider, H., & Hirsch, J. (2012). Neural systems for speech and song in autism. *Brain, 135*, 961–975.

Levin, A. R., Varcin, K. J., O'Leary, H. M., Tager-Flusberg, H., & Nelson, C. A. (2017). EEG power at 3 months in infants at high familial risk for autism. *Journal of Neurodevelopmental Disorders, 9*, 1–13.

Li, D., Karnath, H. O., & Xu, X. (2017). Candidate biomarkers in children with autism spectrum disorder: A review of MRI studies. *Neuroscience Bulletin, 33*, 219–237.

Lindell, A. K., & Hudry, K. (2013). Atypicalities in cortical structure, handedness, and functional lateralization for language in autism spectrum disorders. *Neuropsychology Review, 23*, 257–270.

Lindgren, K. A., Folstein, S. E., Tomblin, J. B., & Tager-Flusberg, H. (2009). Language and reading abilities of children with autism spectrum disorders and specific language impairment and their first-degree relatives. *Autism Research, 2*, 22–38.

Liu, J., Okada, N. J., Cummings, K. K., Jung, J., Patterson, G., Bookheimer, S. Y., ... & Dapretto, M. (2020). Emerging atypicalities in functional connectivity of language-related networks in young infants at high familial risk for ASD. *Developmental Cognitive Neuroscience, 45*, 100814.

Liu, J., Tsang, T., Jackson, L., Ponting, C., Jeste, S. S., Bookheimer, S. Y., & Dapretto, M. (2019). Altered lateralization of dorsal language tracts in 6-week-old infants at risk for autism. *Developmental Science, 22*, e12768.

Liu, J., Tsang, T., Ponting, C., Jackson, L., Jeste, S. S., Bookheimer, S. Y., & Dapretto, M. (2021). Lack of neural evidence for implicit language learning in 9-month-old infants at high risk for autism. *Developmental Science, 24*, e13078.

Lloyd-Fox, S., Blasi, A., Pasco, G., Gliga, T., Jones, E. J. H., Murphy, D. G. M., Elwell, C. E., Charman, T., & Johnson, M. H. (2018). Cortical responses before 6 months of life associate with later autism. *European Journal of Neuroscience, 47*, 1–14.

Lu, C., Qi, Z., Harris, A., Wisman Weil, L., Han, M., Halverson, K., Perrachione, T. K., Kjelgaard, M., Wexler, K., Tager-Flusberg, H., & Gabrieli, J. D. (2016). Shared neuroanatomical substrates of impaired phonological working memory across reading disability and autism. *Biological Psychiatry: Cognitive Neuroscience and Neuroimaging, 1*, 169–177.

Matsuzaki, J., Ku, M., Dipiero, M., Chiang, T., Saby, J., Blaskey, L., Kuschner, E. S., Kim, M., Berman, J. I., Bloy, L., Chen, Y.-h., Dell, J., Liu, S., Brodkin, E. S.,

Embick, D., & Roberts, T. P. (2020). Delayed auditory evoked responses in autism spectrum disorder across the life span. *Developmental Neuroscience, 41,* 223–233.

Matsuzaki, J., Kuschner, E. S., Blaskey, L., Bloy, L., Kim, M., Ku, M., Edgar, J. C., Embick, D., & Roberts, T. P. L. (2019). Abnormal auditory mismatch fields are associated with communication impairment in both verbal and minimally verbal/nonverbal children who have autism spectrum disorder. *Autism Research, 12,* 1225–1235.

McCleery, J. P., Ceponienè, R., Burner, K. M., Townsend, J., Kinnear, M., & Schreibman, L. (2010). Neural correlates of verbal and nonverbal semantic integration in children with autism spectrum disorders. *Journal of Child Psychology and Psychiatry,* 51(3), 277–286.

Messinger, D. S., Young, G. S., Webb, S. J., Ozonoff, S., Bryson, S. E., Carter, A., ... & Zwaigenbaum, L. (2015). Early sex differences are not autism-specific: A Baby Siblings Research Consortium (BSRC) study. *Molecular Autism, 6,* 1–12.

Minnigulova, A., Davydova, E., Pereverzeva, D., Sorokin, A., Tyushkevich, S., Mamokhina, U., ... & Arutiunian, V. (2023). Corpus callosum organization and its implication to core and co-occurring symptoms of autism spectrum disorder. *Brain Structure and Function, 228,* 775–785.

Nagae, L. M., Zarnow, D. M., Blaskey, L., Dell, J., Khan, S. Y., Qasmieh, S., ... & Roberts, T. P. L. (2012). Elevated mean diffusivity in the left hemisphere superior longitudinal fasciculus in autism spectrum disorders increases with more profound language impairment. *American Journal of Neuroradiology, 33,* 1720–1725.

Naigles, L. R., Johnson, R., Mastergeorge, A., Ozonoff, S., Rogers, S. J., Amaral, D. G., & Nordahl, C. W. (2017). Neural correlates of language variability in preschool-aged boys with autism spectrum disorder. *Autism Research, 10,* 1107–1119.

Nair, A., Jalal, R., Liu, J., Tsang, T., McDonald, N. M., Jackson, L., ... & Dapretto, M. (2021). Altered thalamocortical connectivity in 6-week-old infants at high familial risk for autism spectrum disorder. *Cerebral Cortex, 31,* 4191–4205.

Norbury, C. F. (2013). Are you speaking my language? Raising awareness of language learning impairments in developmental psychopathology. *Journal of Child Psychology and Psychiatry, 54,* 705–706.

O'Brien, A. M., Perrachione, T. K., Wisman Weil, L., Sanchez Araujo, Y., Halverson, K., Harris, A., Ostrovskaya, I., Kjelgaard, M., Wexler, K., Tager-Flusberg, H., Gabrieli, J. D., & Qi, Z. (2023). Altered engagement of the speech motor network is associated with reduced phonological working memory in autism. *NeuroImage: Clinical, 37,* 103299.

Okada, N. J., Liu, J., Tsang, T., Nosco, E., McDonald, N. M., Cummings, K. K., Jung, J., Patterson, G., Bookheimer, S. Y., Green, S. A., Jeste, S. S., & Dapretto, M. (2022). Atypical cerebellar functional connectivity at 9 months of age predicts delayed socio-communicative profiles in infants at high and low risk for autism. *Journal of Child Psychology and Psychiatry,* 63(9), 1002–1016.

Olivé, G., Slušná, D., Vaquero, L., Muchart-López, J., Rodríguez-Fornells, A., & Hinzen, W. (2022). Structural connectivity in ventral language pathways characterizes non-verbal autism. *Brain Structure and Function, 227*, 1817–1829.

Pang, E. W., Valica, T., MacDonald, M. J., Taylor, M. J., Brian, J., Lerch, J. P., & Anagnostou, E. (2016). Abnormal brain dynamics underlie speech production in children with autism spectrum disorder. *Autism Research, 9*, 249–261.

Pecukonis, M., Perdue, K. L., Wong, J., Tager-Flusberg, H., & Nelson, C. A. (2021). Exploring the relation between brain response to speech at 6-months and language outcomes at 24-months in infants at high and low risk for autism spectrum disorder: A preliminary functional near-infrared spectroscopy study. *Developmental Cognitive Neuroscience, 47*, 100897.

Peterson, D., Mahajan, R., Crocetti, D., Mejia, A., & Mostofsky, S. (2015). Left-hemispheric microstructural abnormalities in children with high-functioning autism spectrum disorder. *Autism Research, 8*, 61–72.

Phan, L., Tariq, A., Lam, G., Pang, E. W., & Alain, C. (2021). The neurobiology of semantic processing in autism spectrum disorder: An activation likelihood estimation analysis. *Journal of Autism and Developmental Disorders, 51*, 3266–3279.

Pickles, A., Anderson, D. K., & Lord, C. (2014). Heterogeneity and plasticity in the development of language: A 17-year follow-up of children referred early for possible autism. *Journal of Child Psychology and Psychiatry, 55*, 1354–1362.

Pijnacker, J., Geurts, B., Van Lambalgen, M., Buitelaar, J., & Hagoort, P. (2010). Exceptions and anomalies: An ERP study on context sensitivity in autism. *Neuropsychologia*, 48(10), 2940–2951.

Pote, I., Wang, S., Sethna, V., Blasi, A., Daly, E., Kuklisova-Murgasova, M., ... & BASIS Team. (2019). Familial risk of autism alters subcortical and cerebellar brain anatomy in infants and predicts the emergence of repetitive behaviors in early childhood. *Autism Research, 12*, 614–627.

Ribeiro, T. C., Valasek, C. A., Minati, L., & Boggio, P. S. (2013). Altered semantic integration in autism beyond language: A cross-modal event-related potentials study. *Neuroreport*, 24(8), 414–418.

Righi, G., Tierney, A. L., Tager-Flusberg, H., & Nelson, C. A. (2014). Functional connectivity in the first year of life in infants at risk for autism spectrum disorder: An EEG study. *PLoS ONE, 9*, e105176.

Roberts, T. P. L., Bloy, L., Ku, M., Blaskey, L., Jackel, C. R., Edgar, J. C., & Berman, J. I. (2020). A multimodal study of the contributions of conduction velocity to the auditory evoked neuromagnetic response: Anomalies in autism spectrum disorder. *Autism Research, 13*, 1730–1745.

Roberts, T. P., Khan, S. Y., Rey, M., Monroe, J. F., Cannon, K., Blaskey, L., Woldoff, S., Qasmieh, S., Gandal, M., Schmidt, G. L., Zarnow, D. M., Levy, S. E., & Edgar, J. C. (2010). MEG detection of delayed auditory evoked responses in autism spectrum disorders: Towards an imaging biomarker for autism. *Autism Research, 3*, 8–18.

Romeo, R. R., Choi, B., Gabard-Durnam, L. J., Wilkinson, C. L., Levin, A. R., Rowe, M. L., ... & Nelson III, C. A. (2022). Parental language input predicts neuroscillatory patterns associated with language development in toddlers at risk of autism. *Journal of Autism and Developmental Disorders, 52*, 2717–2731.

Ruser, T., Arin, D., Dowd, M., Putnam, S., Winklosky, B., Rosen-Sheidley, B., Tomblin, B., Tager-Flusberg, H., & Folstein, S. (2007). Communicative competence in parents of children with autism and parents of children with specific language impairment. *Journal of Autism and Developmental Disorders, 37*, 1323–1336.

Schumann, C. M., Bloss, C. S., Barnes, C. C., Wideman, G. M., Carper, R. A., Akshoomoff, N., ... & Courchesne, E. (2010). Longitudinal magnetic resonance imaging study of cortical development through early childhood in autism. *Journal of Neuroscience, 30*, 4419–4427.

Schwartz, S., Shinn-Cunningham, B., & Tager-Flusberg, H. (2018). Meta-analysis and systematic review of the literature characterizing auditory mismatch negativity in individuals with autism. *Neuroscience Biobehavioral Reviews, 87*, 106–117.

Schwartz, S., Wang, L., Shinn-Cunningham, B. G., & Tager-Flusberg, H. (2020). Atypical perception of sounds in minimally and low verbal children and adolescents with autism as revealed by behavioral and neural measures. *Autism Research, 13*, 1718–1729.

Seery, A., Tager-Flusberg, H., & Nelson, C. A. (2014). Event-related potentials to repeated speech in 9-month-old infants at risk for autism spectrum disorder. *Journal of Neurodevelopmental Disorders, 6*, 1–12.

Seery, A. M., Vogel-Farley, V., Tager-Flusberg, H., & Nelson, C. A. (2013). Atypical lateralization of ERP response to native and non-native speech in infants at risk for autism spectrum disorder. *Developmental Cognitive Neuroscience, 5*, 10–24.

Shen, M. D., Nordahl, C. W., Young, G. S., Wootton-Gorges, S. L., Lee, A., Liston, S. E., ... & Amaral, D. G. (2013). Early brain enlargement and elevated extra-axial fluid in infants who develop autism spectrum disorder. *Brain, 136*, 2825–2835.

Stanfield, A. C., McIntosh, A. M., Spencer, M. D., Philip, R., Gaur, S., & Lawrie, S. M. (2008). Towards a neuroanatomy of autism: A systematic review and meta-analysis of structural magnetic resonance imaging studies. *European Psychiatry, 23*, 289–299.

Swanson, M. R., Shen, M. D., Wolff, J. J., Elison, J. T., Emerson, R. W., Styner, M. A., ... & Gu, H. (2017). Subcortical brain and behavior phenotypes differentiate infants with autism versus language delay. *Biological Psychiatry: Cognitive Neuroscience and Neuroimaging, 2*, 664–672.

Tager-Flusberg, H., & Joseph, R. M. (2003). Identifying neurocognitive phenotypes in autism. *Philosophical Transactions of the Royal Society of London. Series B: Biological Sciences, 358*, 303–314.

Tager-Flusberg, H., & Kasari, C. (2013). Minimally verbal school-aged children with autism spectrum disorder: The neglected end of the spectrum. *Autism Research, 6*, 468–478.

Talbott, M. R., Nelson, C. A., & Tager-Flusberg, H. (2015). Diary reports of concerns in mothers of infant siblings of children with autism across the first year of life. *Journal of Autism and Developmental Disorders, 45*, 2187–2199.

Taylor, L. J., Maybery, M. T., Grayndler, L., & Whitehouse, A. J. (2014). Evidence for distinct cognitive profiles in autism spectrum disorders and specific language impairment. *Journal of Autism and Developmental Disorders, 44*, 19–30.

Thye, M. D., Bednarz, H. M., Herringshaw, A. J., Sartin, E. B., & Kana, R. K. (2018). The impact of atypical sensory processing on social impairments in autism spectrum disorder. *Developmental Cognitive Neuroscience, 29*, 151–167.

Tierney, A. L., Gabard-Durnam, L., Vogel-Farley, V., Tager-Flusberg, H., & Nelson, C. A. (2012). Developmental trajectories of resting EEG power: An endophenotype of autism spectrum disorder. *PLoS ONE, 7*, e39127.

Toichi, M., & Kamio, Y. (2002). Long-term memory and levels-of-processing in autism. *Neuropsychologia, 40*, 964–969.

Toichi, M., & Kamio, Y. (2003). Long-term memory in high-functioning autism: Controversy on episodic memory in autism reconsidered. *Journal of Autism and Developmental Disorders, 33*, 151–161

Tomblin, B. (2011). Co-morbidity of autism and SLI: Kinds, kin and complexity. *International Journal of Language & Communication Disorders, 46*, 127–137.

Van Dijk, K. R., Sabuncu, M. R., & Buckner, R. L. (2012). The influence of head motion on intrinsic functional connectivity MRI. *Neuroimage, 59*, 431–438.

Wan, C. Y., Marchina, S., Norton, A., & Schlaug, G. (2012). Atypical hemispheric asymmetry in the arcuate fasciculus of completely nonverbal children with autism. *Annals of the New York Academy of Science, 1252*, 332–337.

Wang, Z., Wang, Y., Sweeney, J. A., Gong, Q., Lui, S., & Mosconi, M. W. (2019). Resting-state brain network dysfunctions associated with visuomotor impairments in autism spectrum disorder. *Frontiers in Integrative Neuroscience, 13*, 17.

Wilkinson, C. L., Gabard-Durnam, L. J., Kapur, K., Tager-Flusberg, H., Levin, A. R., & Nelson, C. A. (2020). Use of longitudinal EEG measures in estimating language development in infants with and without familial risk for autism spectrum disorder. *Neurobiology of Language, 1*, 33–53.

Williams, Z. J., Abdelmessih, P. G., Key, A. P., & Woynaroski, T. G. (2021a). Cortical auditory processing of simple stimuli is altered in autism: A meta-analysis of auditory evoked responses. *Biological Psychiatry: Cognitive Neuroscience and Neuroimaging, 6*, 767–781.

Williams, Z. J., Suzman, E., & Woynaroski, T. G. (2021b). Prevalence of decreased sound tolerance (hyperacusis) in individuals with autism spectrum disorder: A meta-analysis. *Ear and Hearing, 42*, 1137–1150.

Yerys, B. E., Jankowski, K. F., Shook, D., Rosenberger, L. R., Barnes, K. A., Berl, M. M., ... & Gaillard, W. D. (2009). The fMRI success rate of children and adolescents: Typical development, epilepsy, attention deficit/hyperactivity disorder, and autism spectrum disorders. *Human Brain Mapping, 30*, 3426–3435.

Zhang, F., Moerman, F., Niu, H., Warreyn, P., & Roeyers, H. (2022). Atypical brain network development of infants at elevated likelihood for autism spectrum disorder during the first year of life. *Autism Research, 15*, 2223–2237.

Zhang, Y., Qin, B., Wang, L., Chen, J., Cai, J., & Li, T. (2022). Sex differences of language abilities of preschool children with autism spectrum disorder and their anatomical correlation with Broca and Wernicke areas. *Frontiers in Pediatrics, 10*, 762621.

Part V

Brain, Language, and Music

16

Neuroplastic Effects of Music Expertise on Speech-Language Processing

Gavin M. Bidelman, Jane A. Brown, Rose Rizzi, and Jessica MacLean

16.1 Introduction

Music is among the most important factors of the human experience. It draws on core perceptual-cognitive processes and similar hierarchical organizational schemes as found in language. Consequently, musicians have been a model for understanding neuroplasticity and its far-reaching transfer effects to perception, action, and cognition. Excellent reviews on this topic can be found elsewhere (Besson et al., 2011; Herholz & Zatorre, 2012; Munte et al., 2002). The current chapter focuses on the perceptual-cognitive benefits that music exerts on the brain with specific reference to the transfer effects it has on *speech* and *language* functions.

While being "musical" no doubt encompasses more than simple hearing or perceptual abilities (e.g., instrumental production, creativity), we focus on individuals who self-identify as a "musician," typically defined in the literature as an amateur with ~10 years of formal instrumental training. This is distinct from notions of "musicality" – the innate, natural, and spontaneous development of widely shared traits, constrained by our cognitive abilities and biology, that underlie the capacity for music (Honing, 2018). Indeed, a non-musician may be highly musical without having any formal music training (Mankel & Bidelman, 2018). Instead, our focus on the categorical definition of a "musician" is consistent with most of the literature on music-induced neuroplasticity reviewed herein, which, for better or worse, has been predominantly cross-sectional. Nevertheless, we attempt to highlight emerging longitudinal training studies that have begun to replicate earlier results and extend cross-sectional findings on music's impact on linguistic processing. We then emphasize commonalities and distinctions in brain plasticity afforded by experience in the speech and music domains, drawing special attention to cross-domain transfer effects (or lack thereof) in how musical training influences linguistic processing and vice versa.

16.2 Perceptual Benefits Relevant to Speech Processing

16.2.1 Psychoacoustics

Music experience has been shown to enrich many auditory perceptual abilities. Musicians outperform their nonmusician peers in temporal processing including the detection and discrimination of small timing changes embedded in rhythmic sequences (Jones & Yee, 1997; Rammsayer & Altenmuller, 2006). Auditory fusion thresholds, a measure reflecting the minimum temporal interval required to distinguish two separate auditory events, also tend to be smaller in musicians (Rammsayer & Altenmuller, 2006), indicating a higher temporal resolving power with musical expertise. These psychophysical benefits may result from musicians developing a more acute, and possibly more adaptive, temporal integration window (Russeler et al., 2001). However, superior temporal abilities might be limited to aspects of timing derived from more immediate perceptual processing. A musician advantage is not observed, for instance, in tasks involving temporal generalization, such as when judging whether a final beat in a regular sequence occurred earlier or later than expected (Lim et al., 2003; Rammsayer & Altenmuller, 2006).

Increased spectral acuity has also been documented as a function of musical training. In these studies, the most widely reported measures involve pitch and frequency discrimination, where behavioral advantages are often reported in terms of difference limens (DLs), that is, the smallest change in frequency/pitch a listener can reliably detect. In general, musically trained listeners achieve DL thresholds that are ~two to four times smaller (i.e., better) than musically naïve listeners for both pure and complex tones (Bidelman et al., 2011; Kishon-Rabin et al., 2001; Micheyl et al., 2006; Spiegel & Watson, 1984). Presumably, these perceptual advantages could be mediated by the higher sensitivity of musicians' cortical (Brattico et al., 2009; Koelsch et al., 1999; Tervaniemi et al., 2005) and subcortical (Bidelman et al., 2011b; Bidelman et al., 2011c) brain activity for subtle manipulations in pitch.

Musicians' better spectrotemporal acuity extends to listening skills outside the direct domain of music. For example, musicians parse and segregate competing signals in complex auditory scenes more effectively than nonmusicians (Munte et al., 2001; Zendel & Alain, 2009). While these and other psychophysical studies demonstrate enhancements in aspects of auditory processing, musicians' perceptual benefits are not exclusive to the auditory domain. Recent studies, for example, suggest that musical expertise might also enhance temporal acuity in the visual modality (Rammsayer et al., 2012; see also Section 16.4, this chapter).

16.2.2 Speech Processing and "Cocktail Party" Listening

Intriguingly, music experience also influences faculties necessary for speech and language processing, representing far transfer of music experience on seemingly unrelated skills. Perceptually, musicians show improved language-specific abilities including better performance in the identification of lexical tones (Lee & Hung, 2008; Wong et al., 2007) and a higher sensitivity in detecting timbral changes in speech (Chartrand & Belin, 2006). Interestingly, these benefits also facilitate speech processing in suboptimal acoustic conditions including noise-degraded listening environments, the so-called cocktail party scenario.

Among the perceptual and speech processing advantages associated with musicianship is the so-called musician advantage in speech-in-noise (SIN) processing (for review see Coffey et al., 2017; Hennessy et al., 2022). Several studies demonstrate musicians outperform nonmusicians in figure-ground perception, as measured in a variety of degraded speech recognition tasks (Anaya et al., 2016; Bidelman & Krishnan, 2010; Brown et al., 2017; Brown & Bidelman, 2022; Clayton et al., 2016; Deroche et al., 2017; Du & Zatorre, 2017; Mankel & Bidelman, 2018; Parbery-Clark et al., 2011; Swaminathan et al., 2015; Torppa et al., 2018; Yoo & Bidelman, 2019). Amateur musicians (~ten years training) are better at identifying and discriminating target speech amidst acoustic interferences including reverberation (Bidelman & Krishnan, 2010) and noise babble (Parbery-Clark, Skoe, & Kraus, 2009). In standardized (audiological) measures of SIN perception (e.g., Hearing in Noise Test (HINT); QuickSIN test) (Killion et al., 2004; Nilsson et al., 1994), musicians also tolerate ~1 dB more noise than their nonmusician peers during degraded speech recognition (Mankel & Bidelman, 2018; Parbery-Clark, Skoe, Lam, et al., 2009; Yoo & Bidelman, 2019; Zendel & Alain, 2012). While this seems small at first glance, a difference of 1–2 dB can equate to over 10–15 percent better speech recognition performance. Similar results have been reported for acoustic degradations to nonspeech sounds (Başkent et al., 2018; Fuller et al., 2014).

Still, several studies have failed to find a musician advantage even with identical SIN tasks (e.g., QuickSIN, HINT) (Boebinger et al., 2015; Escobar et al., 2020; Madsen et al., 2017; Ruggles et al., 2014; Yeend et al., 2017). The failure to replicate could be due to the small nature of this effect and/or unmeasured differences in music aptitude, even among self-reported musicians, that confer perceptual gains in SIN processing (Bidelman & Mankel, 2019; Brown & Bidelman, 2022). Musicians' SIN benefits are also more apparent for difficult speech-on-speech tasks, conditions which require juggling multiple competing talker streams, rather than signal-in-noise detection, per se (Yoo & Bidelman, 2019). This suggests musicians' SIN advantages might be at least partially driven by cognitive factors and/or lesser susceptibility to informational masking (Oxenham & Shera, 2003; Swaminathan et al., 2015; Yoo & Bidelman, 2019).

Bidelman and Yoo (2020) directly tested this premise by measuring speech recognition in musicians and nonmusicians using a multi-talker paradigm that simulated a 3D cocktail party environment in an anechoic chamber. Most studies reporting musician SIN benefits had been limited to headphones, where target and masking signals are presented to the same ear – a highly controlled but arguably artificial listening scenario. The advantage of the Bidelman and Yoo (2020) paradigm was that figure-ground perception could be assessed in a competitive, multi-talker scenario that offered more realistic spatial cues for segregation and engaged the binaural system, two important elements of real-world SIN processing. Results showed that trained musicians were far superior to their nonmusician peers in deciphering speech within this naturalistic cocktail party environment (Figure 16.1). Musicians also had faster

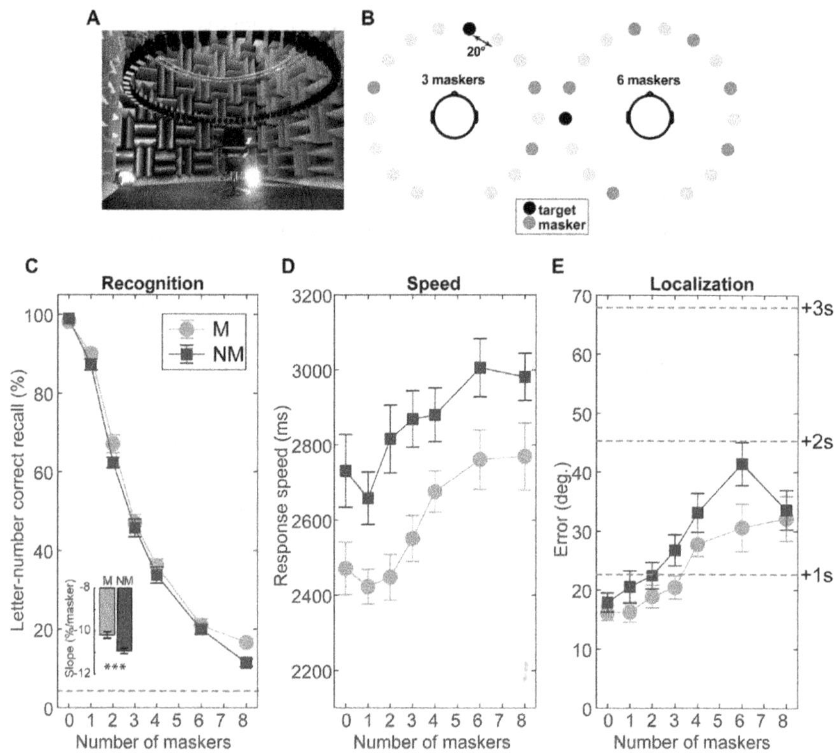

Figure 16.1 Cocktail party listening is superior in musicians. (A) Speech streaming task. Participants were seated in the center of a 16-channel speaker array within an anechoic chamber. Speaker heights were positioned at ear level (~130 cm) during the task with a radial distance of 160 cm to the center of the head and speaker-to-speaker distance of ~20°. (B) Example stimulus presentation (3- and 6-talker conditions). Participants were asked to recall the color, number, and perceived location of target callsign sentences from the coordinate response measure corpus (Bolia et al., 2000). Target location was varied randomly from trial to trial and occurred simultaneously with between zero and eight concurrent masking talkers. (C) Speech recognition declines with increasing masker counts in both groups, but musicians show less performance decrement up to eight interfering talkers (inset). Dotted line = chance performance. (D) Musicians show faster (~200–400 ms) speech recognition speeds than nonmusicians. (E) Both groups localized correctly identified targets within two speakers (<40° error) with better localization in musicians. Error bars = ± 1 s.e.m.

Bidelman, G. M., & Yoo, J. (2020). Musicians show improved speech segregation in competitive, multi-talker cocktail party scenarios. *Frontiers in Psychology, 11*(1927).

and better localization of target speech for up to eight simultaneous talkers and experienced less noise-related decline in performance with increasing masker-talker counts relative to musically naïve listeners. Moreover, these SIN benefits were paralleled in normative measures of figure-ground perception (i.e., QuickSIN test). The findings extended prior studies by demonstrating a relationship between musicianship and cocktail party listening skills (stream segregation).

Still, these and other previous findings do not necessarily resolve whether musical training better enables attending to the target stream, suppressing of background noise, or simultaneous processing of multiple streams. All three mechanisms find evidence in the literature and are probably not mutually exclusive in accounting for the musician SIN advantage (e.g., Bidelman & Yoo, 2020; Munte et al., 2001; Swaminathan et al., 2015; Yoo & Bidelman, 2019; Zendel & Alain, 2013). It is also conceivable that musicians deploy different strategies and/or cognitive resources to achieve their more robust speech-in-noise recognition. For example, as suggested by eye-tracking and pupillometry data (a physiological index of listening effort), musicians exhibit fewer looks to lexical competitors in noise and resolve ambiguity faster than nonmusicians (Kaplan et al., 2021). At the very least, these findings suggest that a musician might solve the cocktail party problem differently than a nonmusician.

Because cocktail party listening draws heavily on a broad range of cognitive faculties (e.g., working memory, attention) and musicianship seems to enhance these functions, advantages in SIN perception might also be somewhat epiphenomenal, secondary to the cognitive benefits induced by musical experience (see Section 16.5 and Andrews and Rammell, Chapter 3, this volume). Indeed, musicians' cocktail party performance is correlated with cognition including attention and working memory (Bidelman & Yoo, 2020; Yoo & Bidelman, 2019) – although we should note group differences between musicians and nonmusicians remain even after partialling out cognitive predictors. Moreover, SIN advantages are also more apparent in older relative to younger adults (Worschech et al., 2021; Zendel & Alain, 2012). Thus, musicians' SIN processing advantage might change with age, becoming more useful in later life when degraded speech perception is most effortful (for review, see Alain et al., 2014).

It is important to note the aforementioned studies on "musician SIN advantages" are nearly entirely cross-sectional. Nevertheless, longitudinal and training experiments have documented improved SIN performance in listeners with short-term musical engagement ranging from a few weeks to a few years of training [one to two years in normal hearing children (Kraus et al., 2014; Slater et al., 2015); twelve weeks in children with hearing loss (Lo et al., 2020); six months in older adults (Zendel et al., 2019)]. Still, future randomized training studies are needed to confirm whether or not the speech-in-noise advantages are causally related to musical training rather than innate predispositions in auditory skills (Mankel & Bidelman, 2018).

16.3 Phonological Awareness and Reading Skills

For thorough examinations of the linguistic benefits of music, the reader is referred to several excellent reviews (Moreno & Bidelman, 2014; Patel, 2011; Patel & Iversen, 2007) and large sample (N=3928) meta-analyses on the topic (Neves et al., 2022). Here, we highlight several key functions of language where musicians have been shown to excel.

16.3.1 Phonological Awareness and Reading

Given its natural foundations in auditory skills, several studies have examined whether the behavioral and cognitive benefits of musical training might transfer to benefit prereading skills, most notably, phonological awareness and sound-to-word mapping. Phonological awareness is the ability to perceive and manipulate the spoken aspects of language and is an early marker of reading and related developmental deficits (Goswami & Bryant, 2016). Anvari and colleagues (2002) conducted a large scale study with four- to five-year-old children and observed a link between music and preliminary reading skills. Their results showed music perception was predictive of phonological awareness and reading development. This finding was confirmed by a later intervention study conducted by Dege and Schwarzer (2011) who tested preschoolers before and after three types of training: music, phonological awareness, and sport. Results indicated that a music program improved phonological skills to a similar degree as traditional phonological training. However, this finding is tempered due to the small sample size of the study (i.e., forty-one children divided across three groups). Nevertheless, converging evidence from more recent studies shows a consistent positive relationship between music and early reading skills (Huss et al., 2011; Tsang & Conrad, 2011) as well as novel word learning (Smit et al., 2022).

Music proficiency involves several factors – rhythm, pitch, tuning, timbre, etc. Rhythm and beat perception seem to play a salient role in phonological awareness, an important component of reading and literacy ability. For example, children with better beat perception and synchronization showed better preliteracy skills and more robust frequency-following responses (FFRs) (Bonacina et al., 2021), a neural potential that represents how well the brain encodes complex sounds (for details, see Section 16.6). Additionally, the neural timing of subcortical brain responses to speech is closely tied to reading abilities. Banai and colleagues (2009) administered a battery of phonological processing and literacy tests in children. Children with better phonological and literacy skills showed FFRs with more precise timing and harmonic encoding, suggesting better sound feature representation in the brain. Presumably, the connection between musical beat synchronization skills and reading might be mediated by general auditory temporal processing abilities (Goswami et al., 2002; Goswami et al., 2013). For example, individuals who are good at discriminating subtle differences in speech rise time might be better equipped to segment speech sequences and distinguish syllable boundaries, which in turn could help foster their vocabulary acquisition and reading (Chung et al., 2021).

Inevitably, the connection between music and reading might be more complex than what is often painted in the literature. Some evidence shows a link between music and only selective aspects of reading skills. Indeed, the impact of early music training on reading skills may only manifest when the learning process consists of complex phoneme-to-grapheme mappings (Moreno et al., 2009). Subsequent longitudinal studies by our group (Moreno, Friesen, et al., 2011) have also assessed phonological awareness and visual-auditory learning (i.e., the ability to map visual symbols onto words). Children receiving music and visual arts training showed similar improvement on measures of phonological awareness, but the children with music training improved significantly more than the art-trained children on measures of visual-auditory learning. These findings illustrate the complex nature of the transfer mechanisms in reading processing and the necessity of conducting more intervention studies to dissociate the specific and/or general aspects of music training's influences.

16.3.2 Sentence Segmentation

Music experience also seems to improve sentence segmentation, or the ability to separate discrete words or phrases in continuous speech (Tierney & Kraus, 2014). At the phrase level, musicians are more successful in detecting both the presence and absence of clause boundaries (Yang et al., 2022). This advantage extends to segmentation on the word level. In a longitudinal study, François and colleagues (2013) compared sentence segmentation abilities between children who took music lessons versus painting lessons (Figure 16.2). Children were exposed to a series of pseudo-words and partial pseudo-words, which were combinations of pseudo-word syllables with lower transitional probabilities (i.e., "nonwords" in the artificial language). Over a two-year period, children in the music training group became more sensitive to the transitional probabilities of the artificial language and thus were more successful in segmenting sentences than children trained in painting. Furthermore, the music group showed a larger difference in the amplitude of their event-related brain potentials (ERPs) for correct versus incorrect pseudo-words, suggesting the behavioral changes were accompanied by underlying neuroplastic changes in brain function.

16.3.3 Syntactic Processing

Music and language are both hierarchically organized; linguistic syntax is analogous to musical harmonic structure (i.e., musical key) (Lerdahl & Jackendoff, 1983; Rohrmeier, 2007). Thus, it is not surprising that the generalized capability to process hierarchical syntactic structure is shared across the two domains, possibly localized (at least in part) in canonical Broca's area

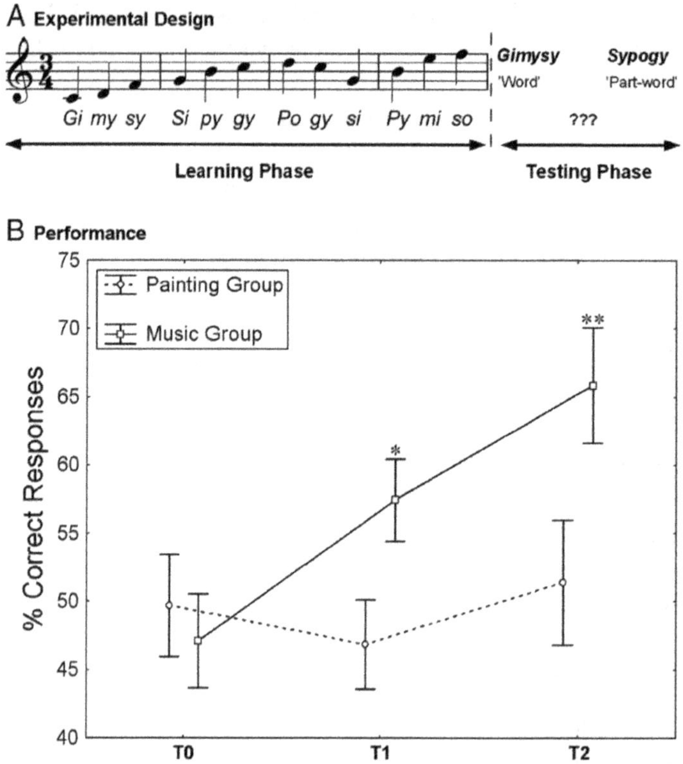

Figure 16.2 Musical training improves speech segmentation skills. Data from François and colleagues (2013) who studied the effects of two years of training (90 min/week in year 1; 45 min in year 2) on behavioral and ERP measures of word segmentation in eight-year-old children. Children were pseudo-randomly assigned to either a music or painting group. (A) Experimental design. During training, children listened to five mins of an artificial sung language built by random concatenation of trisyllabic meaningless words. After familiarization, they were presented two spoken (not sung) items and decided which sounded more familiar. (B) Percentage of correct responses in the three test sessions (T0=baseline pre-training, T1=year 1 post-training, T2=year 2 post-training). No group differences were found before training (T0). Yet, speech segmentation skills improved across testing sessions only in the music group. *p=0.05, **p<0.01.

Data from François, C. F., Chobert, J., Besson, M., & Schön, D. (2013). Music training for the development of speech segmentation. *Cerebral Cortex 23*(9). Reprinted by permission of Oxford University Press.

(Fitch & Martins, 2014; Patel, 2003). Hierarchical processing involves the continuous integration of syntactic information into an existing linguistic (e.g., phrase, sentence) or musical (e.g., phrase, key) representation, as outlined in the Shared Syntactic Integration Resource Hypothesis (SSIRH) between music and language processing (Patel, 2003).

Slevc and colleagues (2009) tested the SSIRH by using garden path sentences (sentences with phrases that disambiguate to a less preferred syntactic structure and thus require reinterpretation of the sentence). Participants read sentences that either manipulated syntactic expectancy with a garden path sentence (e.g., "The attorney advised (that) the defendant was likely to commit more crimes," where "that" is the disambiguating word) or manipulated semantic expectation (e.g., "The boss warned the mailman to watch for angry dogs/pigs when delivering the mail"). Sentences were presented phrase by phrase, and the region of interest – the disambiguating or semantically unexpected word – was paired with a harmonically expected (in-key) or unexpected (out-of-key) chord, which acted as a musical syntax manipulation. In accordance with the SSIRH, harmonically unexpected chords increased reading times only in the garden path sentences but not in the semantic manipulation. The observed interference effects were provocative and suggested both garden path recovery in language and harmonic violations in music utilize the same syntactic processing resources because processing both simultaneously taxed overlapping resources. A follow-up experiment by Perruchet and Poulin-Charronnat (2013) used the same paradigm, but added semantic garden path sentences (e.g., "The old man went to the (river) bank to withdraw his net, which was empty") where the disambiguating word changes the expected meaning, not the grammatical construction. They found that both types of garden path sentences were affected by the harmonically unexpected chord, indicating that the effect is not necessarily semantic but possibly a result of the garden path construction. To reconcile these findings, Slevc and Okada (2015) suggested that the shared resources between music and language processing may not be specific to syntax, but a more general cognitive control, since both music and language require continuous integration of information into expectancies (syntactic, semantic, harmonic, etc.).

16.4 Music-Induced Plasticity beyond the Auditory System: Multisensory Processing

Multisensory integration is the process of combining or binding information from multiple sensory modalities (for review, see Bizley et al., 2016). Audiovisual multisensory integration is best illustrated by the McGurk effect, where visual speech cues (e.g., seeing a talker's lips) influence the auditory input to create illusory speech percepts (McGurk & MacDonald, 1976). Multisensory processing represents a ubiquitous operation in everyday human behaviors that shapes our perception for speech and noncommunicative signals alike. While studies have clearly demonstrated musician advantages in nearly all aspects of auditory processing, there is evidence that musical experience can enhance sensory-perceptual functions outside the realm of audition. Making music is a salient multimodal experience in that it not only engages the auditory system but requires the brain to coordinate different sensory systems in real time. This has led to studies investigating multisensory benefits of musical training including auditory-motor coordination and audiovisual integration.

Indeed, a growing number of behavioral and neuroimaging studies suggest musicians have more refined multisensory integration for the neural encoding and processing of audiovisual signals for both *speech* and *musical* stimuli (Lee & Noppeney, 2011; Lee et al., 2014; Musacchia et al., 2007; Paraskevopoulos et al., 2012). For example, using stimuli in which auditory (speech utterances) and visual (speaker's lips) cues are delivered in a temporally asynchronous manner, several studies have reported higher sensitivity for detecting audiovisual coherence, that is, more restricted integration windows in trained musicians (Lee & Noppeney, 2011; Lee et al., 2014; Paraskevopoulos et al., 2012).

Musicians' multisensory benefits are, however, not restricted to speech and musical stimuli, per se, but may in fact be domain general (Lu et al., 2014). Domain-general multisensory integration can be probed with tasks like the double flash illusion (Foss et al., 2007; Shams et al., 2000). The double flash task requires the perceptual binding of temporally offset auditory and visual cues (i.e., auditory beeps and visual flashes presented either in or out of sync). The paradigm produces a perceptual illusion in that multiple auditory stimuli (beeps) are presented concurrent with only a *single* visual object (flash), yet this presentation induces an illusory perception of multiple flashes (Shams et al., 2000). By parametrically varying the onset asynchrony between auditory and visual events (leads and lags), one can quantify the "temporal window" for fusing audiovisual perceptual objects. Using the double flash illusion paradigm, Bidelman (2016) showed that musically trained individuals were faster and more accurate at processing concurrent audiovisual objects than their nonmusician peers (Figure 16.3). Moreover, musicians showed less susceptibility to the illusion and better resolved the auditory and visual cues, resulting in a ~2–3× shorter temporal binding window. These results suggest that the experience-dependent plasticity of intensive musical training extends beyond the auditory domain and improves the integration of information from multiple sensory systems (audition and vision).

At a mechanistic level, it is conceivable that musicianship changes functional brain organization to enhance connectivity between sensory systems that are highly engaged by music rehearsal (i.e., audition, vision, motor). Indeed, enhanced connectivity in musicians has been observed between auditory and motor cortices (Du & Zatorre, 2017; Grahn & Rowe, 2009), suggesting the potential for increased and/or faster access between sensory modalities with training. Unfortunately, no studies have examined potential music-induced differences in brain connectivity between primary auditory and visual brain regions that would presumably mediate audiovisual binding (cf. Paraskevopoulos et al., 2015). However, the likelihood of perceiving the double flash illusion is highly correlated with white matter connectivity between occipitoparietal regions, the putative ventral/dorsal streams comprising the "what/where" pathways (Kaposvari et al., 2015). This suggests that parallel visual channels play an important role in audiovisual interactions and the temporal binding of disparate cues as required by double-flash percepts (Shams et al., 2000; Shams et al., 2002). It is possible that musicians might show more refined temporal binding of auditory and visual events behaviorally due to increased functional connectivity between the auditory and visual systems.

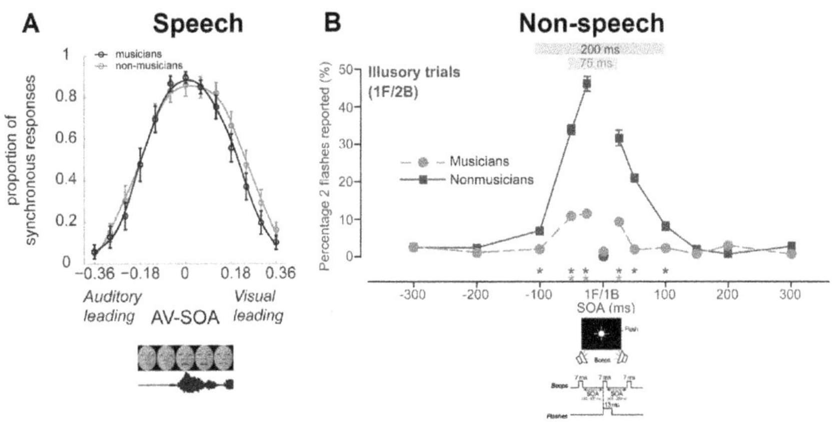

Figure 16.3 Musicians show more refined temporal binding windows for integrating audiovisual information. (A) AV integration psychometric function for *speech* stimuli as reported by Lee and Noppeney (2014). The proportion of synchronous responses is plotted as a function of stimulus onset asynchrony (SOA), that is, the time delay between auditory and visual cues of the talker. Musicians show less susceptibility to offsets over a narrower range of SOAs compared to nonmusicians indicative of superior AV binding. (B) AV integration psychometric function for *nonspeech* stimuli (flash-beeps) as reported by Bidelman (2016). As in speech, musicians show overall less susceptibility to audiovisual asynchrony than nonmusicians and thus a narrow range of AV binding (Ms: 75 ms, NM: 200 ms). *$p<0.05$.

16.3A adapted from Lee, H., & Noppeney, U. (2014). Music expertise shapes audiovisual temporal integration windows for speech, sinewave speech, and music. *Frontiers in Psychology, 5*(868). and 16.3B from Bidelman, G. (2016). Musicians have enhanced audiovisual multisensory binding: Experience-dependent effects in the double-flash illusion. *Experimental Brain Research, 234*(10). Reprinted by permission from Springer Nature.

Relatedly, several neuroimaging studies demonstrate enhancements in non-auditory brain regions (e.g., motor cortex; Elbert et al., 1996), increased auditory-motor coupling (Zatorre et al., 2007), and shorter temporal auditory-motor integration windows in trained musicians (van Vugt & Tillmann, 2014). Conceivably, long-term musical rehearsal and production might also act to improve the prediction of both when and what auditory and visual events are likely to occur as it does in the auditory-motor system (Novembre & Keller, 2014; van Vugt & Tillmann, 2014). As conceived by Novembre and Keller (2014), experience-dependent coupling of perception and actions (or other percepts) might help "scaffold the human ability to represent complex (structured) actions and entrain to multiple agents" (Novembre & Keller, 2014; p.1). Structured actions and multi-agent entrainment are essential in joint musical tasks such as ensemble performance. Coupling an additional sensory modality (e.g., motor activity) during auditory learning has also been shown to increase cortical responses to behaviorally relevant sounds (Lappe et al., 2008; Paraskevopoulos et al., 2012). Thus, in the case of the auditory-motor system, higher multisensory coupling would be advantageous in musical performance as coordination with the motor system could act to increase sensory encoding within the auditory modality. Given that producing music also involves coordinated visual engagement (e.g., reading notes from a score), it stands to reason that intensive musical experience might similarly tighten coupling between auditory and visual cortices. In turn, more precise binding of audiovisual cues would tend to provide higher cognitive brain regions a more successful readout of the speech signal during communicative interactions.

There are several broader implications of music's enhancements to audiovisual processing worth mentioning. First, emerging evidence suggests that the temporal, multisensory binding window might be prolonged in a handful of diffuse disorders including autism, dyslexia, and schizophrenia (e.g., Foss-Feig et al., 2010; Wallace & Stevenson, 2014). Deficits in audiovisual integration and reduced sensitivity to audiovisual asynchrony have also been observed in children with a history of language learning disorders (Kaganovich et al., 2014). In contrast, if musicianship enhances multisensory processes (Bidelman, 2016; Lee & Noppeney, 2011; Lee & Noppeney, 2014; Musacchia et al., 2007; Paraskevopoulos et al., 2012), it stands to reason that musical training might be used as a rehabilitation strategy in the treatment and management of certain disorders to help strengthen face-to-face communication skills (see Section 16.7 for other clinical interventions for improving speech-language outcomes). Second, it is well known that speech intelligibility in noisy listening environments benefits when listeners are able to utilize simultaneous visual cues of the talker (Bidelman et al., 2020; MacLeod & Summerfield, 1987; Navarra & Soto-Faraco, 2007; Sumby & Pollack, 1954). Musicians excel at speech-in-noise processing (see Section 16.2.2). Thus, at least some of musicians' "cocktail party" listening advantages in real-world listening scenarios (i.e., outside the laboratory) might stem from their superiority in combining auditory and visual information. Lastly, some scholars argue that concept knowledge, the building blocks of complex reasoning and linguistic processing, might be better described not in terms of usual abstract/symbolic properties of cognition, but rather, more elementary computations that are coupled within the brain's sensory-motor systems (Gallese & Lakoff, 2005). Under this premise, musicians' advantages in higher-level faculties like speech-language processing might not be due to improved linguistic and/or cognitive inference, per se, but rather, a transfer of benefit from their enhanced multimodal processing.

16.5 Cognitive Benefits of Music

Although cognition encompasses a wide variety of mental processes, it is generally understood as the ability of the brain to learn, process, store, and use new information (Khera & Rangasamy, 2021). Cognition may refer to a suite of different functions including learning, memory, attention, executive function, and intelligence. Many studies have demonstrated that musical training may provide cognitive benefits across these different domains (Bugos & Wang, 2022; Guhn et al., 2020; Jentzsch et al., 2014; Kaganovich et al., 2013; Nie et al., 2022; Sluming et al., 2007; Talamini et al., 2017). Given that these mental processes are critical for speech and language processing, it is useful to review the effects of music on the various elements of cognition.

16.5.1 Intelligence

One such cognitive domain is intelligence (IQ). Intelligence was the first domain to be explored regarding the cognitive benefits of musical training. Given that musical training can begin at a young age, many studies have examined its effects on academics in school-aged children. In a well-controlled seminal intervention study, Schellenberg (2004) followed three groups of children who participated in either weekly music lessons (keyboard or voice), theater instruction, or no lessons at all for duration of one year. Compared with children in the control groups, children in the music groups exhibited greater increases in Wechsler full-scale IQ. While the effect was relatively small (a few percentage points), it nevertheless generalized across IQ subtests, index scores, and a standardized measure of academic achievement. These results were important in that they demonstrated the first causal link between music training and intelligence improvements. Subsequent larger correlational studies helped confirm the relationship between musical training and intelligence (Schellenberg, 2006).

A recent comprehensive (i.e., population-level) study examined the academic performance of 112,916 children in Canada in grades ten and twelve (Guhn et al., 2020). These researchers controlled for language and cultural background, socioeconomic status, and grade seven academic performance. Those students who participated in music had higher exam scores in both math and science, with a stronger advantage for instrumental training than vocal training, suggesting that long-term musical training benefits academic achievement. A meta-analysis by Román-Caballero and colleagues (2022) supported the claim that instrumental musical training was better for investigating benefits of music on cognition, specifically on academic achievement. This study also demonstrated that instrument learning had a positive impact on academic performance and showed individuals who participated in musical training had higher baseline academic and overall cognitive performance.

16.5.2 Working Memory

Another commonly studied aspect of cognition is working memory (WM), or memory used to plan and execute behavior (Cowan, 2008). Improvements in WM due to musical training have been observed in several studies. For example, musicians are able to hold and manipulate information in a short-term memory buffer longer than their nonmusician peers (Chan et al., 1998). Neuroimaging work corroborates these behavioral data, demonstrating increased cortical activation in musicians relative to nonmusicians during WM tasks (Pallesen et al., 2010). Interestingly, recent training studies have demonstrated similar WM improvements in older adults following short-term (six months) individualized piano instruction (Bugos et al., 2007). Thus, repeated musical practice may act to stimulate and reinforce multiple integrated brain networks, including those subserving more general cognitive functions (WM, executive functions). Bugos and colleagues (2007) posit that the enhanced engagement of such networks late into life may serve as an effective cognitive intervention to offset age-related cognitive declines. More thorough discussions of the impact of musical training on the aging brain can be found elsewhere (Alain et al., 2014; Strait & Kraus, 2014).

Nevertheless, there are somewhat conflicting reports on whether the WM benefits of music are restricted to audition; WM is of course a domain general construct that operates across sensory modalities. Some studies have shown improvements in verbal (i.e., auditory), but not visual working memory following formal music training (Brandler & Rammsayer, 2003; Ho et al., 2003; Parbery-Clark, Skoe, Lam, et al., 2009; Strait et al., 2010; Tierney et al., 2008), while others show increased working memory performance for musicians independent of modality (Bidelman et al., 2013; George & Coch, 2011). Our own studies demonstrated enhanced visuospatial processing (Corsi Blocks task) in musicians relative to musically naïve individuals (Bidelman et al., 2013), suggesting the functional benefits of musicianship may not be exclusively auditory in nature. Still, these equivocal results imply that the cognitive benefits of musical experience – at least onto working memory – might be stronger in the auditory than nonauditory domain. Conversely, these inconsistencies might also be explained by subtle differences in musicians' listening histories. For example, instruments like the piano require a continuous mapping and recall of spatial location along the keyboard to execute the correct order of notes. Indeed, an association between musicianship and visuospatial working memory has been observed in trained pianists, who commonly exercise abstract visual rules in their music practice (Bidelman et al., 2013). Nevertheless, the equivocality of these findings highlights the necessity to further examine the role of WM in mediating the broader plastic effects of musical training. Additionally, as is true of most music-related plasticity studies, most of these findings are largely based on cross-sectional comparisons and cannot infer causality between music training and working memory benefits.

In a true longitudinal training study, Nie and colleagues studied long-term effects of instrumental training on working memory (Nie et al., 2022). They assigned students in China to either an instrumental music training group, a second language learning group, or a control group. Auditory WM was measured with a digit span task prior to and after a year of training. They found greater improvement in scores in the backward digit span task condition for the musical training group compared to both the second language learners and control groups, concluding that this music training benefits the central executive system by improving WM (Figure 16.4). These findings are supported by cross-sectional studies (e.g., Yoo & Bidelman, 2019) and meta-

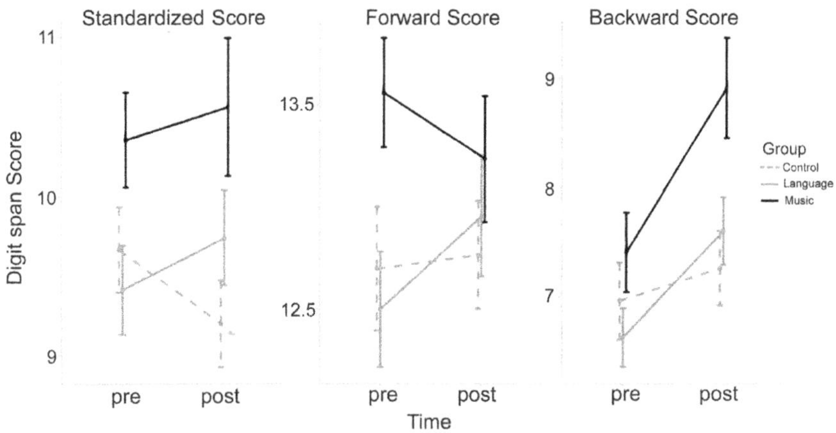

Figure 16.4 Musical training enhances working memory span. Data from Nie and colleagues (2022) examining the effects of one year of training on working memory in n=123 children. Children were assigned to either a music or second language class, or no contact control group. No Group × Time interactions were found in digit span forward or standardized scores. However, the music group showed significant improvements pre- to post-test in digit span *backward* scores (right panel). Adapted from Nie, P., Wang, C., Rong, G., Du, B., Lu, J., Li, S., Putkinen, V., Tao, S., & Tervaniemi, M. (2022). Effects of music training on the auditory working memory of Chinese-speaking school-aged children: A longitudinal intervention study. *Frontiers in Psychology, 12*, 770425.

analyses suggesting musicians perform better on a variety of WM tasks (Talamini et al., 2017).

16.5.3 Executive Processing: Inhibitory Control

Other investigations have examined the relationship between music and another prominent component of executive function: inhibitory control. Bialystok and DePape (2009) used a modified Stroop task to investigate whether extensive music experience leads to enhancements in executive processing. The stimuli involved auditory and linguistic conflict between a word and its pitch (e.g., the word "high" spoken with a low pitch). Musicians outperformed the control group on this task, revealing an influence of music expertise on inhibition processing. Another intervention study examining response inhibition in young children (Moreno, Bialystok, et al., 2011) suggested that after one month of music training, children showed improvements in verbal IQ and (visual) inhibition, as indexed by a go/no-go task. Moreover, they found a positive correlation between improvements in IQ and functional changes at the cortical level during the go/no-go task (i.e., increased P2 magnitude in ERPs). The degree of neural enhancement post-training was associated with an increase in the children's verbal IQ. Recent studies extend these results and support the notion that certain aspects of executive function, particularly the regulation of attentional resources, may play a mediating role between music training, auditory sensory plasticity, and the observed higher-order cognitive skills in musicians (e.g., intelligence; Degé et al., 2011). For further discussion on mediation effects see Moreno and Bidelman (2014).

At the opposite end of the age spectrum, enhanced executive processing from musicianship might have a quasi-protective effect on the brain, increasing cognitive reserve and therefore slowing the mental aging processing in older adults (for discussion, see Alain et al., 2014). Indeed, there is some evidence that musicianship offsets the progressive effects of mild cognitive impairment (Price & Bidelman, 2022). Mechanistically, such improvements could result from better widespread organization of neuronal activity, even in the absence of a task, known as "resting-state connectivity" (Raichle, 2011). Ongoing intrinsic activity of the brain, traditionally considered neural noise, might allow for better coordination and control of sensory-cognitive brain regions subserving complex behaviors. Indeed, resting-state scans have confirmed stronger and more flexible endogenous brain connectivity in trained musicians (Eierud et al., 2023; Hutka et al., 2013; Zamorano et al., 2017). More interestingly, such intrinsic circuits of the "default mode brain network" appear to experience less age-related decline in older musicians relative to nonmusicians (Eierud et al., 2023). Such pervasive interregional communication may aid the coordination, suppression, integration, and control of information in the brain and help account for putative improvements in inhibitory control and executive processes observed in young (Bialystok & DePape, 2009; Moreno, Bialystok, et al., 2011) and older musicians (Lu et al., 2022; Moussard et al., 2016) via both cross-sectional and randomized training studies.

16.5.4 Attention

Attention, as measured via processing speed, also seems to be enhanced by musical training. Jentzsch and colleagues investigated the effects of instrumental musical training on performance monitoring (Jentzsch et al., 2014). They found that with more practice, responses after errors in performance were more efficiently corrected, demonstrating that more musical practice leads to both improved processing speed and executive control. Another study explored attention and neural encoding of complex sounds in musicians through an auditory distraction task (Kaganovich et al., 2013). Through ERP measurements, musicians showed improved early neural encoding of complex sounds and better attention than nonmusicians. Bugos and Wang studied the effects of piano training on cognitive performance in an aging population (Bugos & Wang, 2022). Their participants completed either piano training, computer-assisted cognitive training, or no training over the course of sixteen weeks. They measured cognitive performance, psychosocial well-being, and physiological stress before, directly after, and three months after training. Both piano training and computer training enhanced working memory and processing speed, and piano alone increased verbal fluency skills, self-efficacy, and category switching skills. Additionally, Sluming and colleagues showed a benefit in visuospatial cognition and increased activity in Broca's area stemming from sight reading skills in orchestral musicians (Sluming et al., 2007).

16.5.5 Confounding Factors

Although there is abundant evidence suggesting advantages of music training on cognition, there are other factors to consider, such as personality and baseline musical aptitude, when examining correlations between music and cognitive benefits. For example, Corrigall and colleagues studied whether individual differences in cognition and personality could predict musical training (Corrigall et al., 2013). They found that individual differences played a role in whether a person took music lessons as well as the duration of those lessons. In this study, both cognitive and personality factors were significant predictors of musical experience. Thus, it is difficult to parse out which cognitive benefits may be from individual personality differences, and which are from musical training alone. Another study by Swaminathan and colleagues further investigated this question, studying whether the benefit in intelligence was due to training effects or general music aptitude (Swaminathan et al., 2017). They found that intelligence and music aptitude were positively associated, but the association between music training and intelligence disappeared when controlling for music aptitude. They hypothesized that the association between music training and intelligence may exist because people who are high functioning in multiple cognitive domains may be more likely to have good music aptitude and thus choose to partake in training. Though the benefits of musical training on cognition are difficult to disentangle, compelling evidence strongly suggests benefits to cognition across many domains. In fact, music therapy capitalizes on these proposed enhancements, using music to improve cognitive skills in different populations (see Section 16.7).

16.6 Music Pit against Language Expertise

A large body of empirical evidence suggests that the brain mechanisms governing music and language processing interact and might share an important link with respect to their underlying neurophysiological processing (Bidelman et al., 2011a; Koelsch et al., 2002; Maess et al., 2001; Patel, 2008; Slevc et al., 2009). For example, neuroanatomical regions including Broca's and Wernicke's areas and electrophysiological markers in the ERPs (N400 and P600) typically associated with language-specific operations (e.g., semantic/syntactic processing) are also recruited for processing the melodic and harmonic relationships of music (Koelsch et al., 2002; Maess et al., 2001; Patel et al., 1998). In trained musicians, frontal regions (e.g., BA 47) typically associated with higher-order language comprehension also show activation to the complex metric and rhythmic structures of music (Vuust et al., 2006).

These studies provide evidence for a common neuronal mechanism subserving the temporal coherence found in both domains and demonstrate the intimate coupling between underlying neural processes recruited for language- and music-related processing.

Pitch provides a common window to examine brain plasticity associated with musical training and language experience since it is a shared acoustic parameter found in both domains (Schon et al., 2004). Still, pitch is used very differently in the contexts of music and language, unfolding in discrete, stair-stepped patterns in the former and curvilinear patterns in the latter. The differential use of pitch in music and language has thus allowed the unique opportunity to examine how specific expertise in music vs. tone-language experience differentially shape brain processing for pitch. In this regard, electrophysiological studies examining the frequency-following response (FFR) – an EEG potential putatively generated in the upper brainstem (Bidelman, 2018b; Smith et al., 1975) – have provided recent insights into the sensory encoding and plasticity of pitch. The FFR is a neurophonic potential that closely mirrors the spectrotemporal properties of its evoking stimulus. Indeed, the fidelity of the FFR (a neural potential) is immediately apparent when the response is replayed (sonified) as an acoustic audio stimulus to listeners and they easily hear it as intelligible speech (Bidelman, 2018a; Galbraith et al., 1995; Weiss & Bidelman, 2015).

The FFR studies comparing musicians and nonmusicians reveal enhanced brainstem encoding of pitch-relevant information in trained individuals (Bidelman & Krishnan, 2010; Bidelman et al., 2011; Bidelman et al., 2011a; Musacchia et al., 2007). Additionally, as indicated by shorter, less "jittered" response latencies, musicians' neural activity is also more temporally precise than that of nonmusicians. Musical training therefore not only magnifies the "gain" of subcortical brain activity but also refines it by increasing the temporal precision of the brain's response to musical pitch (Bidelman et al., 2011). Interestingly, these neural indices are correlated with an individual's degree of musical training/experience (Musacchia et al., 2007; Wong et al., 2007) as well as their perceptual abilities (Bidelman et al., 2013; Bidelman et al., 2011c). Such enhanced sensory processing for pitch is presumably why musicians often show highly superior psychophysical threshold, memory, and discrimination of complex pitch and musical melodies (Bidelman et al., 2011; Bidelman et al., 2011b, 2011c; Bidelman et al., 2013).

Similarly, tone-language listeners (e.g. Mandarin, Thai speakers) show stronger and more precise neural tracking in their FFRs to the lexical tones of their native tongue as compared to nontone language listeners (Bidelman et al., 2011a; Krishnan et al., 2005; Krishnan et al., 2009; Krishnan et al., 2010a, 2010b). The neural representation for voice pitch (F0) is often more resilient in these listeners such that it is less prone to interference by noise masking (Krishnan et al., 2010a) or unusually fast rates of F0 change (Krishnan et al., 2010). Thus, like musical pitch for musicians, tone-language speakers show enhanced representation for the types of pitch heard in their everyday language environment. Therefore, both music and language tune the brain to process sounds in an individual's everyday listening environment.

Recognizing the shared brain structures between language and music leads to the provocative question of whether music ability might impact language-related abilities and vice versa. Indeed, the extensive overlap between these domains has led many to posit that musicianship and certain language backgrounds might impact processing in the complementary domain, that is, so-called perceptual-cognitive "transfer effects" (Bialystok & DePape, 2009; Bidelman et al., 2011a; Moreno, 2009). Such *cross-domain* influences have been more extensively reported in the direction from *music-to-language*. For example, musicians demonstrate perceptual enhancements in a myriad of language-specific abilities including phonological processing (Anvari et al., 2002), verbal memory (Chan et al., 1998; Franklin et al., 2008) and verbal intelligence (Moreno, Bialystok, et al., 2011), formant and voice pitch discrimination (Bidelman & Krishnan, 2010), sensitivity to prosodic cues (Thompson et al., 2004), detecting durational cues in speech (Milovanov et al., 2009), degraded speech perception (Bidelman & Krishnan, 2010; Parbery-Clark, Skoe, Lam, et al., 2009), second language proficiency (Marques et al., 2007; Slevc & Miyake, 2006), and lexical tone identification (Delogu et al., 2006, 2010; Lee & Hung, 2008).

Evidence for the opposite direction of transfer, namely *language-to-music* influences, also exists though is arguably more scarce (for discussion, see Bidelman et al., 2013). Still, there is some evidence that Cantonese is associated with higher auditory perceptual performance for music listening. Compared to English-speaking nonmusicians, for example, Cantonese listeners show superior performance on tasks of auditory pitch acuity, music perception, and general cognitive ability (e.g., fluid intelligence, working memory), similar to musicians (Bidelman et al., 2013). The fact that such language-to-music transfer is not observed in other tone languages (e.g., Bidelman et al., 2011c) is probably due to higher demands of Cantonese; the proximity of Cantonese tones is on the order of a semitone (Peng, 2006), that is, 6 percent difference in frequency, which parallels the minimum distance between adjacent pitches found in music.

When directly pitted against one another, do music and linguistic pitch experiences provide similar enhancements to neural encoding? Cross-domain comparisons of brainstem responses reveal overall enhancements in the FFRs (e.g., response strength, pitch tracking accuracy) elicited by either musical or lexical pitch patterns in musicians and tone-language (Mandarin Chinese) speakers alike (Bidelman et al., 2011a, 2011c; Bidelman et al., 2013) (Figure 16.5). Thus, both language and musical experience provide some mutual benefit to the neural extraction of pitch which transfers to the opposite domain (i.e., music→language; language→music).

Yet, the similarities between language and music experience are not as clear cut on closer inspection (Bidelman et al., 2011c; Bidelman et al., 2013). Indeed, auditory pitch representations are subject to a further "tuning"

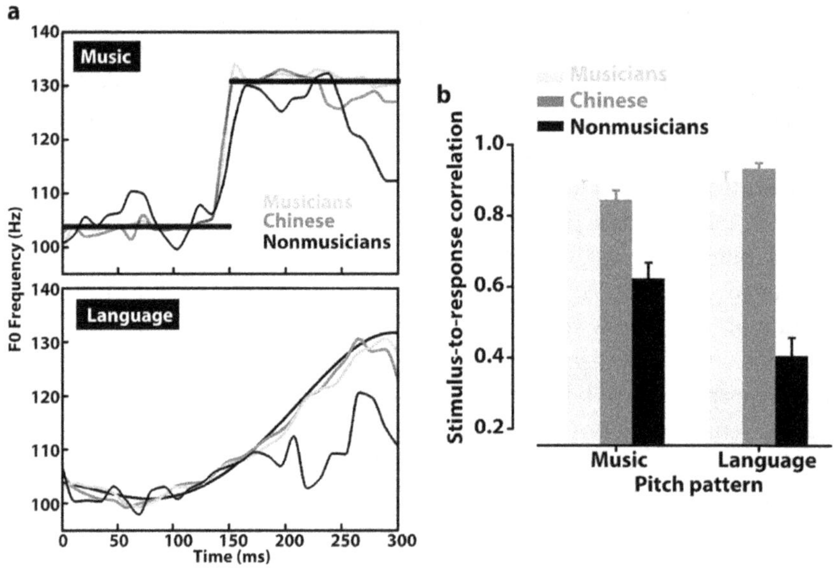

Figure 16.5 FFR neural pitch tracking accuracy for linguistic and musical pitch is enhanced in musicians and tone-language (Chinese) listeners and transfer across domains. (A) Stimulus (thick black) and neural response pitch tracks (shaded lines) extracted from listeners' FFRs. Top, major third musical pitch interval (M3); bottom, Mandarin lexical tone (T2). (B) Stimulus-to-response correlations (i.e., pitch tracking accuracy). Both musicians and Chinese show superior brain tracking of pitch compared to English-speaking nonmusicians, regardless of domain. Moreover, long-term experience with pitch transfers across domains. Musicians are comparable to Chinese in their ability to track lexical tone; and likewise, Chinese are comparable to musicians in their ability to track the musical interval.

Data from Bidelman, G. M., Gandour, J. T., & Krishnan, A. (2011a). Cross-domain effects of music and language experience on the representation of pitch in the human auditory brainstem. *Journal of Cognitive Neuroscience*, *23*(2), 425–434.

according to specific acoustic features encountered in a listener's experience. For example, when presented with an identical gliding pitch stimulus, musicians show enhanced sensory encoding when the pitch pattern intersects discrete notes along the musical scale; tone language speakers, on the other hand, show enhanced sensory encoding during rapidly changing portions of tonal contour (Bidelman et al., 2011a, 2011b). Such "cue weighting" is consistent with each group's unique listening experience and the relative importance of these dimensions to music (Burns & Ward, 1978) and lexical tone perception (Gandour, 1983), respectively. These findings collectively suggest that while both language and musical experience provide some mutual benefit to the neural encoding of auditory information, specific features of the acoustic signal are highlighted in the brain depending on their perceptual salience and function within a listener's domain of expertise. They also suggest that language and music plasticity may not be entirely isomorphic with regard to the sensory encoding of sound (Bidelman et al., 2013). Presumably, this is due to the higher auditory-perceptual demands of music compared to speech (e.g., Peretz & Hyde, 2003).

Experience-dependent enhancements in sensory processing are also not restricted to pitch. In the music domain, brainstem responses (FFRs) are sensitive to individuals' long-term exposure to specific timbres. For example, pianists' responses reflect the timbral characteristics of piano sounds with greater fidelity than those of nonpianists (Strait et al., 2011). Similar timbre-specific enhancements are also observed in the cortex (Pantev et al., 2001). In the language domain, pitch encoding may similarly be influenced by the linguistic status of timbre. For example, the representation of F0 in brain responses (FFRs) is larger when co-occurring with a native as compared to nonnative vowel quality (Krishnan et al., 2011). Such experience-dependent effects suggest that subcortical encoding is also tuned by a person's specific listening history with certain qualities (timbres) of sound. They further suggest that like in psychophysics (see Section 16.2.1), pitch interacts with timbre as the brain transcribes complex sounds. Together, the enhancements observed in both musicians' and tone-language speakers' neural responses to behaviorally relevant sounds indicate experience-dependent neuroplasticity is pervasive, selectively tuning nearly all levels of the hearing brain in accordance with the individual's expertise and listening demands of their environment (for further review, see Moreno & Bidelman, 2014).

16.7 Music-Based Interventions for Improving Speech-Language Outcomes

From a translational standpoint, the neural overlaps in music and language processing highlighted earlier can be leveraged to target clinical areas of need. Music therapy involves tailoring music-based interventions for individuals by a credentialed professional within the context of a therapeutic relationship (American Music Therapy Association, 2005). One approach to music therapy, neurologic music therapy (NMT), utilizes techniques developed from research in music and neuroscience with specific attention to effects of musical elements on neural systems and behaviors (Thaut & Hoemberg, 2014). NMT consists of twenty standardized techniques categorized by targeted clinical need area and proposed therapeutic mechanism (for a review of NMT techniques for speech and language function, see Lee et al., 2019). Below we provide examples of two successful NMT techniques for improving speech-language function and their purported therapeutic mechanisms.

16.7.1 Melodic Intonation Therapy

Melodic intonation therapy (MIT) uses tapping and short, intoned phrases to improve speech output, most classically for individuals with expressive aphasia (for the NMT protocol, see Thaut, Thaut, et al., 2014). The intoned speech features exaggerated prosodic elements of melody and rhythm, with specific content tailored to individual needs. In MIT, the therapist follows a step-by-step protocol which involves guiding the client in left-hand tapping and production of the intoned phrase, gradually leading to the patient's independent production of the phrase in intoned or typical speech.

The therapeutic mechanism of MIT likely involves bypassing damaged left-hemisphere networks such as Broca's area through targeting intact right-hemisphere language-sensitive networks by way of exaggerated prosody (Patel, 2008), use of musical contour (Seger et al., 2013), and contralateral activation through left-hand tapping (Gentilucci & Volta, 2008). Rhythm may be the primary musical mechanism of MIT as it provides a stable stimulus for pacing and entrainment (Stahl et al., 2011; Thaut, McIntosh, et al., 2014). Temporal prediction and motor planning involved in rhythmic entrainment and MIT may provide domain-general neural overlap for speech outcome improvements for individuals with nonfluent aphasia (Shi & Zhang, 2020).

16.7.2 Rhythmic Speech Cueing

Rhythmic speech cueing (RSC) involves production of speech with a steady, rhythmic stimulus to improve aspects of speech such as speech rate, intelligibility, and articulation (Mainka & Mallien, 2014). Example target populations for RSC include individuals experiencing dysarthria (e.g., Thaut et al., 2001), stuttering (Ingham et al., 2012), or apraxia of speech (Brendel & Ziegler, 2008). The RSC protocol involves modulation of speech rate through progressive alteration of the tempo and metric structure of the cued speech, beginning at a tempo slightly below the individual's current speech rate and moving to the desired rate. The clinician guides the individual in entraining to an isochronous stimulus by tapping and producing speech syllables in time with a metronome at a syllabic or metrically patterned level.

Musically, the elements of rhythm, meter, and tempo allow for rate control, temporal prediction, and prosodic cueing that facilitate functional transfer between music and speech perception (Hausen et al., 2013). The therapeutic mechanism of RSC is thought to involve the facilitation of speech-motor networks through rhythmic priming and entrainment (Cason et al., 2015; Mainka & Mallien, 2014; Thaut, 2013). Facilitation of domain-general processing of a widely distributed network involved in rhythm processing, including the basal ganglia, may be particularly helpful for populations with basal ganglia dysfunction, such as Parkinson's disease (Thaut et al., 2001).

16.8 Summary

Music is an enjoyable and engaging activity that recruits nearly all sensory and cognitive systems of the human brain. Musicianship has been shown to improve a variety of functions related to speech and language processing. Musicians show enhanced timing and precision in how their brain encodes the acoustic elements of speech that in turn enables more rapid phonological mapping and word learning relevant to preliteracy skills. Music plasticity also seems to reach beyond the auditory system to tune general cognitive functions (audiovisual integration, working memory, attention) that may similarly improve and maintain communication abilities across the lifespan. Consequently, short-term musical training programs may offer potential benefit to improve or at least offset the difficulties in speech processing experienced by older adults (e.g., Bidelman & Alain, 2015), listeners with cognitive impairments (e.g., Price & Bidelman, 2022), or children with language-based learning delays (e.g., Gross et al., 2010).

Unfortunately, the ever-growing body of work on the linguistic benefits of music is still mainly cross-sectional, comparing listeners with decades of musical training vs. novices. Randomized controlled studies are needed to confirm whether the myriad of benefits reported in these studies are truly caused by music *experience,* and not preexisting factors of the individual (e.g., inherent hearing aptitude, personality traits, or cognitive predispositions; Corrigall et al., 2013; Mankel & Bidelman, 2018). Related, early work on music interventions and rehabilitation is promising in suggesting music activity as an effective means to improve certain speech-language outcomes. Yet, as with any viable clinical treatment, additional studies are needed to determine the correct "dosage" and long-term efficacy of successful music-based interventions.

Notes

Preparation of this chapter was supported by the National Institute on Deafness and Other Communication Disorders of the National Institutes of Health under award number NIH/NIDCD R01DC016267.

References

Alain, C., Zendel, B. R., Hutka, S., & Bidelman, G. M. (2014). Turning down the noise: The benefit of musical training on the aging auditory brain. *Hearing Research, 308*, 162–173.

Anaya, E. M., Pisoni, D. P., & Kronenberger, W. G. (2016). Long-term musical experience and auditory and visual perceptual abilities under adverse conditions. *Journal of the Acoustical Society of America, 140*(3), 2074–2081.

Anvari, S. H., Trainor, L. J., Woodside, J., & Levy, B. A. (2002). Relations among musical skills, phonological processing and early reading ability in preschool children. *Journal of Experimental Child Psychology, 83*(2), 111–130.

Association, American Music Therapy. (2005). What is Music Therapy? AMTA Official Definition of Music Therapy. www.musictherapy.org/about/musictherapy/

Banai, K., Hornickel, J., Skoe, E., Nicol, T., Zecker, S., & Kraus, N. (2009). *Reading and Subcortical Auditory Function Cerebral Cortex* (Vol. 19, pp. 2699–2707). Department of Communication Sciences, Auditory Neuroscience Lab, Northwestern University.

Başkent, D., Fuller, C. D., Galvin, J. J., Schepel, L., Gaudrain, E., & Free, R. H. (2018). Musician effect on perception of spectro-temporally degraded speech, vocal emotion, and music in young adolescents. *Journal of the Acoustical Society of America, 143*(5), EL311–EL316.

Besson, M., Chobert, J., & Marie, C. (2011). Transfer of training between music and speech: Common processing, attention, and memory. *Frontiers in Psychology, 2*, 94.

Bialystok, E., & DePape, A. M. (2009). Musical expertise, bilingualism, and executive functioning. *Journal of Experimental Psychology: Human Perception and Performance, 35*(2), 565–574.

Bidelman, G. M. (2016). Musicians have enhanced audiovisual multisensory binding: Experience-dependent effects in the double-flash illusion. *Experimental Brain Research, 234*(10), 3037–3047.

Bidelman, G. M. (2018a). Sonification of scalp-recorded frequency-following responses (FFRs) offers improved response detection over conventional statistical metrics. *Journal of Neuroscience Methods, 293*, 59–66.

Bidelman, G. M. (2018b). Subcortical sources dominate the neuroelectric auditory frequency-following response to speech. *NeuroImage, 175*, 56–69.

Bidelman, G. M., & Alain, C. (2015). Musical training orchestrates coordinated neuroplasticity in auditory brainstem and cortex to counteract age-related declines in categorical vowel perception. *Journal of Neuroscience, 35*(2), 1240 –1249.

Bidelman, G. M., Brown, B., Mankel, K., & Price, C. N. (2020). Psychobiological responses reveal audiovisual noise differentially challenges speech recognition. *Ear and Hearing, 41*(2), 268–277.

Bidelman, G. M., Gandour, J. T., & Krishnan, A. (2011a). Cross-domain effects of music and language experience on the representation of pitch in the human auditory brainstem. *Journal of Cognitive Neuroscience, 23*(2), 425–434.

Bidelman, G. M., Gandour, J. T., & Krishnan, A. (2011b). Musicians demonstrate experience-dependent brainstem enhancement of musical scale features within continuously gliding pitch. *Neuroscience Letters, 503*(3), 203–207.

Bidelman, G. M., Gandour, J. T., & Krishnan, A. (2011c). Musicians and tone-language speakers share enhanced brainstem encoding but not perceptual benefits for musical pitch. *Brain and Cognition, 77*(1), 1–10.

Bidelman, G. M., Hutka, S., & Moreno, S. (2013). Tone language speakers and musicians share enhanced perceptual and cognitive abilities for musical pitch: Evidence for bidirectionality between the domains of language and music. *PLoS ONE, 8*(4), e60676.

Bidelman, G. M., & Krishnan, A. (2010). Effects of reverberation on brainstem representation of speech in musicians and non-musicians. *Brain Research, 1355*, 112–125.

Bidelman, G. M., Krishnan, A., & Gandour, J. T. (2011). Enhanced brainstem encoding predicts musicians' perceptual advantages with pitch. *European Journal of Neuroscience, 33*(3), 530–538.

Bidelman, G. M., & Mankel, K. (2019). Reply to Schellenberg: Is there more to auditory plasticity than meets the ear?. *Proceedings of National Academy of Sciences of the United States of America [Letter to the Editor], 116*(8), 2785–2786.

Bidelman, G. M., & Yoo, J. (2020). Musicians show improved speech segregation in competitive, multi-talker cocktail party scenarios. *Frontiers in Psychology, 11* (1927), 1–11.

Bizley, J. K., Maddox, R. K., & Lee, A. K. C. (2016). Defining auditory-visual objects: Behavioral tests and physiological mechanisms. *Trends in Neuroscience, 39*(2), 74–85.

Boebinger, D., Evans, S., Rosen, S., Lima, C. F., Manly, T., & Scott, S. K. (2015). Musicians and non-musicians are equally adept at perceiving masked speech. *Journal of the Acoustical Society of America, 137*(1), 378–387.

Bolia, R. S., Nelson, W. T., Ericson, M. A., & Simpson, B. D. (2000). A speech corpus for multitalker communications research. *Journal of the Acoustical Society of America, 107*(2), 1065–1066.

Bonacina, S., Huang, S., White-Schwoch, T., Krizman, J., Nicol, T., & Kraus, N. (2021). Rhythm, reading, and sound processing in the brain in preschool children. *npj Science of Learning, 6*, 1–11.

Brandler, S., & Rammsayer, T. H. (2003). Differences in mental abilities between musicians and nonmusicians. *Psychology of Music, 31*, 123–138.

Brattico, E., Pallesen, K. J., Varyagina, O., Bailey, C., Anourova, I., Jarvenpaa, M., ... Tervaniemi, M. (2009). Neural discrimination of nonprototypical chords in music experts and laymen: An MEG study. *Journal of Cognitive Neuroscience, 21*(11), 2230–2244.

Brendel, B., & Ziegler, W. (2008). Effectiveness of metrical pacing in the treatment of apraxia of speech. *Aphasiology, 22*(1), 77–102.

Brown, C. J., Jeon, E-K., Driscoll, V., Mussoi, B., Deshpande, S. B., Gfeller, K., & Abbas, P. J. (2017). Effects of long-term musical training on cortical auditory evoked potentials. *Ear and Hearing, 38*(2), e74–e84.

Brown, J. A., & Bidelman, G. M. (2022). Familiarity of background music modulates the cortical tracking of target speech at the "cocktail party". *Brain Sciences, 12*(10), 1320.

Bugos, J. A., Perlstein, W. M., McCrae, C. S., Brophy, T. S., & Bedenbaugh, P. H. (2007). Individualized piano instruction enhances executive functioning and working memory in older adults. *Aging and Mental Health, 11*(4), 464–471.

Bugos, J. A., & Wang, Y. (2022). Piano training enhances executive functions and psychosocial outcomes in aging: Results of a randomized controlled trial. *J Gerontology, 77*(9), 1625–1636.

Burns, E. M., & Ward, W. D. (1978). Categorical perception – phenomenon or epiphenomenon: Evidence from experiments in the perception of melodic musical intervals. *Journal of the Acoustical Society of America, 63*(2), 456–468.

Cason, N., Hidalgo, C., Isoard, F., Roman, S., & Schon, D. (2015). Rhythmic priming enhances speech production abilities: Evidence from prelingually deaf children. *Neuropsychology, 29*(1), 102–107.

Chan, A. S., Ho, Y. C., & Cheung, M. C. (1998). Music training improves verbal memory. *Nature, 396*(6707), 128.

Chartrand, J. P., & Belin, P. (2006). Superior voice timbre processing in musicians. *Neuroscience Letters, 405*(3), 164–167.

Chung, W.-L., Jarmulowicz, L., & Bidelman, G. M. (2021). Cross-linguistic contributions of acoustic cues and prosodic awareness to first and second language vocabulary knowledge. *Journal of Research in Reading, 44*(2), 434–452.

Clayton, K. K., Swaminathan, J., Yazdanbakhsh, A., Zuk, J., Patel, A. D., & Kidd, G., Jr. (2016). Executive function, visual attention and the cocktail party problem in musicians and non-musicians. *PLoS ONE, 11*(7), e0157638.

Coffey, E. B. J., Mogilever, N. B., & Zatorre, R. J. (2017). Speech-in-noise perception in musicians: A review. *Hearing Research, 352*, 49–69.

Corrigall, K. A., Schellenberg, E. G., & Misura, N. M. (2013). Music training, cognition, and personality. *Frontiers in Psychology, 4*, 222.

Cowan, N. (2008). What are the differences between long-term, short-term, and working memory? *Progress in Brain Research, 169*, 323–338.

Degé, F., Kubicek, C., & Schwarzer, G. (2011). Music lessons and intelligence: A relation mediated by executive functions. *Music Perception, 29*(2), 195–201.

Degé, F., & Schwarzer, G. (2011). The effect of a music program on phonological awareness in preschoolers. *Frontiers in Psychology, 2*, 124.

Delogu, F., Lampis, G., & Olivetti Belardinelli, M. (2006). Music-to-language transfer effect: May melodic ability improve learning of tonal languages by native nontonal speakers? *Cognitive Processing, 7*(3), 203–207.

Delogu, F., Lampis, G., & Olivetti Belardinelli, M. (2010). From melody to lexical tone: Musical ability enhances specific aspects of foreign language perception. *European Journal of Cognitive Psychology, 22*(1), 46–61.

Deroche, M. L. D., Limb, C. J., Chatterjee, M., & Gracco, V. L. (2017). Similar abilities of musicians and non-musicians to segregate voices by fundamental frequency. *The Journal of the Acoustical Society of America, 142*(4), 1739–1755.

Du, Y., & Zatorre, R. J. (2017). Musical training sharpens and bonds ears and tongue to hear speech better. *Proceedings of the National Academy of Sciences of the United States of America, 114*(51), 13579–13584.

Eierud, C., Michael, A., Banks, D., & Andrews, E. (2023). Resting-state functional connectivity in lifelong musicians. *Psychoradiology, 3*, kkad003.

Elbert, T., Pantev, C., Wienbruch, C., Rockstroh, B., & Taub, E. (1996). Increased cortical representation of the fingers of the left hand in string players. *Science, 270*, 305–307.

Escobar, J., Mussoi, B. S., & Silberer, A. B. (2020). The effect of musical training and working memory in adverse listening situations. *Ear and Hearing, 41*(2), 278–288.

Fitch, W. T., & Martins, M. D. (2014). Hierarchical processing in music, language, and action: Lashley revisited. *Annals of the New York Academy of Science, 1316*, 87–104.

Foss, A. H., Altschuler, E. L., & James, K. H. (2007). Neural correlates of the Pythagorean ratio rules. *Neuroreport, 18*(15), 1521–1525.

Foss-Feig, J. H., Kwakye, L. D., Cascio, C. J., Burnette, C. P., Kadivar, H., Stone, W. L., & Wallace, M. T. (2010). An extended multisensory temporal binding window in autism spectrum disorders. *Experimental Brain Research, 203*(2), 381–389.

François, C., Chobert, J., Besson, M., & Schön, D. (2013). Music training for the development of speech segmentation. *Cerebral Cortex, 23*, 2038–2043.

Franklin, M. S., Sledge Moore, K., Yip, C-Y., Jonides, J., Rattray, K., & Moher, J. (2008). The effects of musical training on verbal memory. *Psychology of Music, 36*(3), 353–365.

Fuller, C. D., Galvin, J. J., 3rd, Maat, B., Free, R. H., & Baskent, D. (2014). The musician effect: Does it persist under degraded pitch conditions of cochlear implant simulations? *Frontiers in Neuroscience, 8*, 179.

Galbraith, G. C., Arbagey, P. W., Branski, R., Comerci, N., & Rector, P. M. (1995). Intelligible speech encoded in the human brain stem frequency-following response. *Neuroreport, 6*(17), 2363–2367.

Gallese, V., & Lakoff, G. (2005). The brain's concepts: The role of the sensory-motor system in conceptual knowledge. *Cognitive Neuropsychology, 22*(3), 455–479.

Gandour, J.T. (1983). Tone perception in Far Eastern languages. *Journal of Phonetics, 11*, 149–175.

Gentilucci, M., & Volta, R. D. (2008). Spoken language and arm gestures are controlled by the same motor control system. *Quarterly Journal of Experimental Psychology, 61*(6), 944–957.

George, E. M., & Coch, D. (2011). Music training and working memory: An ERP study. *Neuropsychologia, 49*(5), 1083–1094.

Goswami, U., & Bryant, P. (2016). *Phonological Skills and Learning to Read*. Routledge.

Goswami, U., Huss, M., Mead, N., Fosker, T., & Verney, J. P. (2013). Perception of patterns of musical beat distribution in phonological developmental dyslexia: Significant longitudinal relations with word reading and reading comprehension. *Cortex, 49*(5), 1363–1376.

Goswami, U., Thomson, J., Richardson, U., Stainthorp, R., Hughes, D., Rosen, S., & Scott, S. K. (2002). Amplitude envelope onsets and developmental dyslexia: A new hypothesis. *Proceedings of the National Academy of Sciences, 99*(16), 10911–10916.

Grahn, Jessica A., & Rowe, J. B. (2009). Feeling the beat: Premotor and striatal interactions in musicians and nonmusicians during beat perception. *The Journal of Neuroscience, 29*(23), 7540–7548.

Gross, W., Linden, U., & Ostermann, T. (2010). Effects of music therapy in the treatment of children with delayed speech development: Results of a pilot study. *BMC Complementary Alternative Medicine, 10*, 39.

Guhn, M., Emerson, S. D., & Gouzouasis, P. (2020). A population-level analysis of associations between school music participation and academic achievement. *Journal of Educational Psychology, 112*(2), 308–328.

Hausen, M., Torppa, R., Salmela, V. R., Vainio, M., & Sarkamo, T. (2013). Music and speech prosody: A common rhythm. *Frontiers in Psychology, 4*, 566.

Hennessy, S., Mack, W. J., & Habibi, A. (2022). Speech-in-noise perception in musicians and non-musicians: A multi-level meta-analysis. *Hearing Research, 416*, 108442.

Herholz, S. C., & Zatorre, R. J. (2012). Musical training as a framework for brain plasticity: Behavior, function, and structure. *Neuron, 76*(3), 486–502.

Ho, Y., Cheung, M., & Chan, A. (2003). Music training improves verbal but not visual memory: Cross sectional and longitudinal explorations in children. *Neuropsychology, 17*(3), 439–450.

Honing, H. (2018). *The Origins of Musicality*. The MIT Press.

Huss, M., Verney, J. P., Fosker, T., Mead, N., & Goswami, U. (2011). Music, rhythm, rise time perception and developmental dyslexia: Perception of musical meter predicts reading and phonology. *Cortex, 47*(6), 674–689.

Hutka, S. A., Bidelman, G. M., & Moreno, S. (2013). Brain signal variability as a window into the bidirectionality between music and language processing: Moving from a linear to a nonlinear model. *Frontiers in Psychology, 4*(984), 1–11.

Ingham, R. J., Bothe, A. K., Wang, Y., Purkhiser, K., & New, A. (2012). Phonation interval modification and speech performance quality during fluency-inducing conditions by adults who stutter. *Journal of Communication Disorders, 45*(3), 198–211.

Jentzsch, I., Mkrtchian, A., & Kansal, N. (2014). Improved effectiveness of performance monitoring in amateur instrumental musicians. *Neuropsychologia, 52*, 117–124.

Jones, M. R., & Yee, W. (1997). Sensitivity to time change: The role of context and skill. *Journal of Experimental Psychology: Human Perception and Performance, 23*, 693–709.

Kaganovich, N., Kim, J., Herring, C., Schumaker, J., Macpherson, M., & Weber-Fox, C. (2013). Musicians show general enhancement of complex sound encoding and better inhibition of irrelevant auditory change in music: An ERP study. *European Journal of Neuroscience, 37*(8), 1295–1307.

Kaganovich, N., Schumaker, J., Leonard, L. B., Gustafson, D., & Macias, D. (2014). Children with a history of SLI show reduced sensitivity to audiovisual temporal asynchrony: An ERP study. *Journal of Speech, Language, and Hearing Research, 57*(4), 1480–1502.

Kaplan, E., Wagner, A. E., Toffanin, P., & Başkent, D. (2021). Do musicians and non-musicians differ in speech-on-speech processing? *Frontiers in Psychology, 12*(281).

Kaposvari, P., Csete, G., Bognar, A., Csibri, P., Toth, E., Szabo, N., ... Kincses, Z. T. (2015). Audio-visual integration through the parallel visual pathways. *Brain Research, 1624*, 71–77.

Khera, T., & Rangasamy, V. (2021). Cognition and pain: A review. *Frontiers in Psychology, 12*.

Killion, M. C., Niquette, P. A., Gudmundsen, G. I., Revit, L. J., & Banerjee, S. (2004). Development of a quick speech-in-noise test for measuring signal-to-noise ratio loss in normal-hearing and hearing-impaired listeners. *Journal of the Acoustical Society of America, 116*(4 Pt 1), 2395–2405.

Kishon-Rabin, L., Amir, O., Vexler, Y., & Zaltz, Y. (2001). Pitch discrimination: Are professional musicians better than non-musicians? *Journal of Basic and Clinical Physiology and Pharmacology*, *12*(2), 125–143.

Koelsch, S., Gunter, T. C., Cramon, D. Y., Zysset, S., Lohmann, G., & Friederici, A. D. (2002). Bach speaks: A cortical "language-network" serves the processing of music. *Neuroimage*, *17*(2), 956–966.

Koelsch, S., Schroger, E., & Tervaniemi, M. (1999). Superior pre-attentive auditory processing in musicians. *Neuroreport*, *10*(6), 1309–1313.

Kraus, N., Slater, J., Thompson, E. C., Hornickel, J., Strait, D. L., Nicol, T., & White-Schwoch, T. (2014). Music enrichment programs improve the neural encoding of speech in at-risk children. *Journal of Neuroscience*, *34*(36), 11913–11918.

Krishnan, A., Gandour, J. T., Ananthakrishnan, S., Bidelman, G. M., & Smalt, C. J. (2011). Linguistic status of timbre influences pitch encoding in the brainstem. *Neuroreport*, *22*(16), 801–803.

Krishnan, A., Gandour, J. T., & Bidelman, G. M. (2010a). Brainstem pitch representation in native speakers of Mandarin is less susceptible to degradation of stimulus temporal regularity. *Brain Research*, *1313*, 124–133.

Krishnan, A., Gandour, J. T., & Bidelman, G. M. (2010b). The effects of tone language experience on pitch processing in the brainstem. *Journal of Neurolinguistics*, *23*, 81–95.

Krishnan, A., Gandour, J. T., Bidelman, G. M., & Swaminathan, J. (2009). Experience-dependent neural representation of dynamic pitch in the brainstem. *Neuroreport*, *20*(4), 408–413.

Krishnan, A., Gandour, J. T., Smalt, C. J., & Bidelman, G. M. (2010). Language-dependent pitch encoding advantage in the brainstem is not limited to acceleration rates that occur in natural speech. *Brain and Language*, *114*(3), 193–198.

Krishnan, A., Swaminathan, J., & Gandour, J. T. (2009). Experience-dependent enhancement of linguistic pitch representation in the brainstem is not specific to a speech context. *Journal of Cognitive Neuroscience*, *21*(6), 1092–1105.

Krishnan, A, Xu, Y, Gandour, J. T., & Cariani, P. (2005). Encoding of pitch in the human brainstem is sensitive to language experience. *Brain Research Cognitive Brain Research*, *25*(1), 161–168.

Lappe, C., Herholz, S. C., Trainor, L. J., & Pantev, C. (2008). Cortical plasticity induced by short-term unimodal and multimodal musical training. *Journal of Neuroscience*, *28*, 9632–9639.

Lee, C. Y., & Hung, T. H. (2008). Identification of Mandarin tones by English-speaking musicians and nonmusicians. *Journal of the Acoustical Society of America*, *124*(5), 3235–3248.

Lee, C.-Y., Lekich, A., & Zhang, Y. (2014). Perception of pitch height in lexical and musical tones by English-speaking musicians and nonmusiciansa. *Journal of the Acoustical Society of America*, *135*(3), 1607–1615.

Lee, H., & Noppeney, U. (2011). Long-term music training tunes how the brain temporally binds signals from multiple senses. *Proceedings of the National Academy of Sciences of the United States of America*, *108*(51), E1441–1450.

Lee, H. L., & Noppeney, U. (2014). Music expertise shapes audiovisual temporal integration windows for speech, sinewave speech and music. *Frontiers in Psychology, 5*(868), 1–9.

Lee, Y. S, Thaut, C., & Santorini, C. (2019). Neurologic music therapy for speech and language rehabilitation. In M. H. Thaut & D. A. Hodges (Eds.), *The Oxford Handbook of Music and the Brain*. Oxford University Press.

Lerdahl, F., & Jackendoff, R. (1983). *A Generative Theory of Tonal Music. MIT Press Series on Cognitive Theory and Mental Representation* (pp. xiv, 368). MIT Press.

Lim, V. K., Bradshaw, J. L., nicholls, M., & Altenmuller, E. (2003). Perceptual differences in sequential stimuli across patients with musician's and writer's cramp. *Movement Disorders, 11*, 1286–1293.

Lo, C. Y., Looi, V., Thompson, W. F., & McMahon, C. M. (2020). Music training for children with sensorineural hearing loss improves speech-in-noise perception. *Journal of Speech, Language, and Hearing Research, 63*(6), 1990–2015.

Lu, J., Moussard, A., Guo, S., Lee, Y., Bidelman, G. M., Moreno, S., ... Alain, C. (2022). Music training modulates theta brain oscillations associated with response suppression. *Annals of the New York Academy of Sciences, 1516*(1), 212–221.

Lu, Y., Paraskevopoulos, E., Herholz, S. C., Kuchenbuch, A., & Pantev, C. (2014). Temporal processing of audiovisual stimuli is enhanced in musicians: Evidence from magnetoencephalography (MEG). *PLoS ONE, 9*(3), e90686.

MacLeod, A., & Summerfield, Q. (1987). Quantifying the contribution of vision to speech perception in noise. *British Journal of Audiology, 21*(2), 131–141.

Madsen, S. M. K., Whiteford, K. L., & Oxenham, A. J. (2017). Musicians do not benefit from differences in fundamental frequency when listening to speech in competing speech backgrounds. *Scientific Reports, 7*(1), 12624.

Maess, B., Koelsch, S., Gunter, T. C., & Friederici, A. D. (2001). Musical syntax is processed in Broca's area: An MEG study. *Nature Neuroscience, 4*(5), 540–545.

Mainka, S., & Mallien, G. (2014). Rhythmic speech cueing (RSC). In M. H. Thaut & V. Hoemberg (Eds.), *Handbook of Neurologic Music Therapy* (pp. 150–160). Oxford University Press.

Mankel, K., & Bidelman, G. M. (2018). Inherent auditory skills rather than formal music training shape the neural encoding of speech. *Proceedings of the National Academy of Sciences of the United States of America, 115*(51), 13129–13134.

Marques, C., Moreno, S., Castro, S. L., & Besson, M. (2007). Musicians detect pitch violation in a foreign language better than nonmusicians: Behavioral and electrophysiological evidence. *Journal of Cognitive Neuroscience, 19*(9), 1453–1463.

McGurk, H., & MacDonald, J. (1976). Hearing lips and seeing voices. *Nature, 264*(5588), 746–748.

Micheyl, C., Delhommeau, K., Perrot, X., & Oxenham, A. J. (2006). Influence of musical and psychoacoustical training on pitch discrimination. *Hearing Research, 219*(1–2), 36–47.

Milovanov, R., Huotilainen, M., Esquef, P. A., Alku, P., Valimaki, V., & Tervaniemi, M. (2009). The role of musical aptitude and language skills in preattentive duration processing in school-aged children. *Neuroscience Letters, 460*(2), 161–165.

Moreno, S. (2009). Can music influence language and cognition? *Contemporary Music Review, 28*(3), 329–345.

Moreno, S., Bialystok, E., Barac, R., Schellenberg, E. G., Cepeda, N. J., & Chau, T. (2011). Short-term music training enhances verbal intelligence and executive function. *Psychological Science, 22,* 1425–1433.

Moreno, S., & Bidelman, G. M. (2014). Examining neural plasticity and cognitive benefit through the unique lens of musical training. *Hearing Research, 308,* 84–97.

Moreno, S., Friesen, D., & Bialystok, E. (2011). Effect of music training on promoting preliteracy skills: Preliminary causal evidence. *Music Perception, 29*(2), 165–172.

Moreno, S., Marques, C., Santos, A., Santos, M., Castro, S. L., & Besson, M. (2009). Musical training influences linguistic abilities in 8-year-old children: More evidence for brain plasticity. *Cerebral Cortex, 19*(3), 712–723.

Moussard, A., Bermudez, P., Alain, C., Tays, W., & Moreno, S. (2016). Life-long music practice and executive control in older adults: An event-related potential study. *Brain Research, 1642,* 146–153.

Munte, T. F., Altenmuller, E., & Jancke, L. (2002). The musician's brain as a model of neuroplasticity. *Nature Reviews Neuroscience, 3*(6), 473–478.

Munte, T. F., Kohlmetz, C., Nager, W., & Altenmuller, E. (2001). Superior auditory spatial tuning in conductors. *Nature, 409*(6820), 580.

Musacchia, G., Sams, M., Skoe, E., & Kraus, N. (2007). Musicians have enhanced subcortical auditory and audiovisual processing of speech and music. *Proceedings of the National Academy of Sciences of the United States of America, 104*(40), 15894–15898.

Navarra, J., & Soto-Faraco, S. (2007). Hearing lips in a second language: Visual articulatory information enables the perception of second language sounds. *Psychological Research, 71*(1), 4–12.

Neves, L., Correia, A. I., Castro, S. L., Martins, D., & Lima, C. F. (2022). Does music training enhance auditory and linguistic processing? A systematic review and meta-analysis of behavioral and brain evidence. *Neuroscience & Biobehavioral Reviews, 140,* 104777.

Nie, P., Wang, C., Rong, G., Du, B., Lu, J., Li, S., ...& Tervaniemi, M. (2022). Effects of music training on the auditory working memory of Chinese-speaking school-aged children: A longitudinal intervention study. *Frontiers in Psychology, 12.*

Nilsson, M., Soli, S. D., & Sullivan, J. A. (1994). Development of the hearing in noise test for the measurement of speech reception thresholds in quiet and in noise. *Journal of the Acoustical Society of America, 95*(2), 1085–1099.

Novembre, G., & Keller, P. E. (2014). A conceptual review on action-perception coupling in the musicians' brain: What is it good for? *Frontiers in Human Neuroscience, 8,* 603.

Oxenham, A. J., & Shera, C. A. (2003). Estimates of human cochlear tuning at low levels using forward and simultaneous masking. *Journal of the Association for Research in Otolaryngology, 4,* 541–554.

Pallesen, K. J., Brattico, E., Bailey, C. J., Korvenoja, A., Koivisto, J., Gjedde, A., & Carlson, S. (2010). Cognitive control in auditory working memory is enhanced in musicians. *PLoS ONE, 5*(6), e11120.

Pantev, C., Roberts, L. E., Schulz, M., Engelien, A., & Ross, B. (2001). Timbre-specific enhancement of auditory cortical representations in musicians. *Neuroreport, 12*(1), 169–174.

Paraskevopoulos, E., Kraneburg, A., Herholz, S. C., Bamidis, P. D., & Pantev, C. (2015). Musical expertise is related to altered functional connectivity during audiovisual integration. *Proceedings of the National Academy of Sciences of the United States of America, 112*(40), 12522–12527.

Paraskevopoulos, E., Kuchenbuch, A., Herholz, S. C., & Pantev, C. (2012). Evidence for training-induced plasticity in multisensory brain structures: An MEG study. *PLoS ONE, 7,* e36534.

Parbery-Clark, A., Skoe, E., & Kraus, N. (2009). Musical experience limits the degradative effects of background noise on the neural processing of sound. *Journal of Neuroscience, 29*(45), 14100–14107.

Parbery-Clark, A., Skoe, E., Lam, C., & Kraus, N. (2009). Musician enhancement for speech-in-noise. *Ear and Hearing, 30*(6), 653–661.

Parbery-Clark, A., Strait, D. L., Anderson, S., Hittner, E., & Kraus, N. (2011). Musical experience and the aging auditory system: Implications for cognitive abilities and hearing speech in noise. *PLoS ONE, 6*(5), e18082.

Patel, A D. (2003). Language, music, syntax and the brain. *Nature Neuroscience, 6,* 674–681.

Patel, A. D. (2008). *Music, Language, and the Brain.* Oxford University Press.

Patel, A. D. (2011). Why would musical training benefit the neural encoding of speech? The OPERA hypothesis. *Frontiers in Psychology, 2,* 142.

Patel, A. D., Gibson, E., Ratner, J., Besson, M., & Holcomb, P. J. (1998). Processing syntactic relations in language and music: An event-related potential study. *Journal of Cognitive Neuroscience, 10*(6), 717–733.

Patel, A. D., & Iversen, J. R. (2007). The linguistic benefits of musical abilities. *Trends in Cognitive Sciences, 11*(9), 369–372.

Peng, G. (2006). Temporal and tonal aspects of Chinese syllables: A corpus-based comparative study of Mandarin and Catonese. *Journal of Chinese Linguistics, 34*(1), 134–154.

Peretz, I., & Hyde, K. L. (2003). What is specific to music processing? Insights from congenital amusia. *Trends in Cognitive Sciences, 7*(8), 362–367.

Perruchet, P., & Poulin-Charronnat, B. (2013). Challenging prior evidence for a shared syntactic processor for language and music. *Psychonomic Bulletin & Review, 20*(2), 310–317.

Price, C. N., & Bidelman, G. M. (2022). Musical experience partially counteracts temporal speech processing deficits in putative mild cognitive impairment. *Annals of the New York Academy of Sciences, 1516*(1), 114–122.

Raichle, M. E. (2011). The restless brain. *Brain Connect, 1*(1), 3–12.

Rammsayer, T., & Altenmuller, E. (2006). Temporal information processing in musicians and nonmusicians. *Music Perception, 24,* 37-48.

Rammsayer, T., Buttkus, F., & Altenmüller, E. (2012). Musicians do better than non-musicians in both auditory and visual timing tasks. *Music Perception, 30,* 85–96.

Rohrmeier, M. (2007). A generative grammar approach to diatonic harmonic structure. Proceedings of the 4th Sound and Music Computing Conference, SMC 2007, 97–100.

Román-Caballero, R., Vadillo, M. A., Trainor, L. J., & Lupiáñez, J. (2022). Please don't stop the music: A meta-analysis of the cognitive and academic benefits of instrumental musical training in childhood and adolescence. *Educational Research Review, 35,* 100436.

Ruggles, D. R., Freyman, R. L., & Oxenham, A. J. (2014). Influence of musical training on understanding voiced and whispered speech in noise. *PLoS ONE, 9*(1), e86980.

Russeler, J., Altenmuller, E., Nager, W., Kohlmetz, C., & Munte, T. F. (2001). Event-related brain potentials to sound omissions differ in musicians and non-musicians. *Neuroscience Letters, 308*(1), 33–36.

Schellenberg, E. G. (2004). Music lessons enhance IQ. *Psychological Science, 15*(8), 511–514.

Schellenberg, E. G. (2006). Long-term positive associations between music lessons and IQ. *Journal of Educational Psychology, 98,* 457–468.

Schon, D., Magne, C., & Besson, M. (2004). The music of speech: Music training facilitates pitch processing in both music and language. *Psychophysiology, 41*(3), 341–349.

Seger, C. A., Spiering, B. J., Sares, A. G., Quraini, S. I., Alpeter, C., David, J., & Thaut, M. H. (2013). Corticostriatal contributions to musical expectancy perception. *Journal of Cognitive Neuroscience, 25*(7), 1062–1077.

Shams, L., Kamitani, Y., & Shimojo, S. (2000). What you see is what you hear. *Nature, 408*(14), 788.

Shams, L., Kamitani, Y., & Shimojo, S. (2002). Visual illusion induced by sound. *Cognitive Brain Research, 14,* 147–152.

Shi, E. R., & Zhang, Q. (2020). A domain-general perspective on the role of the basal ganglia in language and music: Benefits of music therapy for the treatment of aphasia. *Brain and Language, 206,* 104811.

Slater, J., Skoe, E., Strait, D., O'Connell, S., Thompson, E., & Kraus, N. (2015). Music training improves speech-in-noise perception: Longitudinal evidence from a community-based music program. *Behavioural Brain Research, 291,* 244–252.

Slevc, L. R., & Miyake, A. (2006). Individual differences in second-language proficiency: Does musical ability matter? *Psychological Science, 17*(8), 675–681.

Slevc, L. R., Rosenberg, J. C., & Patel, A. D. (2009). Making psycholinguistics musical: Self-paced reading time evidence for shared processing of linguistic and musical syntax. *Psychonomic Bulletin & Review, 16*(2), 374–381.

Slevc, L. R., & Okada, B. M. (2015). Processing structure in language and music: A case for shared reliance on cognitive control. *Psychonomic Bulletin and Review, 22*, 637–652.

Sluming, V., Brooks, J., Howard, M., Downes, J. J., & Roberts, N. (2007). Broca's area supports enhanced visuospatial cognition in orchestral musicians. *Journal of Neuroscience, 27*(14), 3799–3806.

Smit, E. A., Milne, A. J., & Escudero, P. (2022). Music perception abilities and ambiguous word learning: Is there cross-domain transfer in nonmusicians? *Frontiers in Psychology, 13*.

Smith, J. C., Marsh, J. T., & Brown, W. S. (1975). Far-field recorded frequency-following responses: Evidence for the locus of brainstem sources. *Electroencephalography and Clinical Neurophysiology, 39*(5), 465–472.

Spiegel, M. F., & Watson, C. S. (1984). Performance on frequency-discrimination tasks by musicians and nonmusicians. *Journal of the Acoustical Society of America, 766*(6), 1690–1695.

Stahl, B., Kotz, S. A., Henseler, I., Turner, R., & Geyer, S. (2011). Rhythm in disguise: Why singing may not hold the key to recovery from aphasia. *Brain, 134*(Pt 10), 3083–3093.

Strait, D. L., Chan, K., Ashley, R., & Kraus, N. (2011). Specialization among the specialized: Auditory brainstem function is tuned in to timbre. *Cortex, 48*(3), 360–362.

Strait, D. L., & Kraus, N. (2014). Biological impact of auditory expertise across the life span: Musicians as a model of auditory learning. *Hearing Research, 308*, 109–121.

Strait, D. L., Kraus, N., Parbery-Clark, A., & Ashley, R. (2010). Musical experience shapes top-down auditory mechanisms: Evidence from masking and auditory attention performance. *Hearing Research, 261*(1–2), 22–29.

Sumby, W. H., & Pollack, I. (1954). Visual contribution to speech intelligibility in noise. *Journal of the Acoustical Society of America, 26*, 212–215.

Swaminathan, J., Mason, C. R., Streeter, T. M., Best, V., Kidd, G., Jr., & Patel, A. D. (2015). Musical training, individual differences and the cocktail party problem. *Scientific Reports, 5*, 11628.

Swaminathan, S., Schellenberg, E. G., & Khalil, S. (2017). Revisiting the association between music lessons and intelligence: Training effects or music aptitude? *Intelligence, 62*, 119–124.

Talamini, F., Altoè, G., Carretti, B., & Grassi, M. (2017). Musicians have better memory than nonmusicians: A meta-analysis. *PLoS ONE, 12*(10), e0186773.

Tervaniemi, M., Just, V., Koelsch, S., Widmann, A., & Schroger, E. (2005). Pitch discrimination accuracy in musicians vs nonmusicians: An event-related potential and behavioral study. *Experimental Brain Research, 161*(1), 1–10.

Thaut, M. H., McIntosh, G. C., & Hoemberg, V. (2014). Neurobiological foundations of neurologic music therapy: Rhythmic entrainment and the motor system. *Frontiers in Psychology, 5*, 1185.

Thaut, M. H., McIntosh, K. W., McIntosh, G. C., & Hoemberg, V. (2001). Auditory rhythmicity enhances movement and speech motor control in patients with Parkinson's disease. *Functional Neurology, 16*(2), 163–172.

Thaut, M. H., Thaut, C., & McIntosh, K. (2014). Melodic Intonation Therapy (MIT). In M. H. Thaut & V. Hoemberg (Eds.), *The Handbook of Neurologic Music Therapy* (pp. 140–145). Oxford University Press.

Thaut, M. (2013). *Rhythm, music, and the brain: Scientific foundations and clinical applications*: Routledge.

Thaut, M., & Hoemberg, V. (2014). *Handbook of Neurologic Music Therapy*. Oxford University Press.

Thompson, W. F., Schellenberg, E. G., & Husain, G. (2004). Decoding speech prosody: Do music lessons help? *Emotion, 4*, 46–64.

Tierney, A. T., Bergeson, T. R., & Pisoni, D. B. (2008). Effects of early musical experience on auditory sequence memory. *Empirical Musicology Review*, *3*(4), 178–186.

Tierney, A., & Kraus, N. (2014). Auditory-motor entrainment and phonological skills: Precise auditory timing hypothesis (PATH). *Frontiers in Human Neuroscience, 8*.

Torppa, R., Faulkner, A., Kujala, T., Huotilainen, M., & Lipsanen, J. (2018). Developmental links between speech perception in noise, singing, and cortical processing of music in children with cochlear implants. *Music Perception*, *36*(2), 156–174.

Tsang, C. D., & Conrad, N. J. (2011). Music training and reading readiness. *Music Perception*, *29*(2), 157–163.

van Vugt, F. T., & Tillmann, B. (2014). Thresholds of auditory-motor coupling measured with a simple task in musicians and non-musicians: Was the sound simultaneous to the key press? *PLoS ONE*, *9*(2), e87176.

Vuust, P., Roepstorff, A., Wallentin, M., Mouridsen, K., & Ostergaard, L. (2006). It don't mean a thing... Keeping the rhythm during polyrhythmic tension, activates language areas (BA47). *NeuroImage*, *31*(2), 832–841.

Wallace, M. T., & Stevenson, R. A. (2014). The construct of the multisensory temporal binding window and its dysregulation in developmental disabilities. *Neuropsychologia*, *64C*, 105–123.

Weiss, M. W., & Bidelman, G. M. (2015). Listening to the brainstem: Musicianship enhances intelligibility of subcortical representations for speech. *Journal of Neuroscience*, *35*(4), 1687–1691.

Wong, P. C., Skoe, E., Russo, N. M., Dees, T., & Kraus, N. (2007). Musical experience shapes human brainstem encoding of linguistic pitch patterns. *Nature Neuroscience*, *10*(4), 420–422.

Worschech, F., Marie, D., Jünemann, K., Sinke, C., Krüger, T. H. C., Großbach, M., ...& Altenmüller, E. (2021). Improved speech in noise perception in the elderly after 6 months of musical instruction. *Frontiers in Neuroscience*, *15*.

Yang, X., Shen, X., Zhang, Q., Wang, C., Zhou, L., & Chen, Y. (2022). Music training is associated with better clause segmentation during spoken language processing. *Psychonomic Bulletin and Review*, *22*, 1472–1479.

Yeend, I., Beach, E. F., Sharma, M., & Dillon, H. (2017). The effects of noise exposure and musical training on suprathreshold auditory processing and speech perception in noise. *Hearing Research*, *353*, 224–236.

Yoo, J., & Bidelman, G. M. (2019). Linguistic, perceptual, and cognitive factors underlying musicians' benefits in noise-degraded speech perception. *Hearing Research, 377*, 189–195.

Zamorano, A. M., Cifre, I., Montoya, P., Riquelme, I., & Kleber, B. (2017). Insula-based networks in professional musicians: Evidence for increased functional connectivity during resting state fMRI. *Human Brain Mapping, 38*(10), 4834–4849.

Zatorre, R. J., Chen, J. L., & Penhune, V. B. (2007). When the brain plays music: Auditory-motor interactions in music perception and production. *Nature Reviews Neuroscience, 8*(7), 547–558.

Zendel, B. R., & Alain, C. (2009). Concurrent sound segregation is enhanced in musicians. *Journal of Cognitive Neuroscience, 21*(8), 1488–1498.

Zendel, B. R., & Alain, C. (2012). Musicians experience less age-related decline in central auditory processing. *Psychology and Aging, 27*(2), 410–417.

Zendel, B. R., & Alain, C. (2013). The influence of lifelong musicianship on neurophysiological measures of concurrent sound segregation. *Journal of Cognitive Neuroscience, 25*(4), 503–516.

Zendel, B. R., West, G. L., Belleville, S., & Peretz, I. (2019). Musical training improves the ability to understand speech-in-noise in older adults. *Neurobiology of Aging, 81*, 102–115.

17

Music and Language in the Brain

Balancing Domain-Specific and Domain-General Mechanisms

Stephen C. Van Hedger and Urvi Maheshwari

17.1 Introduction

Music is the universal language of mankind

Henry Wadsworth Longfellow

Over the years, the relationship between music and language has been explored by philosophers, poets, musicians, evolutionary biologists, and, most pertinently for the present chapter, cognitive neuroscientists. A common theme emerging from these comparisons is that music and language share remarkable similarities, with music perhaps even representing a kind of universal language (a sentiment captured by the opening quotes to this chapter).

Indeed, the observable links between music and language are hard to ignore. Music and spoken language both make use of complex sound patterns that unfold over time. Both systems are shaped by culture (Cross, 2001), are human activities present in every known culture (McDermott & Hauser, 2005; Mehr et al., 2019), are learned over the course of development (McMullen & Saffran, 2004), and convey meaning to listeners (Slevc & Patel, 2011). Archaeological evidence of bone flutes dating back 50,000 years (Higham et al., 2012; Turk et al., 2020; though see Diedrich, 2015), suggests that music making was an integral part of early cultural development. The existence of music for tens of thousands of years has led some to theorize that music and language were part of the same early communicative system, making use of prosodic changes (i.e., patterns of intonation and stress) to convey affective state (Brown, 2017; Mithen, 2005). Although the evolutionary function of music has been the subject of considerable debate (e.g., Honing et al., 2015; Pinker, 1999), the perspective offered by Brown (2017) suggests that music and language share contemporary similarities because both stem from a common communicative system.

Yet, reductionist comparisons of music and language highlight some of the tenuous links that bridge both domains. For example, despite both music and spoken language consisting of structured sequences of sounds that convey meaning to listeners, the syntactic organization and means of conveying meaning differ considerably across domains. Differences between music and language are perhaps most salient in the domain of semantics. Whereas language can convey concrete, referential meaning through (mostly) arbitrary patterns of sound (e.g., "The tired professor grades the term papers"), music cannot convey a similarly specific referential meaning, even though it might be able to represent a subjective psychological state of the tired professor somewhat adeptly (e.g., via slow, quiet, repetitive sound patterns). This example represents a rather extreme difference in how music and language operate, but it highlights the need for researchers exploring music-language links to consider appropriate levels of analysis to make fruitful cross-domain comparisons (cf. Asano & Boeckx, 2015). This is critical to gaining traction on the question of how distinct or overlapping neural representations of these systems are. Moreover, understanding the relative distinctiveness of both systems has broader implications. For example, it provides insight into whether speech is a special cognitive ability, or whether it might be explained under a broader auditory framework that includes music (Peretz & Hyde, 2003; Pinker & Jackendoff, 2005).

In the present chapter, we compare music and language from a cognitive neuroscience lens. We begin by highlighting the apparent similarities in how these domains are organized (Section 17.2), in terms of acoustic (Section 17.2.1), syntactic (Section 17.2.2), semantic (Section 17.2.3) features. However, throughout this discussion, we highlight common pitfalls within this area of research, which involve attempts to apply the lens of one domain too myopically to the other domain. We then turn away from the specific features of both music and language to examine broader cognitive mechanisms involved in acquiring and maintaining these systems of knowledge (Section 17.3). We focus on the developmental constraints in learning music and language (Section 17.3.1), the role of implicit learning mechanisms in pattern extraction across domains (Section 17.3.2), and the conditions under which training in one domain generalizes to the other domain (Section 17.3.3). We conclude (Section 17.4) by summarizing the current state of the scientific literature and commenting on how the comparison of music and language lends insight into fundamental questions within cognitive neuroscience.

17.2 Featural Similarities between Music and Language

The featural similarities between music and language have become the subject of considerable research in psychology and cognitive neuroscience. Understanding the featural elements of both music and language is a necessary step in assessing the degree to which these domains have distinct or shared processing in the brain. In this section, we deconstruct these featural similarities across music and language, from a description of the sound patterns that comprise both systems (Section 17.2.1), to the ways in which sounds are organized hierarchically (Section 17.2.2), and how these organized sound patterns convey meaning to listeners (Section 17.2.3).

17.2.1 Acoustic Features of Music and Language

Both music and spoken language are acoustic signals that unfold over time. Music and speech signals interact with the same auditory organs and are not inherently differentiated at the level of the cochlear partition. Yet, music and speech are (generally) not confusable signals to most listeners, despite following the same sensory pathway and aggregating as a single waveform if heard simultaneously. How, then, do listeners clearly separate these signals to appreciate both as separable auditory objects?

Although a comprehensive description of the acoustic similarities and differences between music and speech is beyond the scope of this chapter, both spectral (related to the frequency components of sounds) and temporal (related to how sounds unfold over time) factors differentiate music from speech. Music typically contains discrete (stable) pitches and isochronous rhythmic patterns that allow for the extraction of a "beat" (e.g., Patel, 2003). By contrast, speech typically contains larger, more continuous pitch changes and rhythmic patterns that are not as strongly periodic (Peelle & Davis, 2012; Zatorre & Baum, 2012). Music and speech also differ in their reliance on fine-grained (spectral) information versus temporal (amplitude envelope) information (Smith et al., 2002). These findings can be contextualized by considering the demands of music and speech comprehension. In speech, temporal differences on the order of tens of milliseconds can entirely change the meaning of the signal – for example, in differentiating stop consonants like /ba/ and /pa/ (e.g., Miller & Volaitis, 1989). Music unfolds more slowly (Ding et al., 2017) and places greater demands on spectral resolution. For example, most listeners can readily detect when elements of a musical performance are "out of tune" relative to one another, even when these differences are under one semitone, corresponding to about a 5.95 percent change in frequency (Larrouy-Maestri et al., 2019). By contrast, pitch changes in spoken language are on the order of half an octave (i.e., 50 percent changes in pitch), with precise tuning or intervallic information not playing a critical role in understanding (Patel, Peretz, et al., 1998)

These acoustic differences between music and speech might partly underlie the observed differences in brain lateralization across domains. Although music and speech are not well differentiated early in the auditory pathway, including in the primary auditory cortex where tonotopic organization is largely preserved (e.g., Humphries et al., 2010; Zatorre et al., 1992), a consistent finding over decades of neuroimaging research has found that language processing tends to be left lateralized, whereas music processing tends to be right lateralized (Tervaniemi & Hugdahl, 2003). Given that the left hemisphere has been found to be more temporally sensitive (extracting information over short timescales), whereas the right hemisphere has been found to be more spectrally sensitive (extracting information over longer timescales), the lateralization of music and speech may in part be due to acoustic differences (Zatorre et al., 2002).

One approach trying to examine the extent to which observed neural differences between music and speech might be acoustically driven focuses on how *song*, relative to non-sung speech, is represented in the brain (e.g., Norman-Haignere et al., 2015). Although this approach reduces the spectro-temporal richness that characterizes music to a single instrument (the voice), it addresses the role of acoustic features in differentiating music and speech representations, as both signals are produced by the same source. To highlight the leverage gained by using song to assess acoustic differences in speech and music processing, Albouy and colleagues (2020) played participants a cappella songs (with lyrics) using a functional magnetic resonance imaging (fMRI) paradigm. When songs were degraded in the temporal domain, speech perception was impaired, yet music perception was preserved; and when songs were degraded in the spectral domain, music perception was impaired, and speech perception was preserved. Critically, the processing of these degraded sounds in terms of song or speech was highly lateralized: the neural decoding of speech was dependent on activity patterns in the left auditory cortex, and the neural decoding of music was dependent on activity patterns in the right auditory cortex. By degrading the same sounds to selectively impair speech and music perception, the researchers provide strong evidence that the hemispheric asymmetry found between speech and music can be explained by specialization of left and right auditory cortex for temporal and spectral analysis, respectively.

Given that both song and speech make use of the same vocal and articulatory mechanisms, this opens the possibility that these signals might not be clearly differentiable at an acoustic level. For example, the "speech-to-song" illusion is an effect where an acoustic signal, clearly heard as speech initially, transforms to song as a function of repetition (Deutsch et al., 2008). This illusion provides a promising window for distinguishing acoustic factors from music and speech representations in the brain, as the same acoustic stimulus can serve as both speech and music. A significant amount of research has been conducted to examine the acoustic factors that make some speech segments more likely to transform into song points to acoustic factors, such as the pitch salience and pitch stability of the utterance, as well as the extent to which repetition leads to perceptual grouping of sounds that accentuates metrical information, (e.g., stressed and unstressed beats; Falk et al., 2014). This work examining speech and song has found that speech – given its greater variability in terms of pitch changes and rhythmic variation – sometimes appears to incidentally contain sufficient pitch stability and sufficient metrical patterns to transform to song, blurring the distinction between music and speech at an acoustic level.

The speech-to-song illusion thus offers a valuable window into the neural processing of both music and spoken language, as it allows for the manipulation of domain (music versus speech) while controlling for acoustic factors. Using fMRI, Tierney and colleagues (2013) played participants excerpts from an audiobook. Some of these excerpts transformed into songs when repeated (i.e., the speech-to-song illusion), whereas others did not (control stimuli). This design allowed the researchers to examine the changes in blood-oxygenation-level-dependent (BOLD) responses in the brain as a function of hearing acoustically similar stimuli as either speech or song. The authors found several additional areas that increased in BOLD activation in response to hearing song (versus speech), including areas implicated in pitch processing (e.g., anterior and posterior superior temporal gyrus), and areas implicated in audio-motor integration (e.g., supramarginal gyrus and inferior frontal gyrus). The authors conclude that these differences in "song" versus "speech" stimuli are unlikely to be the result of acoustic factors, but rather reflect the increased demands that song perception places on pitch processing and motor processing. Interestingly, the findings from this study were entirely asymmetrical – perception of song led to increased BOLD responses in multiple pitch- and motor-sensitive areas, but the control stimuli that did not transform did not lead to increased BOLD responses in any area. These findings suggest that perceiving a stimulus as music might place greater processing demands in a myriad of brain areas, which has been proposed as one of the ways in which musical training might lead to benefits in speech processing (e.g., see Patel, 2011 and also Section 17.3.3).

Music and spoken language are quite similar at an acoustic level. Both contain complex harmonic information, a perceptible fundamental frequency, and periodic and quasi-periodic rhythmic patterns as they unfold over time. Yet, musical sounds tend to have more stable pitches and more regular rhythms that allow for the perception of a metrical beat. In earlier parts of the auditory pathway (including the earliest cortical processing of sounds in the primary auditory cortex), music and speech are not clearly distinguishable. However, methodological advances and clever experimental paradigms (e.g., using the speech-to-song illusion) have demonstrated differentiation of music and speech signals in areas implicated in pitch and motor processing. The existence of speech selective and music selective areas of the brain, even when controlling for acoustic factors, suggests that these systems have domain-specific representations, even if they share substantial overlap in earlier auditory processing areas.

17.2.2 Syntactic Features of Music and Language

The acoustic signatures of both music and language are not static, but rather unfold over time. An important consideration when comparing how music and language are represented in the brain is to examine the ways in which these sounds are structured. Both music and language are governed by rules that constrain how smaller units (e.g., notes, words) are combined into larger structural units or sequences (e.g., phrases, sentences). Both music and language have *syntax*, broadly defined as the set of rules that dictate how discrete elements of a system are combined hierarchically into larger units or sequences (e.g., Adger, 2015).

Even though both language and music are governed by syntactic rules, a pitfall in this area of research is to examine these relationships using a myopic lens (e.g., see Asano & Boeckx, 2015 for a similar argument). Although music and language have a hierarchical structure, trying to identify surface-level parallels across domains is problematic for several reasons (Lerdahl, 2013). First, hierarchical organization can occur at multiple levels of analysis within each domain (e.g., the phonological, morphological, and sentential level in language; the harmonic and rhythmic levels in music), and it is unclear how the "units" of each domain should be compared. Second, syntax is intricately tied to meaning, particularly within music (Asano & Boeckx, 2015). For example, *harmonic syntax* (the rules governing how tonal sequences unfold over time) generates strong expectations in listeners, and violating these expectations is one of the ways in which music can represent affective meaning (e.g., Steinbeis et al., 2006). Similarly, in language, syntax is used to determine and convey meaning, perhaps most saliently in a process called syntactic bootstrapping. On these accounts, language learners might acquire semantic knowledge (e.g., verb meanings) from syntactic knowledge, such as using grammatical cues (e.g., tense markings, morphology) and discourse structure (Brown, 1957; Gleitman, 1990; Jakobson, 1965; Landau & Gleitman, 2009; Naigles, 1990; Naigles & Swensen, 2007) to narrow their hypothesis space about the meanings of words. Given the radically different ways in which music and language convey meaning (see Section 17.2.3), this intertwining of syntax and semantics complicates direct and literal comparisons of musical and linguistic syntax. Consequently, much of the research examining syntactic relationships between music and language focuses on the *processing* of syntax in both domains, rather than on literal translations between musical and linguistic syntactic units.

How is syntax processed in the brain? In language, syntactic processing has been dissociated from semantic processing using a variety of methodologies, including electroencephalography (EEG; Neville et al., 1991) and fMRI (Friederici, 2003). Early neuroimaging work in this area found evidence for the involvement of "Broca's Area" in the processing of syntax in language (Embick et al., 2000). However, the specific role of Broca's Area in the context of the classic Broca–Wernicke–Lichtheim–Geschwind model of language (e.g., Geschwind, 1970) has been more recently reexamined, in part because it is not clearly marked anatomically (Tremblay & Dick, 2016) and appears to dynamically change functionality depending on coactivation of other related networks (Hagoort, 2014). As such, it is more appropriate to characterize syntax processing as being subserved by a broader network than Broca's Area, with the left inferior frontal gyrus and anterior superior temporal gyrus critically implicated in syntactic processing (Grodzinsky & Friederici, 2006).

Intriguingly, musical syntax has been argued to have substantial shared processing overlap with linguistic syntax (Patel, 2003). Converging evidence from several neuroscientific methodologies, including EEG (Patel, Gibson, et al., 1998), magnetoencephalography (MEG; Maess et al., 2001), and fMRI (e.g., Tillmann et al., 2003) highlight the similarities in both the time course of processing musical and linguistic syntax, as well as the presumed activated networks (including the inferior frontal gyri and "Broca's Area"). These findings have been incorporated into the shared syntactic integration resource hypothesis (SSIRH; Patel, 2003), which posits that the online processing of both music and language syntax draws upon a common syntactic processor in the brain. Although the SSIRH might initially appear to be at odds with double dissociations found between music and language – specifically, amusia without aphasia (Peretz et al., 1994) and aphasia without amusia (Luria et al., 1965), the hypothesis states that both domains are representationally distinct (i.e., have domain-specificity), and only draw upon common processing resources when integrating elements into larger syntactic structures.

One prediction that stems from the SSIRH is that syntactic processing demands in music should interfere with processing demands in language (or vice versa) – a finding that has been supported in the literature (e.g., Fedorenko et al., 2009; Slevc et al., 2009). Neuroimaging (fMRI) investigations of these interference effects have found that the processing of musical and linguistic syntax results in interactive effects in the left inferior frontal gyrus – consistent with the SSIRH and the view that the left inferior frontal gyrus is critically involved in the online processing of syntax, regardless of domain (Kunert et al., 2015). In further support of the SSIRH, individuals with Broca's aphasia were found to exhibit syntactic processing difficulties for music, with these difficulties not easily attributable to low-level deficits in pitch perception or short-term memory (Patel et al., 2008).

The SSIRH has received considerable support from converging methodologies. However, the primary challengers to this hypothesis have questioned whether the behavioral and neuroscientific results are most parsimoniously explained under a "syntactic processor" framework, as opposed to a more general cognitive framework invoking attentional and memory processes. This has been shown using "garden path" sentences (i.e., sentences that are grammatically correct, but bias an individual toward the wrong interpretation and thus require online reevaluation and reanalysis). Work supporting the SSIRH found that syntactic garden path sentences (e.g., "The old man the boat") interfere with the processing of musical syntax (Slevc et al., 2009). However, follow-up work found that this interference is not specific to syntactic processing, as *semantic* garden path sentences (e.g., "The old man went to the bank to withdraw the net") result in the same patterns of interference on musical syntax (Perruchet & Poulin-Charronnat, 2013). These results suggest that it is the nature of garden path sentences – which broadly place demands on attention and working memory to reevaluate and correctly interpret – rather than syntactic processing demands specifically, that might be most parsimoniously explaining shared processing between musical and linguistic structure. This challenge to the SSIRH can be integrated into a broader research literature suggesting that Broca's Area might be better explained in terms of broader (working) memory processes in sequence understanding (Fiebach et al., 2005; Stromswold et al., 1996). In addition, musical syntax has also been explained in terms of auditory short-term memory processes (Bigand et al., 2014). These findings do not challenge the empirical findings used in support of the SSIRH; rather, they suggest that an alternative explanation of the results might involve broader auditory memory and attentional processes over specific syntactic processing.

The relationship between musical and linguistic syntax is thus complex and actively debated. At first glance, the similarities between the two domains appear numerous (e.g., both involve the sequencing of smaller units into hierarchical structures), which has undoubtedly contributed to the notion of music as a kind of language. However, the presumed similarities between musical and linguistic syntax depends on the level of analysis which one applies to syntactic structures in both domains. The most promising approach in this area of research has been to focus on the *processing* of syntactic structures in both domains, rather than on deriving a common taxonomy for describing syntactic units. To this end, the SSIRH has become a highly influential framework for understanding syntactic processing in both music and language. Although the SSIRH has been supported in many contexts, using converging methodologies to show overlapping neural networks (e.g., inferior frontal gyrus) and behavioral interference effects in processing musical and linguistic syntax, one outstanding question in this area of research is whether these findings are best explained via broader auditory attention and memory mechanisms.

17.2.3 Semantic Features of Music and Language

Both music and language are communicative systems that convey meaning to listeners. As such, they can be considered under the framework of semantics (e.g., Crystal, 1981). In this section, we first highlight the key differences in how musical and linguistic meaning operate. We then turn to some promising approaches to assess the extent to which semantic processing might overlap in musical and linguistic domains.

Musical meaning is largely derived from its intramusical structure (e.g., Lerdahl & Jackendoff, 1983; Meyer, 1956; Narmour, 1990), which establishes (and sometimes violates) listeners' expectations in service of aesthetic or emotional meaning (Steinbeis et al., 2006). Under this framework, musical meaning is *self-referential* – that is, determined by musical structure, which blurs the distinction between musical syntax and musical meaning (e.g., Asano & Boeckx, 2015). Although music can certainly take on meaning beyond its own internal structure (for example, externally associating the repeating two note leitmotif from the "Jaws" soundtrack, composed by John Williams, with a shark or with an ominous event more broadly), these "extra-musical" associations have been treated as a more ancillary source of meaning compared to meaning generated from the musical structure itself (Meyer, 1956).

In contrast, linguistic meaning is derived largely from its ability to symbolically reference ideas outside of the internal structure of language.[1] In describing the "design features" (key properties) of language, Hockett (1958, 1960) describes how language is differentiated from other forms of communication (e.g., nonhuman animal communication) and details several elements of language that other forms of communication do not possess (although see Wacewicz & Żywiczyński, 2015 for a modern critique; also see Andrews, 2014, 2019 for a perspective from cognitive neuroscience). Of particular importance for the present chapter is the design feature of *arbitrariness* – that is, the referential units in language (e.g., spoken or written forms of the word "dog") are arbitrarily associated with the concept (in this case, the concept of a furry, four-legged animal). The arbitrariness of language is thought to be a critical feature of the generativity of linguistic meaning, as well as expressing abstract concepts (e.g., liberty), as there could not be a sound or symbol that would "naturally" map onto this construct (Lupyan & Winter, 2018).

However, the notion of *arbitrariness* as a design feature of language has been challenged for decades (e.g., see Bolinger, 1949; Jakobson, 1995). Specifically, there exist contexts in which language-meaning mappings are *iconic*, in which the symbol (e.g., the sound of a word) maps onto its referent in a nonarbitrary manner. Examples of these nonarbitrary mapping include onomatopoeia (e.g., the words *buzz* or *hiss* mimic acoustic features of their referents), as well as less obvious iconic relationships between words and their meanings, such as the association between *sl-* sounds and negative properties, such as *slime, sludge, slum,* and *slur* (Perniss et al., 2010). One of the most prominent examples of nonarbitrary mappings in language is the "maluma/takete" effect (also referred to as the "bouba/kiki" effect), in which sounds like "maluma" or "bouba" map onto rounded shapes, whereas sounds like "takete" or "kiki" map onto to spiky shapes (Köhler, 1929; Ramachandran & Hubbard, 2001). The bouba/kiki effect replicates across cultures and writing systems (Ćwiek et al., 2022). Moreover, this iconic mapping – also called *sound symbolism* (e.g., Nygaard et al., 2009) – occurs in real-world instances (i.e., is not limited to pseudowords like "bouba" or "kiki") (Sidhu et al., 2021), more frequently than would be expected by chance, and facilitates word learning in children (Monaghan et al., 2014).

These iconic mappings between linguistic sounds and their referents open up the possibility that language and music might share more fundamental ways of conveying meaning, especially given the case that spoken language and music have a number of acoustic similarities (e.g., see Section 17.2.1). The foundations for these sound symbolic mappings have been found in prelinguistic infants, which suggests that the foundation for this sound-meaning mapping might be derived from broader statistical properties of the environment or even innate synesthetic correspondences (Imai & Kita, 2014; Walker et al., 2010). Moreover, individuals have been shown to spontaneously use this analog acoustic channel of expression, independent from the propositional structure of language, to convey referential information about an object. For example, individuals modulate their speaking rate to convey information about an object's speed and modulate their fundamental frequency to convey information about an object's vertical movement (Shintel et al., 2006). Subsequent work demonstrated that underscoring of spoken sentences with either accelerating or decelerating musical motifs influenced individuals' perceptual interpretations of the sentence content (e.g., picturing a galloping horse if the music was accelerating and a resting horse if the music was decelerating), suggesting that these nonarbitrary mappings interact across domains, in a conceptually similar manner to the SSIRH but for semantic processing (Hedger et al., 2013).

How might these nonarbitrary associations between sound and meaning in both music and language be similarly represented in terms of neural activity? In EEG paradigms, semantic violations have reliably elicited a specific ERP component – the N400 – making this a promising methodology and biomarker for addressing this question. The N400 is a negative peaking electrophysio logical event peaking about 400 milliseconds after a stimulus onset and is consistently implicated in semantic – but not syntactic – violations (Kutas & Hillyard, 1980). For example, if one were presented with the written sentence "She takes her coffee with cream and floor," one word at a time, one would expect an N400 ERP following the word "floor" as it represents a semantic violation given the preceding context.

Subsequent research on the N400 has shown that it is elicited by semantic violations that extend beyond language (see Kutas & Federmeier, 2011 for a review), including the processing of pictures (Ganis et al., 1996), mathematical symbols (Niedeggen & Rösler, 1999), and nonlinguistic environmental sounds (Uddin et al., 2018). These findings, which associate the N400 with violations of meaning more broadly, suggest that music might also exhibit N400 responses. Although early work in this area did not find any evidence that violations in music elicited N400s, musical violations were marked by different ERP signatures such as the P300 (Besson & Macar, 1987). Critically, however, this work operationalized musical expectancy violations via intramusical violations (altering notes within a scale or melody), not meaning derived from sound symbolic mappings.

Koelsch and colleagues (2004) tested whether music might elicit an N400 by examining semantic processing in music more in line with the ideas of sound symbolism. In this paradigm, individuals were presented with written words (e.g., "wide"), which could be preceded by a related or an unrelated prime, in either language or music. In language, the related prime was a spoken sentence that was semantically related to the target word (e.g., "The gaze wandered into the distance"), whereas the unrelated prime was a spoken sentence incongruent with the target word (e.g., "The manacles allow only little movement"). In music, the related prime was an excerpt that was selected to reference the target word through nonarbitrary acoustic-meaning mappings (e.g., playing an excerpt from Opus 54 by Richard Strauss, which depicts wideness through instrumental spacing set in open position), whereas the unrelated prime was an excerpt incongruent with the target word (e.g., playing an excerpt from *E-minor piece for accordion* by Heikki Valpola, which depicts narrowness through closely spaced chords). The critical questions of this research were (1) does a musical prime result in an N400 if incongruent with the subsequent word, and (2) if so, how do the electrophysiological components of this ERP differ between linguistic and musical primes? The results demonstrated that, in this capacity, music elicited an N400 and moreover, the signature of the N400 (e.g., its latency, its magnitude) was statistically comparable to the N400 elicited by linguistic primes. The latter point is particularly important given that N400 amplitudes can index the strength of semantic priming and violation (Kutas & Federmeier, 2011). Although the findings of Koelsch and colleagues (2004) are important in demonstrating that music can convey sufficient specificity in referential meaning to elicit N400 responses, the findings should not be interpreted as evidence that music represents meaning in the same manner as language (e.g., see Slevc & Patel, 2011). In other words, even if music conveys meaning through sound symbolic mappings, this is inherently coarser in terms of communicative specificity compared to language, which can make use of both sound symbolic and arbitrary sound-to-meaning mappings.

Although sound symbolic mappings provide one means of comparing meaning across language and music, paralinguistic (nonlexical) cues in speech provide a particularly strong point of comparison across domains. In speech, speakers manipulate prosody – the stress, timing, and intonation of speech – to convey information to listeners. Prosody is often discussed in two realms: affective (emotional) prosody, which conveys the emotional state of the speaker (Pichon & Kell, 2013), and linguistic prosody, which provides structural disambiguation, such as differentiating a question from a statement through pitch (Bolinger, 1983). Prosody is often described as the "music" of speech (Cole, 2015), with affective prosody providing some of the clearest links between meaning in music and language, given that both signals are considered to convey emotional information.

One question that arises in comparing prosody and music is whether both systems make use of similar acoustic manipulations to convey emotional state. In other words, for a given emotional state (e.g., sadness), do the acoustic signatures in speech and music resemble one another? Answering this question is complicated by the fact that both music and language are culturally constrained; thus, the answer to the question may depend on how language and music are being defined in a particular sociocultural context. The extent to which emotional categories are universally represented or recognized across musical and linguistic systems is actively debated (e.g., Fritz et al., 2009; Paulmann & Uskul, 2014; Smit et al., 2022; see also Nettl, 1999 for an alternative account on the universality of meaning in music). Nevertheless, within a Western music framework, research has addressed whether there might be consistent acoustic cues that are used in both systems to convey emotion.

In Western music, "basic" emotions such as happiness and sadness have been associated with consistent psychoacoustic properties. Happiness is often represented in terms of major scale tonality, a relatively fast tempo, and relatively loud dynamics, whereas sadness is often represented in terms of minor scale tonality, a relatively slow tempo, and relatively soft dynamics (e.g., Juslin, 2000). In speech, these general acoustic properties also reflect happiness and sadness in emotional prosody, possibly because these cues covary with arousal (e.g., Curtis & Bharucha, 2010; Scherer, 1995), which has been used to further bolster the conjecture that both music and language evolutionarily stemmed from a protolinguistic, affective communicative system (Mithen, 2005).

Associations between emotional prosody and music perception have also been used to investigate and ultimately challenge the domain-specificity of musical and linguistic representations. As mentioned previously, a double dissociation between music and language has been reported (aphasia without amusia and amusia without aphasia). Yet, given the substantial overlap in how emotion is conveyed in both music and emotional prosody, it is possible that individuals with *amusia* might show impairments in emotional prosody perception, highlighting shared processing across domains. This was verified empirically by Thompson and colleagues (2012). In their design, participants with amusia were compared to control participants without amusia in terms of decoding semantically neutral sentences (e.g., "The broom is in the closet and the book is on the desk") spoken with six different intended emotional categories (happy, sad, tender, irritated, afraid, and no emotion). Although the two groups performed comparably for the no-emotion condition, the listeners with amusia were significantly impaired relative to the control group in identifying the other emotion categories. These findings are consistent with neuroimaging studies, which demonstrate substantial overlap in the processing of emotion in speech and music (e.g., Escoffier et al., 2013), and suggest that music and language convey emotional information via shared networks.

Comparing how music and language convey meaning is inextricably tied to how meaning is defined in both systems. In language, the arbitrary association between symbols and their referents allows for the composition of limitless and novel meanings. In contrast, musical meaning is often determined through intramusical structure, blurring the line between semantic and syntactic representations. However, there are several promising means of comparing meaning in both music and language. First, both music and language can convey meaning using sound symbolic processes, with music showing comparable electrophysiological signatures of semantic violations as language under these circumstances. Second, emotional prosody in language has clear acoustic parallels to music, and both systems appear to convey emotional meaning via similar psychoacoustic properties. These findings suggest that although semantic representations in music and language might be distinct under many circumstances, the two systems exhibit overlap in processing when meaning is inextricably tied to the properties of the sounds (i.e., when sound-to-meaning mappings are nonarbitrary).

17.3 Development of Language and Music

A common theme throughout this chapter is that language and music both draw on the auditory system in complex and sometimes overlapping ways. However, the interaction between these domains is not limited to featural parallels along acoustic, syntactic, and semantic dimensions. Rather, the cross-modal exchange between music and language is intertwined in development across the lifespan, drawing on both domain-specific mechanisms and domain-general developmental processes. In this section, we review the role of critical periods in the development of music and language (Section 17.3.1), the implicit mechanisms that underpin learning processes within these domains (Section 17.3.2), and the ways in which training in one domain can influence processing in the other domain (Sections 17.3.3 and 17.3.4). Throughout this section, we highlight the ways in which auditory experience is shaped in the brain and represented in the world through music and language.

17.3.1 Early Development: Critical Periods

The notion of a critical period for language acquisition – that is, a language is acquired most naturally and accurately within a clearly defined developmental window – is widely accepted (see Lenneberg, 1967; Penfield, 1965). The term "critical period" is often used interchangeably with the term "sensitive period" but conveys an important distinction: whereas critical period suggests that windows of acquisition are sharply defined, sensitive period suggests that humans experience gradual shifts in sensitivity to stimuli in their environment. Here, we use the term critical period for consistency, but also suggest that windows for language and music acquisition are flexible.

Several aspects of language are acquired within the first few months of life (e.g., Johnson & Newport, 1989; Jusczyck & Bertoncini, 1988; Mehler et al., 1988; see for review, Dahaene-Lambertz et al., 2008). In fact, the critical period might even begin before birth since newborn infants show preference for languages they hear *in-utero* (Mehler et al., 1988). Within the first few days of birth, infants can discriminate between languages with different rhythmic structure (Dehaene-Lambertz et al. 2006; Nazzi et al., 1998; Querleu et al. 1988), identify and categorize abstract phonemes (Dehaene-Lambertz & Pena, 2001), and by four months of age, even distinguish their native language from rhythmically similar languages (Nazzi and Ramus, 2003). However, at around four to six years old, children's proficiency in detecting such differences declines rapidly, plateauing around adulthood (Johnson & Newport, 1989,

1991; Newport, 1990; see for review: Meisel, 2013). Although the specific onset and offset times of the critical period for language acquisition continues to be debated, one view is held consistent: proficiency in a given language is strongly related to the age of exposure to that language (e.g., Newport, 1990; Johnson & Newport, 1989, 1991; Oyama, 1976).

By contrast, evidence for a critical period for the acquisition of musical abilities is less widespread. The strongest claims for a critical period in music are grounded in the phenomenon of absolute pitch (AP), which is the ability to name or produce a musical note without a reference (e.g., Deutsch, 2013). Critical periods in AP are supported in principle by several converging findings, including (1) associating the relative incidence of AP with the age at which individuals began musical instruction (Bachem, 1955), (2) training studies that demonstrate superior AP learning among children compared to adults (Crozier, 1997), and (3) pharmacological interventions, suggested to "reopen" critical periods in animal models, leading to enhanced AP learning among adults (Gervain et al., 2013).

Another line of evidence to support critical periods in AP involves the intersection of musical and linguistic processing. Specifically, one theory of AP acquisition is that early experience attending to pitch in a linguistic context (e.g., in the case of tonal languages, such as Mandarin, Cantonese, and Vietnamese) facilitates the development of AP if musical lessons are begun relatively early, when a presumed critical period is still open (Deutsch et al., 2004). For example, Deutsch and colleagues (2006) compared AP prevalence in two groups of first-year music conservatory students – those who either spoke a tonal language (Mandarin) or a nontonal language (English) – and found a greater prevalence of AP in Mandarin- compared to English-speaking students. From these findings, it seems that we might acquire aspects of music in the same way as learning a first language. In fact, a global study conducted by Liu and colleagues (2023) involving upward of 30,000 adult native speakers across nineteen tonal languages and nearly half a million native speakers of other, atonal languages found that tonal language speakers were better at discriminating melodies, but worse at beat perception relative to atonal language speakers. The authors concluded that linguistic experience might shape the perception of music. However, the notion of critical periods in AP has been recently challenged by claims that some adults can acquire AP with sufficient training (Van Hedger et al., 2019; Wong et al., 2020). These findings parallel recent claims in language research – for example, suggesting that developing a native-like accent in a second language is possible under specific conditions beyond a critical period (Dollmann et al., 2020). Overall, these findings may still be consistent with a critical period framework, in which plasticity to acquire new skills is higher within a specified period of development but may persist to some degree across the lifespan (Newport, 2006).

Such parallels between language and music might prompt us to think of the ways in which these domains draw on similar cognitive capacities and experiences early in development. Several aspects of language learning, including accent development, language production, comprehension of morphology and syntax, as well as processing speed and accuracy are facilitated within critical periods during infancy and childhood. Some evidence for this comes from second language learning. For example, Korean- and Chinese-speaking adults make significant errors in grammatical judgments of English with age. Moreover, age of exposure also impacts the way language is represented in the brain. Additionally, ERP and fMRI studies show strong left hemisphere activation for their native language in bilingual adults (Dehaene et al., 1997; Perani et al., 1996; Yetkin et al., 1996). When a second language is learned, patterns of brain activation nearly overlap with those for the first or native language. For late language learners however, neural organization is less lateralized and more variable across individuals (Dehaene et al., 1997; Kim et al., 1997; Yetkin et al., 1996). Several cross-linguistic studies show domain-specific contrasts in first and second language learning (e.g., Neville et al., 1997; Weber-Fox & Neville, 1996) providing evidence for a critical period for acquiring phonological and grammatical patterns of language learning.

A closer look at the role of speech perception in language learning shows that a language-specific pattern of listening accelerates language acquisition over the first two-and-a-half years of life. This is evidenced by a phenomenon of *perceptual narrowing* with increased exposure to the native, over other languages between six and twelve months of age - the critical period for phonetic learning (Oyama, 1976). Werker and Tees (1984) demonstrated this phenomenon of perceptual narrowing in a study investigating infants' ability to discriminate English from other (i.e., Hindi and Salish) consonants. English-speaking infants showed a reduced ability to discriminate between Hindi and Salish consonant contrasts at twelve compared to six months of age, even though native Hindi- and Salish-learning infants continued to effectively discriminate between them. Subsequent evidence from behavioral and neurological studies supports the connection between speech perception and language acquisition. Kuhl and colleagues (2005; 2006) used both behavioral and ERP measures to show that infants' ability to discriminate native versus non-native speech at seven months of age differentially predicts later language outcomes at two years of age: infants better at discriminating native speech showed faster advancements in language acquisition. These results were later extended to show discrimination between English and Spanish (Rivera-Gaxiola et al., 2005, as well as Finnish and a non-native Russian contrast (Silvén et al., 2014), among other languages.

Corresponding evidence of perceptual narrowing is also found in music, as musical experience influences children's perception of musical scales. Lynch and colleagues (1990) tested the ability to notice mistunings in melodies based on Western and non-Western (i.e., Javanese pelog) scales in English-speaking infants and adults, finding that six-month-old infants were similarly able to perceive Western and non-Western scales, whereas adults were better at perceiving Western scales. Follow-up experiments conducted by Lynch and

Eilers (1992) found that six-month-old infants were less perceptually acculturated than one-year-olds, suggesting that young infants have an equipotentiality for detecting differences between musical scales, but show a perceptual narrowing with age.

Given these similarities in learning language and music during early childhood, it might be the case that increased overall auditory experience facilitates perception. Under this view, perceptual narrowing observed during this critical period for phonetic learning is malleable. One example of this malleability comes from research examining infants' ability to discriminate lexical tones, that is, pitch patterns used to contrast word meaning in some languages. Infants learning nontonal languages find it harder to discriminate lexical tones by nine months of age than their tonal language learning counterparts (Mattock et al., 2008; Mattock & Burnham, 2006), but these differences can be shifted through additional auditory experience. For example, a study by Kuhl and colleagues (2003) showed that nine-month-old English-learning infants' ability to discriminate speech contrasts in Mandarin increased after greater exposure to a Mandarin speaker. English-learning infants exposed to Mandarin showed an ability to discriminate speech contrasts on a par with Mandarin-learning infants. Similar research using EEG measures also demonstrated neural sensitivity to speech contrasts (Conboy and Kuhl, 2011; Lytle et al., 2018). An intervention using music, rather than a tonal language, showed a similar shift in children's neural processing of non-native speech contrasts and musical rhythm (Zhao & Kuhl, 2016, 2022). Zhao and colleagues (2022) explored whether music intervention reversed the observed decline of early sensory encoding of non-native speech with gains in native linguistic experience. Seven-to-ten-month-old English-learning infants received a music intervention (i.e., increased auditory experience) and their frequency-following response (FFR) to Mandarin lexical tones was measured longitudinally. The no-intervention group showed a marked decline in FFR pitch-tracking accuracy to Mandarin lexical tones, whereas the music-intervention group showed no such decline. From these findings, it seems that language and music both influence infants' speech encoding within a presumed critical period of development.

17.3.2 Domain-General Learning Mechanisms: Statistical Learning

Newborn infants receive various types of information from their physical and social environments. To make sense of the world, these young learners must keep track of different regularities in their environment and learn complex rules that organize their environment, without prior knowledge about which information is important. One way in which infants do this is through statistical learning, a domain-general learning mechanism by which infants become sensitive to different patterns in their environment implicitly, without external feedback or reinforcement (Saffran et al., 1996; Saffran & Kirkham, 2018).

Initially, statistical learning was limited in scope, conceptualized solely as a computational process of tracking syllable patterns to support segmentation

(Saffran et al., 1996a, 1996b). However, subsequent research has shown that statistical learning is domain-general, extending beyond speech to recognizing regularities within other environmental inputs. Infants can detect statistical patterns within streams of continuous speech by the time they are eight months old.

Similar to speech, listeners are sensitive to musical input from their environments (McMullin & Saffran, 2004). Young learners discover higher-order structures to organize individual units of language (e.g., words and syntax) and music (e.g., melodic and harmonic sequences) in an implicit manner (Fiser & Aslin, 2001; Tillmannet al., 2000). Not only can infants detect regularities in speech, but they also can do the same for musical tones by the time they are eight months old (Saffran et al., 1999). Eventually, listeners can implicitly learn several musical features, including timbre and melodic pitch relationships, from their listening environments (see Ettlinger et al., 2011; Rohrmeier & Rebuschat, 2012). Under the statistical learning hypothesis, the brain not only calculates and organizes the transitional probabilities (TPs) of sequential phenomena (words, tones, etc.) from the environment, but also updates this acquired statistical knowledge in line with new environmental inputs through the lifespan (Daikoku et al., 2017). Differences in brain responses as a result of statistical learning may then depend on the previously acquired knowledge (e.g., Furl et al., 2011; Yumoto et al., 2005).

Evidence for experience and expectancy-related differences in statistical learning can be found in both language and music. In language, continuous speech has both universal regularities and language-specific statistical patterns. For example, adjacent units (syllables) in language have a high transitional probability of co-occurrence across languages (Saffran et al., 1996a, 1996b; Swingley, 2005) that can be implicitly segmented are found across languages, and therefore be leveraged by infant and adult language learners. By contrast, linguistic properties like lexical stress are language-specific, and therefore, not readily available for all language learners (e.g., Sohail & Johnson, 2016). Adult learners might bootstrap language-specific patterns (lexical stress, phonotactics, etc.) onto universal elements such as TPs to identify word boundaries in continuous speech (e.g., Benitez & Saffran, 2021). For example, English-speaking adults given brief, unguided exposure to a foreign language (Italian) in the form of podcasts demonstrate initial word-form learning, compared to adults with no exposure to the foreign language (Alexander et al., 2022), which suggests that second-language learners might be sensitive to the characteristics of different word features (TPs, phonetic patterns, etc.) of a novel language after some exposure to that language. In music, some evidence of expectancy-related learning comes from research conducted with musically trained individuals. For example, adult musicians perform better on statistical learning tasks and are generally better at learning the implicit statistical structures of sung language compared to nonmusicians (François and Schön, 2011, 2014; Vasuki et al., 2016). In an ERP study, musicians showed a larger N1 familiarity effect compared to their nonmusician counterparts, which suggested that they are better at recognizing statistical patterns. Moreover, ERPs recorded during the familiarization of a sung language showed that the N400 learning curve saturates earlier in musicians, which suggested that they are also faster at detecting statistical regularities in sung language compared to their nonmusician counterparts (François and Schön, 2014). Similarly, a longitudinal study conducted with eight-year-old children found that children learning music were better at identifying statistical regularities in auditory streams compared to children who were not formally trained in music (François et al., 2014). Moreover, subsequent research by Vasuki and colleagues (2017) showed that musically trained children were better at melody discrimination, rhythm discrimination, frequency discrimination, in addition to auditory statistical learning, compared to their nontrained counterparts; and the triplet onset stimulus used in their statistical learning paradigm elicited larger responses in musically trained compared to untrained children.

Into adulthood, both language and music demonstrate effects of *entrenchment*, wherein the efficacy of learning the statistical regularities of novel language and music sequences depends on prior representations developed within both domains. In language, variance in performance on a novel statistical learning task could be partly explained by an individual's previous language background – specifically, the extent to which the novel sequences aligned with the individual's native language (Siegelman et al., 2018). In music, tonal sequences that did not adhere to a conventional tuning standard were judged to sound "correct" if they incidentally contained a greater proportion of common intervals found in Western music (perfect fourths and fifths), even though the presence of these intervals was not indicative of the structure of the novel sequences (Van Hedger et al., 2022). These parallels across language and music suggest that as listeners develop increasingly rich representations in both domains, subsequent implicit learning efficacy interacts with these previously developed representations.

Language and music both rely at least partially on statistical learning, a domain-general mechanism for learning and processing novel information. Learners might be using this mechanism for acquiring and maintaining properties of both language (Peretz et al., 2012; Romberg & Saffran, 2010; Saffran, 2001; Saffran, 2003) as well as music (Furl et al., 2011; Saffran, 2003; Saffran et al., 1999). The literature on statistical learning shows us that language and music are not simply parallel domains with surface level similarities, but rather, draw on shared domain-general processes that shape auditory experience, with developed representations influencing the efficacy of statistical learning into adulthood.

17.3.3 Training and Transfer between Domains

So far, we have suggested language and music are distinct modalities, comparable not just by surface-level features, but also related by a common auditory sensory experience and maintained by shared cognitive learning mechanisms. We can now turn our attention to the ways in which the modalities interact with each other in cross-modal exchange. One view holds that the prevalence of higher-order structural similarities indicates that music draws from language in unequal measure (Pinker, 1997) or perhaps, even that language is a product of musical experience (Darwin, 2007/1874). Another view challenges the idea that these domains are linked at a structural level (e.g., Federenko, Behr, et al. 2011; Federenko, McDermott, et al., 2012). While music and language have differentiable representations at the neural level, overlap in the sensory and cognitive systems employed by these modalities allows room for the possibility that experience and training in one domain might facilitate performance in another domain.

Evidence for training and transfer effects from language to music come from research conducted with linguistically and musically trained samples. In one study, Bidelman and colleagues (2013) compared adult Cantonese-speaking nonmusicians to English-speaking trained musicians and English-speaking controls on pitch discrimination and pitch memory tasks. They found that Cantonese-speaking participants outperformed English-speaking controls on all aspects of pitch perception and discrimination. By contrast, Cantonese speakers performed similarly to their English-speaking musically trained counterparts, suggesting that pitch expertise, arising from either musical training or tonal language experience, is associated with lower sensitivity to pitch discrimination, higher tonal memory and melodic discrimination. Moreover, the authors found a significant relationship between behavioral measures of pitch ability and length of exposure to the tonal language or musical training. In another study, Elmer and colleagues (2011) conducted an event-related fMRI study to examine whether long-term language training facilitates the discrimination of nonverbal stimulus attributes in a nonverbal auditory discrimination task, and addressed whether functional transfer effects might support behavioral performance by comparing professional simultaneous interpreters to controls. Their results suggested that discrimination of target stimuli was associated with differential BOLD responses in the frontoparietal regions of the two groups, even though behavioral results showed no significant effects. Language training seemed to modulate brain activity in regions associated with top-down regulation of auditory functions, such as auditory attention and categorization. These findings are in line with the idea that overall auditory experience plays a crucial role in language processing.

Parallel evidence for music to language transfer comes from different levels of processing: multiple studies have examined acoustic features (e.g., Chartrand & Belin, 2006; Bidelman & Krishnan, 2010; Weiss & Bidelman, 2015), cognitive factors (e.g., Anvari et al., 2002; Moreno et al., 2009), and executive function (e.g., Bialystok & DePape, 2009). Moreover, the impact of musical training on language acquisition and development has been explored in the context of neural encoding of speech in noise (Strait et al., 2009; Strait & Kraus, 2011), phonological awareness (e.g., Tsang & Conrad, 2011), as well as vocabulary and reading comprehension (e.g., Corrigal & Trainor, 2011;

Moreno, Bialystok et al., 2011). Musical training studies corresponding to the ones discussed above reliably show that musical training facilitates linguistic performance. For example, Magne and colleagues (2006) found that musically trained children detected incongruities in music and language better than their musically untrained counterparts. Musically trained children showed early negative components in music and late positive components in language, whereas no components were present in their untrained counterparts. These results are in line with the finding that musically trained individuals show enhanced neural encoding of speech sounds in the auditory brainstem compared to their untrained counterparts (Kraus & Chanrasekaran, 2010).

Similar to language-to-music transfer, enhanced performance in language is also correlated with length and degree of exposure to musical training (e.g., Besson et al., 2011; Musacchia et al., 2007; Strait et al., 2009). Although these studies relied on expert populations, longitudinal training experiments have yielded similar results. For example, Moreno and colleagues (2009, 2011a, 2011b) conducted a series of longitudinal studies to explore whether musical training facilitated language processing in young children. Eight-year-old children given musical training showed improvements in EEG correlates of speech-related pitch processing after receiving six months of training, compared to matched controls. Behaviorally, four-to-six-year-old children showed similar enhancements in verbal intelligence after participating in an intensified version of the musical training program conducted over twenty days (Moreno, Bialystok et al., 2011). These findings illustrate that a degree of plasticity may be acquired through exposure to auditory activities in the form of language or music.

17.3.4 Mechanisms Underlying Transfer: Insights from Cognitive Neuroscience

A central question emerging from examining the developmental learning trajectories of music and language, including how training in one domain can transfer to the other domain, is: *what exactly is being learned*? Understanding the perceptual, cognitive, and neural processes by which the brain generalizes and transfers information from one domain (e.g., music) to another (e.g., language) requires us to conceptualize the two domains as relying on shared auditory and attentional systems. Moreover, given recent evidence that lifelong musicianship is associated with preserved white matter tract integrity in individuals, which may promote healthy aging (Andrews et al., 2021), understanding the proposed mechanisms of transfer has implications for designing interventions to improve psychological and neurobiological health across the lifespan.

One explanation for the cross-modal interaction between language and music comes from the OPERA hypothesis (Patel, 2011), which states that, neural areas recruited in music and language overlap (O) when partaking in a learning process that involves precision (P), emotional engagement (E), repetition (R), and attentional focus (A). According to this theory, cross-modal transfer occurs because largely overlapping cortical and subcortical brain regions are recruited when the acoustic features of speech and music are encoded. Increased precision of neural processing through the components of this model (i.e., emotional engagement, repetition, and attention) make the acoustic information more salient to the listener. Cross-modal transfer is possible when acoustic features of music or language are encoded with a high degree of precision with increased exposure (e.g., tonal language) and training (e.g., formal musical training). Training allows learners to give their attention to the acoustic features of music and language, and this enhanced attention in turn mediates the overlapping neural networks involved in music and language processing.

Building on the OPERA hypothesis, Moreno and Bidelman (2014) proposed a different model of cross-modal transfer. Under this view, the neural networks determining the transfer of auditory information across music and language are affected by two dimensions. First, is a sensory-cognitive dimension that distinguishes low-level auditory-sensory information from high-level domain-general processes that support language. Whereas some training might lead to experience-dependent plasticity, other training might impact cortical plasticity, leading to overarching changes in language processing and executive function. A second dimension along which training and transfer occurs is the near-far dimension. A near-transfer involves seeing the development of more precise listening skills (relative pitch discrimination, improvements in pattern detection, etc.) because of repeated exposure to auditory patterns. By contrast, a far-transfer occurs when the music training benefits auditory sensory encoding in speech production and language acquisition. The benefit of musical training is not just a product of length and intensity of training, but also the degree to which the training targets higher-order cognitive skills.

Both these accounts shed some light on the ways in which underlying neural and cognitive mechanisms underlying music and language might be facilitating a cross-modal transfer between these domains. The developmental trajectory of these cognitive domains, discussed in earlier parts of the chapter, also lends insights into the possibilities and restraints on transfers between them (cf. discussion of sensory periods: White et al., 2013). Some skills (AP, detecting unfamiliar sounds, etc.) emerge early, and may be restricted by a developmental timeline. However, it also seems true that both abilities interact beyond these critical periods, and might contribute to learning and development (e.g., second language acquisition, musical pattern detection) within the other domain.

A third account of transfer across domains is that training plays a relatively small role, with preexisting differences accounting for most of the shared variance in language and musical abilities (e.g., Swaminathan & Schellenberg, 2019; see also Figure 17.1). For example, Swaminathan and Schellenberg (2020) found that speech abilities were positively correlated with musical abilities and IQ among six-to-nine-year-old children; however, musical training

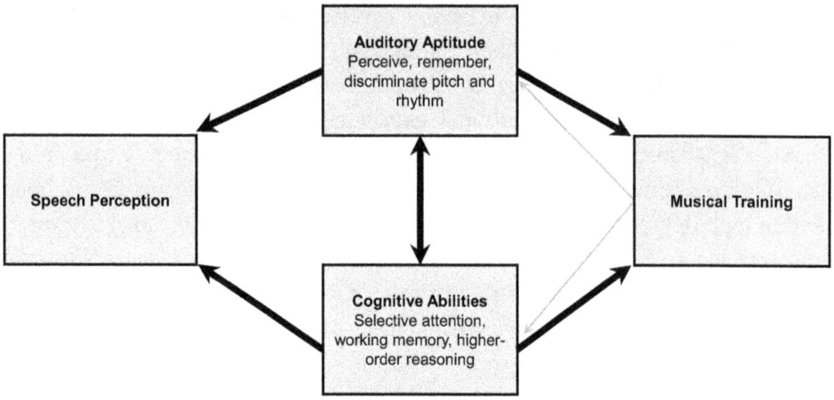

Figure 17.1 Potential paths explaining observed associations between musical training and spoken language perception.

Note: The width and shading of the arrows represent pathway strength, with thicker and darker arrows representing stronger connections. The figure outlines two primary ways through which musical training could be associated with speech perception. The first, non-causal path is due to preexisting auditory and cognitive abilities (middle), which then influence an individual's tendency to continue with musical training and also relate to an individual's spoken language abilities. The second, causal path shows how musical training could improve both auditory and cognitive abilities, which then in turn could improve speech perception.

Figure modified and expanded upon from Swaminathan and Schellenberg (2019).

was not a significant predictor of language abilities. Among adults, recent work has reported cases of "musical sleepers" – individuals without formal musical training who nevertheless demonstrate perceptual acuity and electrophysio logical signatures of sound encoding (e.g., FFRs) that are comparable to individuals with extensive musical training (Mankel & Bidelman, 2018). In line with the preexisting difference account, research has also demonstrated that personality factors (e.g., openness) are associated in both children and adults with the likelihood of taking music lessons and persisting with training (Corrigall et al., 2013). Although these findings do not inherently reject the notion that training in one domain may transfer to another domain, they strongly caution against the use of quasi-experimental designs (e.g., comparing trained musicians to nonmusicians) in making causal claims about the nature of training and transfer.

Through a discussion of (1) the developmental constraints on acquiring music and language, (2) the domain-general learning mechanisms involved in learning music and language, and (3) the mechanisms underlying cross-modal transfer between music and language in this section, we suggest that learning in these domains is both an implicit, bottom-up process that relies on auditory-sensory input to engage overlapping neural networks, and an explicit, top-down process that requires precise encoding of information to facilitate cross-modal transfer between the two domains. To better understand the underlying mechanisms that allow for training and transfer, longitudinal studies with random assignment to condition are critical, given the reports that preexisting auditory processing and personality factors tend to covary with musical training.

17.4 Conclusion

Comparisons between music and language have been the focus of research in psychology and neuroscience for several decades. Throughout this chapter, we demonstrate how music and language overlap in terms of psychoacoustic properties, recruit shared neural networks in processing syntactically organized sequences, and can use nonarbitrary cues to convey meaning. Given these surface-level similarities between language and music, it seems tempting, and even practical to think of the two domains as related and parallel domains. However, applying too fine-grained a level of analysis, such as focusing solely on featural cues, restricts the way we think about the interactions between these domains. For example, language and music are organized hierarchically at multiple levels of analysis, ranging from the phonological to morphological, and sentential level in language, and from harmonic to rhythmic levels in music. Seeing these solely as hierarchical processes limits us from comparing the units (morphology, rhythm, etc.) in a meaningful way. Conversely, looking for functional equivalence between the two domains might lend itself to miscalibrating the true underlying purpose and function of these skills. For example, meaning is conveyed differently in both music and language: it differs not just in transparency, but also content. It is difficult to compare, and therefore, can be dangerous to equate across modalities. Music and language are not just parallel domains that share surface-level features and should be treated and compared at an appropriate level of analysis.

Despite these pitfalls that inevitably arise from featural comparisons of music and language, studying the cognitive and perceptual processes underlying these domains can provide insight into the ways in which music and language interact and draw on each other. Both modalities are maintained by domain-general learning mechanisms, with similar constraints on developmental trajectories of learning (e.g., critical periods) and shared neurocognitive systems for supporting implicit learning via statistical regularities in the environment. Moreover, training in one domain often leads to enhanced skills in the other domain, showing evidence for a complex training and transfer mechanism. Applying an appropriate level of analysis demonstrates that music and language interact in sophisticated ways and can provide insight into shared (domain-general) versus distinct (domain-specific) neural and cognitive mechanisms.

Notes

1 Several related accounts have proposed alternative and complementary sources of linguistic meaning, such as sensorimotor systems (e.g., Gallese & Lakoff,

2005), embodied perception and action (e.g., Gibbs, 2003), metaphoric images and schemas (e.g., Glucksberg & Keysar, 1990; Murphy, 1996), as well as nonreferential sources of musical meaning including embodiment (e.g., Cox, 2001), and culture (e.g., Shepard & Wicke, 1997). For the purposes of our discussion, however, we are focusing on the similarities and differences in language and music from the human ability to derive meaning by mapping sounds to words and words to their referents (Section 17.2.1) and bootstrapping our hypotheses about the meanings of words, sentences, and musical sequences from syntax (Section 17.2.2).

References

Adger, D. (2015). Syntax: Syntax. *Wiley Interdisciplinary Reviews: Cognitive Science*, *6*(2), 131–147. https://doi.org/10.1002/wcs.1332

Albouy, P., Benjamin, L., Morillon, B., & Zatorre, R. J. (2020). Distinct sensitivity to spectrotemporal modulation supports brain asymmetry for speech and melody. *Science*, *367*(6481), 1043–1047. https://doi.org/10.1126/science.aaz3468

Alexander, E., Van Hedger, S. C., & Batterink, L. J. (2023). Learning words without trying: Daily second language podcasts support word-form learning in adults. *Psychonomic Bulletin & Review*, 30(2), 751–762. https://doi.org/10.3758/s13423-022-02190-1

Andrews, E. (2014). *Neuroscience and Multilingualism*. Cambridge University Press.

Andrews, E. (2019). Cognitive neuroscience and multilingualism. In J. W. Schwieter & M. Paradis (Eds.), *The Handbook of the Neuroscience of Multilingualism* (1st ed., pp. 19–47). Wiley. https://doi.org/10.1002/9781119387725.ch2

Andrews, E., Eierud, C., Banks, D., Harshbarger, T., Michael, A., & Rammell, C. (2021). Effects of lifelong musicianship on white matter integrity and cognitive brain reserve. *Brain Sciences*, *11*(1), 67. https://doi.org/10.3390/brainsci11010067

Anvari, S. H., Trainor, L. J., Woodside, J., & Levy, B. A. (2002). Relations among musical skills, phonological processing, and early reading ability in preschool children. *Journal of Experimental Child Psychology*, 83(2), 111–130. https://doi.org/10.1016/S0022-0965(02)00124-8

Asano, R., & Boeckx, C. (2015). Syntax in language and music: What is the right level of comparison? *Frontiers in Psychology*, *6*. https://doi.org/10.3389/fpsyg.2015.00942

Bachem, A. (1955). Absolute pitch. *The Journal of the Acoustical Society of America*, *27*(6), 1180–1185. https://doi.org/10.1121/1.1908155

Benitez, V. L., & Saffran, J. R. (2021). Two for the price of one: Concurrent learning of words and phonotactic regularities from continuous speech. *PLoS ONE*, 16(6), e0253039. https://doi.org/10.1371/journal.pone.0253039

Besson, M., Chobert, J., & Marie, C. (2011). Transfer of training between music and speech: Common processing, attention, and memory. *Frontiers in Psychology*, 2, 94. https://doi.org/10.3389/fpsyg.2011.00094

Besson, M., & Macar, F. (1987). An event-related potential analysis of incongruity in music and other non-linguistic contexts. *Psychophysiology, 24*(1), 14–25. https://doi.org/10.1111/j.1469-8986.1987.tb01853.x

Bialystok, E., & DePape, A. M. (2009). Musical expertise, bilingualism, and executive functioning. *Journal of Experimental Psychology: Human Perception and Performance, 35*(2), 565. https://doi.org/10.1037/a0012735

Bidelman, G. M., Hutka, S., & Moreno, S. (2013). Tone language speakers and musicians share enhanced perceptual and cognitive abilities for musical pitch: Evidence for bidirectionality between the domains of language and music. *PLoS ONE,* 8(4), e60676. https://doi.org/10.1371/journal.pone.0060676

Bidelman, G. M., & Krishnan, A. (2010). Effects of reverberation on brainstem representation of speech in musicians and non-musicians. *Brain Research, 1355,* 112–125. https://doi.org/10.1016/j.brainres.2010.07.100

Bidelman, G. M., & Yoo, J. (2020). Musicians show improved speech segregation in competitive, multi-talker cocktail party scenarios. *Frontiers in Psychology, 11.* www.frontiersin.org/articles/10.3389/fpsyg.2020.01927

Bigand, E., Delbé, C., Poulin-Charronnat, B., Leman, M., & Tillmann, B. (2014). Empirical evidence for musical syntax processing? Computer simulations reveal the contribution of auditory short-term memory. *Frontiers in Systems Neuroscience, 8.* https://doi.org/10.3389/fnsys.2014.00094

Bolinger, D. (1949). The sign is not arbitrary. *Boletín del Instituto Caro y Cuervo (= Thesaurus), 5,* 52–62.

Bolinger, D. (1983). Intonation and gesture. *American Speech, 58*(2), 156. https://doi.org/10.2307/455326

Brown, R. (1957). Linguistic determinism and the part of speech. *The Journal of Abnormal and Social Psychology, 55*(1), 1–5. https://doi.org/10.1037/h0041199

Brown, R. (1973). *A First Language: The Early Stages.* Harvard University Press.

Brown, S. (2017). A joint prosodic origin of language and music. *Frontiers in Psychology, 8,* 1894. https://doi.org/10.3389/fpsyg.2017.01894

Chartrand, J. P., & Belin, P. (2006). Superior voice timbre processing in musicians. *Neuroscience Letters,* 405(3), 164–167. https://doi.org/10.1016/j.neulet.2006.06.053

Cole, J. (2015). Prosody in context: A review. *Language, Cognition and Neuroscience, 30*(1–2), 1–31. https://doi.org/10.1080/23273798.2014.963130

Conboy, B. T., & Kuhl, P. K. (2011). Impact of second-language experience in infancy: Brain measures of first-and second-language speech perception. *Developmental Science,* 14(2), 242–248. https://doi.org/10.1111/j.1467-7687.2010.00973.x

Corrigall, K. A., Schellenberg, E. G., & Misura, N. M. (2013). Music training, cognition, and personality. *Frontiers in Psychology, 4.* https://doi.org/10.3389/fpsyg.2013.00222

Corrigall, K. A., & Trainor, L. J. (2011). Associations between length of music training and reading skills in children. *Music Perception,* 29(2), 147–155. https://doi.org/10.1525/mp.2011.29.2.147

Cox, A. (2001). The mimetic hypothesis and embodied musical meaning. *Musicae Scientiae, 5*(2), 195–212. https://doi.org/10.1177/102986490100500204

Cross, I. (2001). Music, cognition, culture, and evolution. *Annals of the New York Academy of Sciences, 930*(1), 28–42. https://doi.org/10.1111/j.1749-6632.2001.tb05723.x

Crozier, J. B. (1997). Absolute pitch: Practice makes perfect, the earlier the better. *Psychology of Music, 25*(2), 110–119. https://doi.org/10.1177/0305735697252002

Crystal, D. (1981). Semantics. In G. E. Arnold, F. Winckel, & B. D. Wyke (Eds.), *Clinical Linguistics* (Vol. 3, pp. 131–191). Springer Vienna. https://doi.org/10.1007/978-3-7091-4001-7_5

Curtis, M. E., & Bharucha, J. J. (2010). The minor third communicates sadness in speech, mirroring its use in music. *Emotion, 10*(3), 335–348. https://doi.org/10.1037/a0017928

Ćwiek, A., Fuchs, S., Draxler, C., Asu, E. L., Dediu, D., Hiovain, K., ... & Winter, B. (2022). The bouba/kiki effect is robust across cultures and writing systems. *Philosophical Transactions of the Royal Society B: Biological Sciences, 377*(1841), 20200390. https://doi.org/10.1098/rstb.2020.0390

Darwin, C. (2007/1874). *The Descent of Man: The Concise Edition.* Penguin.

Dehaene, S., Dupoux, E., Mehler, J., Cohen, L., Paulesu, E., Perani, D., ... & Le Bihan, D. (1997). Anatomical variability in the cortical representation of first and second language. *Neuroreport*, 8(17), 3809–3815.

Dehaene-Lambertz, G., Hertz-Pannier, L., & Dubois, J. (2006). Nature and nurture in language acquisition: Anatomical and functional brain-imaging studies in infants. *Trends in Neurosciences*, 29(7), 367–373. https://doi.org/10.1016/j.tins.2006.05.011

Dehaene-Lambertz, G., Hertz-Pannier, L., Dubois, J., & Dehaene, S. (2008). How does early brain organization promote language acquisition in humans? *European Review*, 16(4), 399–411. doi:10.1017/S1062798708000513

Dehaene-Lambertz, G., & Pena, M. (2001). Electrophysiological evidence for automatic phonetic processing in neonates. *Neuroreport*, 12(14), 3155–3158.

Deutsch, D. (2013). Absolute pitch. In *The Psychology of Music* (pp. 141–182). Elsevier. https://doi.org/10.1016/B978-0-12-381460-9.00005-5

Deutsch, D., Henthorn, T., & Dolson, M. (2004). Absolute pitch, speech, and tone language: Some experiments and a proposed framework. *Music Perception, 21*(3), 339–356. https://doi.org/10.1525/mp.2004.21.3.339

Deutsch, D., Henthorn, T., Marvin, E., & Xu, H. (2006). Absolute pitch among American and Chinese conservatory students: Prevalence differences, and evidence for a speech-related critical period. *The Journal of the Acoustical Society of America*, 119(2), 719–722. https://doi.org/10.1121/1.2151799

Deutsch, D., Lapidis, R., & Henthorn, T. (2008). The speech-to-song illusion. *The Journal of the Acoustical Society of America, 124*(4), 2471–2471. https://doi.org/10.1121/1.4808987

Diedrich, C. G. (2015). "Neanderthal bone flutes": Simply products of Ice Age spotted hyena scavenging activities on cave bear cubs in European cave bear dens. *Royal Society Open Science, 2*(4), 140022. https://doi.org/10.1098/rsos.140022

Ding, N., Patel, A. D., Chen, L., Butler, H., Luo, C., & Poeppel, D. (2017). Temporal modulations in speech and music. *Neuroscience & Biobehavioral Reviews, 81*, 181–187. https://doi.org/10.1016/j.neubiorev.2017.02.011

Dollmann, J., Kogan, I., & Wei?mann, M. (2020). Speaking accent-free in L2 beyond the critical period: The compensatory role of individual abilities and opportunity structures. *Applied Linguistics, 41*(5), 787–809. https://doi.org/10.1093/applin/amz029

Elmer, S., Meyer, M., Marrama, L., & Jäncke, L. (2011). Intensive language training and attention modulate the involvement of fronto-parietal regions during a non-verbal auditory discrimination task. *European Journal of Neuroscience, 34*(1), 165–175. https://doi.org/10.1111/j.1460-9568.2011.07728.x

Embick, D., Marantz, A., Miyashita, Y., O'Neil, W., & Sakai, K. L. (2000). A syntactic specialization for Broca's area. *Proceedings of the National Academy of Sciences, 97*(11), 6150–6154. https://doi.org/10.1073/pnas.100098897

Escoffier, N., Zhong, J., Schirmer, A., & Qiu, A. (2013). Emotional expressions in voice and music: Same code, same effect? *Human Brain Mapping, 34*(8), 1796–1810. https://doi.org/10.1002/hbm.22029

Falk, S., Rathcke, T., & Dalla Bella, S. (2014). When speech sounds like music. *Journal of Experimental Psychology: Human Perception and Performance, 40*(4), 1491–1506. https://doi.org/10.1037/a0036858

Fedorenko, E., Behr, M. K., & Kanwisher, N. (2011). Functional specificity for high-level linguistic processing in the human brain. *Proceedings of the National Academy of Sciences*, 108(39), 16428–16433. https://doi.org/10.1073/pnas.1112937108

Fedorenko, E., McDermott, J. H., Norman-Haignere, S., & Kanwisher, N. (2012). Sensitivity to musical structure in the human brain. *Journal of Neurophysiology*, 108(12), 3289–3300. https://doi.org/10.1152/jn.00209.2012

Fedorenko, E., Patel, A., Casasanto, D., Winawer, J., & Gibson, E. (2009). Structural integration in language and music: Evidence for a shared system. *Memory & Cognition, 37*(1), 1–9. https://doi.org/10.3758/MC.37.1.1

Fiebach, C. J., Schlesewsky, M., Lohmann, G., von Cramon, D. Y., & Friederici, A. D. (2005). Revisiting the role of Broca's area in sentence processing: Syntactic integration versus syntactic working memory. *Human Brain Mapping, 24*(2), 79–91. https://doi.org/10.1002/hbm.20070

Fiser, J., & Aslin, R. N. (2001). Unsupervised statistical learning of higher-order spatial structures from visual scenes. *Psychological Science*, 12(6), 499–504. https://doi.org/10.1111/1467-9280.003

François, C., Jaillet, F., Takerkart, S., & Schön, D. (2014). Faster sound stream segmentation in musicians than in nonmusicians. *PLoS ONE*, 9(7), e101340. https://doi.org/10.1371/journal.pone.0101340

François, C., & Schön, D. (2014). Neural sensitivity to statistical regularities as a fundamental biological process that underlies auditory learning: The role of musical practice. *Hearing Research*, 308, 122–128. https://doi.org/10.1016/j.heares.2013.08.018

Francois, C., & Schön, D. (2011). Musical expertise boosts implicit learning of both musical and linguistic structures. *Cerebral Cortex*, 21(10), 2357–2365. https://doi.org/10.1093/cercor/bhr022

Friederici, A. D. (2003). The role of left inferior frontal and superior temporal cortex in sentence comprehension: Localizing syntactic and semantic processes. *Cerebral Cortex, 13*(2), 170–177. https://doi.org/10.1093/cercor/13.2.170

Fritz, T., Jentschke, S., Gosselin, N., Sammler, D., Peretz, I., Turner, R., Friederici, A. D., & Koelsch, S. (2009). Universal recognition of three basic emotions in music. *Current Biology, 19*(7), 573–576. https://doi.org/10.1016/j.cub.2009.02.058

Furl, N., Kumar, S., Alter, K., Durrant, S., Shawe-Taylor, J., & Griffiths, T. D. (2011). Neural prediction of higher-order auditory sequence statistics. *Neuroimage*, 54(3), 2267–2277. https://doi.org/10.1016/j.neuroimage.2010.10.038

Gallese, V., & Lakoff, G. (2005). The brain's concepts: The role of the sensory-motor system in conceptual knowledge. *Cognitive Neuropsychology, 22*(3–4), 455–479. https://doi.org/10.1080/02643290442000310

Ganis, G., Kutas, M., & Sereno, M. I. (1996). The search for "common sense": An electrophysiological study of the comprehension of words and pictures in reading. *Journal of Cognitive Neuroscience, 8*(2), 89–106. https://doi.org/10.1162/jocn.1996.8.2.89

Gervain, J., Vines, B. W., Chen, L. M., Seo, R. J., Hensch, T. K., Werker, J. F., & Young, A. H. (2013). Valproate reopens critical-period learning of absolute pitch. *Frontiers in Systems Neuroscience, 7.* https://doi.org/10.3389/fnsys.2013.00102

Geschwind, N. (1970). The organization of language and the brain: Language disorders after brain damage help in elucidating the neural basis of verbal behavior. *Science, 170*(3961), 940–944. https://doi.org/10.1126/science.170.3961.940

Gibbs, R. W. (2003). Embodied experience and linguistic meaning. *Brain and Language, 84*(1), 1–15. https://doi.org/10.1016/S0093-934X(02)00517-5

Gleitman, L. (1990). The structural sources of verb meanings. *Language Acquisition, 1*(1), 3–55.

Glucksberg, S., & Keysar, B. (1990). Understanding metaphorical comparisons: Beyond similarity. *Psychological Review, 97*(1), 3–18. https://doi.org/10.1037/0033-295X.97.1.3

Grodzinsky, Y., & Friederici, A. D. (2006). Neuroimaging of syntax and syntactic processing. *Current Opinion in Neurobiology, 16*(2), 240–246. https://doi.org/10.1016/j.conb.2006.03.007

Hagoort, P. (2014). Nodes and networks in the neural architecture for language: Broca's region and beyond. *Current Opinion in Neurobiology, 28,* 136–141. https://doi.org/10.1016/j.conb.2014.07.013

Hedger, S. C., Nusbaum, H. C., & Hoeckner, B. (2013). Conveying movement in music and prosody. *PLoS ONE, 8*(10), e76744. https://doi.org/10.1371/journal.pone.0076744

Higham, T., Basell, L., Jacobi, R., Wood, R., Ramsey, C. B., & Conard, N. J. (2012). Testing models for the beginnings of the Aurignacian and the advent of figurative art and music: The radiocarbon chronology of Gei?enklösterle. *Journal of Human Evolution, 62*(6), 664–676. https://doi.org/10.1016/j.jhevol.2012.03.003

Hockett, C. F. (1958). A course in modern linguistics. *Language Learning, 8*(3–4), 73–75. https://doi.org/10.1111/j.1467-1770.1958.tb00870.x

Hockett, C. F. (1960). The origin of speech. *Scientific American, 203*(3), 88–96. https://doi.org/10.1038/scientificamerican0960–88

Honing, H., ten Cate, C., Peretz, I., & Trehub, S. E. (2015). Without it no music: Cognition, biology and evolution of musicality. *Philosophical Transactions of the Royal Society B: Biological Sciences, 370*(1664), 20140088. https://doi.org/10.1098/rstb.2014.0088

Humphries, C., Liebenthal, E., & Binder, J. R. (2010). Tonotopic organization of human auditory cortex. *NeuroImage, 50*(3), 1202–1211. https://doi.org/10.1016/j.neuroimage.2010.01.046

Imai, M., & Kita, S. (2014). The sound symbolism bootstrapping hypothesis for language acquisition and language evolution. *Philosophical Transactions of the Royal Society B: Biological Sciences, 369*(1651), 20130298. https://doi.org/10.1098/rstb.2013.0298

Jakobson, R. (1965). Quest for the essence of language. *Diogenes*, 13(51), 21–37. https://doi.org/10.1177/039219216501305103

Jakobson, R. (1995). *On Language* (ed. L. R. Waugh). Harvard University Press.

Johnson, J. S., & Newport, E. L. (1989). Critical period effects in second language learning: The influence of maturational state on the acquisition of English as a second language. *Cognitive Psychology*, 21(1), 60–99. https://doi.org/10.1016/0010-0285(89)90003-0

Jusczyk, P. W., & Bertoncini, J. (1988). Viewing the development of speech perception as an innately guided learning process. *Language and Speech*, 31(3), 217–238. https://doi.org/10.1177/002383098803100301

Juslin, P. N. (2000). Cue utilization in communication of emotion in music performance: Relating performance to perception. *Journal of Experimental Psychology: Human Perception and Performance, 26*(6), 1797–1812. https://doi.org/10.1037/0096-1523.26.6.1797

Kim, K. H., Relkin, N. R., Lee, K. M., & Hirsch, J. (1997). Distinct cortical areas associated with native and second languages. *Nature*, 388(6638), 171–174. https://doi.org/10.1038/40623

Koelsch, S., Kasper, E., Sammler, D., Schulze, K., Gunter, T., & Friederici, A. D. (2004). Music, language and meaning: Brain signatures of semantic processing. *Nature Neuroscience, 7*(3), 302–307. https://doi.org/10.1038/nn1197

Köhler, W. (1929). *Gestalt Psychology: An Introduction to New Concepts in Modern Psychology*. Liveright.

Kraus, N., & Chandrasekaran, B. (2010). Music training for the development of auditory skills. *Nature Reviews Neuroscience*, 11(8), 599–605. https://doi.org/10.1038/nrn2882

Kuhl, P. K., Conboy, B. T., Padden, D., Nelson, T., & Pruitt, J. (2005). Early speech perception and later language development: Implications for the "critical period". *Language Learning and Development*, 1(3–4), 237–264. https://doi.org/10.1080/15475441.2005.9671948

Kuhl, P. K., Stevens, E., Hayashi, A., Deguchi, T., Kiritani, S., & Iverson, P. (2006). Infants show a facilitation effect for native language phonetic perception between 6 and 12 months. *Developmental Science*, 9(2), F13–F21. https://doi.org/10.1111/j.1467-7687.2006.00468.x

Kuhl, P. K., Tsao, F. M., & Liu, H. M. (2003). Foreign-language experience in infancy: Effects of short-term exposure and social interaction on phonetic learning. *Proceedings of the National Academy of Sciences*, 100(15), 9096–9101. https://doi.org/10.1073/pnas.1532872100

Kunert, R., Willems, R. M., Casasanto, D., Patel, A. D., & Hagoort, P. (2015). Music and language syntax interact in Broca's Area: An fMRI study. *PLoS ONE, 10*(11), e0141069. https://doi.org/10.1371/journal.pone.0141069

Kutas, M., & Federmeier, K. D. (2011). Thirty years and counting: Finding meaning in the N400 component of the event-related brain potential (ERP). *Annual Review of Psychology, 62*(1), 621–647. https://doi.org/10.1146/annurev.psych.093008.131123

Kutas, M., & Hillyard, S. A. (1980). Reading senseless sentences: Brain potentials reflect semantic incongruity. *Science, 207*(4427), 203–205. https://doi.org/10.1126/science.7350657

Landau, B., & Gleitman, L. R. (2009/1985). *Language and Experience: Evidence from the Blind Child* (Vol. 8). Harvard University Press.

Landau, B., Gleitman, L. R., & Landau, B. (2009). *Language and Experience: Evidence from the Blind Child* (Vol. 8). Harvard University Press.

Larrouy-Maestri, P., Harrison, P. M. C., & Müllensiefen, D. (2019). The mistuning perception test: A new measurement instrument. *Behavior Research Methods, 51*(2), 663–675. https://doi.org/10.3758/s13428-019-01225-1

Lenneberg, E. H. (1967). *Biological Foundations of Language.* John Wiley & Sons.

Lerdahl, F. (2013). Musical syntax and its relation to linguistic syntax. In M. A. Arbib (Ed.), *Language, Music, and the Brain* (pp. 257–272). The MIT Press. https://doi.org/10.7551/mitpress/9780262018104.003.0010

Lerdahl, F., & Jackendoff, R. (1983). *A Generative Theory of Tonal Music* (Repr.). MIT Press.

Liu, J., Hilton, C. B., Bergelson, E., & Mehr, S. A. (2023). Language experience predicts music processing in a half-million speakers of fifty-four languages. *Current Biology, 33*(10), 1916–1925.

Lupyan, G., & Winter, B. (2018). Language is more abstract than you think, or, why aren't languages more iconic? *Philosophical Transactions of the Royal*

Society B: Biological Sciences, 373(1752), 20170137. https://doi.org/10.1098/rstb.2017.0137

Luria, A. R., Tsvetkova, L. S., & Futer, D. S. (1965). Aphasia in a composer. *Journal of the Neurological Sciences, 2*(3), 288–292. https://doi.org/10.1016/0022-510X(65)90113-9

Lynch, M. P., & Eilers, R. E. (1992). A study of perceptual development for musical tuning. *Perception & Psychophysics,* 52(6), 599–608. https://doi.org/10.3758/BF03211696

Lynch, M. P., Eilers, R. E., Oller, D. K., & Urbano, R. C. (1990). Innateness, experience, and music perception. *Psychological Science,* 1(4), 272–276. https://doi.org/10.1111/j.1467-9280.1990.tb00213.x

Lytle, S. R., Garcia-Sierra, A., & Kuhl, P. K. (2018). Two are better than one: Infant language learning from video improves in the presence of peers. *Proceedings of the National Academy of Sciences,* 115(40), 9859–9866. https://doi.org/10.1073/pnas.1611621115

Maess, B., Koelsch, S., Gunter, T. C., & Friederici, A. D. (2001). Musical syntax is processed in Broca's area: An MEG study. *Nature Neuroscience, 4*(5), 540–545. https://doi.org/10.1038/87502

Magne, C., Schön, D., & Besson, M. (2006). Musician children detect pitch violations in both music and language better than nonmusician children: Behavioral and electrophysiological approaches. *Journal of Cognitive Neuroscience,* 18(2), 199–211. https://doi.org/10.1162/jocn.2006.18.2.199

Mankel, K., & Bidelman, G. M. (2018). Inherent auditory skills rather than formal music training shape the neural encoding of speech. *Proceedings of the National Academy of Sciences, 115*(51), 13129–13134. https://doi.org/10.1073/pnas.1811793115

Mattock, K., & Burnham, D. (2006). Chinese and English infants' tone perception: Evidence for perceptual reorganization. *Infancy,* 10(3), 241–265. https://doi.org/10.1207/s15327078in1003_3

Mattock, K., Molnar, M., Polka, L., & Burnham, D. (2008). The developmental course of lexical tone perception in the first year of life. *Cognition,* 106(3), 1367–1381. https://doi.org/10.1016/j.cognition.2007.07.002

McDermott, J., & Hauser, M. (2005). The origins of music: Innateness, uniqueness, and evolution. *Music Perception, 23*(1), 29–59. https://doi.org/10.1525/mp.2005.23.1.29

McMullen, E., & Saffran, J. R. (2004). Music and language: A developmental comparison. *Music Perception, 21*(3), 289–311. https://doi.org/10.1525/mp.2004.21.3.289

Mehler, J., Jusczyk, P., Lambertz, G., Halsted, N., Bertoncini, J., & Amiel-Tison, C. (1988). A precursor of language acquisition in young infants. *Cognition,* 29(2), 143–178. https://doi.org/10.1016/0010-0277(88)90035-2

Mehr, S. A., Singh, M., Knox, D., Ketter, D. M., Pickens-Jones, D., Atwood, S., ... & Glowacki, L. (2019). Universality and diversity in human song. *Science, 366*(6468), eaax0868. https://doi.org/10.1126/science.aax0868

Meisel, J. M. (2013). *Language Acquisition and Change: A Morphosyntactic Perspective*. Edinburgh University Press.

Meyer, L. B. (1956). *Emotion and Meaning in Music* (Paperback ed., [Nachdr.]). University of Chicago Press.

Miller, J. L., & Volaitis, L. E. (1989). Effect of speaking rate on the perceptual structure of a phonetic category. *Perception & Psychophysics, 46*(6), 505–512. https://doi.org/10.3758/BF03208147

Mithen, S. J. (2005). *The Singing Neanderthals: The Origins of Music, Language, Mind and Body*. Weidenfeld & Nicolson.

Monaghan, P., Shillcock, R. C., Christiansen, M. H., & Kirby, S. (2014). How arbitrary is language? *Philosophical Transactions of the Royal Society B: Biological Sciences, 369*(1651), 20130299. https://doi.org/10.1098/rstb.2013.0299

Moreno, S., Bialystok, E., Barac, R., Schellenberg, E. G., Cepeda, N. J., & Chau, T. (2011a). Short-term music training enhances verbal intelligence and executive function. *Psychological Science*, 22(11), 1425–1433. https://doi.org/10.1177/095679761141699

Moreno, S., & Bidelman, G. M. (2014). Examining neural plasticity and cognitive benefit through the unique lens of musical training. *Hearing Research*, 308, 84–97. https://doi.org/10.1016/j.heares.2013.09.012A

Moreno, S., Friesen, D., & Bialystok, E. (2011b). Effect of music training on promoting preliteracy skills: Preliminary causal evidence. *Music Perception*, 29(2), 165–172. https://doi.org/10.1525/mp.2011.29.2.165

Moreno, S., Marques, C., Santos, A., Santos, M., Castro, S. L., & Besson, M. (2009). Musical training influences linguistic abilities in 8-year-old children: More evidence for brain plasticity. *Cerebral Cortex*, 19(3), 712–723. https://doi.org/10.1093/cercor/bhn120A

Murphy, G. L. (1996). On metaphoric representation. *Cognition, 60*(2), 173–204. https://doi.org/10.1016/0010-0277(96)00711-1

Musacchia, G., Sams, M., Skoe, E., & Kraus, N. (2007). Musicians have enhanced subcortical auditory and audiovisual processing of speech and music. *Proceedings of the National Academy of Sciences*, 104(40), 15894–15898. https://doi.org/10.1073/pnas.0701498104

Naigles, L. (1990). Children use syntax to learn verb meanings. *Journal of Child Language, 17*(2), 357–374.

Naigles, L. R., & Swensen, L. D. (2007). Syntactic supports for word learning. In E. Hoff & M. Shatz (Eds.), *Blackwell Handbook of Language Development* (pp. 212–231). Blackwell Publishing. https://doi.org/10.1002/9780470757833.ch11

Narmour, E. (1990). *The Analysis and Cognition of Basic Melodic Structures: The Implication-Realization Model*. University of Chicago Press.

Nazzi, T., Bertoncini, J., & Mehler, J. (1998). Language discrimination by newborns: Toward an understanding of the role of rhythm. *Journal of Experimental Psychology: Human Perception and Performance*, 24(3), 756.

Nazzi, T., & Ramus, F. (2003). Perception and acquisition of linguistic rhythm by infants. *Speech Communication*, 41(1), 233–243. https://doi.org/10.1016/S0167-6393(02)00106-1

Nettl, B. (1999). An ethnomusicologist contemplates universals in musical sound and musical culture. In N. L. Wallin, B. Merker, & S. Brown (Eds.), *The Origins of Music* (pp. 463–472). The MIT Press. https://doi.org/10.7551/mitpress/5190.003.0032

Neville, H., Nicol, J. L., Barss, A., Forster, K. I., & Garrett, M. F. (1991). Syntactically based sentence processing classes: Evidence from event-related brain potentials. *Journal of Cognitive Neuroscience, 3*(2), 151–165. https://doi.org/10.1162/jocn.1991.3.2.151

Neville, H. J., Coffey, S. A., Lawson, D. S., Fischer, A., Emmorey, K., & Bellugi, U. (1997). Neural systems mediating American Sign Language: Effects of sensory experience and age of acquisition. *Brain and Language, 57*(3), 285–308. https://doi.org/10.1006/brln.1997.1739

Newport, E. (2006). Language development, critical periods in. In L. Nadel (Ed.), *Encyclopedia of Cognitive Science* (p. s00506). John Wiley & Sons, Ltd.https://doi.org/10.1002/0470018860.s00506

Newport, E. L. (1990). Maturational constraints on language learning. *Cognitive Science*, 14(1), 11–28. https://doi.org/10.1016/0364-0213(90)90024-Q

Niedeggen, M., & Rösler, F. (1999). N400 effects reflect activation spread during retrieval of arithmetic facts. *Psychological Science, 10*(3), 271–276. https://doi.org/10.1111/1467-9280.00149

Norman-Haignere, S., Kanwisher, N. G., & McDermott, J. H. (2015). Distinct cortical pathways for music and speech revealed by hypothesis-free voxel decomposition. *Neuron, 88*(6), 1281–1296. https://doi.org/10.1016/j.neuron.2015.11.035

Nygaard, L. C., Cook, A. E., & Namy, L. L. (2009). Sound to meaning correspondences facilitate word learning. *Cognition, 112*(1), 181–186. https://doi.org/10.1016/j.cognition.2009.04.001

Oyama, S. (1976). A sensitive period for the acquisition of a nonnative phonological system. *Journal of Psycholinguistic Research, 5*, 261–283. https://doi.org/10.1007/BF01067377

Patel, A. D. (2003). Language, music, syntax and the brain. *Nature Neuroscience, 6*(7), 674–681. https://doi.org/10.1038/nn1082

Patel, A. D. (2011). Why would musical training benefit the neural encoding of speech? The OPERA hypothesis. *Frontiers in Psychology, 2*. https://doi.org/10.3389/fpsyg.2011.00142

Patel, A. D., Gibson, E., Ratner, J., Besson, M., & Holcomb, P. J. (1998). Processing syntactic relations in language and music: An event-related potential study. *Journal of Cognitive Neuroscience, 10*(6), 717–733. https://doi.org/10.1162/089892998563121

Patel, A. D., Iversen, J. R., Wassenaar, M., & Hagoort, P. (2008). Musical syntactic processing in agrammatic Broca's aphasia. *Aphasiology, 22*(7–8), 776–789. https://doi.org/10.1080/02687030701803804

Patel, A. D., Peretz, I., Tramo, M., & Labreque, R. (1998). Processing prosodic and musical patterns: A neuropsychological investigation. *Brain and Language, 61*(1), 123–144. https://doi.org/10.1006/brln.1997.1862

Paulmann, S., & Uskul, A. K. (2014). Cross-cultural emotional prosody recognition: Evidence from Chinese and British listeners. *Cognition and Emotion, 28*(2), 230–244. https://doi.org/10.1080/02699931.2013.812033

Peelle, J. E., & Davis, M. H. (2012). Neural oscillations carry speech rhythm through to comprehension. *Frontiers in Psychology, 3*. https://doi.org/10.3389/fpsyg.2012.00320

Penfield, W. (1965). Conditioning the uncommitted cortex for language learning. *Brain*, 88(4), 787–798.

Perani, D., Dehaene, S., Grassi, F., Cohen, L., Cappa, S. F., Dupoux, E., ... & Mehler, J. (1996). Brain processing of native and foreign languages. *NeuroReport – International Journal for Rapid Communications of Research in Neuroscience*, 7(15), 2439–2444.

Peretz, I., & Hyde, K. L. (2003). What is specific to music processing? Insights from congenital amusia. *Trends in Cognitive Sciences, 7*(8), 362–367. https://doi.org/10.1016/S1364-6613(03)00150-5

Peretz, I., Kolinsky, R., Tramo, M., Labrecque, R., Hublet, C., Demeurisse, G., & Belleville, S. (1994). Functional dissociations following bilateral lesions of auditory cortex. *Brain, 117*(6), 1283–1301. https://doi.org/10.1093/brain/117.6.1283

Peretz, I., Saffran, J., Schön, D., & Gosselin, N. (2012). Statistical learning of speech, not music, in congenital amusia. *Annals of the New York Academy of Sciences*, 1252(1), 361–366. https://doi.org/10.1111/j.1749-6632.2011.06429.x

Perniss, P., Thompson, R. L., & Vigliocco, G. (2010). Iconicity as a general property of language: Evidence from spoken and signed languages. *Frontiers in Psychology, 1*. https://doi.org/10.3389/fpsyg.2010.00227

Perruchet, P., & Poulin-Charronnat, B. (2013). Challenging prior evidence for a shared syntactic processor for language and music. *Psychonomic Bulletin & Review, 20*(2), 310–317. https://doi.org/10.3758/s13423-012-0344-5

Pichon, S., & Kell, C. A. (2013). Affective and sensorimotor components of emotional prosody generation. *The Journal of Neuroscience, 33*(4), 1640–1650. https://doi.org/10.1523/JNEUROSCI.3530-12.2013

Pinker, S. (1997). Words and rules in the human brain. *Nature*, 387(6633), 547–548. https://doi.org/10.1038/42347

Pinker, S. (1999). *How the Mind Works* (1. publ). Norton.

Pinker, S., & Jackendoff, R. (2005). The faculty of language: What's special about it? *Cognition, 95*(2), 201–236. https://doi.org/10.1016/j.cognition.2004.08.004

Querleu, D., Renard, X., Versyp, F., Paris-Delrue, L., & Crèpin, G. (1988). Fetal hearing. *European Journal of Obstetrics & Gynecology and Reproductive Biology*, 28(3), 191–212. https://doi.org/10.1016/0028-2243(88)90030-5

Ramachandran, V. S., & Hubbard, E. M. (2001). Synaesthesia: A window into perception, thought and language. *Journal of Consciousness Studies, 8*(12), 3–34.

Rivera-Gaxiola, M., Silva-Pereyra, J., & Kuhl, P. K. (2005). Brain potentials to native and non-native speech contrasts in 7- and 11-month-old American infants. *Developmental Science*, 8(2), 162–172. https://doi.org/10.1111/j.1467-7687.2005.00403.x

Romberg, A. R., & Saffran, J. R. (2010). Statistical learning and language acquisition. *Wiley Interdisciplinary Reviews: Cognitive Science*, 1(6), 906–914. https://doi.org/10.1002/wcs.78

Saffran, J. R. (2001). Words in a sea of sounds: The output of infant statistical learning. *Cognition*, 81(2), 149–169. https://doi.org/10.1016/S0010-0277(01)00132-9

Saffran, J. R. (2003). Musical learning and language development. *Annals of the New York Academy of Sciences*, 999(1), 397–401. https://doi.org/10.1196/annals.1284.050

Saffran, J. R., Aslin, R. N., & Newport, E. L. (1996a). Statistical learning by 8-month-old infants. *Science*, 274(5294), 1926–1928. https://doi.org/10.1126/science.274.5294.1926

Saffran, J. R., Johnson, E. K., Aslin, R. N., & Newport, E. L. (1999). Statistical learning of tone sequences by human infants and adults. *Cognition*, 70(1), 27–52. https://doi.org/10.1016/S0010-0277(98)00075-4

Saffran, J. R., & Kirkham, N. Z. (2018). Infant statistical learning. *Annual Review of Psychology*, 69, 181–203. https://doi.org/10.1146/annurev-psych-122216-011805

Saffran, J. R., Newport, E. L., & Aslin, R. N. (1996b). Word segmentation: The role of distributional cues. *Journal of Memory and Language*, 35(4), 606–621. https://doi.org/10.1006/jmla.1996.0032

Scherer, K. R. (1995). Expression of emotion in voice and music. *Journal of Voice*, *9*(3), 235–248. https://doi.org/10.1016/S0892-1997(05)80231-0

Shepherd, J., & Wicke, P. (1997). *Music and Cultural Theory*. Polity Press.

Shintel, H., Nusbaum, H. C., & Okrent, A. (2006). Analog acoustic expression in speech communication. *Journal of Memory and Language*, *55*(2), 167–177. https://doi.org/10.1016/j.jml.2006.03.002

Sidhu, D. M., Westbury, C., Hollis, G., & Pexman, P. M. (2021). Sound symbolism shapes the English language: The maluma/takete effect in English nouns. *Psychonomic Bulletin & Review*, *28*(4), 1390–1398. https://doi.org/10.3758/s13423-021-01883-3

Siegelman, N., Bogaerts, L., Elazar, A., Arciuli, J., & Frost, R. (2018). Linguistic entrenchment: Prior knowledge impacts statistical learning performance. *Cognition*, *177*, 198–213. https://doi.org/10.1016/j.cognition.2018.04.011

Silvén, M., Voeten, M., Kouvo, A., & Lundén, M. (2014). Speech perception and vocabulary growth: A longitudinal study of Finnish-Russian bilinguals and Finnish monolinguals from infancy to three years. *International Journal of Behavioral Development*, 38(4), 323–332. https://doi.org/10.1177/0165025414533748

Slevc, L. R., & Patel, A. D. (2011). Meaning in music and language: Three key differences. *Physics of Life Reviews*, *8*(2), 110–111. https://doi.org/10.1016/j.plrev.2011.05.003

Slevc, L. R., Rosenberg, J. C., & Patel, A. D. (2009). Making psycholinguistics musical: Self-paced reading time evidence for shared processing of linguistic and musical syntax. *Psychonomic Bulletin & Review*, *16*(2), 374–381. https://doi.org/10.3758/16.2.374

Smit, E. A., Milne, A. J., Sarvasy, H. S., & Dean, R. T. (2022). Emotional responses in Papua New Guinea show negligible evidence for a universal effect of major versus minor music. *PLoS ONE, 17*(6), e0269597. https://doi.org/10.1371/journal.pone.0269597

Smith, Z. M., Delgutte, B., & Oxenham, A. J. (2002). Chimaeric sounds reveal dichotomies in auditory perception. *Nature, 416*(6876), 87–90. https://doi.org/10.1038/416087a

Sohail, J., & Johnson, E. K. (2016). How transitional probabilities and the edge effect contribute to listeners' phonological bootstrapping success. *Language Learning and Development, 12*(2), 105–115. https://doi.org/10.1080/15475441.2015.1073153

Steinbeis, N., Koelsch, S., & Sloboda, J. A. (2006). The role of harmonic expectancy violations in musical emotions: Evidence from subjective, physiological, and neural responses. *Journal of Cognitive Neuroscience, 18*(8), 1380–1393. https://doi.org/10.1162/jocn.2006.18.8.1380

Strait, D. L., & Kraus, N. (2011). Can you hear me now? Musical training shapes functional brain networks for selective auditory attention and hearing speech in noise. *Frontiers in Psychology*, 2, 113. https://doi.org/10.3389/fpsyg.2011.00113

Strait, D. L., Kraus, N., Skoe, E., & Ashley, R. (2009). Musical experience and neural efficiency – effects of training on subcortical processing of vocal expressions of emotion. *European Journal of Neuroscience*, 29(3), 661–668. https://doi.org/10.1111/j.1460-9568.2009.06617.x

Stromswold, K., Caplan, D., Alpert, N., & Rauch, S. (1996). Localization of syntactic comprehension by positron emission tomography. *Brain and Language, 52*(3), 452–473. https://doi.org/10.1006/brln.1996.0024

Swaminathan, S., & Schellenberg, E. G. (2019). Music training and cognitive abilities: Associations, causes, and consequences. In M. H. Thaut & D. A. Hodges (Eds.), *The Oxford Handbook of Music and the Brain* (pp. 644–670). Oxford University Press. https://doi.org/10.1093/oxfordhb/9780198804123.013.26

Swaminathan, S., & Schellenberg, E. G. (2020). Musical ability, music training, and language ability in childhood. *Journal of Experimental Psychology: Learning, Memory, and Cognition, 46*(12), 2340–2348. https://doi.org/10.1037/xlm0000798

Swingley, D. (2005). Statistical clustering and the contents of the infant vocabulary. *Cognitive Psychology*, 50(1), 86–132. https://doi.org/10.1016/j.cogpsych.2004.06.001

Tervaniemi, M., & Hugdahl, K. (2003). Lateralization of auditory-cortex functions. *Brain Research Reviews, 43*(3), 231–246. https://doi.org/10.1016/j.brainresrev.2003.08.004

Thompson, W. F., Marin, M. M., & Stewart, L. (2012). Reduced sensitivity to emotional prosody in congenital amusia rekindles the musical protolanguage hypothesis. *Proceedings of the National Academy of Sciences, 109*(46), 19027–19032. https://doi.org/10.1073/pnas.1210344109

Tierney, A., Dick, F., Deutsch, D., & Sereno, M. (2013). Speech versus song: Multiple pitch-sensitive areas revealed by a naturally occurring musical illusion. *Cerebral Cortex, 23*(2), 249–254. https://doi.org/10.1093/cercor/bhs003

Tillmann, B., Bharucha, J. J., & Bigand, E. (2000). Implicit learning of tonality: A self-organizing approach. *Psychological Review*, 107(4), 885. https://doi.org/10.1037/0033-295X.107.4.885

Tillmann, B., Janata, P., & Bharucha, J. J. (2003). Activation of the inferior frontal cortex in musical priming. *Cognitive Brain Research, 16*(2), 145–161. https://doi.org/10.1016/S0926-6410(02)00245-8

Tremblay, P., & Dick, A. S. (2016). Broca and Wernicke are dead, or moving past the classic model of language neurobiology. *Brain and Language, 162*, 60–71. https://doi.org/10.1016/j.bandl.2016.08.004

Tsang, C. D., & Conrad, N. J. (2011). Music training and reading readiness. *Music Perception*, 29(2), 157–163. https://doi.org/10.1525/mp.2011.29.2.157

Turk, M., Turk, I., & Otte, M. (2020). The neanderthal musical instrument from Divje Babe I Cave (Slovenia): A critical review of the discussion. *Applied Sciences, 10*(4), 1226. https://doi.org/10.3390/app10041226

Uddin, S., Heald, S. L. M., Van Hedger, S. C., & Nusbaum, H. C. (2018). Hearing sounds as words: Neural responses to environmental sounds in the context of fluent speech. *Brain and Language, 179*, 51–61. https://doi.org/10.1016/j.bandl.2018.02.004

Van Hedger, S. C., Heald, S. L. M., & Nusbaum, H. C. (2019). Absolute pitch can be learned by some adults. *PLoS ONE, 14*(9), e0223047. https://doi.org/10.1371/journal.pone.0223047

Van Hedger, S. C., Johnsrude, I. S., & Batterink, L. J. (2022). Musical instrument familiarity affects statistical learning of tone sequences. *Cognition, 218*, 104949. https://doi.org/10.1016/j.cognition.2021.104949

Vasuki, P. R. M., Sharma, M., Demuth, K., & Arciuli, J. (2016). Musicians' edge: A comparison of auditory processing, cognitive abilities and statistical learning. *Hearing Research*, 342, 112–123. https://doi.org/10.1016/j.heares.2016.10.008

Vasuki, P. R. M., Sharma, M., Ibrahim, R., & Arciuli, J. (2017). Statistical learning and auditory processing in children with music training: An ERP study. *Clinical Neurophysiology*, 128(7), 1270–1281. https://doi.org/10.1016/j.clinph.2017.04.010

Wacewicz, S., & Żywiczyński, P. (2015). Language evolution: Why Hockett's design features are a non-starter. *Biosemiotics, 8*(1), 29–46. https://doi.org/10.1007/s12304-014-9203-2

Walker, P., Bremner, J. G., Mason, U., Spring, J., Mattock, K., Slater, A., & Johnson, S. P. (2010). Preverbal infants' sensitivity to synaesthetic cross-modality correspondences. *Psychological Science, 21*(1), 21–25. https://doi.org/10.1177/0956797609354734

Weiss, M. W., & Bidelman, G. M. (2015). Listening to the brainstem: Musicianship enhances intelligibility of subcortical representations for speech. *Journal of Neuroscience*, 35(4), 1687–1691. https://doi.org/10.1523/JNEUROSCI.3680-14.2015

Werker, J. F., & Tees, R. C. (1984). Cross-language speech perception: Evidence for perceptual reorganization during the first year of life. *Infant Behavior and Development*, 7(1), 49–63. https://doi.org/10.1016/S0163-6383(84)80022-3

White, E. J., Hutka, S. A., Williams, L. J., & Moreno, S. (2013). Learning, neural plasticity and sensitive periods: Implications for language acquisition, music training and transfer across the lifespan. *Frontiers in Systems Neuroscience*, 7, 90. https://doi.org/10.3389/fnsys.2013.00090

Wong, Y. K., Lui, K. F. H., Yip, K. H. M., & Wong, A. C.-N. (2020). Is it impossible to acquire absolute pitch in adulthood? *Attention, Perception, & Psychophysics*, *82*(3), 1407–1430. https://doi.org/10.3758/s13414-019-01869-3

Yetkin, O., Yetkin, F. Z., Haughton, V. M., & Cox, R. W. (1996). Use of functional MR to map language in multilingual volunteers. *American Journal of Neuroradiology*, 17(3), 473–477.

Yumoto, M., Uno, A., Itoh, K., Karino, S., Saitoh, O., Kaneko, Y., ... & Kaga, K. (2005). Audiovisual phonological mismatch produces early negativity in auditory cortex. *Neuroreport*, 16(8), 803–806.

Zatorre, R. J., & Baum, S. R. (2012). Musical melody and speech intonation: Singing a different tune. *PLOS Biology*, *10*(7), e1001372. https://doi.org/10.1371/journal.pbio.1001372

Zatorre, R. J., Belin, P., & Penhune, V. B. (2002). Structure and function of auditory cortex: Music and speech. *Trends in Cognitive Sciences*, *6*(1), 37–46. https://doi.org/10.1016/S1364-6613(00)01816-7

Zatorre, R. J., Evans, A., Meyer, E., & Gjedde, A. (1992). Lateralization of phonetic and pitch discrimination in speech processing. *Science*, *256*(5058), 846–849. https://doi.org/10.1126/science.1589767

Zhao, T. C., & Kuhl, P. K. (2016). Musical intervention enhances infants' neural processing of temporal structure in music and speech. *Proceedings of the National Academy of Sciences*, 113(19), 5212–5217. https://doi.org/10.1073/pnas.1603984113

Zhao, T. C., Llanos, F., Chandrasekaran, B., & Kuhl, P. K. (2022). Language experience during the sensitive period narrows infants' sensory encoding of lexical tones: Music intervention reverses it. *Frontiers in Human Neuroscience*, 16, 941853. https://doi.org/10.3389/fnhum.2022.941853

Part VI

New Directions and Perspectives

18

Hippocampal Dependent Memory Supports Communication and Language

Implications for Cognitive-Communication Disorders

Melissa C. Duff and Natalie V. Covington

18.1 Language and Memory: Understanding of Their Interactions and Interdependencies

Language and memory have historically been studied apart as unique cognitive abilities with distinct neural substrates. This separation was supported, in part, by early lesion studies and demonstrations of double dissociations in language and memory. For example, while damage restricted to the left hemisphere perisylvian language areas commonly results in aphasia, it does not produce a co-occurring amnesia (e.g., Broca, 1865). Likewise, damage confined to the hippocampus and medial temporal lobes produces anterograde amnesia but not a concomitant aphasia (e.g., Scoville & Milner, 1957). These early lesion studies were revolutionary in the mapping of brain-behavior relations and in advancing the neurobiology of language and of memory. Early lesion studies also highlighted the value of patient studies more broadly as a key method in empirical cognitive neuroscience investigations of language and memory. The main take away from these early lesion studies was that language and memory were largely independent of one another, both functionally and neuroanatomically.

Cognitive neuroscience research over the past several decades, however, now suggests that there is not a simple one-to-one mapping of a single cognitive ability to a discrete brain region (e.g., Sherman et al., 2023). Rather, multiple neural systems can contribute in concert to a particular cognitive ability. Conversely, a single brain region can make contributions to multiple cognitive functions. Accordingly, we have come to view cognition as being organized around large-scale neural networks. For example, the language network is composed of different regions located across the brain that together support language processes (Friederici & Gierhan, 2013). The neural network approach can be combined with patient studies and the lesion method to test both local and network hypotheses about the language network. These hypotheses address how, and under what conditions, brain regions outside the left hemisphere perisylvian language areas may be recruited and coactivate with the language network to meet the various demands of language processing (Fedorenko & Thompson-Schill, 2014). These methods also inform models of how the language network changes as a function of brain damage and recovery (Kiran et al., 2019; Wilson et al., 2019).

Since the earliest lesion studies, there is also growing recognition that disruptions in language and memory can occur along a continuum depending on the nature and extent of neuropathology and specific task demands. Indeed, while frank[1] presentation of aphasia and amnesia have historically been considered categorically distinct disorders that can be differentiated from other forms of language and memory deficits in specificity and severity (e.g., semantic dementia; age-related memory decline), language and memory disruptions are common symptoms that cut across numerous disease categories. For example, hippocampal dysfunction and declarative memory impairment are features of a number of clinical populations (e.g., traumatic brain injury (TBI), Alzheimer's disease (AD), schizophrenia, autism, depression, developmental language disorder (DLD)) and deficits can vary in severity both within and across populations. It has remained a challenge, however, to capture the degree of language and memory disruption that is frequent in conditions outside of the aphasia and amnesia as many neuropsychological tests were designed to diagnose these distinct and frank disorders rather than to be sensitive to mild and moderate disruptions or to individual differences.

Emerging from the neural network perspective and increased efforts to capture a range of language and memory ability, there is now a growing body of evidence suggesting that language and memory are more heavily intertwined than originally purported, can be impaired in tandem, and may even rely on shared cognitive and neural mechanisms (Banjac et al., 2021; Blank et al., 2016; Bonhage et al., 2015; Covington & Duff, 2016; Piai et al., 2016; Pu et al., 2020; Roger et al., 2022). Patient studies have been highly informative in these advances. For example, even though damage to the left hemisphere perisylvian areas does not cause frank amnesia, individuals with aphasia can have co-occurring disruptions in memory (e.g., Mayer & Murray, 2012; Vallila-Rohter & Kiran, 2013). Likewise, individuals with hippocampal damage and amnesia do not have frank aphasia but they do have deficits in language (e.g., Duff & Brown-Schmidt, 2012, 2017; MacKay et al, 1998). Cutting across various theoretical and methodological approaches and patient populations, these findings offer novel insights into the interactions and interdependencies of language and memory. Additionally, their significance extends to the theoretical and clinical realms, as they have considerable implications for the neurobiology of language and memory and for the assessment and treatment of populations where co-occurring disorders of language and memory are common.

18.2 Communication and Memory: Implications for Cognitive-Communication Disorders

Over the past two decades, our research group, with our collaborators, has been examining the language use and communicative interactions of individuals with hippocampal damage and severe declarative memory impairment (see Duff & Brown-Schmidt, 2012, 2017; Rubin et al., 2014 for previous reviews). The study of individuals with hippocampal amnesia provides a compelling opportunity to directly test the role of a given memory system to language and to generate a biologically and psychologically plausible account of the contributions of memory to language processes and language use. Our work has focused on testing the proposal that the same processes by which the hippocampal declarative memory system creates and flexibly integrates representations across diverse sources of information in the formation of new memories, and maintains representations online to be evaluated and used in service of behavioral performance, are the same processes necessary for meeting many of the demands of language processing and use.

When we began this work, we were particularly interested in the role of hippocampal dependent declarative memory in communicative and discursive contexts of language use (e.g., Duff et al., 2006, 2007, 2008, 2009; Kurczek & Duff, 2011). This was, in part, because of the connection to the literature on cognitive-communication disorders. Cognitive-communication disorders refer to difficulty with any aspect of communication secondary to impairments in cognition including attention, memory, and/or executive function (ASHA, 1987, 2005). In the field of communication sciences and disorders, and in the study of neurogenic language disorders more broadly, the term cognitive-communication disorder is used to differentiate the disruptions in *communication*, secondary to impairments in cognition (e.g., memory, attention), from the disruptions in *language* observed in the aphasias due to impairments in the linguistic system (Holland, 1982). Initial evidence for a distinction between communication and language came from studies demonstrating that aphasia secondary to left hemisphere perisylvian lesions produced utterance level impairments (i.e., language at the level of word, phrase, and sentence level comprehension and production) while the ability to manage discourse and conversation (i.e., communication and the use of language above the level of the utterance) was a relative strength. The opposite pattern was observed in populations we now refer to as having cognitive-communication disorders such as individuals with TBI or AD. In TBI, for example, disruptions are frequently observed in communication, at the levels of discourse and conversation, while the rapid processing of words and individual utterances are preserved (or at least as measured by standardized aphasia batteries and offline language tasks). In this way, the historical distinction between communication and language, and between cognitive-communication and aphasia, has been helpful in demonstrating broad distinctions between categories of language disorders and types of language tasks and language demands.

From this history, the study of cognitive-communication disorders has largely focused on eliciting and analyzing narratives and conversation (utterances above the level of words, phrases, and sentences as these microlinguistic abilities are thought to be intact) and relating discourse-level performance to aspects of cognition thought to underlie discourse ability (e.g., memory, attention, executive function) (Cherney et al., 1998; Snow et al., 1998; Rousseaux et al., 2010). These studies are critical in documenting the presence and severity of communication impairments in these populations. However, populations with cognitive-communication disorders typically have impairments to *multiple* cognitive domains and neural substrates, given the diffuse nature of neuropathology of these conditions. This prevents researchers and clinicians from attributing any observed deficit in a specific aspect of communication to a particular cognitive domain or neural substrate. For example, it is challenging to conclude that impairments in cohesion (observed in TBI and AD) are due to working memory deficits secondary to frontal lobe damage, which has been suggested, if the study population also has impairments in declarative memory, attention, and executive functions as a result of widespread neural pathology (Kurczek & Duff, 2011, 2012). Indeed, efforts to establish a link between aspects of discourse (e.g., cohesion) and cognition (e.g., executive functioning, working memory) have yielded only moderate and inconsistent correlations (Snow et al., 1998; Youse & Coelho, 2005).

We approached the study of communication and discourse in individuals with hippocampal amnesia as a rare opportunity to examine the specific contribution of memory to communication and to advance mechanistic accounts of cognitive-communication disorders. As we will describe below, this work has illuminated the impact of isolated hippocampal dependent memory impairments on communication. Those occasions when individuals with hippocampal amnesia are successful in using language are as informative as their failures in mapping the role of multiple memory systems to language use and communication. Patterns of successes and failures in these patients also help flesh out a theory of the contribution and coordination of multiple cognitive and neural systems to the larger language network.

While this work has its roots in understanding the role of memory in interactive and communicative contexts, we have also documented disruptions in more basic aspects of language in individuals with hippocampal amnesia including at the word, phrase, and sentence level across offline and online language-processing tasks. These findings are striking given the historical and theoretical distinctions between communication and language and the purported independence between language and memory. As we will review below, this research, linking both communication and language to hippocampal dependent memory, suggests that hippocampal dependent memory contributes to different levels of language processing and use and that we can observe graded differences in the contribution of this form of memory depending on the demands of a particular language function. In doing so, this research demonstrates how promiscuously the hallmark processing features of the hippocampus are called upon by a variety of cognitive domains including, but not limited to, language. This research raises questions about the reality and utility of the historical distinction between communication and language in defining cognitive-communication disorders, as we show deficits that cut across both communication and language in individuals with hippocampal amnesia.

In this chapter, we briefly review the history and rationale for our proposal for a critical role of hippocampal dependent memory in communication and language and the methodological approach we have taken to conduct this work. We next present data from studies of individuals with hippocampal amnesia documenting patterns of success and impairment across various domains of communication and language. Finally, we discuss the implications of this work for theories describing the language-memory interface and for the treatment and rehabilitation of language and memory disorders. An overarching goal of this chapter is to update our previous reviews of disruptions in language and communication in individuals with amnesia (Duff & Brown-Schmidt, 2012, 2017; Rubin et al., 2014) and to situate this work in the context of the literature on cognitive-communication disorders.

18.3 Hippocampal Dependent Memory Supports to Language and Communication

We are certainly not the first research group to propose a relationship between language and memory. In fact, there have been numerous attempts to link aspects of memory to the particular properties of language or demands of language processing. Much of this work has focused on linking the demands of language processing to working memory and/or executive control processes associated with prefrontal cortex mechanisms (e.g., Novick et al., 2005). Other proposals have focused on the role of semantic memory in providing support for language processing in terms of maintaining enduring representations of word meanings and knowledge (e.g., Kutas & Federmeier, 2000; Federmeier et al., 2002). Yet, with the exception of a limited role in the acquisition of new vocabulary (e.g., Ullman et al., 1997; Ullman, 2004), the hippocampal dependent declarative memory system has largely been ignored as a candidate mechanism for meeting the demands of language use and processing. Historically, this made good sense.

The traditional view of the hippocampal dependent declarative memory system was that it contributed exclusively to long-term memory. Indeed, short-term memory was thought to be intact in individuals with hippocampal amnesia because they performed within normal limits on neuropsychological tests of working memory like digit span. Hippocampal amnesia is defined as a severe and selective deficit of memory leaving other cognitive domains, including language, intact (O'Connor et al., 1995; O'Connor & Verfaellie, 2002). Finally, although the hippocampus is critical for acquiring new semantic knowledge, semantic information was thought to become independent of the hippocampus over time via neocortical consolidation processes (McClelland et al., 1995).

Over the past two decades, a number of discoveries have expanded the breadth of hippocampal functionality (see Hannula & Duff, 2017) and have significant implications for theories of language use and processing. First, evidence started to emerge that challenged the view of hippocampal declarative memory as contributing exclusively to long-term memory. Empirical work with patients with hippocampal damage and amnesia revealed deficits in declarative memory when there are minimal delays, and even when there were no delays at all (i.e., when all the necessary information was immediately available and over a time course similar to working memory) (Barense et al., 2007; Hannula et al., 2006; Shrager et al., 2008). Moreover, fMRI studies provided converging evidence showing hippocampal activation for relational learning over similarly short delays (e.g., Hannula & Ranganath, 2009; Ranganath & D'Esposito, 2001). These findings suggested that new hippocampal dependent representations are available rapidly enough to influence ongoing processing when: new information is perceived; old information is retrieved; and representations are held online to be evaluated, manipulated, integrated, and used in service of behavioral performance. Although the initial work in this area came primarily from studies using visuo-spatial stimuli, the strong implication was that the hallmark flexibility and integration of hippocampal dependent representations are deployed and rapidly available when information is processed in an ongoing fashion. These new findings suggested to us that performance of patients with hippocampal amnesia on language tasks would be impaired when sufficient demands were placed on the processing capabilities of the hippocampus and hippocampal dependent representations, even on the shortest of timescales (i.e., during online language processing).

Second, research revealed that the hippocampus participates in a neural network that supports the flexible and creative (re)construction and use of mental representation for remembering the past, imagining the future, simulating the thoughts of others, and creativity (Addis & Schacter, 2012; Buckner & Carroll, 2007; Duff et al., 2013; Hassabis et al., 2007; Kurczek et al., 2015). Researchers started to propose that the processes that enable a person to combine and recombine individual elements of real or imagined experience may also serve flexible and creative cognition including support of imagination, decision making, navigation, social cognition, attentional processing, and language (Aly & Turk-Browne, 2017; Duff & Brown-

Schmidt, 2012; Laurita & Spreng, 2017; Rubin et al., 2014; Schacter et al., 2017; Schlichting & Preston, 2017). Notably, language use is generative, involving the flexible and creative use of words and thoughts to (re)create, (re)purpose, and (re)contextualize utterances for specific partners and communicative contexts. If the hippocampus contributes to flexible and creative (re)construction and use of mental representations, we argued such contributions may also extend to supporting creative flexibility in discourse and interactional language use.

Finally, research demonstrating hippocampal contributions to implicit memory, or learning below consciousness, began to accumulate across methods (e.g., fMRI, lesion studies) (see Hannula & Greene, 2012 for a review). In the memory literature, these findings were striking given the long history of using consciousness as a defining feature to distinguish among memory systems (i.e., declarative/explicit memory required consciousness for encoding and retrieval whereas nondeclarative (procedural)/implicit memory did not). Research demonstrating hippocampal contributions to statistical learning and sequence learning tasks further expanded evidence for hippocampal contributions to implicit and unconscious processing (see Hannula & Greene, 2012 for a review). Statistical learning and rule-based sequence learning are critical to the acquisition and use of language (e.g., Saffran et al., 1996; Conway & Christiansen, 2005). While historically linked to the nondeclarative memory system and to the basal ganglia, research linking the hippocampus to the unconscious learning of statistics and sequences raised the possibility that it may also play a role in various rule-based language functions (Covington et al., 2018; Schapiro et al., 2014).

These advances in the memory literature expanded the breadth of hippocampal functionality and extended the reach of the hippocampus beyond the domain of memory. The expansion in our understanding of the processing capabilities of the hippocampus have significant implications for psychological and biological theories of communication and language and provided the foundation for our proposal that the hippocampal dependent memory system plays a critical role in meeting many of the demands of language processing and use (Duff & Brown-Schmidt 2012, 2017; Rubin et al., 2014). The advances reviewed above also impacted the terms used to reference memory systems and tasks. Given research demonstrating that the hippocampus contributes to memory processing in the moment (on a timescale of working memory) and to memory and learning of statistics and sequences (previously considered the domain of nondeclarative/procedural memory), some have argued that the term "declarative memory" no longer provides an exclusive term for the memory functions supported by the hippocampus. The term "hippocampal dependent memory" has since become more common although declarative memory is still used in the context of neuropsychological testing measuring episodic memory performance. Given that our work spans this time of change in terminology, we will use hippocampal dependent declarative memory and hippocampal dependent memory interchangeably to be consistent with the wording in our original papers. These advances have also resulted in a move away from referring to tasks as implicit or explicit as such labels no longer reliably map to a specific form of memory (Ryan, 2012).

18.3.1 Methodological Approach: The Lesion Method and Individuals with Hippocampal Amnesia

The study of individuals with hippocampal damage and amnesia provides a unique and compelling opportunity to examine the specific contribution of a particular form of memory to aspects of communication and language. That said, it is challenging to form a group of participants appropriate for these studies given that it is rare to find individuals who have bilateral damage to the hippocampus necessary to produce a severe memory deficit while not being so extensive as to significantly damage other neural structures and/or produce widespread cognitive impairment. Sample sizes for our studies have ranged between three and nine participants and varied due to participant availability and, over time, participants being excluded due development of secondary conditions. All participants were adults when they acquired amnesia, and the majority have four-year college degrees while two completed an associate's degree.

The two primary etiologies for amnesia in our studies are anoxia (e.g., cardiac arrest; status epilepticus) and herpes simplex encephalitis (HSE). Structural magnetic resonance imaging (MRI) is used to confirm bilateral hippocampal damage and to rule out significant damage in other lobes of the brain (Allen et al., 2006). For participants with anoxia, hippocampal damage is disproportionate, even when compared with other medial temporal lobe regions and the remainder of the temporal lobe (gray and white matter). Figure 18.1 displays magnetic resonance scans of some of our participants with hippocampal amnesia from anoxia (1606, 1846, 2144, 2363, and 2571) and HSE (1951 and 2308). As can be seen, patients with HSE have more extensive bilateral medial temporal lobe damage that includes the hippocampus but also extends into the temporal lobe on one side. Despite the more extensive temporal lobe damage, neither of these participants have aphasia or semantic dementia (e.g., 2308 has extensive left temporal lobe damage but is left-handed).

Following the historical view that amnesia is a severe and selective memory impairment, neuropsychological testing is used to confirm a diagnosis of amnesia. Conventionally, a 20-point disparity between performance on a standardized test of general intelligence (e.g., Wechsler Adult Intelligence Scale) and a standardized test of memory (e.g., Wechsler Memory Scales) is used to confirm and diagnose amnesia (O'Connor et al., 1995). All our participants meet this criterion. While all our participants have severe to profound deficits on standardized tests of declarative/episodic memory (e.g., Auditory Verbal Learning Test), they perform within normal limits on measures of intelligence, attention, executive function, language, and perception. They also perform normally on perceptual and motor measures of nondeclarative memory (Cavaco et al., 2004).

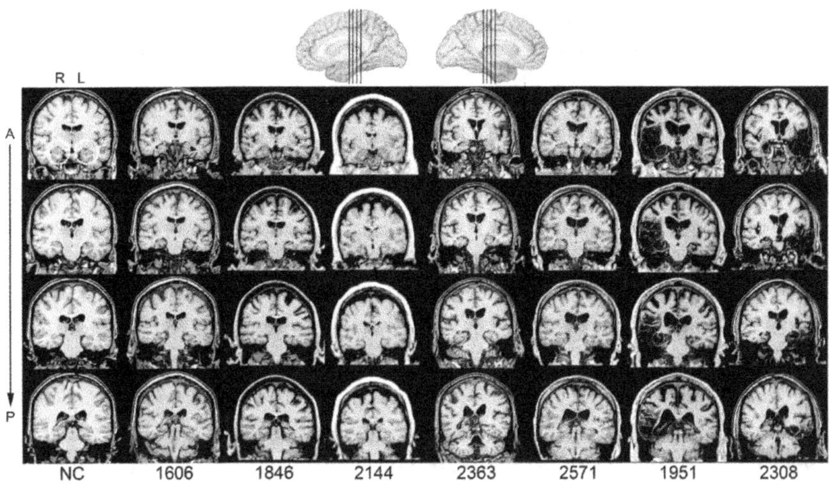

Figure 18.1 Magnetic resonance scans of hippocampal patients. Images are coronal slices through four points along the hippocampus from T1-weighed scans. Volume changes can be noted in the region of the hippocampus bilaterally for patients 1606, 1846, 2144, 2363, and 2571 and significant bilateral damage to the medial temporal lobes (MTL), including hippocampus, can be observed for 1951 and 2308. R = right; L = left; A = anterior; P = Posterior; NC = a neurotypical comparison brain.

From Hilverman, C., Cook, S. W., & Duff, M. C. (2016). Hippocampal declarative memory supports gesture production: Evidence from amnesia. *Cortex, 85*, 25–36. Copyright Elsevier 2016.

We take different approaches to testing the specificity of our findings. We first analyze the data by comparing the performance of the participants with amnesia to a group of demographically matched non-brain injured comparison participants. We then remove the participants with HSE from the analysis to make sure that our results are not driven by these two participants who have more extensive lesions. In all our studies, our results do not change based on the inclusion of these two participants. In our review below, we will include all participants with amnesia in a single group as etiology and lesion size do not influence results. We also frequently use a brain-damaged comparison group with bilateral damage to a neural region outside of the medial temporal lobes and who do not have memory impairments. A brain-damaged comparison (BDC) group allows us to make claims about hippocampal specificity in the observed disruptions in communication and language as opposed to capturing the effects of brain damage more broadly. For this, we have studied individuals with bilateral ventromedial prefrontal cortex (vmPFC) damage. Individuals with bilateral vmPFC damage are an ideal comparison group because they do not have damage to the medial temporal lobes or hippocampus, and they do not have memory impairments.

Below, we review our work on the contributions of hippocampal dependent memory to communication and language. Consistent with our goal to situate this work in the context of the literature on cognitive-communication disorders, we separate our findings into studies examining hippocampal contributions to communication (i.e., communication as using language above the level of the utterance) and to language (i.e., language as word, phrase, and sentence level comprehension and production).

18.4 Contributions of Hippocampal Dependent Memory to Communication

18.4.1 Hippocampal Amnesia and Common Ground Processes

During a conversation, we form assumptions about what our communication partners believe, what they know, and what they do not know. These mental representations of what is mutually known between conversational partners is called common ground and it serves as the critical backdrop for successful communication by guiding language production and comprehension (Clark, 1996; Clark & Marshall, 1978; Wilkes-Gibbs & Clark, 1992). Once conversational partners establish some common ground (e.g., what language to speak, what information can already be assumed as shared, and what information cannot), they are on solid footing for proceeding with the conversation (Brown-Schmidt & Duff, 2016). While it seems obvious that memory processes must be involved in the representation and use of common ground, little is known about which memory systems support encoding and retrieving the necessary information about what other individuals do and do not know. Clark (1996) asserted that common ground requires conversational partners to keep track of events, to recall events, and to continuously update the record of new events. If Clark is correct, then hippocampal amnesia should impair common ground processes in conversation.

In the first study on common ground and amnesia, we asked individuals with hippocampal amnesia to acquire and use referential labels for a set of abstract figures while interacting with a familiar communication partner (friend, spouse) (Duff et al., 2006). The study followed previous work by Clark and Wilkes-Gibbs (1986) who had pairs of neurotypical young adults perform a referential communication task in which the pairs repeatedly described a set of abstract tangram figures to each other. In the original paradigm, Clark and Wilkes-Gibbs found that through the process of communicating about the figures, the pairs got much faster at referencing the figures and had developed concise referential labels (e.g., "the ice skater"). We asked whether the same acquisition of novel labels could be achieved in the absence of declarative memory for the interactions in which the labels were developed. In a modified version of the referential communication task, individuals with hippocampal amnesia played the role of Director, and were seated across from their familiar partner who played the role of Matcher. The Director and Matcher each had a board with twelve numbered spaces and a set of twelve cards displaying the abstract tangram figures. On the table between them was a low barrier preventing a view of the others' cards (Figure 18.2). The participants with amnesia always played the role of Director, communicating to the Matcher how to arrange their cards so that at the end of the trial the

Figure 18.2 Set up for referential communication task.

cards were placed in the same numbered spaces on each board. The pairs communicated freely across twenty-four trials, each of which involved sorting the same twelve tangram figures into a different random order.

Despite severe declarative memory impairments, the individuals with hippocampal amnesia demonstrated robust learning, arriving at increasingly concise and stable labels (e.g., the transition of "this one looks like it has wings, like an angel" on an early trial to just "angel" on a later trial) that facilitated rapid and efficient communication with their partner (Duff et al., 2006). In fact, the rate of learning exhibited by participants with amnesia (measured by the reduction in time and words necessary to complete each trial), did not differ from that of comparison participants. This learning would seem to constitute an example of intact common ground processes, suggesting that at least some forms of common ground can be supported outside of explicit recollection and declarative memory. This interpretation opens the possibility that nondeclarative mechanisms (which are intact in this population) play some role in acquiring and representing information in these communicative interactions. On the other hand, these findings do not indicate whether the participants with amnesia represented these labels as partner-specific; knowledge that is jointly and uniquely shared with their partners (and that would be unknown to a new partner). Partner specificity is a critical component of common ground knowledge and without it, the individuals with amnesia may have instead developed egocentric representations of these labels (i.e., labels they would use with any partner).

In a secondary analysis of these referential communication sessions, we examined the referential form of the labels used by the participants to determine if they used definite reference to signal to their partner that the referent was part of shared knowledge (Duff et al., 2011a). Neurotypical speakers tend to signal when information is shared, using a definite reference, such as *The* Viking ship, and indefinite reference, *A* Viking ship, when it is not (see Issacs & Clark, 1987). The neurotypical comparison pairs overwhelmingly marked shared labels with definite reference (90 percent) whereas the participants with amnesia used definite reference significantly less often (56 percent). That is, even though the participants with amnesia had described the tangram figures multiple times and were using concise labels, they did not consistently mark these labels as part of common ground through definite reference. Even during the final round of matching – the 24th time they had described the figures – the participants with amnesia still used indefinite references, as if they were seeing the figures and deriving the labels for the very first time. By contrast, comparison Directors nearly always used definite references to refer to the figures after the first trial.

This study together with the previous 2006 study suggested a potential bifurcation in the memory systems that support common ground: concise labels can be developed and consistently used in conversation with a familiar partner in the absence of declarative memory, but declarative memory is required to signal (via use of a definite reference) that those same labels are part of shared knowledge. Interestingly, if the roles are reversed and the familiar partners are the Directors and the individuals with hippocampal amnesia are the Matchers (six months later with a new set of tangram figures), the familiar partners also use indefinite articles when communicating to their partners with amnesia (Duff et al., 2011b). Documenting the use of indefinite references by the familiar partners who have intact language and intact memory when communicating with their partner with amnesia further points to the flexibility and complex interaction of language and memory. It further suggests the familiar partners (who are well aware of their partner's significant memory impairment) were not confident that the labels they were developing during the game were maintained as shared knowledge by their family members with amnesia. In terms of hippocampal specificity, we note that damage to vmPFC does not disrupt the formation or expression of common ground as measured by the development of concise labels or the use of definite reference (Gupta et al., 2012).

While the ability of individuals with amnesia to acquire and successfully use novel labels for the tangram figures demonstrates impressive new learning in this communicative context, the data on the use of definite reference use suggests that the individuals with amnesia may not have encoded in memory that these labels were part of common ground with their partner. To directly test partner specificity, we had participants with amnesia describe a set of tangram figures five times with an initially unfamiliar partner and then a new unfamiliar partner would arrive later in the experiment (Yoon et al., 2017). More specifically, after each trial the initial partner (Partner 1) would leave the room for a few minutes and come back in to do another trial. This occurred across the first five trials. On the critical trial, trial six, either Partner 1 came back to complete the task (same-Matcher Condition) or instead of Partner 1 returning, a new unfamiliar partner (Partner 2) came in to complete the task (different-Matcher Condition).

If individuals with hippocampal amnesia can form partner-specific common ground, we would expect them to produce short, concise descriptions during their interactions with Partner 1, with whom they have common ground for the labels. When talking to Partner 2 (the new unfamiliar partner), participants with amnesia should produce longer descriptions to accommodate the new partner's lack of knowledge for the labels. If, however, partner specific common ground cannot be established in the presence of severe declarative memory impairment, individuals with amnesia should continue to use the concise labels acquired with Partner 1 when communicating with Partner 2 (i.e., the knowledge would not be partner specific and they would use the newly acquired labels with any communication partner (Partner 2) not only the partner with whom that knowledge is shared (Partner 1)). We found that individuals with amnesia can form partner-specific common ground representations. Like non-injured comparison participants, individuals with amnesia successfully tailored their language to the knowledge shared with their conversational partner – their common ground. Participants with amnesia produced concise descriptions of the tangram images for Partner 1 and then on the critical trial, provided more descriptive, longer expressions for Partner 2 (Figure 18.3).

These findings demonstrate remarkable sparing in amnesia of the acquisition and use of partner-specific knowledge that underlies common ground. Moreover, they also illuminate the complex memory demands of common ground. That partner-specific representations can be acquired in the absence of intact hippocampal declarative memory fits with other demonstrations of individuals with amnesia acquiring person-specific information (e.g.,

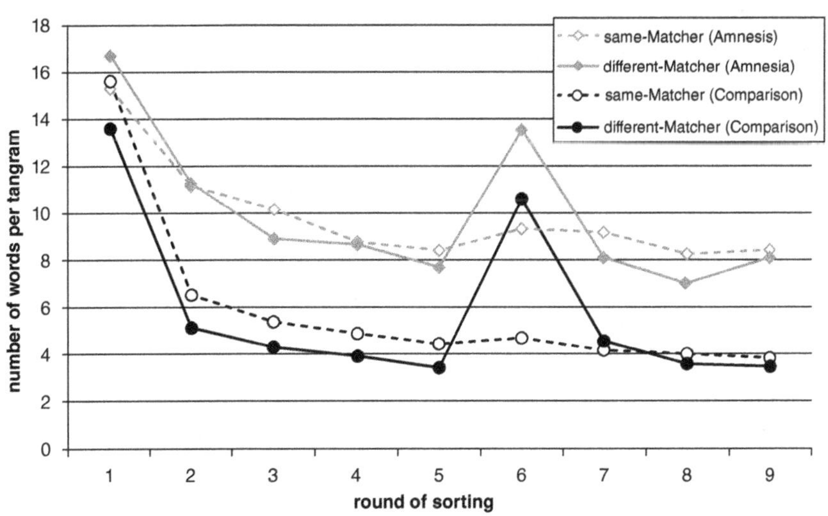

Figure 18.3 Average number of words used by participants to describe each image in the same-Matcher and different-Matcher conditions.

Figure from Yoon, S. O., Duff, M. C., & Brown-Schmidt, S. (2017). Learning and using knowledge about what other people do and don't know despite amnesia. *Cortex, 94.*

personality attributes, accent, political preferences; Coronel et al., 2012; Trude et al., 2014; Tranel & Damasio, 1993).

Common ground status is represented in memory, and multiple memory systems play complementary roles in support of common ground processes (Brown-Schmidt & Duff, 2016). The study of common ground has been taken up by researchers studying language and memory to better understand communication in populations such as Alzheimer's disease (Duff et al., 2013; Paek & Yoon, 2021), traumatic brain injury (Gupta & Duff, 2016), and aphasia (Doedens et al., 2021; Hengst, 2003). This work examining common ground processing across different populations, where deficits in language and memory are common, is an important step in understanding the patterns of performance across different cognitive-communication disorders. An interesting future direction would be to examine how common ground is represented beyond dyadic interactions, in group and multiparty interactions (e.g., Yoon & Brown-Schmidt, 2018), which are the routine communicative contexts of families, education, work, and social participation.

18.4.2 Hippocampal Amnesia and Creative and Flexible Discourse

Language is generative. Language generativity requires creativity and flexibility at the level of the utterance but also in the way speakers play with language to craft specific utterances to meet the dynamic contextual demands of communication across time, places, and speakers. Hippocampal dependent memory has a generative and creative capacity underlying its role in creating, imagining, and (re)combining elements of experience into mental representations, as well as for their flexible and novel use (Cohen et al., 1997; Eichenbaum & Cohen, 2001; Rubin & Cohen, 2017). The generativity and the compositionality afforded by the hippocampus has been linked to the evolution of language (Corballis, 2019) and to creative thinking (e.g., Duff et al., 2013; Madore & Schacter, 2014). We have proposed that hippocampal dependent memory directly supports discursive creativity and flexibility across partners and discourse contexts (Duff et al., 2007; Duff & Brown-Schmidt, 2012).

In our first study examining the creative and flexible use of discourse we examined the conversational use of reported speech (Duff et al., 2007). Reported speech is a discourse practice in which speakers represent, or reenact, words or thoughts from other times and/or places (e.g., *If I ever have kids I'm going to tell them, "You don't have to agree with me but you can't be mean to me;" John said he would be here at six; I can't believe she said that*) (Tannen, 1989). We coded instances of reported speech across discourse samples collected using the mediated discourse elicitation protocol (Hengst & Duff, 2007) where participants have a conversation, tell narratives, describe pictures, and explain how to perform tasks with an experimenter. We found that in the conversational interactions of individuals with hippocampal amnesia there were only half as many instances of reported speech ($M = 30.3$) as there were in the conversations of demographically matched non-injured comparison

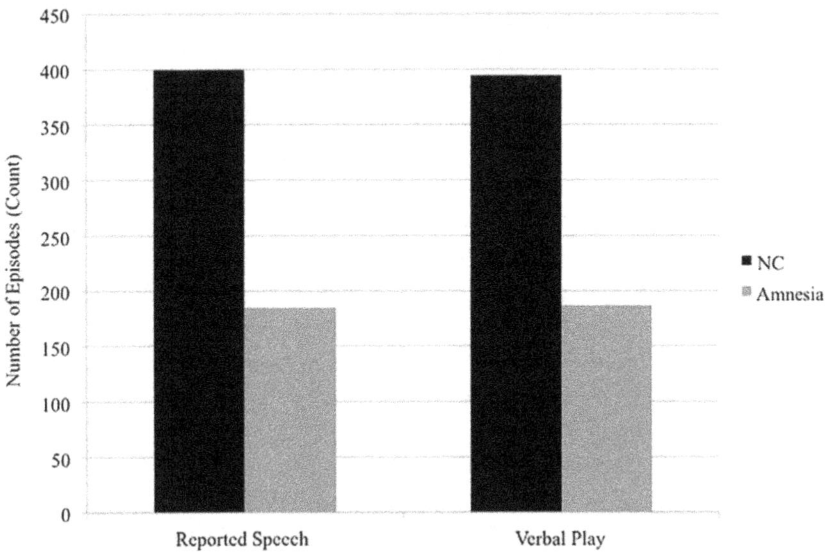

Figure 18.4 Reported speech and verbal play. In conversational interactions with a clinician, patients with hippocampal amnesia produce significantly fewer episodes of reported speech (185) than comparison participants (400). In the interactions with a familiar partner while completing trials of a collaborative referencing task, participants with amnesia produced significantly fewer episodes of verbal play (187) than comparison participants (395). Data presented are group totals for participants with amnesia (Amnesia) and neurotypical comparison (NC) participants. Data of interactional partners (clinician; familiar communication partner) are not presented.

Figure from Rubin, R. D., Watson, P. D., Duff, M. C., & Cohen, N. J. (2014). The role of the hippocampus in flexible cognition and social behavior. *Frontiers in Human Neuroscience, 8*, 742.

participants (M = 61.5) (Figure 18.4). This reduction in reported speech was evident across all time periods; individuals with amnesia used fewer instances of reported speech for events since the onset of their amnesia and even when producing detailed and vivid memories from their remote past. The reduction in reported speech was also independent of session length; sessions with the participants with amnesia were not significantly shorter, on average, than comparison sessions. Rather, reported speech seems to place high demands on hippocampal dependent memory as its use requires access to autobiographical experiences as well as the ability to flexibly and creatively generate unique combinations of the reconstructed elements (what details to represent, what details to omit to meet the specific interactional goals of this telling, on this occasion, with this communication partner). We propose that any discursive practices that place high demands on the processing of the hippocampus would similarly be disrupted.

We have also examined verbal play, the creative use of sounds and meanings of words through the use of puns, voices and sound effects, teasing, and telling funny stories (Crystal, 1998). Verbal play is ubiquitous in everyday conversations. We coded instances of verbal play in the face-to-face interactions of participants as they completed the collaborative referencing task described above. We found that individuals with hippocampal amnesia produced significantly fewer verbal play episodes (Total episodes = 187) than the comparison participants (Total = 395) (Figure 18.4; Duff et al., 2009). In addition to the quantitative group differences in verbal play there were also qualitative differences. In contrast to comparison participants, verbal play episodes produced by the participants with amnesia were rote and repetitive. A common occurrence was for a participant with amnesia to reproduce the same joke, nearly verbatim, multiple times in the same conversation. In addition, these episodes were less richly and skillfully deployed (i.e., fewer productions combining verbal, prosodic, and gestural resources). In the interactions with an individual with amnesia, collaborative playful themes, where each participant contributes to the verbal play, were not sustained across stretches of interaction or returned to in subsequent interactions. These findings suggest that the ability to creatively and flexibly deploy the communicative and cognitive resources necessary to meet the moment-to-moment demands of interactional discourse is impaired following hippocampal damage. We note hippocampal specificity in meeting these demands as bilateral vmPFC damage does not impair reported speech (Duff et al., 2015) or verbal play (Gupta et al., 2012).

Language is generative, creative, and flexible. We have demonstrated that hippocampal dependent memory supports the creative and flexible use of language across discourse histories, contexts, and communication partners. We predict that any aspect of language use that places high demands on creative and flexible uses of language should recruit the hippocampus. We also note that researchers and clinicians routinely lament the challenge of detecting cognitive-communication disorders in clinical and research settings. This is most likely due to an overreliance on tasks (e.g., monologic narratives) and measures (e.g., t-units, story grammar) that are stripped of the rich communicative contexts and functions of everyday language. By using designs that place high demands on contextual language use and processing, we have been able to reveal disruptions in communication that more closely mirror the types of communication breakdown individuals with cognitive-communication disorders report outside of the lab or clinic.

18.4.3 Hippocampal Amnesia and Co-Speech Gestures

When people talk, they gesture. Influential theories of gesture argue that language and gesture arise from the same cognitive (linguistic) system (e.g., McNeill, 1992), a theoretical stance with which we agree. In the study of cognitive-communication disorders, however, gesture is often considered as part of the larger set of nonverbal behaviors that contribute to communication but are supralinguistic. Consistent with that tradition, we review work on hippocampal contributions to co-speech gestures here in the section on communication.

Gesture has communicative benefits for the speaker and the listener, as well as broader cognitive benefits including facilitating memory and learning (see

Clough & Duff, 2020 for review). We have been particularly interested in the links between gesture and memory and the examination of which memory systems may support the various communicative and cognitive functions of gesture. Examining co-speech gestures in individuals with hippocampal amnesia provides a window into the cognitive and neural links between language, gesture, and memory and provides a foundation for understanding co-speech gestures in other populations where memory impairments are common.

Gestures are tightly linked to spoken language semantically (McNeill, 1992), pragmatically (Kelly et al., 1999), and temporally (Habets et al., 2011). However, gestures do not simply communicate information redundant with spoken language. Rather, gestures, by communicating visually and motorically, can convey information not present in spoken language (McNeill, 1992). Moreover, the information conveyed in gesture is often unique to the gesture modality and not reflected in what is said (e.g., features of actions or physical space) (Cassell et al., 1999). These properties suggest that gesture may reflect content that is qualitatively different from that in spoken language, potentially reflecting types of memory representations that are not encoded in, or accessible to, speech. For example, when giving driving directions, a speaker may list in words the turns that should be made, mentioning landmarks and cardinal directions. Concurrently, speakers produce hand gestures that *show* the listener the way, reflecting on their hands the representation of the physical/spatial route that the speaker has in their mind.

Gesture has features that would suggest it is supported by nondeclarative memory. For example, gesture is action; gesture's benefits come from *producing* and *experiencing* gesture. This feature of gesture is reminiscent of the connection between some forms of procedural memory and the motor system, and is also similar to characterizations of nondeclarative memory as resulting directly from experience (Eichenbaum & Cohen, 2001; Reber et al., 1996). These shared properties suggest a link between gesture and nondeclarative memory.

Gesture also has features that would suggest a link to hippocampal dependent memory. When we communicate, we draw on rich mental representations to describe past events or speculate about future plans. The hippocampus plays a critical role in relational binding and flexibility for the reconstruction and recreation of richly detailed, multimodal, mental representations of experience (Eichenbaum & Cohen, 2001) and may support the representation we draw on when communicating. We also adjust our speech and gestures based on the knowledge of the listener and/or the assumed common ground among communication partners. For example, when communicating novel information to a listener, speakers produce gestures that are more complex and informative (Gerwing & Bavelas, 2004) and higher in the visual field (Hilliard & Cook, 2016) than when communicating shared or known information to a listener. As we described above, hippocampal dependent memory plays a critical role in some aspects of common ground processing. When we

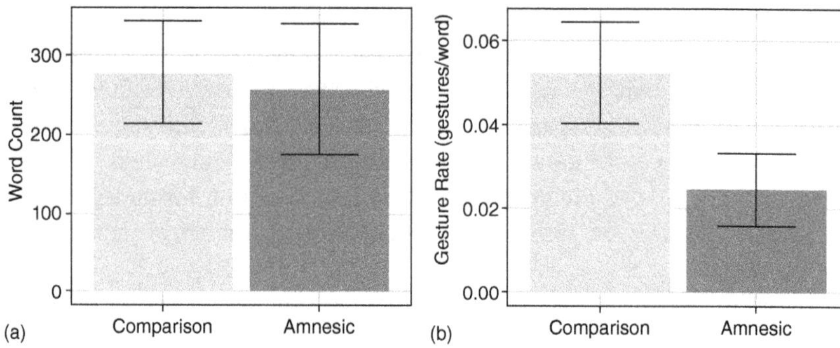

Figure 18.5 (A) Word count by participant group across all discourse topics. The amount of speech produced was highly variable both within and across the different narratives. (B) Gesture rate by group across all discourse narratives. Participants with hippocampal amnesia gestured at a lower rate than comparison participants. Error bars represent 95 percent confidence intervals.
Figure from Hilverman, C., Cook, S. W., & Duff, M. C. (2016). Hippocampal declarative memory supports gesture production: Evidence from amnesia. *Cortex, 85*, 25–36. Copyright Elsevier 2016.

began examining co-speech gestures, an open question was as if hippocampal dependent representations were supporting the tailoring of communication via gesture.

In the first study in this line of work, we examined the spontaneous gesture production of participants with amnesia as they produced narratives in response to four prompts (e.g., most frightening experience, how to make a sandwich) (Hilverman et al., 2016). We found that even though the participants with amnesia and neurotypical peers produced a similar number of words in their narratives, the participants with amnesia had a reduced gesture rate, producing significantly fewer gestures per word than the neurotypical peers (Figure 18.5). It is important to note that hippocampal amnesia does not produce motor impairments that would interfere with the physical ability to produce gestures. Moreover, for neurotypical comparison participants, the number of episodic features in a narrative was a positive predictor of gesture rate; participants gestured more as they provided more episodic detail. This relationship between episodic details and gesture production was not present in the participants with amnesia. These findings suggested a link between gesture production and the mental representations supported by hippocampal dependent memory.

We have also examined gesture comprehension in amnesia. Listeners integrate information from co-speech gestures with information in spoken language. For example, when hearing the utterance, “A piece of the log flew up and hit Carl in the face” while viewing a gesture directed at the nose, listeners tend to later report that the log hit Carl in the nose (information only in gesture) rather than in the face (information only in speech). We tested the possibility that the hippocampus, known for its role in relational memory and information integration, plays a role in integrating information in speech and gesture (Hilverman et al., 2018a). We asked participants to watch videos of an experimenter telling narratives that included hand gestures that contained information not in speech. Participants were asked to retell the narratives and we examined their retellings for the presence of information from gesture. Consistent with their severe memory impairment, individuals with amnesia recalled significantly fewer narrative features overall. Yet, their retellings included features that contained information that had been uniquely present in gesture in amounts that did not differ from the comparison group. Thus, hippocampal declarative memory is not necessary for gesture-speech integration over short timescales (i.e., immediate retellings).

Other studies demonstrate that the spontaneous co-speech gestures produced by individuals with amnesia are communicative and uniquely reveal information about their knowledge and experiences. As discussed above, when people share knowledge (common ground) with a communication partner, there is evidence of that shared knowledge in their speech: individuals produce fewer words in their referencing expressions and increase their use of definite references. When neurotypical speakers share common ground with a listener, they also attenuate their gestures producing fewer total gestures that are also lower in the visual field (Hilliard & Cook, 2016). Recall that in the collaborative referencing task in which participants accrue incremental common ground with a partner across multiple interactions around a matching game, people with amnesia arrived at increasing concise spoken labels with their partners (Duff et al., 2006), but they did not consistently use definite references like comparison participants (Duff et al., 2011a). In gesture, participants with amnesia signal common ground status in their gestures like comparison participants by producing fewer visible gestures above the barrier over the course of the task as shared knowledge increases (Hilverman et al., 2019). These findings suggested that gestures can reflect common ground status during conversation in the absence of explicit knowledge.

More recently we examined how individuals adapt their speech and gesture to meet the unique listener needs of a child vs an adult. Participants were asked to describe how to complete everyday tasks (e.g., how to tie a shoe), first to a live adult listener then, after a delay, to an imagined child listener. Although individuals with amnesia did not increase their total number of words and instructional steps for the child listener, they did produce representational gestures at significantly higher rates for the imagined child compared to the adult listener (Clough et al., 2022). Participants with amnesia and the comparison participants did not differ in the number of words produced or the number of steps given. Thus, if we had only examined spoken language for evidence of adaptation to listener knowledge and mental states, we would have missed that the adaptation in individuals with amnesia was happening in gesture and not speech. We also note that in this study individuals with amnesia gestured at similar frequencies to neurotypical peers, suggesting that hand gesture can be a meaningful communicative resource, in some contexts, even in the case of severe declarative memory impairment.

These studies provide evidence that speech and gesture can dissociate and, under certain conditions or task demands, may be differentially supported by distinct memory systems. Potential connections between nondeclarative memory and gesture (see Klooster et al., 2015) raise the possibility that individuals with declarative memory impairment may benefit from presentation of information in gesture to enhance communication and new learning. Indeed, preliminary evidence suggests that gestures can facilitate improved memory in a word learning task (Hilverman et al., 2018b). We asked participants with amnesia to learn novel word-object associations while producing gesture, observing gesture, or without gesture. Consistent with their profound memory impairment, participants with hippocampal amnesia were unable to freely recall the labels when tested. However, they demonstrated above chance recognition memory for labels when they produced gestures during encoding. This line of work offers exciting new directions leveraging gestures to facilitate new learning in individuals with memory impairment. More broadly, the study of gesture, language, and memory represents a novel research space for populations with cognitive-communication disorders which has largely been unexplored (Clough & Duff, 2020) but also holds promise for advancing the characterization and remediation of cognitive-communication disorders.

18.5 Contributions of Hippocampal Dependent Memory to Language

18.5.1 Hippocampal Amnesia and Online Language Processing

Language processing is incremental. Spoken language unfolds over time and many words and phrases are, at least temporarily, ambiguous. To meet the demands of incremental language processing, listeners must generate, integrate, and maintain information in real-time to create meaning. Processes associated with the prefrontal cortex, such as cognitive control and working memory, have long been linked with incremental language processing (e.g., MacDonald et al. 1992; Novick et al., 2005). Findings from behavioral and neuroimaging studies demonstrating recruitment and processing of hippocampal dependent memory over very short delays, and even no delays at all, have significant implications for mechanistic accounts of language processing and use. Applying these findings to the domain of language, we argued that new hippocampal dependent representations are available rapidly enough to influence ongoing language processing, making it a candidate mechanism for meeting the demands of incremental and online language processing (Duff & Brown-Schmidt, 2012). Findings suggesting recruitment and processing of hippocampal dependent memory in the moment, or on the same timescale as working memory, also had significant implications for conceptualizing language processing in individuals with hippocampal amnesia. The previous assumption was that individuals with isolated memory impairments have intact language processing and can follow and understand the thread of a conversation (Lackner, 1974; Milner et al., 1968). Thus, the consequence of a memory impairment on language was thought to be limited to remembering the content of what was said for later, after a significant delay, and was not thought to extend to how language was processed and understood in the moment.

We began testing the role of hippocampal dependent memory to online language processing by examining referential processing. Establishing and maintaining reference is a central component of language processing, as much of what we talk about involves referring to entities in the world around us. Referential processing is also an interesting target for studying online language processes as individuals must maintain a representation of the unfolding discourse history and work to resolve ambiguity as language unfolds in real time.

In our first study in this line of work, we assessed the role of hippocampal declarative memory in the use of common ground during online referential ambiguity resolution (Rubin et al., 2011) as representations of common ground are relevant to language production and comprehension in conversation. We created situations where linguistic ambiguity was eliminated if common ground information was taken into account. We had individuals with amnesia sit across a table from an experimenter, both of whom were in front of a computer that displayed a 3-D rendering of nine cubbyholes. Each cubbyhole contained a picture of an object, but, critically, not all the pictures were visible to both partners. Objects that were visible to both the participant and the experimenter were considered to be in common ground. Objects that were only visible to either the participant or the experimenter were considered to be in their respective privileged ground. Three of the nine objects were in common ground, three objects were in the participants' privileged ground, and the final three objects were in the experimenter's privileged ground. We monitored the eye movements of the participants as they engaged in the task. We were particularly interested in the eye fixations that the participants with amnesia made to the objects in the different cubbyholes as they interpreted the experimenter's sentences. On critical trials, displays contained two identical objects (e.g., two ducks). In one condition, common ground was cued visually; one duck was in an open cubbyhole (common ground), and the other duck was in the participant's privileged ground (i.e., designed to appear as if the experimenter could not view it). In another condition, both ducks were in the participants' privileged ground. The experimenter would then bring one duck into common ground verbally by asking, *What's in your bottom left cubby*? The participant's response ("a duck") would bring that object into common ground. Then, immediately, or after a brief delay (~40 seconds), the experimenter gave the critical instruction, *Look at <u>the</u> duck*, during which the participant's eye movements were monitored. Linguistic ambiguity (which duck was being referenced) can be eliminated by taking into account common ground information.

Like comparison participants, the participants with amnesia were able to rapidly use common ground when common ground was cued visually. The participants with amnesia could also successfully use verbal common ground when there was no delay. However, when there was a short (~40 seconds) delay between establishing common ground, and the ambiguous reference (e.g., *Look at the duck*), individuals with amnesia were equally likely to look at either of the two ducks. There was no evidence they knew that one of the ducks had just been brought into common ground less than a minute earlier. We also observed that even in trials when the target was successfully fixated, participants with amnesia were more likely than comparison participants to glance at the competitor object. This lingering competition raised questions about the distinctness of the representations in the participants with amnesia, and suggested that the hippocampus may influence the online resolution of competition during language processing (Rubin et al., 2011).

We have also examined the ability of individuals with hippocampal amnesia to understand a personal pronoun such as "he" or "she" when its referent appeared in a previous sentence. Personal pronouns can be highly ambiguous, and successful interpretation of pronouns typically draws on representations of the discourse history, built up using contextual information as well as past utterances. In the following example, the pronoun "he" in the second sentence is typically interpreted as referring to the subject of the first sentence:

David is playing the viola for *Rob* as the sun is shining overhead. *He* is wearing a burgundy shirt, and it looks like it will be a nice day.

While the pronoun itself is highly ambiguous, the subject of the prior sentence, "David," who was also mentioned first in that sentence, is likely to be the intended referent. Indeed, in a variant of the visual world paradigm, Arnold and colleagues (2000) (also see Kaiser et al., 2009) find that in situations like this, neurotypical young adults show a preference to fixate the subject of the first sentence, for example, "David," when interpreting the ambiguous pronoun. Using a similar paradigm, we asked whether this ability to use information about the relative prominence of two candidate referents introduced in a previous sentence, and are both depicted as copresent in a visual scene, is preserved or impaired in individuals with hippocampal amnesia (Kurczek et al., 2013). Even though these two-sentence stories were short (only about ten seconds long), participants with amnesia showed significant impairment relative to comparison participants. When interpreting the critical pronoun, for example, "he," the participants with amnesia only showed a slight tendency to fixate the first-mentioned character (e.g., "David") over the second-mentioned character (e.g., "Rob"). When gender provided an additional cue to the intended referent (e.g., "David is playing the viola for Natalie ... he..."), the participants with amnesia were more successful, but not nearly at the levels of comparison participants. Here again we studied a group of individuals with bilateral vmPFC damage as a brain-damaged comparison group and they performed similarly to neurotypical comparison participants.

These two studies point to a key role of hippocampal dependent memory in online language comprehension, specifically when there are high demands on resolving ambiguity. These data suggest that individuals with memory impairment may struggle with referential processing in daily conversation. In particular, impairments in amnesia seem to arise when information must be integrated across sentences or across temporal boundaries between elements of discourse. Although making information available in the immediate environment promotes some success, it does not raise performance to the level of the comparison participants without hippocampal damage and memory impairment. These findings provide initial insights into how the hippocampus might contribute to online sentence processing. Future work investigating how information from distinct temporal, physical, and speaker contexts are integrated in the moment and drawn upon to guide language production and comprehension is warranted.

A long-term goal of our work is to inform the development of effective, evidence-based interventions for individuals with memory impairment and cognitive-communication disorders. Building on the findings that hippocampal damage disrupts referential processing in individuals with amnesia (Kurczek & Duff, 2011; Kurczek et al., 2013), we wondered if performance could be improved by increasing the relative salience of a reference through repetition. For example, common communication breakdowns in populations where hippocampal damage and memory impairment are hallmark (e.g., Alzheimer's disease) have led to recommendations and expert opinions to caregivers for verbatim repetition of the most salient information and even avoiding the use of pronouns altogether (e.g., Azuma & Bayles, 1997; Ripich et al., 1998). This recommendation, however, lacks a strong evidentiary basis.

To test if we could improve performance through repetition, we asked participants to listen to short narratives and we monitored their eye movements while they looked at a related scene (Covington et al., 2020). Each narrative introduced two same-gender referents and then referred back to one of these referents with a pronoun. Because both potential referents were of the same gender, the pronoun was always ambiguous. The key manipulation was whether the first-mentioned referent was made more prominent through re-mention. A sample narrative can be found in Covington et al. (2020:174, Table 1).

If increasing salience of the first-mentioned referent through repetition improves pronoun resolution in patients with hippocampal amnesia, we thought this might suggest that the discourse representation is not entirely lost in amnesia, but rather weakened. Such a finding would also offer support for repetition as a candidate communicative intervention. If, however, patients with hippocampal amnesia were unable to take advantage of repetition, we thought this might point to a highly degraded, or lost, representation of the discourse history and offer little support for repetition as an effective communication strategy. What we found was that increasing the salience of the first-mentioned referent through repetition supported interpretation of pronouns; participants with hippocampal amnesia demonstrated improved pronoun processing following repetition. However, this effect was attenuated compared to comparison participants and did not raise the performance of individuals with amnesia to that of the comparison participants. See Covington et al. (2020: 179, Table 6).

Future studies should continue examining the role of compensatory strategies on language processing and use in individuals with memory impairment. Indeed, experimental designs aimed at boosting language performance in participants with hippocampal amnesia provide a unique opportunity to test additional hypotheses about the status of memory representations used in support of language processing and the instantiation of memory representations across distinct temporal, physical, and speaker contexts. Applying similar designs in other clinical populations with cognitive-communication disorders will help test clinical recommendations that have not yet been examined. This will contribute to a needed evidence base for communicative interventions tailored to individuals with memory impairment.

18.5.2 Hippocampal Amnesia and Statistical Learning

Statistical learning refers to the process of extracting underlying patterns from the environment, based on statistical regularities present in the stimulus input. Statistical learning has been shown to occur in humans from infancy through adulthood (e.g., Saffran et al., 1996; Conway & Christiansen, 2005) and is of particular interest to language researchers. Statistical learning is considered a potential mechanism in language acquisition including in the support of word segmentation and word learning from continuous speech (Graf Estes et al., 2007; Saffran et al., 1996), leveraging of adjacent and nonadjacent dependencies for syntax acquisition (Gómez, 2002; Saffran & Wilson, 2003), and tracking of phonotactic, orthographic, and morphological regularities (Chambers et al., 2003; Pacton et al., 2005).

While statistical learning has been well characterized behaviorally, less is known about its neural correlates. Given descriptions of statistical learning as an unconscious, implicit process, and the long-standing links between statistical learning and other implicit learning paradigms, such as artificial grammar learning and serial reaction time tasks (Kim et al., 2009; Perruchet & Pacton, 2006), initial accounts of the neural substrates underlying statistical learning emphasized contributions of the basal ganglia (Evans et al., 2009; Kim et al., 2009; Perruchet & Pacton, 2006). Consistent with recent neural network approaches to cognition, however, neuroimaging studies suggest that statistical learning is not the purview of a single neural system but rather the product of multiple systems working in parallel (e.g., Karuza et al., 2013; Sherman et al., 2023; Turk-Browne et al., 2009).

Of particular interest to our work on language and memory is evidence linking statistical learning to a neural network that includes the medial temporal lobes and the hippocampus. Turk-Browne and colleagues measured neural responses to structured versus random stimulus input in a visual statistical learning task. The study confirmed involvement of the basal ganglia in statistical learning but also extended the network supporting statistical learning to the medial temporal lobes and hippocampus (Turk-Browne et al., 2009). Hippocampal involvement in statistical learning fits with our understanding of the functionality of the hippocampus. For example, statistical learning requires the ability to encode and track the relations between individual stimuli, and a hallmark processing feature of the hippocampus is its ability to support the rapid binding of arbitrarily related elements that make up a scene or event and their temporal, spatial, and interactional relations (Eichenbaum & Cohen, 2001). Furthermore, evidence demonstrating hippocampal contributions to unconscious processing of relational binding in the moment (Hannula & Greene, 2012) position the hippocampus as a potential contributor to statistical learning.

The most compelling initial evidence for a link between hippocampus and statistical learning came from a study demonstrating impaired statistical learning in a single patient with extensive medial temporal lobe damage (Schapiro et al,, 2014). This patient had bilateral hippocampal damage as well as more extensive damage to surrounding the medial temporal lobe cortex and left anterior temporal lobe as a result of herpes simplex encephalitis. The patient failed to show statistical learning across stimuli modalities including shapes, syllables, scenes, and tones. The authors concluded that the medial temporal lobes are critical in statistical learning but acknowledged that the specific role of the hippocampus could not be determined given the extensive damage to the medial temporal lobes.

We conducted a study to test the necessity of the hippocampus in statistical learning by replicating methods from Schapiro and colleagues (2014) in a larger group of patients who have either localized damage to the hippocampus or broader medial temporal lobe damage (Covington et al., 2018). Patients with hippocampal damage consistently showed less learning overall compared with comparison participants (Figure 18.6). This result was consistent with the emerging evidence for hippocampal contributions to statistical learning. Interestingly, lesion size did not reliably predict performance. However, patients with hippocampal damage were not uniformly at chance and demonstrated above-chance performance in some task variants. We concluded that these results suggested that the hippocampus is necessary for statistical learning levels achieved by most neurotypical comparison participants, but significant hippocampal pathology alone does not abolish the capacity for statistical learning.

Evidence linking statistical learning and the hippocampus leads to the possibility that the hippocampus contributes to a broader range of

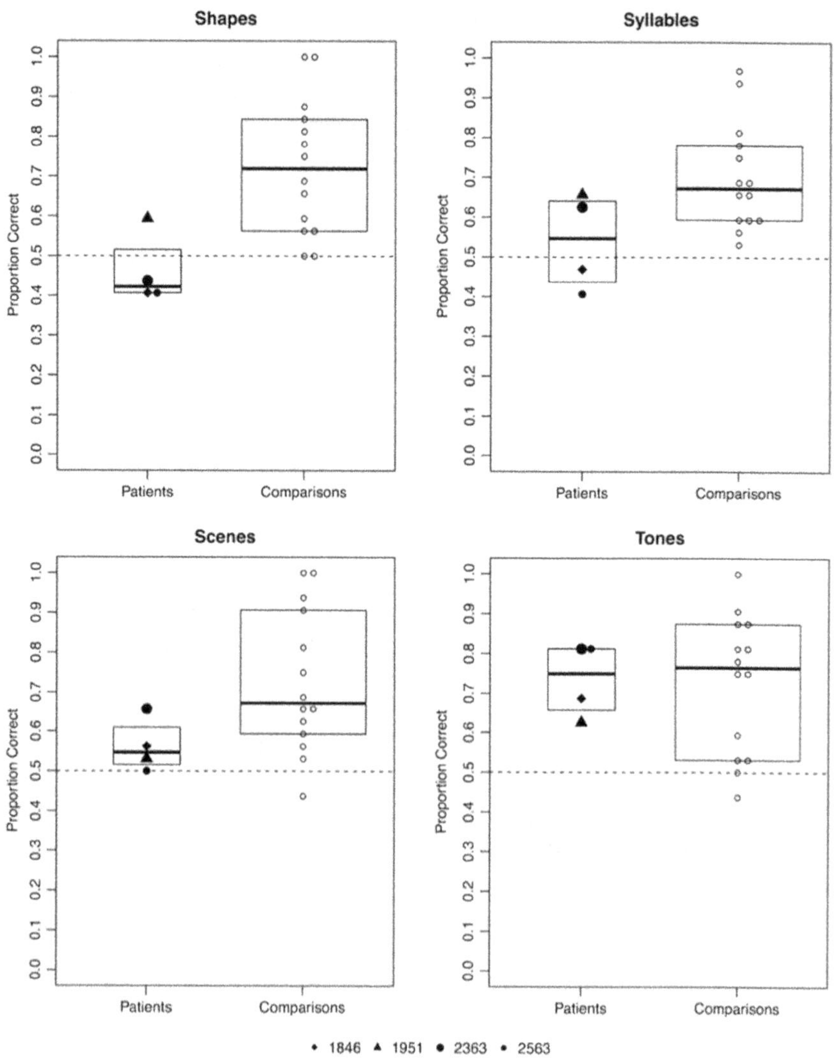

Figure 18.6 Data from Experiment 1; proportion correct across four versions of the statistical learning task by patients with amnesia and non-injured comparison participants. Boxplots indicate the group level interquartile range.

Figure from Covington, N. V., Brown-Schmidt, S., & Duff, M. C. (2018). The necessity of the hippocampus for statistical learning. *Journal of Cognitive Neuroscience, 30*(5), 680–697. Reprinted by permission of The MIT Press.

grammatical functions than has been examined to date. More work investigating the role of the hippocampus in statistical learning and its implications for language are needed. The link between statistical learning and the hippocampus may also lead to new mechanistic accounts of language deficits across different clinical populations. For example, individuals with developmental language disorder, which is characterized by deficits in morphosyntactic components of language that are thought to be acquired through statistical learning (Evans et al., 2009), demonstrate significant correlations between hippocampal volume and language measures (Lee et al., 2013; also see Lee et al., 2020).

Integrity of statistical learning ability may also be an important factor in language intervention and rehabilitation. In individuals with aphasia in the acute phase of recovery, left hippocampal volume predicted long-term language outcomes over and above what was predicted by lesion size and location and acute language deficits information (Schevenels et al., 2022). The researchers also reported that visuomotor statistical learning was correlated with acute hippocampal gray and white matter suggesting that left hippocampal integrity may be a potential marker of language recovery after stroke, perhaps through its capacity for statistical learning. Thus, not only does research linking the hippocampus to language promise to lead to biologically and psychologically plausible models of the role of memory in language use and processing, but it may also hold the key to better predictors of language outcome and interventions for neurogenic language disorders. Finally, the hippocampus may play a long-term role in maintaining information about statistical regularities beyond initial acquisition. We studied collocates (i.e., words that tend to co-occur with high statistical regularity) in individuals with amnesia. Participants were asked to judge acceptable collocates, for example, to endorse that "sudden noise" is permissible in English while "sudden doctor" is not. Individuals with amnesia performed significantly worse than non-injured and brain-damaged comparison participants with vmPFC damage at making these judgments (Klooster & Duff, 2015). This result suggests that the hippocampus may play a long-term role in maintaining information about the relations among words, possibly through its role in statistical learning.

18.5.3 Hippocampal Amnesia and Semantic Representation Over Time

The role of the hippocampus in new semantic learning (i.e., the acquisition of words, facts, and general knowledge) is well established (see Duff et al., 2020 for review). The traditional view, though, has been that this role is time limited as semantic representations were thought to become independent of the hippocampus over time via neocortical consolidation (McClelland et al., 1995). Evidence for this view came from data from individuals with hippocampal amnesia who did not have aphasia or semantic dementia, who performed within normal limits on neuropsychological measures of vocabulary knowledge and naming (Kensinger et al., 2001) and who perform similarly to neurotypical comparison participants on experimental tasks to assess remote word knowledge like naming or matching a label with a phrase, definition, or sentence (Gabrieli et al., 1988; Manns et al., 2003; Verfaellie et al., 2000). Taken together, these data led to the conclusion that remote semantic memory (information acquired before the onset of hippocampal damage) was intact in amnesia.

Many standardized neuropsychological tests, however, were developed to capture frank disorders such as aphasia or semantic dementia and likely lack the sensitivity to detect individual differences in naming or semantic knowledge. The same is likely true for many of the experimental tasks used to examine semantic memory in individuals with amnesia. For example, previous studies showed participants a picture of a common object such as an apple and asked participants to name it, or match the label apple to a definition, or determine if A-P-P-L-E is a real English word. Tasks like this are certainly useful in identification of severe deficits in semantic memory or naming but are unlikely to capture deficits that may fall along a broader continuum of ability.

In our first study of remote semantic memory in amnesia, we examined how much information is associated with highly familiar words that were likely acquired long before the onset of amnesia (Klooster & Duff, 2015). The tasks included a word associates test (identifying synonyms and common collocates; mentioned above), a word senses task (name all the senses of a word; e.g., lemon can be a fruit, a color, a defective automobile) and a word features task (name all of the features of a word; e.g., lemon tastes sour, is native to Asia, used in tea). Participants with amnesia performed significantly worse than neurotypical and brain-damaged comparison groups, on all three measures of word knowledge. For example, patients with amnesia generated, on average, only half the number of features for common words (e.g., shirt) as either comparison group. This deficit was even evident despite showing no differences from comparison participants on self-reported rates of familiarity of words used in the word features and senses tasks (scoring familiarity with a particular word on a nine-point scale). Importantly, the fact that the patients knew these words (i.e., had high familiarity ratings), suggests that they likely would have performed like comparison participants with traditional measures (e.g., naming) that only assess surface level semantic knowledge. Using tasks and measures that assess semantic richness, or depth of semantic knowledge, individuals with hippocampal amnesia perform significantly worse than comparison groups suggesting their remote semantic memory is impoverished.

The findings from Klooster and Duff (2015) suggested to us that the hippocampus may play a long-term role in semantic memory beyond acquisition to include maintaining semantic representations across the lifetime. Consistent with views of word learning as a protracted process unfolding over days, weeks, months, and even years (Carey, 2010; McMurray et al., 2012), we proposed that the hippocampus, in its capacity for relational binding (Cohen & Eichenbaum, 1993; Eichenbaum & Cohen, 2001), reconsolidation (McKenzie & Eichenbaum, 2011), and flexible integration of information (Zeithamova & Preston, 2010), is well situated to meet the demands of maintaining and updating semantic memory as more information is associated with a word or concept over time and with experience (e.g., Laszlo & Federmeier, 2011; McGregor et al., 2002). From this perspective, the hippocampus may strengthen and enrich existing semantic representations as words are retrieved and used in novel contexts.

We were next interested in differentiating if the findings from Klooster and Duff (2015) reflected a deficit in enriching semantic representations over time or if individuals with amnesia were losing semantic knowledge (i.e., language attrition). The best method for testing this would be to systematically probe semantic knowledge in participants with hippocampal amnesia from the onset of their amnesia. This type of study, for our group at least, is not possible as our participants had been living with their memory impairment for about twenty years, on average (range of fourteen–thirty-four years). The approach we took instead was to recruit a new group of comparison participants. In contrast to our previous work where we matched the participants with amnesia to a comparison group who were the same age as the participants with amnesia at the time of testing (chronological age (CA)), we next recruited comparison participants who were matched to each patient's age of onset (AoO) of their brain damage and resulting amnesia. Thus, this new comparison group was the same age as the patients with amnesia were when they incurred their brain damage and amnesia. Participants were also matched on years of education. We used the tasks as in Klooster and Duff (2015) with the AoO group including the word features task, word senses tasks, and word associates test.

We found that the older CA comparison group performed significantly better than the younger AoO group on all tasks suggesting that with age, neurotypical individuals add new features, senses, and associates to existing semantic representations (Figure 18.7; Klooster et al., 2020). With many more years of experience and use, the older CA group had significantly more information associated with each lexical/conceptual item even though all the stimuli used in the study were

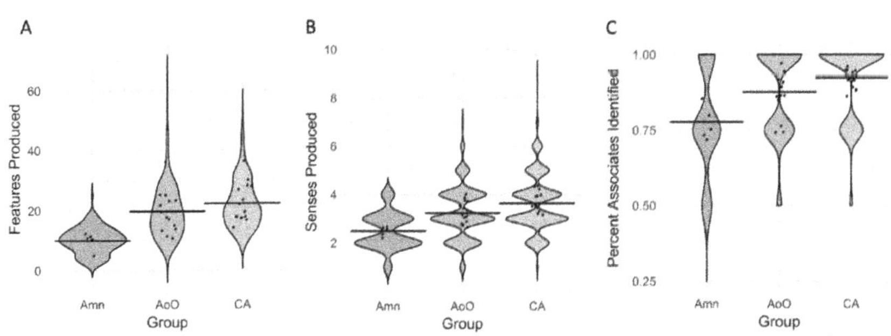

Figure 18.7 Data from participants with hippocampal amnesia (Amn), younger neurotypical comparisons matched to the patients' age of onset (AoO), and older neurotypical comparisons matched to the participants with amnesia's current age (CA). Group means are represented by the black bars. Participant means are represented with black dots. The violin plots (gray shapes) represent the probability density of all observations with wider areas indicating a greater number of observations at that value within that group. Panel A, Word Features Task. Panel B, Word Senses Task. Panel C, Word Associates Task.

Figure adapted from Klooster, N. B., Tranel, D., & Duff, M. C. (2020). The hippocampus and semantic memory over time. *Brain and Language, 20.*

rated as highly familiar to all participants. Thus, the previously reported difference in performance between the participants with amnesia and their chronological age-matched comparison participants becomes all the more striking as it highlights the amount of semantic knowledge the participants with amnesia would have likely acquired over time if not for their hippocampal pathology. This finding provides evidence for the notion that hippocampal damage and amnesia interferes with the ability to add to or enrich semantic representations over time. However, we also found that the group of comparison participants matched to the age the patients were at the time of onset (AoO) of hippocampal damage and amnesia also performed significantly better than the participants with amnesia. This finding raises the possibility that hippocampal damage is associated with some degree of language attrition over time. While longitudinal data from participants with hippocampal pathology are required to test the interpretation that semantic memory is lost over time, such a proposal warrants further study as it could offer key insights into language attrition associated with aberrant aging and neurodegenerative diseases which are frequently associated with changes in the structure and function of the hippocampus.

These results linking hippocampal damage and amnesia to deficits in remote semantic memory led us to reexamine naming ability in individuals with amnesia using a more extensive set of items than are typically used in standardized naming tests or experimental tasks of naming. We conducted a naming task similar to previous studies in which patients with amnesia and neurotypical comparison participants viewed color photographs of items and were instructed to provide a name for the picture (Hilverman & Duff, 2021). Unlike previous naming studies that all contained fewer than 100 images, we used 1,458 items from the Bank of Standardized Stimuli (BOSS) database (Brodeur et al., 2014) that varied across a range of lexical features such as imageability, frequency, and familiarity. By using a wide range of image-word pairs, even subtle differences in naming between participants with amnesia and non-injured comparison participants may be detected. Unlike previous tests of naming in this population, we found that participants with amnesia were less likely than comparison participants to correctly name the objects that they viewed. Furthermore, patients with amnesia were more likely to provide a general label for an object (e.g., *bird* for a *cardinal*) than the comparison participants. Some of the errors that the participants with amnesia made were surprising. For example, one participant with amnesia was the only participant to not correctly name a pair of skis (referring to them as *two paint brushes*). Prior to the onset of his amnesia, he was an avid skier and worked for many years at a ski resort. Another participant with amnesia was the only participant to not correctly name a lizard (referring to it as *rhinoceros*) despite living in a tropical climate and regularly seeing lizards in the wild. Finally, another participant with amnesia was the only participant to not correctly name a syringe (referring to it as *thermometer*) despite working for almost two decades in a medical doctor's office, with one of her primary duties being stocking the supplies. Using a wider range of materials to examine naming ability, we detected evidence for disruptions in naming following hippocampal amnesia.

Examination of language production also reveals disruptions as individuals with amnesia use words rated as less semantically rich relative to comparison participants. In this study, we analyzed the features of words used when participants described events, both past and imagined (Hilverman et al., 2017). Features of words reflect characteristics of what the word describes (e.g., a word's imageability measures the degree to which the word invokes an image in one's mind). Although participants with amnesia are known to produce significantly fewer episodic details in their descriptions of events (Race et al., 2011), the specific words that are used are not necessarily related to the number of episodic details; similar representations can be communicated with the same number of episodic details but using words that vary considerably in their imageability and concreteness. For example, an individual might say, "I was on a jet ski on a nice summer day and water was hitting my face as I went across the lake" or "I was riding a jet ski on a bright summer day and water was spraying my face as I sped across the lake". In both cases, the number of episodic details is the same, but the imageability and concreteness of the words used are much greater in the second example. We found that participants with hippocampal amnesia used words that were significantly less imageable than the comparison participants (Hilverman et al., 2017). This was found even when controlling for the number of overall features in the narrative and word frequency. This finding fits with data from Heyworth and Squire (2019) who found that in narrative recollections of a guided walk, individuals with amnesia used higher-frequency and less concrete words than controls. Thus, even in semi-naturalistic speaking contexts, individuals with amnesia use language that is semantically impoverished.

Individuals with hippocampal damage and amnesia demonstrate disruptions in remote semantic knowledge and use of semantic representations across a range of tasks and language contexts. It is difficult to disentangle the role the hippocampus might play in maintaining existing semantic representations (i.e., the process of strengthening representations and/or protecting against attrition) versus its role in new semantic learning (i.e., the process of acquiring new semantic representations and adding new information to existing representations). That said, there may not be a real distinction, functionally or mechanistically. Maintaining previously acquired semantic representations requires engagement or use. The act of retrieving and producing known semantic representations is in itself a form of learning that changes the nature and strength of those representations as well as the relations among other lexical items (Oppenheim et al., 2007, 2010). Semantic knowledge is also dynamic as individuals update and add to their knowledge of lexical and conceptual information across time. Even in experimental contexts conceived of as purely new learning (i.e., pseudo words and novel objects), the new information mingles with existing knowledge as phonological, orthographic, and conceptual neighbors are activated, held online, and compared. Thus, acquiring a new semantic representation and maintaining an existing representation are both forms of learning that may place similar demands on the hippocampus (see Duff et al., 2020; Hilverman & Duff, 2021 for further discussion).

Taken all together, these findings challenge the historical view that remote semantic memory is intact in amnesia and that semantic memory becomes independent of the hippocampus over time. Our work also suggests that the hippocampus may play a long-term role in maintaining, strengthening, and enriching semantic representations over time. An important implication of these findings is that there are likely more far-reaching disruptions in semantic memory representation and use than previously documented in populations where hippocampal pathology is common.

18.6 Conclusions and Future Directions

The study of individuals with hippocampal damage and amnesia has provided a compelling opportunity to directly test the role of a specific form of memory to communication and language. Our work has focused on testing the proposal that the same processes by which the hippocampal dependent memory system forms and subsequently retrieves new memories to be used in service of behavioral performance, are the same processes necessary for meeting many of the demands of language processing and use (Duff & Brown-Schmidt, 2012, 2017). We have linked hippocampal dependent memory to both communication and language and revealed graded contributions of this form of memory depending on the demands of the particular communicative or language function.

Converging methodological approaches provide additional support for our proposal and extend the evidence base in exciting and novel directions. For example, research using intracranial recordings from depth electrodes placed in the hippocampus of patients with intractable epilepsy demonstrates that the hippocampus contributes to retrieving the arbitrary associations between objects and their names required for naming (Hamamé et al., 2014; also see Jafarpour et al., 2017) and to the encoding, tracking, and building of semantic relations across words during online sentence processing (Piai et al., 2016). These findings provide direct physiological evidence for hippocampal contributions to language and have led to new proposals on the role of the hippocampus in predictive language processing (Bonhage et al., 2015; Piai et al., 2016). Neuroimaging work has revealed a language-memory network and researchers have investigated the reorganization of this network in temporal lobe epilepsy where hippocampal dysfunction is a hallmark feature (Banjac et al., 2021; Roger et al., 2022). These new directions, together with our work, are advancing our understanding of the neurobiology of language by characterizing the contribution of hippocampal dependent memory to language and communication and the role of the hippocampus in the language network (Blank et al., 2016).

Research documenting the interaction and interdependencies of language and memory also has implications for mechanistic accounts and clinical management of populations where co-occurring disorders of language and memory are common (e.g., aphasia, TBI, AD, DLD, schizophrenia). First, our research suggests that hippocampal damage or dysfunction is a risk factor for a range of language and communicative disruptions even in the absence of frank disorders of amnesia or aphasia. Given how prevalent hippocampal dysfunction is across a range of developmental and adult-onset disorders, future research is needed to develop assessments that have the sensitivity to detect a range of language and communicative ability and the specificity to detect hippocampal dependent language and communication impairments. Second, linking hippocampal dependent memory to language and communication holds particular translational promise and significance given the success of interventions known to improve the structure and function of the hippocampus in children, young and older adults, and disordered populations. For instance, hippocampal dependent memory benefits from a range of lifestyle and noninvasive treatments such as neuromodulation (Wang et al., 2014), nutrition (Monti et al., 2014), sleep (Stickgold, 2005), and physical activity interventions (Erickson et al., 2011), which, offered alone or in combination with behavioral treatments, could hold tremendous promise for improving language and communicative outcomes. Indeed, we recently found that individuals with moderate to severe TBI have significant word learning deficits relative to non-injured peers but individuals with and without TBI both remembered more words when they slept after learning (Morrow et al., 2023). Future research is warranted to investigate whether interventions aimed at improving the structure and function of the hippocampus improve language and communication outcomes.

Finally, our work showing deficits in individuals with hippocampal amnesia encompassing both communication and language raises questions about the reality and utility of the historical distinction between communication and language in defining cognitive-communication disorders as distinct from other language disorders such as aphasia. This distinction was based, in part, on the observation that populations we now refer to as having cognitive-communication disorders such as individuals with TBI or AD have disruptions in communication, at the levels of discourse and conversation, while the rapid production and processing of words and individual utterances are preserved. However, much of what we know about language ability in populations with cognitive-communication disorders come from performance on standardized aphasia batteries and offline language tasks. In fact, we know little about language functions such as online language processing, statistical learning, or semantic representation in populations with cognitive-communication disorders. We speculate this lack of basic knowledge is a consequence of the historical distinction between language and communication and the assumption that such language abilities should be preserved. We note that such language abilities were also assumed to be intact in individuals with hippocampal amnesia until empirically tested with designs and materials with sufficient sensitivity to detect impairment. While this distinction was helpful in demonstrating broad differences between categories of language disorders, we suggest it is arbitrary and fails to capture the meaningful patterns of spared and impaired communication and language functions associated with different neurogenic disorders. Moving forward, to better understand the interactions and interdependencies of language and memory and to advance language and communicative outcomes, it may be more fruitful to focus on the underlying mechanisms of shared communication and language deficits (e.g., hippocampal dependent memory) and on methods to detect a range of language and memory abilities rather than on the arbitrary distinction between communication and language. Such an approach might be particularly impactful for the characterization and clinical management of the numerous disorders that have in common disruption of hippocampal dependent memory and communication and language functions.

This work was supported by NIDCD R01DC017926 to Melissa C. Duff

Notes

1 Following de Marchena & Miller (2017), we use the term "frank" from the medical literature to mean a distinctive behavioral presentation that is clinically evident and unmistakable.

References

Addis, D. R., & Schacter, D. L. (2012). The hippocampus and imagining the future: Where do we stand? *Frontiers in Human Neuroscience*, *5*, 173. https://doi.org/10.3389/fnhum.2011.00173

Allen, J., Tranel, D., Bruss, J., & Damasio, H. (2006). Correlations between regional brain volumes and memory performance in anoxia. *Journal of Clinical and Experimental Neuropsychology*, *28*, 457–476.

Aly, M., & Turk-Browne, N. B. (2017). How hippocampal memory shapes, and is shaped by, attention. In D. Hannula & M. C. Duff (Eds.), *The Hippocampus from Cells to Systems: Structure, Connectivity, and Functional Contributions to Memory and Flexible Cognition* (pp. 369–403). Springer International Publishing AG.

American Speech-Language-Hearing Association. (1987). Role of speech-language pathologists in the habilitation and rehabilitation of cognitively impaired individuals. Reprinted from American Speech-Language-Hearing Association Vol. 29, pp. 53–55.

American Speech-Language-Hearing Association. (2005). Roles of speech-language pathologists in the identification, diagnosis, and treatment of individuals with cognitive-communication disorders: Position statement. www.asha.org/policy

Arnold, J., Eisenband, J., Brown-Schmidt, S., & Trueswell, J. (2000). The rapid use of gender information: Evidence of the time course of pronoun resolution from eyetracking. *Cognition, 76,* 813–826.

Azuma T., & Bayles, K. (1997). Memory impairments underlying language difficulties in dementia. *Topics in Language Disorders, 18*(1), 58–71.

Banjac, S., Roger, E., Pichat, C., Cousin, E., Mosca, C., Lamalle, L., Krainik, A., Kahane, P., & Baciu, M. (2021). Reconfiguration dynamics of a language-and-memory network in healthy participants and patients with temporal lobe epilepsy. *NeuroImage: Clinical, 31,* 102702.

Barense, M. D., Gaffan, D., & Graham, K. S. (2007). The human medial temporal lobe processes online representations of complex objects. *Neuropsychologia, 45*(13), 2963–2974.

Blank, I. A., Duff, M. C., Brown-Schmidt, S., & Fedorenko, E. (2016). Expanding the language network: Domain-specific hippocampal recruitment during high-level linguistic processing. www.biorxiv.org/content/10.1101/091900v1

Bonhage, C. E, Mueller, J. L, Friederici, A. D., & Fiebach, C. J. (2015). Combined eye tracking and fMRI reveals neural basis of linguistic predictions during sentence comprehension. *Cortex, 68,* 33–47.

Broca, P. (1865). Sur la faculte du langage articule. *Bull. Soc. Anthropol., 6,* 337–393.

Brodeur, M. B., Guérard, K., & Bouras, M. (2014) Bank of Standardized Stimuli (BOSS) phase II: 930 new normative photos. *PLoS ONE, 9*(9), e106953. https://doi.org/10.1371/journal.pone.0106953

Brown-Schmidt, S., & Duff, M.C. (2016). Memory and common ground processes in language use. *Topics in Cognitive Science, 8*(4), 722–736.

Buckner, R. L., & Carroll, D. C. (2007). Self-projection and the brain. *Trends in Cognitive Science, 11,* 49–57.

Carey, S. (2010). Beyond fast mapping. *Language Learning and Development, 6*(3), 184–205.

Cassell, J., McNeill, D., & McCullough, K.-E. (1999). Speech-gesture mismatches: Evidence for one underlying representation of linguistic and nonlinguistic information. *Pragmat. Cogn. 7,* 1–34.

Cavaco, S., Anderson, S. W., Allen, J. S., Castro-Caldas, A., & Damasio, H. (2004). The scope of preserved procedural memory in amnesia. *Brain, 127*(8), 1853–67.

Chambers, K. E., Onishi, K. H., & Fisher, C. (2003). Infants learn phonotactic regularities from brief auditory experience. *Cognition, 87,* B69–B77.

Cherney, L. R., Shadden, B. B., & Coelho, C. A. (1998). *Analyzing Discourse in Communicatively Impaired Adults Gaithersburg, MAQ2*. Aspen Publishers.

Clark, H. H. (1996). *Using Language*. Cambridge University Press.

Clark, H. H., & Marshall, C. R. (1978). Reference diaries. In D. L. Waltz (Ed.), *Theoretical Issues in Natural Language Processing*, vol 2. (pp. 57–63) Association for Computing Machinery.

Clark, H. H., & Wilkes-Gibbs, D. (1986). Referring as a collaborative process. *Cognition, 22*, 1–39.

Clough, S., & Duff, M. C. (2020). The role of gesture in communication and cognition: Implications for understanding and treating neurogenic communication disorders. *Frontiers in Human Neuroscience, 14*. https://doi.org/10.3389/fnhum.2020.00323

Clough, S., Hilverman, C., Brown-Schmidt, S., & Duff, M. C. (2022). Evidence of audience design in amnesia: Adaptation in gesture but not speech. *Brain Sciences, 12*(8), 1082. https://doi.org/10.3390/brainsci12081082

Cohen, N. J., & Eichenbaum, H. (1993). *Memory, Amnesia, and the Hippocampal System*, vol. 3. MIT Press.

Cohen, N. J., Poldrack, R. A., & Eichenbaum, H. (1997). Memory for items and memory for relations in the procedural/declarative memory framework. *Memory, 5*(1–2), 131–78.

Conway, C. M., & Christiansen, M. H. (2005). Modality-constrained statistical learning of tactile, visual, and auditory sequences. *Journal of Experimental Psychology. Learning, Memory, and Cognition*, 31(1), 24–39. https://doi.org/10.1037/0278-7393.31.1.24

Corballis, M. C. (2019). Language, memory, and mental time travel: An evolutionary perspective. *Frontiers in Human Neuroscience, 13*, 217. https://doi.org/10.3389/fnhum.2019.00217

Coronel, J. C., Duff, M. C., Warren, D. E., Gonsalves, B. D., Tranel, D., & Cohen, N. J. (2012). Remembering and voting: Theory and evidence from amnesic patients. *American Journal of Political Science, 56*(4), 837–848.

Covington, N. V., Brown-Schmidt, S., & Duff, M. C. (2018). The necessity of the hippocampus for statistical learning. *Journal of Cognitive Neuroscience, 30*(5), 680–697.

Covington, N. V., & Duff, M. C. (2016). Expanding the language network: Direct contributions from hippocampus. *Trends in Cognitive Sciences, 20*(12), 869–870.

Covington, N. V., Kurczek, J., Duff, M. C., & Brown-Schmidt, S. (2020). The effect of repetition on pronoun resolution in patients with memory impairment. *Journal of Clinical and Experimental Neuropsychology, 42*(2), 171–184.

Crystal, D. (1998). *Language Play*. University of Chicago Press.

de Marchena, A., & Miller, J. (2017), "Frank" presentations as a novel research construct and element of diagnostic decision-making in autism spectrum disorder. *Autism Research, 10*, 653–662. https://doi.org/10.1002/aur.1706

Doedens, W., Bose, A., Lambert, L., & Meteyard, L. (2021). Face-to-face communication in aphasia: The influence of conversation partner familiarity on a collaborative communication task. *Frontiers in Communication, 6*, 574051. https://doi.org/10.3389/fcomm.2021.574051

Duff, M. C., & Brown-Schmidt, S. (2012). The hippocampus and the flexible use and processing of language. *Frontiers in Human Neuroscience, 6*.https://doi.org/10.3389/fnhum.2012.00069

Duff, M. C., & Brown-Schmidt, S. (2017). Hippocampal contributions to language use and processing. In D. Hannula & M. C. Duff (Eds.), *The Hippocampus from Cells to Systems: Structure, Connectivity, and Functional Contributions to Memory and Flexible Cognition* (pp. 503–536). Springer International Publishing.

Duff, M. C., Covington, N., Hilverman, C., & Cohen, N. J. (2020). Semantic memory and the hippocampus: Revising, reaffirming, and extending the reach of their critical relationship. *Frontiers in Human Neuroscience, 13*, 471. https://doi.org/10.3389/fnhum.2019.00471

Duff, M. C., Gallegos, D., Cohen, N. J. & Tranel, D. (2013). Learning in Alzheimer's disease is facilitated by social interaction and common ground. *Journal of Comparative Neurology, 521*(18), 4356–4369.

Duff, M. C., Gupta, R., Hengst, J., Tranel, D., & Cohen, N. J. (2011a). The use of definite references signals declarative memory: Evidence from hippocampal amnesia. *Psychological Science, 22*(5), 666–673.

Duff, M. C., Hengst, J., Gupta, R., Tranel, D. & Cohen, N. J. (2011b). Distributed impact of cognitive-communication impairment: Disruptions in the use of definite references when speaking to individuals with amnesia. *Aphasiology, 25*(6–7), 675–687.

Duff, M. C., Hengst, J., Tengshe, C., Krema, A., Tranel, D., & Cohen, N. J. (2008). Hippocampal amnesia disrupts the flexible use of procedural discourse in social interaction. *Aphasiology, 22*(7, 8), 866–880.

Duff, M. C., Hengst, J., Tranel, D., & Cohen, N. J. (2006). Development of shared information in communication despite hippocampal amnesia. *Nature Neuroscience, 9*(1), 140–146.

Duff, M. C., Hengst, J., Tranel, D., & Cohen, N. J. (2007). Talking across time: Using reported speech as a communicative resource in amnesia. *Aphasiology, 21*(6, 7, 8), 702–716.

Duff, M. C., Hengst, J., Tranel, D., & Cohen, N. J. (2009). Hippocampal amnesia disrupts verbal play and the creative use of language in social interaction. *Aphasiology, 23*(7), 926–939.

Duff, M. C., Kurczek, J., & Miller, M. (2015). Use of reported speech in the communicative interactions of individuals with ventromedial prefrontal cortex damage. *Journal of Interactional Research in Communication Disorders, 6*(1), 97–114.

Duff, M. C., Kurczek, J., Rubin, R., Cohen, N. J., & Tranel, D. (2013). Hippocampal amnesia impairs creative thinking. *Hippocampus, 23*(12), 1143–1149.

Eichenbaum, H., & Cohen, N. J. (2001). *From Conditioning to Conscious Recollection: Memory Systems of the Brain.* Oxford University Press.

Erickson, K. I., Voss, M. W., Prakash, R. S., Basak, C., Szabo, A., Chaddock, L., Kim, J. S., et al. (2011). Exercise training increases size of hippocampus and improves memory. *PNAS, 108*(7), 3017–3022. https://doi.org/10.1073/pnas.1015950108

Evans, J. L., Saffran, J. R., & Robe-Torres, K. (2009). Statistical learning in children with specific language impairment. *Journal of Speech, Language, and Hearing Research, 52,* 321–335.

Gómez, R. L. (2002). Variability and detection of invariant structure. *Psychological Science, 13,* 431–436.

Graf Estes, K., Evans, J. L., Alibali, M. W., & Saffran, J. R. (2007). Can infants map meaning to newly segmented words? Statistical segmentation and word learning. *Psychological Science, 18,* 254–260.

Federmeier, K., McLennan, D., De Ochoa, E., & Kutas, M. (2002). The impact of semantic memory organization and sentence context information on spoken language processing by younger and older adults: An ERP study. *Psychophysiology, 39*(2), 133–146.

Fedorenko, E., & Thompson-Schill, S. L. (2014). Reworking the language network. *Trends in Cognitive Science, 18*(3), 120–126.

Friederici, A. D., & Gierhan, S. M. (2013). The language network. *Current Opinions in Neurobiology, 23*(2), 250–254.

Gabrieli, J. D. E., Cohen, N. J., & Corkin, S. (1988). The impaired learning of semantic knowledge following bilateral medial temporal-lobe resection. *Brain Cognition, 7,* 157–177.

Gerwing, J., & Bavelas, J. (2004). Linguistic influences on gesture's form. *Gesture, 4,* 157–195.

Gupta, G. R., & Duff, M. C. (2016). Incorporating principles of the collaborative contextualized intervention approach with the empirical study of learning and communication in traumatic brain injury. *Aphasiology, 30*(12), 1461–1482.

Gupta, R., Tranel, D., & Duff, M. C. (2012). Ventromedial prefrontal cortex damage does not impair the development and use of common ground in social interaction: Implications for cognitive theory of mind. *Neuropsychologia, 25*(2), 137–146.

Habets, B., Kita, S., Shao, Z., Ozyurek, A., & Hagoort, P. (2011). The role of synchrony and ambiguity in speech-gesture integration during comprehension. *Journal of Cognitive Neuroscience, 23*(8), 1845–1854.

Hamamé, C. M., Alario, F. X., Llorens, A., Liégeois-Chauvel, C., & Trébuchon-Da Fonseca, A. (2014). High frequency γ activity in the left hippocampus predicts visual object naming performance. *Brain Language, 135,* 104–114.

Hannula, D., & Duff, M. C. (Eds.). (2017). *The Hippocampus from Cells to Systems: Structure, Connectivity, and Functional Contributions to Memory and Flexible Cognition.* Springer International Publishing.

Hannula, D. E., & Greene, A. (2012). The hippocampus reevaluated in unconscious learning and memory: At a tipping point? *Frontiers in Human Neuroscience, 6,* 80. https://doi.org/10.3389/fnhum.2012.00080

Hannula, D. E., & Ranganath, C. (2009). The eyes have it: Hippocampal activity predicts expression of memory in eye movements. *Neuron, 63*(5), 592–599.

Hannula, D. E., Tranel, D., & Cohen, N. J. (2006). The long and the short of it: Relational memory impairments in amnesia, even at short lags. *Journal of Neuroscience, 26*(32), 8352–8359.

Hassabis, D., Kumaran, D., Vann, S. D., & Maguire, E. A. (2007). Patients with hippocampal amnesia cannot imagine new experiences. *PNAS, 104*, 1726–1731.

Hengst, J. (2003). Collaborative referencing between individuals with aphasia and routine communication partners. *Journal of Speech, Language and Hearing Research, 46*, 831–848.

Hengst, J. & Duff, M. C. (2007). Clinicians as communication partners: Developing a mediated discourse elicitation protocol. *Topics in Language Disorders, 27*, 36–47.

Heyworth, N., & Squire, L. (2019). The nature of recollection across months and years after medial temporal lobe damage. *PNAS, 116* (10), 4619–4624.

Hilliard, C., & Cook, S. W. (2016). Bridging gaps in common ground: Speakers design their gestures for their listeners. *Journal of Experimental Psychology: Learning, Memory and Cognition, 42*, 91–103.

Hilverman, C., Brown-Schmidt, S., & Duff, M. C. (2019). Gesture height reflects common ground status even in patients with amnesia. *Brain & Language, 190*, 31–37.

Hilverman, C., Clough, S. A., Duff, M. C., & Cook, S. W. (2018a). Patients with hippocampal amnesia successfully integrate gesture and speech. *Neuropsychologia, 117*, 332–338.

Hilverman, C., Cook, S. W., & Duff, M. C. (2016). Hippocampal declarative memory supports gesture production: Evidence from amnesia. *Cortex, 85*, 25–36.

Hilverman, C., Cook, S. W., & Duff, M. C. (2017). The influence of the hippocampus and declarative memory on word use: Patients with amnesia use less imageable words. *Neuropsychologia, 106*, 179–186.

Hilverman, C., Cook, S. W., & Duff, M. C. (2018b). Hand gestures support word learning in patients with hippocampal amnesia. *Hippocampus, 28*, 406–415.

Hilverman, C., & Duff, M. C. (2021). Evidence of impaired naming in patients with hippocampal amnesia. *Hippocampus, 31*(6), 612–626.

Holland, A. (1982). When is aphasia aphasia? The problem of closed head injury. Paper presentation at the Clinical Aphasiology Conference. http://aphasiology.pitt.edu/746/

Issacs, E. A., & Clark, H. H. (1987). References in conversation between experts and novices. *Journal of Experimental Psychology: General, 116*, 26–37.

Jafarpour, A., Piai, V., Lin, J. J., & Knight, R. T. (2017). Human hippocampal pre-activation predicts behavior. *Scientific Reports, 7*(1), 5959.

Kaiser, E., Runner, J. T., Sussman, R. S., & Tanenhaus, M. K. (2009). Structural and semantic constraints on the resolution of pronouns and reflexives. *Cognition, 112*, 55–80.

Karuza, E. A., Newport, E. L., Aslin, R. N., Starling, S. J., Tivarus, M. E., & Bavelier, D. (2013). The neural correlates of statistical learning in a word segmentation task: An fMRI study. *Brain and Language, 127*, 46–54.

Kelly, S. D., Barr, D. J., Church, R. B., & Lynch, K. (1999). Offering a hand to pragmatic understanding: The role of speech and gesture in comprehension and memory. *Journal of Memory and Language, 40*, 577–592.

Kensinger, E. A., Ullman, M. T., & Corkin, S. (2001). Bilateral medial temporal lobe damage does not affect lexical or grammatical processing: Evidence from the amnesic patient H.M. *Hippocampus, 11*, 347–360.

Kim, R., Seitz, A., Feenstra, H., & Shams, L. (2009). Testing assumptions of statistical learning: Is it long-term and implicit? *Neuroscience Letters, 461*, 145–149.

Kiran, S., Meier, E., & Johnson, J. (2019). Neuroplasticity in aphasia: A proposed framework of language recovery. *Journal of Speech, Language, and Hearing Research, 62*(11), 3973–3985.

Klooster, N., Cook, S., Uc, E., & Duff, M. C. (2015). Gestures make memories, but what kind? Patients with impaired procedural memory display disruptions in gesture production and comprehension. *Frontiers in Human Neuroscience, 8*, 1054. https://doi.org/10.3389/fnhum.2014.01054

Klooster, N., & Duff, M. C. (2015). Remote semantic memory is impoverished in hippocampal amnesia. *Neuropsychologia, 79*(Part A), 42–52.

Klooster, N., Tranel, D., & Duff, M. C. (2020). Hippocampus and semantic memory across time. *Brain and Language, 210*. https://doi.org/10.1016/j.bandl.2019.104711

Kurczek, J., Brown-Schmidt, S., & Duff, M. C. (2013). Hippocampal contributions to language: Evidence of referential processing deficits in amnesia. *Journal of Experimental Psychology: General, 142*(4), 1346–1354.

Kurczek, J., & Duff, M. C. (2011). Cohesion, coherence, and declarative memory: Discourse patterns of patients with hippocampal amnesia. *Aphasiology, 25*, (6–7), 700–712.

Kurczek, J., & Duff, M. C. (2012). Intact discourse cohesion and coherence following bilateral ventromedial prefrontal damage. *Brain and Language, 123*(3), 222–227.

Kurczek, J., Wechsler, E., Ahuja, S., Jensen, U., Cohen, N., Tranel, D., & Duff, M. C. (2015). Differential contributions of hippocampus and medial prefrontal cortex to self-projection and self-referential processing. *Neuropsychologia, 73*, 116–126.

Kutas, M., & Federmeier, K. (2000). Electrophysiology reveals semantic memory use in language comprehension. *Trends in Cognitive Science, 4*(12), 463–470.

Lackner, J. R. (1974). Observations on the speech processing capabilities of an amnesic patient: Several aspects of H.M.'s language function. *Neuropsychologica, 12*, 199–207.

Laszlo, S., & Federmeier, K. D. (2011). The N400 as a snapshot of interactive processing: Evidence from regression analyses of orthographic neighbor and lexical associate effects. *Psychophysiology, 48*(2), 176–186.

Laurita, A., & Spreng, R. N. (2017). The hippocampus and social cognition. In D. Hannula & M. C. Duff (Eds.), *The Hippocampus from Cells to Systems: Structure,*

Connectivity, and Functional Contributions to Memory and Flexible Cognition (pp. 537–558). Springer International Publishing.

Lee, J. C., Nopoulos, P. C., & Tomblin, J. B. (2013). Abnormal subcortical components of the corticostriatal system in young adults with DLI: A combined structural MRI and DTI study. *Neuropsychologia, 51*(11), 2154–2161.

Lee, J. C., Nopoulos, P. C., & Tomblin, J. B. (2020). Procedural and declarative memory brain systems in developmental language disorder (DLD). *Brain and Language, 205*, 104789. https://doi.org/10.1016/j.bandl.2020.104789

MacDonald, M. C., Just, M. A., & Carpenter, P. A. (1992). Working memory constraints on the processing of syntactic ambiguity. *Cognitive Psychology, 24*, 56–98.

MacKay, D. G., Burke, D. M., & Stewart, R. (1998). H.M. revisited: Relations between language comprehension, memory, and the hippocampal system. *Journal of Cognitive Neuroscience, 10*, 377–394.

Madore, K. P., & Schacter, D. L. (2014). An episodic specificity induction enhances means-end problem solving in young and older adults. *Psychology and Aging, 29*, 913–924.

Manns, J. R., Hopkins, R. O., & Squire, L. R. (2003). Semantic memory and the human hippocampus. *Neuron, 38*, 127–133.

Mayer, J. F., & Murray, L. L. (2012). Measuring working memory deficits in aphasia. *Journal of Communication Disorders, 45*(5), 325–339.

McClelland, J. L., McNaughton, B. L., & O'Reilly, R. C. (1995). Why there are complementary learning systems in the hippocampus and neocortex: Insights from the successes and failures of connectionist models of learning and memory. *Psychological Review, 102*, 419–457.

McGregor, K. K., Newman, R. M., Reilly, R. M., & Capone, N. C. (2002). Semantic representation and naming in children with specific language impairment. *Journal of Speech, Language, and Hearing Research, 45*(5), 998–1014.

McKenzie, S., & Eichenbaum, H. (2011). Consolidation and reconsolidation: Two lives of memories? *Neuron, 71*(2), 224–233.

McMurray, B., Horst, J. S., & Samuelson, L. K. (2012). Word learning emerges from the interaction of online referent selection and slow associative learning. *Psychological Review, 119*, 831–877.

McNeill, D. (1992). *Hand and Mind: What Gestures Reveal About Thought.* The University of Chicago Press.

Milner, B., Corkin, S., & Teuber, H. L. (1968). Further analysis of the hippocampal amnesic syndrome: A 14-year follow up study of H.M. *Neuropsychologia, 6*, 215–234.

Monti, J. M., Baym, C. L., & Cohen, N. J. (2014). Identifying and characterizing the effects of nutrition on hippocampal memory. *Advances in Nutrition, 5*(3), 337S–43S. https://doi.org/10.3945/an.113.005397

Morrow, E., Mayberry, L., & Duff, M. C. (2023). The growing gap: A study of sleep, encoding, consolidation of new words in chronic traumatic brain injury.

Neuropsychologia, 184, 108518. https://doi.org/10.1016/j.neuropsychologia.2023.108518

Novick, J. M., Trueswell, J. C., & Thompson-Schill, S. L. (2005). Cognitive control and parsing: Reexamining the role of Broca's area in sentence comprehension. *Cognitive, Affective, & Behavioral Neuroscience, 5,* 263–281.

O'Connor, M., & Verfaellie, M. (2002). The amnesic syndrome. In A. D. Baddeley, M. D. Kopelman, & B. A. Wilson (Eds.), *The Handbook of Memory Disorders* (2nd ed, pp.145–166). John Wiley & Sons Ltd.

O'Connor, M., Verfaellie, M., & Cermak, L. S. (1995). Clinical differentiation of amnesic subtypes. In A. D. Baddeley, B. A. Wilson, & F. N. Watts (Eds.), *The Handbook of Memory Disorders* (pp. 53–80). John Wiley & Sons Ltd.

Olson, I. R., Page, K., Moore, K. S., Chatterjee, A., & Verfaellie, M. (2006). Working memory for conjunctions relies on the medial temporal lobe. *Journal of Neuroscience, 26*(17), 4596–4601.

Oppenheim, G. M., Dell, G. S., & Schwartz, M. F. (2007). Cumulative semantic interference as learning. *Brain and Language, 103*(1–2), 175–176.

Oppenheim, G. M., Dell, G. S., & Schwartz, M. F. (2010). The dark side of incremental learning: A model of cumulative semantic interference during lexical access in speech production. *Cognition, 114*(2), 227–252.

Pacton, S., Fayol, M., & Perruchet, P. (2005). Children's implicit learning of graphotactic and morphological regularities. *Child Development, 76,* 324–339.

Paek, E. J., & Yoon, S. O. (2021). Partner-specific communication deficits in individuals with Alzheimer's Disease. *American Journal of Speech-Language Pathology, 11*(30)(1S), 76–390.

Perruchet, P., & Pacton, S. (2006). Implicit learning and statistical learning: One phenomenon, two approaches. *Trends in Cognitive Sciences, 20,* 233–238.

Piai, V., Anderson, K. L., Lin, J. J., Dewar, C., Parvizi, J., Dronkers, N. F., & Knight, R. T. (2016). Direct brain recordings reveal hippocampal rhythm underpinnings of language processing. *PNAS, 113,* 11366–11371.

Pu, Y., Cheyne, D., Sun, Y., & Johnson, B. W. (2020). Theta oscillations support the interface between language and memory. *NeuroImage, 215,* 116782.

Race, E., Keane, M. M., & Verfaellie, M. (2011). Medial temporal lobe damage causes deficits in episodic memory and episodic future thinking not attributable to deficits in narrative construction. *The Journal of Neuroscience, 31*(28), 10262–10269. https://doi.org/10.1523/JNEUROSCI.1145-11.2011

Ranganath, C., & D'Esposito, M. (2001). Medial temporal lobe activity associated with active maintenance of novel information. *Neuron, 31*(5), 865–873.

Reber, P. J., Knowlton, B. J., & Squire, L. R. (1996). Dissociable properties of memory systems: Differences in the flexibility of declarative and nondeclarative knowledge. *Behavioral Neuroscience, 110*(5), 861–871.

Ripich, D. N., Ziol, E., & Lee, M. M. (1998). Longitudinal effects of communication training on caregivers of persons with Alzheimer's disease. *Clinical Gerontologist, 19*(2), 37–55.

Roger, E., Banjac, S., de Schotten, M. T., & Baciu, M. (2022). Missing links: The functional unification of language and memory (L∪M). *Neuroscience & Biobehavioral Reviews, 133*, 104489.

Rousseaux, M., Vérigneaux, C., & Kozlowski, O. (2010). An analysis of communication in conversation after severe traumatic brain injury. *European Journal of Neurology, 17*, 922–929.

Rubin, R., Brown-Schmidt, S., Duff, M. C., Tranel, D., & Cohen, N. J. (2011). How do I remember that I know you know that I know? *Psychological Science, 22*(12), 1574–1582.

Rubin, R., & Cohen, N. J. (2017). Memory, relational representations, and the long reach of the hippocampus. In D. Hannula & M. C. Duff (Eds.), *The Hippocampus from Cells to Systems: Structure, Connectivity, and Functional Contributions to Memory and Flexible Cognition* (pp. 337–368). Springer International Publishing.

Rubin, R., Watson, P., Duff, M. C., & Cohen, N. J. (2014). The role of the hippocampus in flexible cognition and social behavior. *Frontiers in Human Neuroscience, 8*. https://doi.org/10.3389/fnhum.2014.00742

Ryan, J. (2012) Reconsidering the use of "explicit" and "implicit" as terms to describe task requirements, *Cognitive Neuroscience, 3*(3–4), 211–212.

Saffran, J. R., Aslin, R. N., & Newport, E. L. (1996). Statistical learning by 8-month-old infants. *Science, 274*, 1926–1928.

Saffran, J. R., & Wilson, D. P. (2003). From syllables to syntax: Multilevel statistical learning by 12-month-old infants. *Infancy, 4*, 273–284.

Schacter, D., Addis, D., & Szpunar, K. (2017). Escaping the past: Contributions of the hippocampus to future thinking and imagination. In D. Hannula & M. C. Duff (Eds.), *The Hippocampus from Cells to Systems: Structure, Connectivity, and Functional Contributions to Memory and Flexible Cognition* (pp. 439–466). Springer International Publishing.

Schapiro, A. C., Gregory, E., Landau, B., McCloskey, M., & Turk-Browne, N. B. (2014). The necessity of the medial temporal lobe for statistical learning. *Journal of Cognitive Neuroscience, 26*, 1736–1747.

Schevenels, K., Michiels, L., Lemmens, R., De Smedt, B., Zink, I., & Vandermosten, M. (2022). The role of the hippocampus in statistical learning and language recovery in persons with post stroke aphasia. *Neuroimage: Clinical, 36*, 103243.

Schlichting, M., & Preston, A. (2017). The hippocampus and memory integration: Building knowledge to navigate future decisions. In D. Hannula & M.C. Duff (Eds.), *The Hippocampus from Cells to Systems: Structure, Connectivity, and Functional Contributions to Memory and Flexible Cognition* (pp. 405–438). Springer International Publishing.

Scoville, W. B., & Milner, B. (1957). Loss of recent memory after bilateral hippocampal lesions. *Journal of Neurology, Neurosurgery, and Psychiatry, 20*, 11–21. https://doi.org/10.1136/jnnp.20.1.11

Sherman, B. E., Turk-Browne, N. B., & Goldfarb, E. V. (2023). Multiple Memory Subsystems: Reconsidering Memory in the Mind and Brain. *Perspectives on Psychological Science, 0*(0). https://doi.org/10.1177/17456916231179146

Shrager, Y., Levy, D. A., Hopkins, R. O., & Squire, L. R. (2008). Working memory and the organization of brain systems. *The Journal of Neuroscience: The Official Journal of the Society for Neuroscience, 28*(18), 4818–4822.

Snow, P., Douglas, J. M., & Ponsford, J. (1998). Conversational discourse abilities following severe traumatic brain injury: A follow-up study. *Brain Injury, 12*, 911–935.

Stickgold, R. (2005). Sleep-dependent memory consolidation. *Nature, 437*(7063), 1272–1278. https://doi.org/10.1038/nature04286

Tannen, D. (1989). *Talking Voices: Repetition, Dialogue, and Imagery in Conversational Discourse.* Cambridge University Press.

Tranel, D., & Damasio, A. R. (1993.) The covert learning of affective valence does not require structures in hippocampal system or amygdala. *Journal of Cognitive Neuroscience, 5*, 79–88.

Trude, A., Duff, M. C., & Brown-Schmidt, S. (2014). Talker-specific learning in amnesia: Insight into mechanisms of adaptive speech perception. *Cortex, 54*, 117–123.

Turk-Browne, N. B., Scholl, B. J., Chun, M. M., & Johnson, M. K. (2009). Neural evidence of statistical learning: Efficient detection of visual regularities without awareness. *Journal of Cognitive Neuroscience, 21*, 1934–1945.

Ullman, M. T. (2004). Contributions of memory circuits to language: The declarative/procedural model. *Cognition, 92*, 231–270.

Ullman, M. T., Corkin, S., Coppola, M., Hickok, G., Growdon, J. H., Koroshetz, W. J., Pinker, S. (1997). A neural dissociation within language: Evidence that the mental dictionary is part of declarative memory and that grammatical rules are processed by the procedural system. *Journal of Cognitive Neuroscience, 9*(2), 266–276.

Vallila-Rohter, S., & Kiran, S. (2013). Non-linguistic learning and aphasia: Evidence from a paired associate and feedback-based task. *Neuropsychologia, 51*(1), 79–90.

Verfaellie, M., Koseff, P., & Alexander, M. P. (2000). Acquisition of novel semantic information in amnesia: Effects of lesion location. *Neuropsychologia, 38*, 484–492.

Wang, J. X., Rogers, L. M., Gross, E. Z., Ryals, A. J., Dokucu, M. E., Brandstatt, K. L., Hermiller, M. S., & Voss, J. L. (2014). Targeted enhancement of cortical-hippocampal brain networks and associative memory. *Science*, 345(6200), 1054–1057.

Warren, D. E., Duff, M. C., Tranel, D., & Cohen, N. J. (2011). Observing degradation of visual representations over short intervals when medial temporal lobe is damaged. *Journal of Cognitive Neuroscience, 23*(12), 3862–3873.

Wilkes-Gibbs, D., & Clark, H. H. (1992). Coordinating beliefs in conversation. *Journal of Memory and Language, 31*, 183–194.

Wilson, S., Eriksson, D., Yen, M., Demarco, A., Schneck, S., & Lucanie, J. (2019). Language mapping in aphasia. *Journal of Speech Language and Hearing Research, 62*(11), 3937–3946.

Yoon, S., & Brown-Schmidt, S. (2018). Aim low: Mechanisms of audience design in multiparty conversation. *Discourse Processes, 55*, 566–592.

Yoon, S. O., Duff, M. C., Brown-Schmidt, S. (2017). Learning and using knowledge about what other people do and don't know despite amnesia. *Cortex, 94*, 164–175.

Youse, K. M., & Coelho, C. A. (2005). Working memory and discourse production abilities following closed-head injury. *Brain Injury, 19*, 1001–1009.

Zeithamova, D., & Preston, A. R. (2010). Flexible memories: Differential roles for medial temporal lobe and prefrontal cortex in cross-episode binding. *The Journal of Neuroscience, 30*(44), 14676–14684.

Index

For EU product safety concerns, contact us at Calle de José Abascal, 56–1°,
28003 Madrid, Spain or eugpsr@cambridge.org.

www.ingramcontent.com/pod-product-compliance
Lightning Source LLC
Chambersburg PA
CBHW081404061125
35092CB00005B/301
9781009202282